Tactical Missile Propulsion

Tactical Missile Propulsion

Edited by
Gordon E. Jensen
United Technologies
San Jose, California

David W. Netzer
Naval Postgraduate School
Monterey, California

Volume 170
PROGRESS IN ASTRONAUTICS AND AERONAUTICS
A Volume in the AIAA Tactical Missile Series

Paul Zarchan, Editor-in-Chief
Charles Stark Draper Laboratory, Inc.
Cambridge, Massachusetts

Published by the
American Institute of Aeronautics and Astronautics, Inc.
1801 Alexander Bell Drive, Reston, Virginia 20191-4344

Second Printing.

ISBN 1-56347-118-3

Table of Contents

Preface

Paraphrasing Dickens, "it is the best of times and it is the worst of times" for the tactical missile propulsion system designer. With the ongoing development of new and advanced propulsion systems, there are numerous potential solutions for any particular application. In the past, the designer was essentially limited to solid and liquid propellant rockets or turbine engines for tactical missile propulsion; the choice of system being dictated by the speed, range, and cost requirements that were set for the missile. Today, although enhanced performance solid rockets are still a mainstay for many applications, the significant advances that have been made with other propulsion devices can also be utilized. This is especially true with respect to liquid-fuel, solid-fuel, and ducted-rocket ramjets and scramjets, low-cost and expendable turbine engines, advanced liquid propellant and hybrid rockets, and advanced combined cycle propulsion systems. The current situation has created an almost bewildering number of choices for the designer to consider.

In contrast, overall performance requirements, environmental considerations, cost and reliability constraints, shrinking budgets, and myriad other considerations sorely tax today's designer. Energy management for longer range and higher intercept velocities complicate the design, increase costs, and can compromise reliability. Low-visibility plume and low-observable missile requirements require special attention and, also, generally compromise the performance of the system. Insensitive munitions (IM) requirements similarly have an impact on the designer's options, while further complicating the design and increasing costs. Reduced time to target requirements necessitate the consideration of advanced types of high-Mach number engines that may not have the full database that a designer needs to come to a definitive answer. Ambiguous answers make selecting a "best system" nearly impossible. New long-range national goals for significant cost reductions, reliability improvements, as well as performance improvements also are being formulated, e.g., in the United States the integrated high payoff rocket propulsion technology (IHPRPT) program. Such efforts, while supporting important new technology development, will make the designer's job even more challenging.

The objectives of this volume are to provide today's designer with a summary of the advances in potential propulsion systems, as well as to provide a discussion of major design and selection considerations. Authors have been chosen for their demonstrated knowledge in their respective fields to ensure as complete and up to date a summary as possible.

Gordon E. Jensen
David W. Netzer
July 1996

Chapter 1

Introduction

Gordon E. Jensen*
United Technologies Corporation, San Jose, California 95151
and
David W. Netzer†
Naval Postgraduate School, Monterey, California 93043

THE purpose of this volume, which is part of the AIAA Tactical Missile Series of AIAA's Progress in Astronautics and Aeronautics, is to provide today's designer with a summary of the advances in potential propulsion systems, and to provide discussions of design and selection considerations. The 13 chapters following this one serve to accomplish that purpose. Chapter 2 was prepared by Frederick S. Billig of Johns Hopkins University. He presents a methodology for defining the performance and operating characteristics of potential tactical propulsion systems. Rockets, ramjets, scramjets, and turbojets are analyzed with respect to the key figures of merit of range or "F" pole, travel time to target, and velocity and miss distance at intercept or impact. The results are shown to depend on the number of stages utilized and the chosen trajectory. In addition to addressing the differences in performance and operating characteristics, development requirements for the different systems are discussed. The chapter concludes with specific design procedures for ramjet powered missiles, including inlet and combustor considerations.

Chapter 3 by Ben Wilson of the U.S. Army Missile Command provides a summary of the state of development of liquid rockets. Whereas liquid rockets have been less frequently used in recent tactical motors, the controllability of the thrust-time profile for energy management and advances in gelled fuels and oxidizers for insensitive munitions (IM) compatibility makes the liquid rocket a very viable candidate for many new applications. The purpose of this chapter is to provide sufficient insight into system requirements, the source of these requirements, and the general technology available for tactical liquid propulsion systems to permit the designer to achieve the optimum propulsion system for a specific weapon system. The chapter addresses monopropellant engines, bipropellant engines, and advanced gel systems that can meet today's stringent requirements.

---

*Manager, Advanced Projects & Systems Engineering, Chemical Systems Division.

†Professor, Department of Aeronautics and Astronautics.

As solid propellant rockets are still the primary choice for many propulsion systems, a major portion of the volume addresses the different aspects of solid rocket motor design and the various components. A summary of solid motor design considerations is provided by Alain Davanas of SNPE Défense Espace. He begins with a discussion of the overall design approach, which requires consideration of general, functional, and interface requirements. A description of a typical solid propellant rocket motor follows, and then a discussion of how the propulsion and structural performances are determined. This leads directly into a general method for the design and optimization of the motor. Considerable detail is devoted to the selections of propellant composition and grain configuration. A major section is then devoted to ballistic analysis and grain design, using preliminary design methods. The next major section is concerned with the general procedure for grain design optimization and computer aided preliminary design.

Solid propellant grain structural design and service life analysis are discussed by Barbara Marsh and Tom Duerr of the U.S. Army Missile Command in Chapter 5. The emphasis of the chapter is upon case bonding and propellant grain structural requirements. A careful consideration of loads, thermal, acceleration, transportation and handling, cure shrinkage, as well as pressurization, is provided. In addition to design considerations, test methods to evaluate the different loads and the characteristics of the propellant are provided.

Design methods and characteristics of different tactical motor nozzle designs are provided by William Prescott of the Thiokol Corp. and Michael Mocha, formerly of the Thiokol Corp. Key nozzle design parameters and the impact of operating conditions are related to various potential nozzle configurations. A critical part of the design sequence are the consideration and trades of different materials comprising the nozzle assembly. In addition to the discussion of the factors to be considered in the materials selection process, the designer is provided with the basic design steps required to optimize the design. The design of thrust vector systems and actuations systems is presented in the final section of Chapter 6.

Tactical case designs, materials, and fabrication methods are discussed by Geoff Thorp and M. J. Chase of the British Aerospace, Royal Ordnance Rocket Motor Division. Motor case design considerations, materials selection, and different fabrication techniques are thoroughly addressed in this chapter. Particular attention is given to meeting insensitive munition requirements with the case design. Finally, the author presents a specific rocket motor case design as an illustration of the steps to be undertaken during design.

Energy management, plume visibility, and insensitive munitions are the three major issues facing the solid rocket motor designer. In Chapters 8 and 9 Andrew Victor of Victor Technologies discusses plume and insensitive munitions (IM) requirements and potential solutions for different solid rocket propellants and various motor designs. Chapter 8 on plumes begins with an overview of the tactical missile plume signature problem; types, effects, detectability, and modeling. After a brief historical overview, specific operational considerations are discussed, e.g., physical and radiation interactions. Technical considerations are then discussed, including propellant composition effects, plume flowfields, and the various types of plume signatures. The chapter closes with some design guidelines.

IM has taken on considerable significance in recent years, especially within the NATO community. The IM chapter begins with an overview of how and why the present IM requirements have evolved. This is followed by a discussion of operational considerations (munitions qualifications and hazard threats) and technical considerations (energetic behavior of propellants and explosives). A detailed section is devoted to design guidelines, including mitigation systems and propellant selection. The final section of the chapter provides information that is very helpful for the application of IM criteria, e.g., "simple analytical methods for munition hazard assessment."

Turbine engines and ramjet engines are presented in the following four chapters. J. M. Stricker of Wright Laboratories summarizes design considerations and methodology for gas turbine engines. Turbine engines are well established for long-range, low-velocity applications such as the cruise missile. Advanced materials offer potential usage for supersonic engines, whereas low-cost turbine engines offer a viable alternative for smart munition applications. The capability of the turbine to loiter offers significant systems advantages as well as high-specific impulse performance. In addition, the turbine can be combined with other propulsion systems to meet extended range and high-speed mission requirements. The chapter begins with an overview of the evolution of gas turbine engines for tactical missiles, including data for specific engines. On-design ideal and real cycle analyses are then presented in some detail for both turbojet and turbofan engines. This is followed by a discussion of how the off-design performance is determined and utilizes typical compressor and turbine maps and combustor performance characteristics. Guidelines are provided for cycle selection and design considerations for each of the major engine components (centrifugal and axial compressors, fuel systems, combustors, turbines, exhaust nozzles, and controls) are detailed. A final section is devoted to the important area of system integration.

Ramjet engines are not new and have been successfully used in the past. First-generation designs featured twin-pod, aft-mounted engines. One of these, the Boeing Michigan Air Research Center (BOMARC), has been retired from service for over 20 years, whereas the British Bloodhound is in its fourth decade of service. The Chinese HY-3/C101 uses first-generation technology, although it was not deployed until the 1980s. Two second-generation ramjets, the British Sea Dart and the USSR SA-4 were deployed in the 1960s. These feature aft-mounted tandem boosters that are staged at takeover. The Sea Dart proved to be very effective in the Falklands War. Third-generation ramjets feature integral-rocket-ramjet engine designs. The first third-generation missile was the Soviet SA-6 deployed in the 1970s. Whereas the earlier missiles featured liquid fuels, the SA-6 was based on the use of a fixed-flow, fuel-rich solid gas generator. Third-generation ducted rockets with integral boosters and variable flow gas generators have received considerable attention in recent years. The U.S. Air Force is developing the variable flow ducted rocket (VFDR) as a potential propulsion system for long range air-to-air missile applications, while the United States and France, the United States and Germany, and the United States and Japan are investigating other ducted rocket designs for other applications. John Leingang, formerly of the Wright Laboratories, addresses the issues facing the ducted-rocket-ramjet designer. An overview of the various ducted rocket (DR) configurations is first presented and is followed by a discussion of both metallized and nonmetallized fuels. An entire section is devoted to the factors that affect whether a ducted rocket or a liquid-fuel-ramjet would be selected for a particular mission that has a required value of delivered performance. The performance and geometric differences between the fixed-flow DR, the variable-flow DR, and the "Rustique" self-throttling DR are then discussed. This overview is followed by a procedure for performance analysis and a detailed sample problem.

France developed and deployed the world's first liquid-fuel, integral-rocket ramjet in the 1980 time period. The Aer Sol Moyenne Port (ASMP) missile is the highest performance missile in the world of equivalent size and weight. The air-launched ASMP missile uses dual two-dimensional aft-mounted inlets. This missile provides a 300–400% range increase over an equivalent size solid rocket motor. Improvements offer additional range potential. Appropriately, J. M. Laurent of CELERG in France presents the methodology for liquid ramjet engine design. This chapter begins with a review of the available liquid fuels, their energetic performances, and selection criteria. This is followed by a discussion of one of the most difficult components in the design, the fuel system: pressurization methods, combustion chambers and the associated flame stabilization and combustion stability requirements.

In recent years there has been considerable interest in solid fuel ramjets (SFRJ). The U.S. armed services have conducted a number of programs to develop solid fuel technology and considerable interest has also been shown in other countries. The "wooden round" features of the solid-fuel ramjet offer a number of IM advantages, whereas the self-throttling capability of the engine make it competitive with liquid-fuel ramjet engines. Steve Leisch of CSD and Dave Netzer of the Naval Postgraduate School present design methods for the solid fuel ramjet. The chapter begins with a general comparison of the obtainable performance of the SFRJ with that of a solid rocket and with a discussion of typical SFRJ configurations. This is followed by a discussion of the unique combustion environment of the SFRJ and the types of fuels that are utilized. Factors important in the design process are then considered: combustion efficiency, flammability limits, and fuel regression rate. The equations needed in the design process are summarized, and then a design procedure is presented. Bypass and nonbypass configurations are considered as well as the effects of off-design point operation, inlet behavior, and booster integration. The chapter concludes with an overview of a typical SFRJ development process.

The final chapter by Larry Fink of the Boeing Corporation addresses high-Mach number, combined-cycle propulsion systems. In contrast to such combination engines as the BOMARC engine, which used a solid rocket booster in the fuselage, and pod-mounted ramjet engines for the sustain flight, or integral-rocket-ramjets, which use a booster in the ramjet combustor to achieve takeover followed by ramjet operation, combined cycle propulsion refers to the combination of different propulsion elements in a single engine with cycle-process interactions between elements. The motivation for combined cycle engines is to provide higher performance than would be achievable with separate engines.

As stated initially, the lot of the propulsion design engineer is not easy. Hopefully the contents of this volume will provide some help in meeting specific design goals for new systems.

Chapter 2

Tactical Missile Design Concepts

Frederick S. Billig*
Johns Hopkins University, Applied Physics Laboratory, Laurel, Maryland 20723

Nomenclature

A_p = surface area of propellant
a = acceleration or constant in burn-rate expression, Eq. (4)
C_L = lift coefficient
c_p = specific heat at constant pressure
D = drag
d = diameter
ER = equivalence ratio
F = force
g = gravitational constant
h = enthalpy
I_{sp} = specific impulse
L = lift
m = mass
$\dot{m}$ = mass-flow rate
n = exponent in burn-rate expression, Eq. (4)
p = pressure
q = dynamic pressure
R = gas constant
T = temperature
u = velocity
W = weight
x = down-range distance
z = altitude
t = time
ϵ = structural coefficient
θ = flight-path angle
ν = kinematic viscosity
Π = product function

*Chief Scientist and Associate Head, Aeronautics Dept., Johns Hopkins Road.

ρ = density
ψ = elevation angle

Subscripts

b = burnout
c = combustion
ex = nozzle exit
f = final
i = initial or inlet
L = payload
max = maximum value
n = nth stage
p = propellant
R = reference
s = structure
t = total or stagnation
0 = freestream or initial

Superscript

$*$ = nozzle throat

I. Introduction

THE selection of a propulsion system to meet a mission requirement is arduous and often controversial. Typically, the engineer embarks on the conceptual design endeavor with the aspiration of defining the system with the best capability. Generally, the answer is ambiguous with respect to meeting the customer's requirements. Candidate A is more capable in some respects, candidate B in others. Cost often is the ultimate discriminate. Nonetheless, the methodology for defining the performance and operating characteristics of vehicles powered by different propulsion systems is the crucial element in the selection process.

The principal missions for tactical missiles are surface-to-air or air-to-air weapons for defense of the fleet, ground assets, or aircraft; surface, subsurface, or air-launched antiship or ground-attack weapons; and antiballistic weapons.

Key figures of merit of the effectiveness of these missiles are range or "F pole," travel time to target, and velocity and miss distance at intercept or impact.

Long range provides greater standoff distances which may be crucial to survivability, a large footprint for greater coverage of targets, loiter capability for search when there is uncertainty in the location of the target, and the capability to interdict weapons carriers prior to the release of submunitions. Reduced travel time to target (or range/block speed) reduces the translation distance from the initially defined point of intercept of mobile targets, increases the "kill" probability and/or reduces the number of missiles needed in a "shoot-look-shoot" battle scenario, and increases penetration capability. High velocity at intercept or impact increases kill probability and greatly complicates countermeasure tactics.

With the exception of launch from permanent ground-based installations, both the volume and weight of the vehicle are constrained. Consequently, the delivered thrust/unit weight of the propellants, the weight of the propulsion systems components, and the density of the propellants impact the selection process. Methods for quantitatively appraising these issues will be discussed.

Figure 1 illustrates the three basic types of propulsion system for tactical missiles: the solid rocket, the ramjet, and the turbojet. In the rocket, the propellant is burned in a combustion chamber to produce products of combustion at high pressure (30–200 atm) and high temperature (4500–7200°F). The burned gases are accelerated to high velocity as they pass through the exhaust nozzle. Thrust is produced by the high pressure acting on the internal surfaces of the motor.

Instead of explicitly defining this pressure field, Newton's law, $F = ma$, which establishes the equivalency between force and momentum change, is used to determine the thrust from the momentum of the exhaust gases as they leave the nozzle, i.e.,

$$F = \dot{m}u_{\text{ex}} = \dot{m}[2(h_t - h_{\text{ex}})]^{1/2} \approx \dot{m}(2h_t)^{1/2} = \left[2\dot{m}^2 c_p T_t\right]^{1/2} \tag{1}$$

where $\dot{m}$ is the propellant flow rate, subscript t is the stagnation condition in the rocket motor chamber, and all other terms are as defined in the nomenclature. Thus, to produce thrust efficiently; i.e., a large force for a given flow rate, it is necessary to use propellants that produce gases at high temperatures.

For convenience in handling, rocket-powered tactical missiles generally use solid rather than liquid propellants (Fig. 1a). The propellant is composed of a fuel and oxidizer (which may be a single chemical compound or a mixture of several

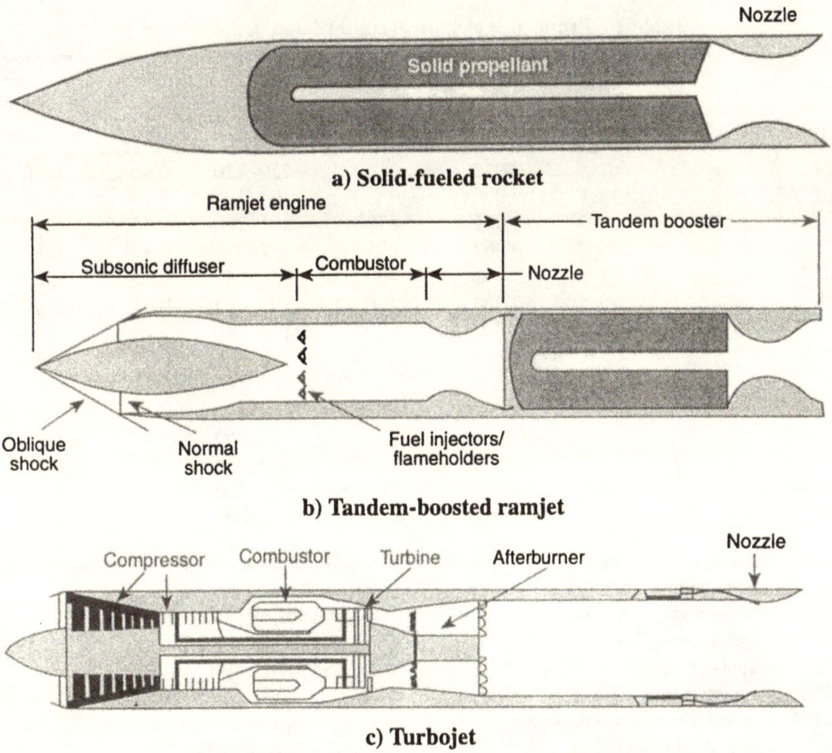

Fig. 1 Propulsion systems for tactical missiles.

ingredients and often contains ignition and burning rate modifiers, catalysts, and inhibitors), and additives to enhance the mechanical and rheological properties (binders, plasticizers, etc.). Typical fuels are organic resins and plastics with or without metal powder additives, typical oxidizers are nitrates or perchlorates; typical homogeneous monopropellants are the organic nitrates containing fuel and oxidizer in the same substance. Table 1 gives the properties of several typical rocket propellants.

The ramjet (Fig. 1b) differs from the rocket in that air is brought from the atmosphere into the engine to serve as the oxidizer for burning the fuel. The specific impulse, $I_{sp} = F/\dot{m}$, of the ramjet is greater than the rocket since it does not have to carry the weight of the oxidizer. However, there is the additional weight of the inlet and loss of available volume for propellants due to the air ducting. Moreover, the thermodynamic cycle of the ramjet requires compression of the air by slowing it down in the engine inlet prior to heat addition. Such compression is insufficient to produce thrust unless the air speed is greater than about Mach 0.9; i.e., the speed is 0.9 times the speed of sound. Thus, the ramjet must be launched from an aircraft or brought up to this speed or beyond by a booster engine, generally a rocket. The booster can either be a separate stage that is jettisoned after propellant exhaustion, as shown in Fig. 1b, or be integrated with the ramjet, i.e., an integral-rocket-ramjet. In the latter case, the rocket propellant fills the ramjet combustion chamber during the boost phase, ejectable covers close off the inlet to the combustor, and a rocket throat is used inside the larger ramjet

Table 1 Properties of typical solid rocket propellants

Propellant	Metal content weight, %	Density, lb/in.^{3}	Flame temperature, °F	I_{sp}, s	Burning rate, in./s	Pressure exponent, n, Eq. (4)
DB	0	0.058	4100	220–230	0.45	0.30
DB/AP/Al	20–21	0.965	6500	260–265	0.78	0.40
DB/AP-HMX/Al	20	0.065	6700	265–270	0.55	0.49
XLDB/AP	19	0.067	6060	269	0.35	0.50
HMX/Al	0	0.061	4600	230–240	0.45	0.38
PVC/AP	21	0.064	5600	260–265	0.45	0.35
PVC/AP/Al	0	0.062	4700	230–240	0.35	0.43
PS/AP	3	0.062	5000	240–250	0.31	0.33
PS/AP/Al	16–20	0.064	5400–6000	260–265	0.27	0.15
PU/AP/Al	19	0.067	6060	269	0.60	0.50
NEPE	16	0.064	5800	260–263	0.55	0.33
PBAN/AP/Al	15–17	0.064	5600–5800	260–265	0.45	0.40
CTPB/AP/Al	4–17	0.067	5600–5800	260–265	0.40	0.40
HTPB/AP/Al	14	0.064	5400–6000	260–265	0.32	0.35
PBAA/AP/Al	—	—	—	—	—	—

Al = aluminum
AP = ammonium perchlorate
CTPB = carboxy-terminated polybutadiene
DB = double base
HMX = cyclotetramethylene tetranitramine
HTPB = hydroxy-terminated polybutadiene
NEPE = nitrate-ester plasticizer
PBAA = polybutadiene-acrylic acid polymer
PBAN = polybutadiene-acrylic acid-acrylonitrile
PS = polysulfide
PU = polyurethane
PVC = polyvinyl chloride
XLDB = cross-linked double base

nozzle. After the rocket propellant has been spent and the missile has reached its minimum operating speed, inlet port covers are ejected to permit external air to be captured by the engine inlets and the rocket nozzle insert is jettisoned.

For speeds between Mach 1 and about Mach 5, the air is compressed by means of a series of shock waves that terminate in a normal shock. The flow into the combustor is subsonic and the engine cycle is called a "conventional" subsonic-combustion ramjet. At speeds greater than about Mach 5, a more efficient cycle results if the inlet air is only partially compressed and enters the combustor with a speed greater than Mach 1. This cycle is called a supersonic combustion ramjet. In the subsequent discussion, the dual-mode ram-scramjet, a single device that can operate with either subsonic or supersonic flow in combustor, will be described.

The third basic engine cycle is the turbojet (Fig. 1c). A rotating air compressor is added ahead of the combustor to circumvent the low-speed limitation of the ramjet, and a turbine to drive this compressor is added behind the combustor. The work that the turbine must provide can be produced by burning only about 25% of the oxygen; i.e., the equivalence ratio in the combustor $ER \approx 0.25$, so an afterburner can be placed downstream of the turbine to increase the thrust for acceleration. For the turbojet to compete, the added weight of the rotating machinery must be more than balanced by a reduced requirement for fuel. At speeds below about Mach 3, the Carnot efficiency of the cycle is higher due to the pressure rise of the compressor and fuel is saved. However, at speeds of about Mach 3, the wave compression in the inlet provides the needed compression and the compressor becomes superfluous. Indeed, the adiabatic compression in the inlet raises the temperature to a point where only a relatively small amount of heat could be added in the combustor without damage to the turbine blades from overheating.

Many variations and combinations of these basic cycles have been used; typical examples are shown in Fig. 2. Figure 2a is a solid-fueled, air-ducted rocket with an integral-rocket booster. In the configuration shown, multiple axisymmetric, aft-entry air inlets are used. Variations include half-round and two-dimensional inlets, perhaps with inward-turning compression surfaces. Depending on the thrust required and the constraints that may be imposed by the stowage of the missile, one, two, or four air inlets can be used. After the booster propellant has been expended and the small rocket nozzle and air inlet covers have been ejected, the sustain propellant is ignited to produce a hot fuel-rich exhaust that is burned to completion with the captured air in the main combustor. As contrasted to the ramjet where a sheet-metal can or metal-baffle obstructions can be used to stabilize the flame, flame stabilization in this design occurs in recirculation zones that are generated in the region of the air-intake ports (as for all integral-rocket ramjets). Since the solid propellant contains some oxidizer, the specific impulse, $I_{sp} = F/\dot{m}$, is lower than for the pure ramjet. However, the relatively high density of the solid fuel provides a high-packaging efficiency that is an important consideration in volume-limited situations.

Figure 2b is an integral rocket, liquid-fueled, supersonic-combustion ramjet. A "chin" inlet is shown that takes advantage of a favorable forebody compression field at angle of attack in the plane of symmetry. On the other hand, bank-to-turn rather than skid-to-turn control is required, which can have detrimental effects on the guidance system accuracy, especially during terminal maneuvers. Rather than a single ejectable nozzle, multiple nozzles are shown, reducing volume at the expense of weight and complexity. When the rocket propellant is consumed and the nozzles and inlet port covers have been ejected, liquid fuel is injected from

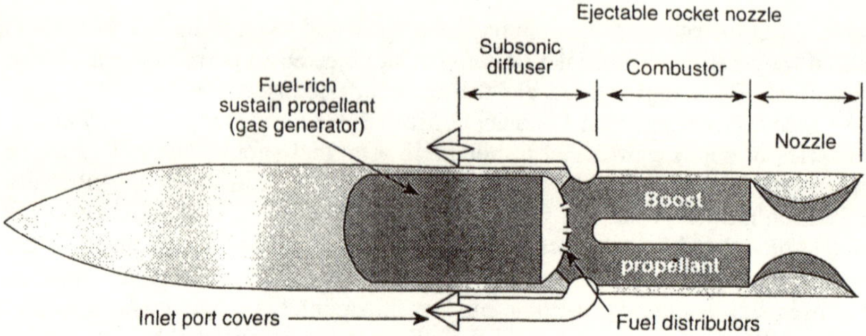

a) Solid-fueled ducted rocket

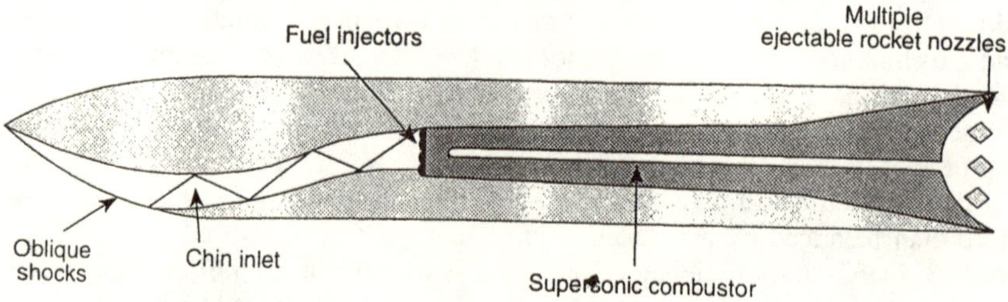

b) Integral rocket liquid-fueled supersonic combustion ramjet

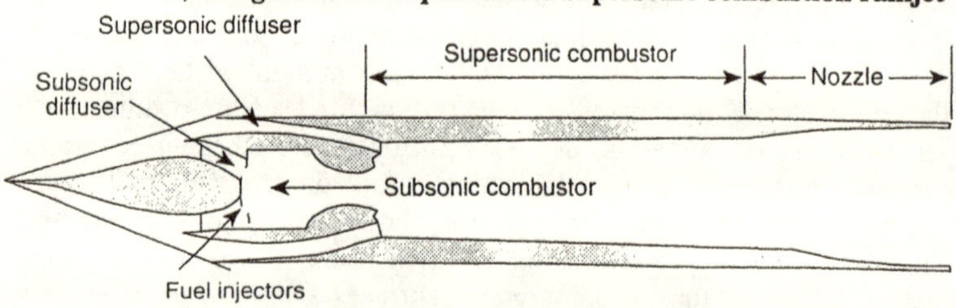

c) Liquid-fueled dual-combustor ramjet

Fig. 2 Variants of basic propulsion cycles.

the walls into the combustor unencumbered by flame holders that would produce undesired shock losses in the supersonic flow. Since the flow is supersonic, the residence time in the combustor is very short, typically 1–2 ms. Therefore, to burn conventional fuels, either the flight Mach number must be high, i.e., above Mach 5, or a highly reactive fuel or fuel blend, such as ethyl decaborane or a hydrocarbon with 20% pentaborane, must be used on special flame holding, or a special flame holding scheme must be used.

An alternative technique to reactive fuels is to inject simultaneously a highly reactive oxidizer, e.g., chlorine trifluoride (ClF_3), with a conventional fuel at the expense of simplicity, increased safety hazard, and loss in specific impulse.

An engine concept that bridges the gap between the supersonic and hypersonic speed ranges, with logistically acceptable fuels, is the dual-combustor ramjet, shown in Fig. 2c. A portion of the airflow captured by the inlet is diverted to a hot-gas generator which, in fact, is a "dump-type" subsonic combustor. To simplify the fuel control and injection system, all of the fuel is injected into the gas generator. Consequently, this combustor must operate considerably in excess of stoichiometric when high thrust is demanded. The hot fuel-rich combustion products pass into the main supersonic afterburner to complete the heat release by reacting with additional air.

Both this engine cycle and the scramjet (Fig. 2b) can, in fact, be dual-mode engines; they can operate in either the subsonic or the supersonic combustion mode. At low-flight Mach number and high-equivalence ratio ER, [$ER =$ (fuel/air)/(fuel/air) at stoichiometric conditions], a normal shock is located in the combustor entrance and heat release begins in subsonic flow. Heat addition occurs simultaneously with pressure drop in a diverging duct that accelerates the flow through the sonic point to supersonic conditions at the combustor exit. The price that is paid for the versatility in operation over a wide speed range is the complexity and weight of the split-flow inlet compression system and gas generator.

The particular choices of inlet type (e.g., chin, axisymmetric nose, or side mounted), nozzle type (single or multiple), and booster type (tandem or integral) in the configurations shown in Fig. 2 were arbitrary. Indeed, nearly every possible combination of these components has been examined for one or more applications. Moreover, Fig. 1a depicts only the simple single-stage rocket. As the subsequent discussion will show, in some applications the optimum rocket missile may either have two stages or a propellant designed to burn intermittently in two or more pulses.

II. Selection of Propulsion Cycle

Selection of the optimum propulsion cycle for a given application can be an arduous task. Generally, many cycles can be eliminated by a cursory examination, but the final selection often requires a detailed composite design. Even then, the selection may be ambiguous; one cycle may be better for a particular mission but another may have the versatility to carry out several missions. There may be a tradeoff between cost and performance, etc.

Two relatively simple expressions can provide some insight into the factors that ultimately lead to the selection of the candidate engine cycles. The first is the Breguet range equation which holds for the simple case of cruising at constant velocity u with the lift L equal to the weight W, viz.,

$$R = uI_{sp}(L/D)\ln(W_i/W_f) \tag{2}$$

where R is the range, I_{sp} is the propellant specific impulse, and W_i and W_f are the weights at the beginning and end of the cruise phase, respectively.

The second expression is for the velocity change of a vehicle during the climb and acceleration phase, viz.,

$$u_b - u_0 = gI_{sp}\ln\frac{W_0}{W_b} - \int_0^{t_b}\left(g\sin\theta + \frac{D}{W}\right)dt \tag{3}$$

where W_0 is the initial weight $(W_S + W_P + W_L)$, W_S is the structural weight, W_P is the weight of the propellant, W_L is the payload weight, $W_b = W_S + W_L$ is the burnout weight, g is the acceleration due to gravity, and θ is the local flight-path angle with respect to the horizontal.

In both expressions, performance (either the range or the accelerative ability) is directly related to the specific impulse I_{sp}. Typical values of I_{sp} for the propulsive cycles depicted in Figs. 1 and 2 are shown in Fig. 3. The values are based on the use of conventional hyrocarbon fuels in the airbreathing engines. The turbojet is

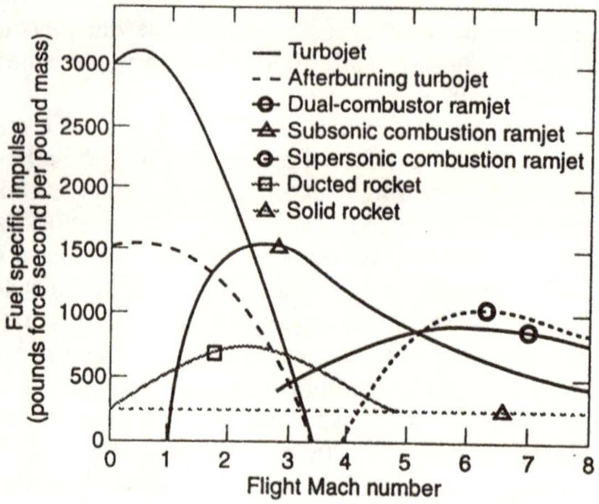

Fig. 3 Fuel specific impulse for various missile propulsion systems.

the most efficient for speeds up to Mach 2.5, which has led to its near universal use as an aircraft power plant. If relatively high thrust is required, the turbojet operates with an afterburner and I_{sp} is significantly lower.

Most missile applications require high thrust. As a result, the high cost of the turbomachinery for an expendable engine eliminates the turbojet from consideration unless long range is the dominant requirement and high acceleration and high speed are not mandatory. The current generation of cruise missiles (Harpoon and Tomahawk) is in this category. In the supersonic to moderate hypersonic speed range, the ramjet is attractive despite the booster requirement to bring it up to speed. The choice among subsonic combustion, supersonic combustion, or dual-combustor ramjets depends on the velocity requirement for the mission as well as cost and logistic considerations, as noted earlier.

As shown in Fig. 3, the rocket's specific impulse is only 20–30% of that of a ramjet. On the other hand, the mass fraction of propellant for a rocket for a tactical missile is typically 50–70% of the initial weight as compared to 25–30% for the ramjet. Thus, the logarithmic terms in Eqs. (2) and (3) are larger by a factor of 2.5–2.8 for the rocket. Since the product of the two terms is significantly lower for the rocket, this generally eliminates it from consideration if long range during cruise is required [Eq. (2)]. However, during rapid acceleration the second term in Eq. (3) can be significant. This term is minimized when dt is small and drag is not too large, which corresponds to a relatively high average thrust-to-weight ratio (F/W) of about 10. By tailoring the grain design and using a relatively large exit nozzle throat, it is possible to design a rocket with F/W of about 10 which is essentially independent of altitude. High-speed ramjets can be designed with F/W of about 10 near sea level. But since the thrust is proportional to the ambient pressure, at an altitude of 50,000 ft, F/W is approximately 1 and at 100,000 ft, about 0.1. Thus, if the missile must climb to high altitude while accelerating rapidly and a long cruise range is not required, the rocket may be the preferred engine.

To develop this argument further and to give some insight into the design of propulsion systems, it is necessary to examine families of climb-out and cruise trajectories. For a given mission objective and constraints inherent to the composite design of the missile, e.g., structural and control limitations, the climb–out trajectory can be optimized by the introduction of lateral aerodynamic force and modulation of the thrust. Lateral force which affects the missile attitude θ can only be generated at the expense of induced drag, so the terms within the integrand of Eq. (3) are coupled. Moreover, drag increases with the square of velocity and decreases exponentially with altitude Z. Thus, rapid acceleration reduces dt but not necessarily $(D/W)dt$. No general solution has been found for Eq. (3), nor is there a rigorous procedure for optimization of the climb out. However, considerable attention has been directed toward suboptimal schemes[1,2] that use high-speed computations to obtain apparent optimal trajectories for particular missile configurations and missions.

The interdependence of the missile composite design and the optimization of the trajectory will become apparent as particular mission requirements are examined. Typical trajectories for the first generic class of missions to be examined are depicted in Fig. 4. In this case, the missile is to be launched vertically ($\theta \approx 90$ deg) from a canister, after which it will climb, accelerate, and intercept a target at an altitude Z_A. The criterion for selecting the best propulsion system is maximum velocity at point A, which in general corresponds to minimum flight time.

Consider first the case of a solid propellant rocket. Let us hypothesize that it will be possible to select propellant grains and specify grain shapes and nozzle sizes that will produce constant flow rates of the propellants. Trajectories can now be flown for different flow rates to determine whether flight time is dependent on them. Curve I in Fig. 4 corresponds to a high flow rate, a short duration turn from the vertical, and a zero-lift ascent trajectory, i.e., no normal aerodynamic force is

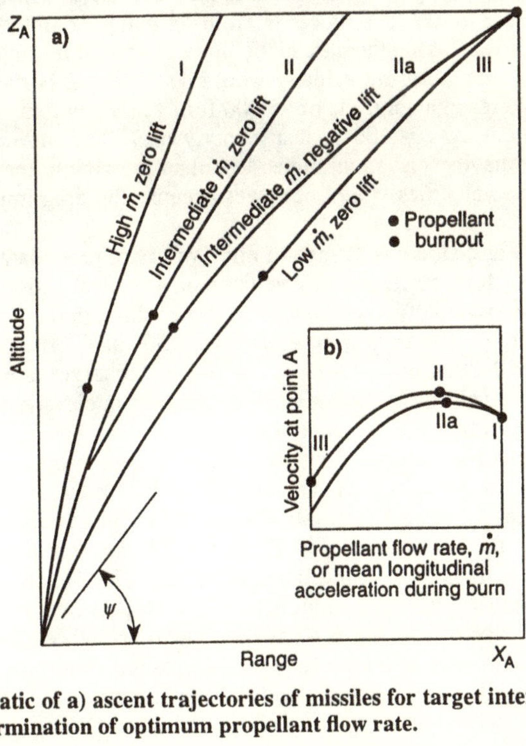

Fig. 4 Schematic of a) ascent trajectories of missiles for target intercept near point A, and b) determination of optimum propellant flow rate.

applied. Thus, the vehicle flies ballistically. Maximum velocity is reached when the propellant burns out, and the missile coasts with some slowdown from burnout to altitude Z_A.

Curve II is for a lower flow rate. Burnout occurs later at a higher altitude and the trajectory "falls over" farther from the vertical because the velocity during the early phase of the climb out is lower than for curve I. Curve III is for the lowest flow rate covered in this example. Burnout occurs nearer X_A, Z_A, and the trajectory falls over somewhat farther.

By introducing negative aerodynamic lift, it is possible to force trajectories I and II to pass through X_A and Z_A also, as typified by curve IIa for the intermediate value of flow rate. Similarly, trajectories II and III could be forced to pass through the terminal point of trajectory I by introducing positive lift.

Regardless of whether a comparison is made for ballistic ascent trajectories or for lifting trajectories forced to pass through the same point in space, there exists a propellant flow rate that yields a maximum mean velocity, as shown in Fig. 4b. The optimum trajectory corresponds to the case where the integrand in Eq. (3) is minimized. This can best be understood by examining the limiting case of $\dot{m} \to \infty$, which from Eq. (3) would produce the maximum velocity immediately after launch. However, the vehicle would experience rapid slowdown due to the high drag at low altitude. With a lower flow rate, the missile would climb through the dense lower atmosphere at a lower velocity with lower drag impulse, taking the same time to arrive at an intermediate altitude but with higher velocity.

As an example, consider the case of a rocket weighing 4000 lb at liftoff, of which two-thirds is propellant that produces an average I_{sp} of 260 s (pound force seconds per pound mass). The rocket is 18 in. in diameter and 230 in. long and has a low-drag nose. Taking the unrealistic limiting case of instantaneous burning of the propellant, the burn-out velocity would be 9190 fps. For a vertical climb, the rocket would reach an altitude of 95,300 ft in 11.99 s with a velocity of 7500 fps and an altitude of 204,100 ft at a velocity of 7000 fps. For ascent trajectories to targets having elevation angles less than the zenith, the corresponding altitudes at these velocities would be lower because the drag impulse would be higher.

As shown in Fig. 5, there is a range of finite values of propellant flow rates that yield a higher Z_A for a corresponding velocity on a vertical climb. For clarity, the entire $\dot{m} = \infty$ curve is not included. Curves drawn through the loci of points-on-ascent trajectories corresponding to velocities of 7500 and 7000 fps are shown for flow rates from 60–200 lbm/s. It is clear that only when targets are near the zenith are high flow-rate values desirable. For targets below an elevation angle of about 34 deg, the optimal flow rate is about 80 lbm/s.

For the case with a constant flow rate of 80 lb/s, the propellant burn time is 33.3 s and the burn-out velocities vary from 7260 fps for targets at low elevation angles to 7858 fps for targets at elevation angles of about 35 deg. For an elevation angle of 90 deg, the burn-out velocity drops to 7770 fps, which again shows the interaction of the terms within the integrand in Eq. (3). This range of burn-out velocities corresponds to a mean longitudinal acceleration during burning, u_b/gt_b, of 6.76–7.32 g where t_b is the time at burnout. Although each missile design and mission would have a unique flow-rate value, the value that yields a mean acceleration of 5–10 g is typical of a wide variety of cases.

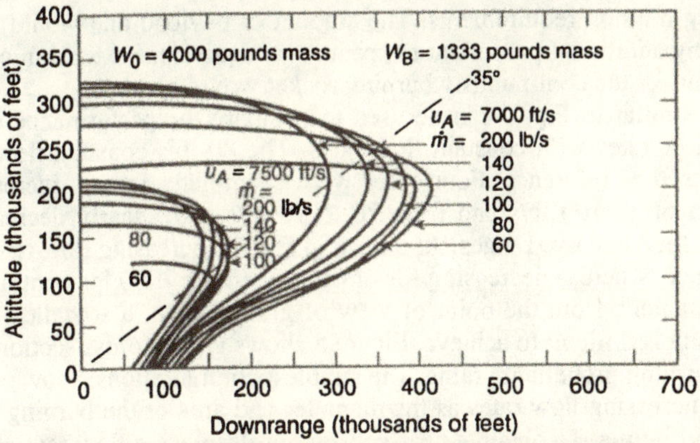

Fig. 5 Effect of propellant flow rate on ascent trajectories.

Granting that a constant flow rate of 80 lb/s is a reasonable design point from which departures can be made to examine other effects, the case of nonconstant flow rate can be studied. Two cases are considered: a constant flow rate with an intermediate period without burning and a linear variation of flow rate with time. Figure 6 shows the effect of interrupting the burning of the propellant for periods of 5–40 s once the vehicle reaches a velocity of 5000 fps. For target elevation angles of 13 deg, pulses up to 30 s lead to increases in performance, i.e., the $u_A = 7000$ fps contour is moved farther from the launch point. For example, at an altitude of 161,000 ft, the $u_A = 7000$-fps contour shifts from a downrange of 415,000 ft to a downrange of 620,000 ft. During the 30-s pulse, the missile slows from 5000 fps to 4600 fps before the second burn period. Lengthening the pulse beyond 30 s causes too great a slowdown to move the contour further; indeed, if the pulse is long enough, the vehicle cannot reach 7000 fps. The choice of 5000 fps for the point at which the zero-flow rate pulse is initiated is about optimum for extending the $u_A = 7000$ fps contour for pulse lengths of 30 s. Had the pulse been delayed until $u_A = 6000$ fps, only about half of the gain would have been realized. Whether the added complexity of the motor design to accommodate pulsing is warranted would depend on how crucial the additional performance is

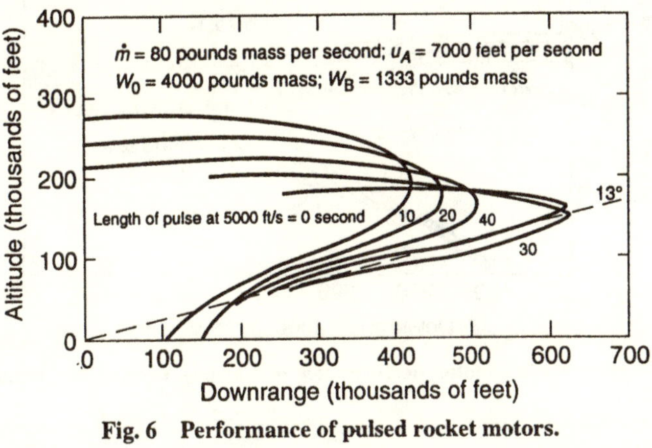

Fig. 6 Performance of pulsed rocket motors.

in meeting mission requirements. The amount of payload that would have to be replaced by additional propellant to produce an equivalent extension of the 7000 fps contour for the continuously burning rocket would be 85 lb.

A plot similar to Fig. 6 can be used to compare the performance of linearly varying flow rates with constant flow rates. The 80-lb/s constant flow-rate case is again used as reference. Four cases were examined: linearly increasing flow rates from 60 to 100 lb/s and from 70 to 90 lb/s and linearly decreasing flow rates over the same two ranges. As shown in Fig. 7, increasing flow rates degrade performance, whereas decreasing flow rates extend the 7000 fps contour.

Unfortunately, from the point of view of grain design, a so-called regressive burning rate is difficult to achieve. Figure 8 shows typical cross sections of some radially burning propellant grains. The simplest configurations, shown in Fig. 8a, result in increasing flow rates as the diameter and area of the burning surface(s) increase with time. To obtain a near-constant or decreasing flow rate, more complicated internal shapes are required, as shown in Fig. 8b, introducing problems of high-stress concentrations in corners and the possibility of producing longitudinal slivers of propellant that could be sheared off and expelled from the rocket chamber without burning.

An end burning grain could also produce a constant flow rate, but the required burn rates for the propellant would have to be increased significantly. For a flow rate of 80 lb/s, the volumetric burning rate for a propellant having a density of 0.067 lb/in.^{3} would be 1194 in.^{3}/s. For a grain having a diameter of about 17.5 in., this would require a linear burning rate of about 5 in./s, about an order of magnitude larger than can be obtained with the propellants shown in Table 1. The addition of longitudinal metallic wires in the propellant offers the possibility of increasing the burning rate, perhaps to the level required.

The internal star designs have surface areas ranging from 1 to 5 *dl*, where *d* and *l* are the overall diameter and length of the grains. Assuming that grain length is two-thirds of the 230-in. missile length, the corresponding range in required burning rates would be 0.09–0.45 in./s which is in the range that can be obtained with the Table 1 propellants.

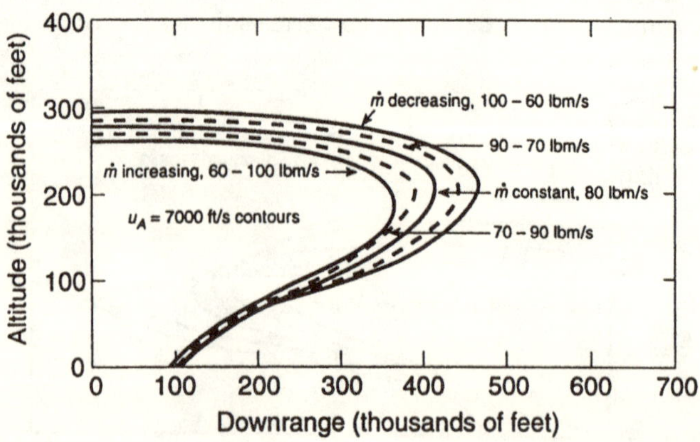

Fig. 7 Effect on rocket motor performance of linearly varying the weight flow of propellant.

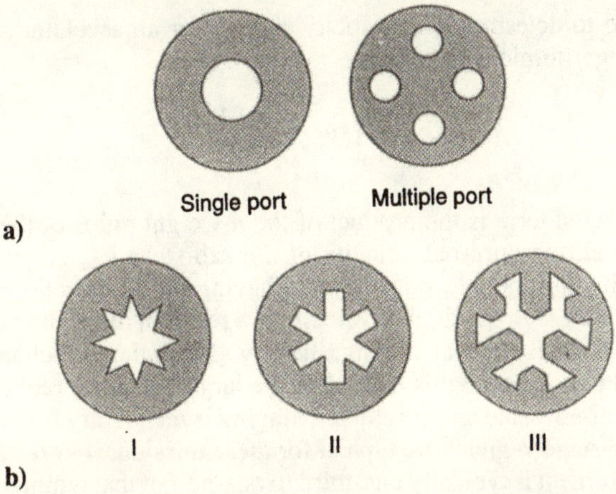

Fig. 8 Cross sections of typical solid propellant grains: a) progressive burning-rate designs, and b) near-neural (I) and regressive (II and III) burning-rate designs.

The burning rate of most propellants is in accordance with St. Robert's law, viz.,

$$\dot{m} = aA_p\rho_p P_c^n \tag{4}$$

where the lead coefficient a is constant for a given grain temperature. Combining this expression with the equation for flow through a sonic throat, viz.,

$$\dot{m} = P_c A^* u^* / RT^* \tag{5}$$

shows that for a fixed nozzle, i.e., A^* constant and propellant temperature (T^*, u^* constant),

$$A_p \approx (\dot{m})^{1-n} \tag{6}$$

The asterisks refer to conditions at the throat of the nozzle. To obtain a regressive burning rate of from 100 to 60 lbm/s, the range of percentage reductions in A_p would be from 29 to 54% for the propellants listed in Table 1. The internal star design shown on the right in Fig. 8 would have an A_p that decreases by about 60%. Either this or a somewhat less complex shape could presumable meet the regressive rate selected in the example.

There is a mitigating effect that was not considered in this example, viz., that the structural weight of the linearly varying-propellant-flow engine would be greater than that of the constant-propellant-flow engine because the design maximum pressure would be greater. Consequently, most missile designs tend to use burning grain designs that generate gases at approximately a uniform rate.

The trend of these results suggests that for two-stage rocket designs, the flow rate of the first stage should be greater than that of the second stage. To establish this point, it is necessary to examine the differences in performances for two-stage vehicles where flow rate is varied for the two stages. A complete examination of this issue is an arduous task, however, some exemplary results can be discussed that show the general trends that would results from a more thorough investigation. Classical theory (e.g., Ref. 3) shows that for vehicles having multiple stages, Eq. (3)

must be revised to determine the velocity change for an accelerating vehicle. In this case, the logarithmic term becomes

$$\ell n\left[\prod_{1}^{n}(W_{oi}/W_{bi})\right] \tag{7}$$

where the bracketed term is the product of the n weight ratios of the i individual stages. In general, the burn-out velocity of the nth stage u_{bn} is greater than the burn-out velocity for a single-stage vehicle having the same ratio of payload to initial weights, W_L/W_O. The higher velocity is a result of discarding the structural weight of the burn-out stage at staging, thereby saving the propellant that would have been used to accelerate this mass. For the large velocities required for long-range ballistic missiles and space vehicles, staging is mandatory for high W_L/W_O. Three- and four-stage vehicles are typical for these missions. For tactical missiles, the velocity increment is typically one-third to one-half of that required for ballistic missiles, and no more than two stages are required.

Analytical solutions (e.g., Ref. 3) to maximize the burn-out velocity for particular propulsive and structural characteristics have been found that show, for the idealized case of drag-free/gravity-free climb out of a vehicle having equal I_{sp} and structural coefficients $\epsilon = W_{si}/(W_{si} + W_{pi})$ for all stages, each stage should have the same weight fraction and, therefore, produce the same velocity increment as every other stage. Here, W_{pi} is the weight of propellant in the ith stage. When drag, gravity, varying I_{spi}, and ϵ_i are considered, the optimum u_{bi} for rocket powered vehicles occurs when the velocity increment of the stage increases with each succeeding stage. Figure 9 shows that optimum performance for two-stage rockets is obtained when the velocity increment of the first stage is 44–46% of the total Δu.

The same trend would also be observed when the second stage is an air-breathing missile. However, this cannot be fully exploited with the fixed geometry engines that are generally used in tactical missiles. Usually it is not possible to design a fixed geometry engine (i.e., the inlet area and the throat of the exhaust nozzle

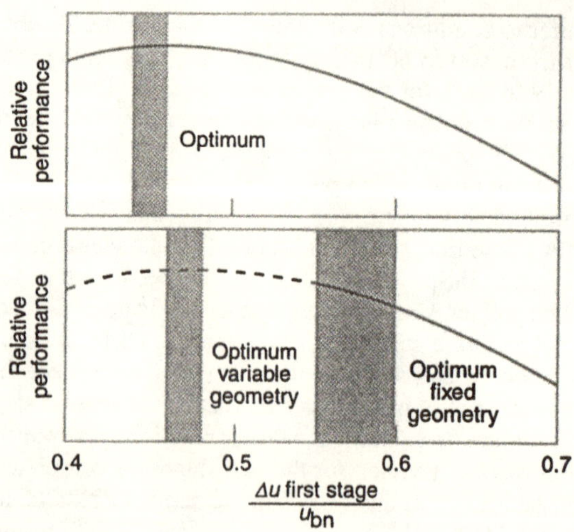

Fig. 9 Optimal staging of tactical missiles: a) two-stage rocket, and b) rocket booster, air-breathing second stage.

cannot be varied) that will operate over a velocity range below 55–60% of the maximum velocity, as shown in Fig. 9b. If variable geometry is practical, the velocity increment of the first stage can be reduced to about 46–48% for optimal performance.

A 4000-lb missile with similar characteristics as before, except that it has two stages rather than one, can now be re-examined. If a structural coefficient $\epsilon_i = 0.1$ is taken for the rocket motor case, not including the nozzle and thrust-vector control system, 1480 lb of propellant should be packaged in the first stage to obtain $u_{b1} \approx 0.45 u_{b2}$. At first-stage burn out, the total weight would be 2520 lb, of which 214 lb would be dropped (164 lb of rocket-motor case and about 50 lb of nozzle, thrust-vector control, and stage clamping devices). The second stage would have an initial weight of 2306 lb, including 1153 lb of propellant, 122 lb of structure, and 20 lb of nozzle. Thus, both missiles would have effective payload weights of 1006 lb if it is assumed that the nozzle thrust-vector control of the one-stage vehicle is 30 lb. The effective payload includes the warhead and guidance radome, the forebody, and all other components of the engine other than the rocket motor.

Comparing the performance of one- and two-stage rockets for altitude values of 120,000 ft or lower shows that the $u_A = 7000$-fps curve for the two-stage rocket would extend about 20,000 ft farther when the flow rates are 80 lb/s for both stages. This increment would about double if the mass-flow rates were optimized, i.e., 140 and 50 lb/s for the respective stages. Again, whether the added cost and complexity would be warranted for a defense missile targeted at low-elevation angle is questionable. On the other hand, for targets at high-elevation angles, a two-stage vehicle would be capable of reaching nearly double the range at the same burn-out velocity for a given elevation angle.

The mission depicted in Fig. 10 will be used to compare the rocket with two different rocket boosted air-breathing missiles. In this scenario, an incoming missile is targeted at a land-based site that is relatively near to an interceptor launch site and search radar. In an equivalent scenario, the target could be an aircraft carrier whose defense is being provided by an Aegis ship located nearby. The Aegis is equipped with the radar and houses interceptor missiles in a matrix in the vertical launch system. At time 0 the search radar detects the incoming missile. Shortly thereafter at time 1, sufficient tracking data have been processed to permit launching the interceptor in the approximate azmuthal plane. Stage separation occurs at time 3. Peak velocity of the second stage is reached at time 4. Rocket powered vehicles would slow down due to drag and gravity whereas the air breather would throttle back to maintain constant velocity to the point of fuel exhaustion. At time 5, onboard sensors acquire the target with a position uncertainty depicted by the cross section of the shaded cone and a corresponding region of possible intercept. From time 5 to the target intercept at time 7, the homing guidance system redirects the interceptor to minimize the miss distance. Higher velocity between time 5 and 7 minimizes time to target and miss distance.

Figure 11 compares the capability of rockets with air breathers for the area-defense mission of Fig. 10. Whereas the domain of velocity values previously discussed was equal or greater than 7000 fps, more meaningful direct comparisons of rockets and air breathers are made at velocities of 4000–6000 fps. A typical subsonic combustion-ramjet interceptor would have a cruise velocity of 4000 fps. For

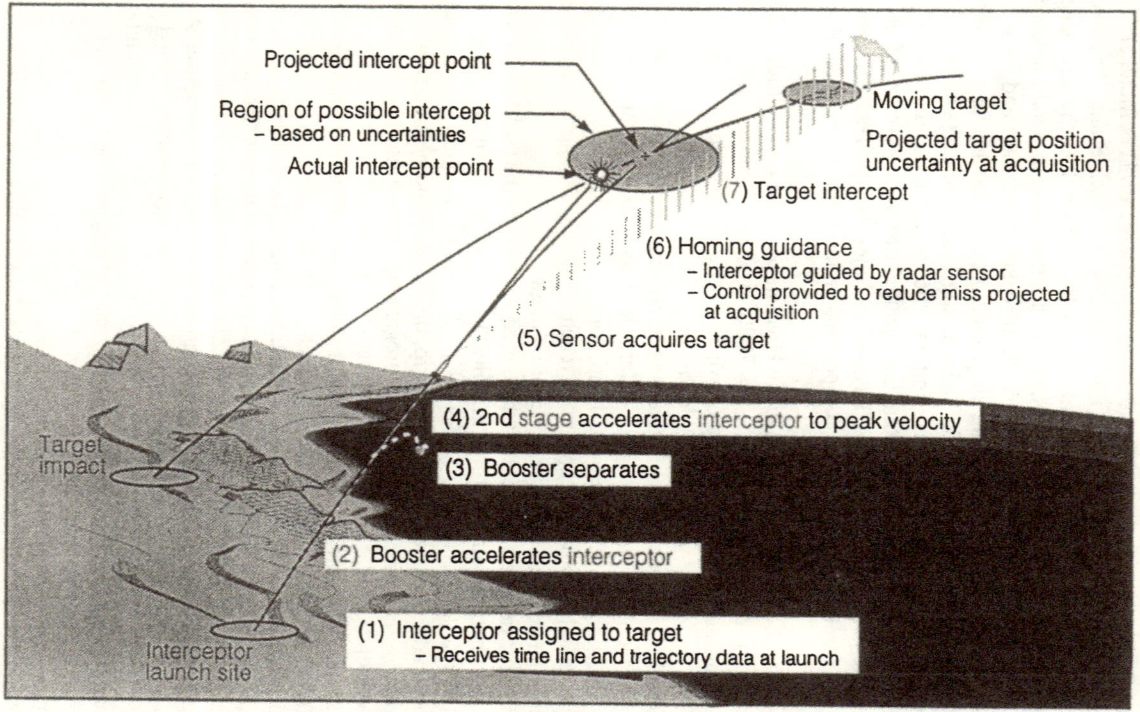

Fig. 10 Scenario of an interceptor providing area defense from incoming targets.

velocities greater than 4000 fps, the dual-mode, ram-scramjet typified in Fig. 2c is the optimal air-breathing engine cycle. The dual-mode, ram-scramjet can reach velocities greater than 6000 fps, but structural problems become formidable and materials choices are quite limited. Two approaches are possible to compare the rockets and air breathers in the range of 4000–6000 fps. The first is to compare an air breather capable of reaching $u_A = 6000$ fps and then slowing down to 4000-fps after fuel exhaustion. The second is to compare a family of air breathers, each designed for a particular maximum value. For velocity values less than 6000 fps, the latter approach would show better performance for the air breathers because more of the total weight of the system could be put in the second stage. For example, to meet the u_{b1} criterion shown in Fig. 9 for a fixed geometry air breather capable of reaching $u_A = 6000$ fps would require a booster weighing about 1960 lb, whereas for $u_A = 4000$ fps missile, the booster weight would be about 1259 lb. Again, these designs are for total weights of 4000 lb, and many assumptions have been made regarding structural weights and other factors.

The procedures used to optimize the air-breather stages and trajectories are complex, and the reader is referred to more detailed discussions of the subject in the following chapters and elsewhere (e.g., Ref. 4). Suffice it to say that the geometry of the engine, the modulation of thrust during climb out, and the variation of normal force all enter into the optimization.

In Fig. 11, the down-range scale has been compressed. The rocket is the preferred engine for all elevation angles above 10 deg. The region for the rocket is terminated since at high altitude, control of the missile would have to be provided by a means other than aerodynamic surfaces. The 160,000-ft limit shown here corresponds to a dynamic pressure q of 45 lb/ft^2 for a velocity of 6000 fps. Control at lower dynamic pressure would have to be provided by auxiliary thrusters; this greatly complicates the missile design. The air breather is preferred for all velocities that can be reached at $Z_A < Z_{A\,\max}$ prior to fuel exhaustion, since the air breather still has fuel remaining to maintain a high velocity, whereas the rocket is coasting. As shown, the air breather operating envelope is terminated at fuel exhaustion and at $Z_A < Z_{A\,\max}$, which corresponds to the altitude above which stable combustion cannot be obtained. The air breathers would also be capable of coasting to higher

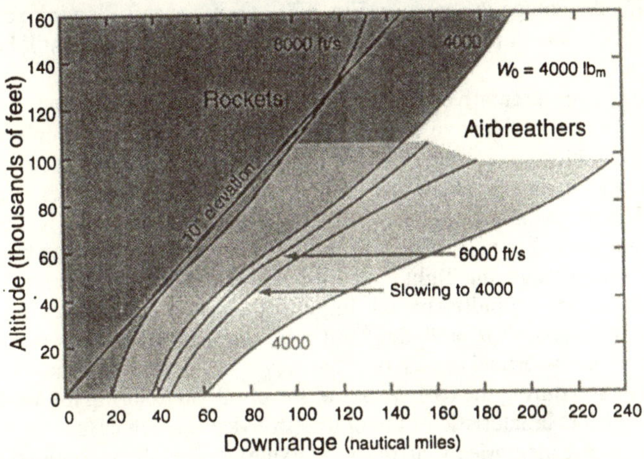

Fig. 11 Comparison of rockets and air breathers for area-defense mission.

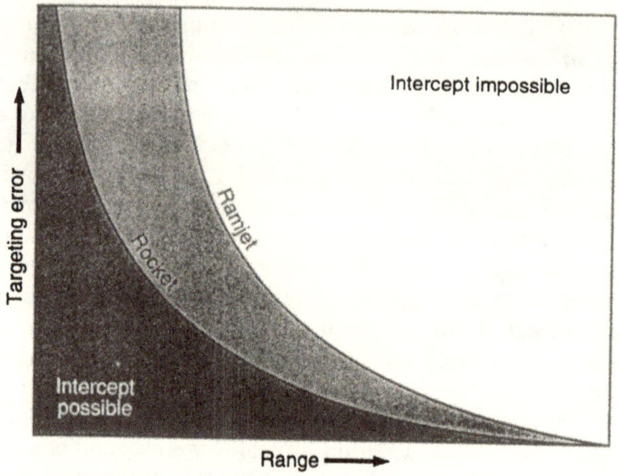

Fig. 12 Effect of targeting error on intercept capability.

altitudes with a corresponding loss in speed. Thus, intercepts could be made with the air breathers at altitudes above $Z_{A\,\max}$.

In the region covered by both the rockets and the air breathers, the latter would be more effective. At relatively long range from intercept when the onboard sensor acquires the target (point 5 in Fig. 10), its position is defined; therefore, the known targeting error is zero. As the end game proceeds, the guidance system on the interceptor perceives a targeting error from measurements of the relative position of the objects. Commands are given to redirect the interceptor to minimize the miss distance, taking into account the predicted velocity changes of both the target and interceptor. Unless the target is nonmaneuvering and on a collision course identical to the interceptor, the trajectories will at best intersect at a point. For crossing targets, the ability to obtain a direct hit for a kinetic-kill weapon or to minimize the miss distance with a proximity fused warhead is dependent on the maneuver capability and the velocity of the interceptor. The air-breathing engine which is powered during the end game has a distinct advantage over a coasting rocked. This advantage is shown in Fig. 12. At any range following acquisition, the air breather can contend with a much larger targeting error and still be capable of intercept.

Trajectories representative of a second generic class of missions are shown in Fig. 13. Here the objective is to obtain maximum range by accelerating and climbing to an altitude Z_c and cruising in near-horizontal flight from c to d. As in the previous example, the rocket has a maximum speed greater than the air breather. The propellant is consumed relatively early in the climb out, i.e., 33.3 s into a climb out of about 100 s, and the vehicle decelerates following burn out. To obtain maximum range, the flight path during climb out and approach to the cruise condition must be optimally shaped. In general, the optimal path corresponds to launch near the vertical ($\theta = 90$ deg) and continual turn down (negative lift) up to the apogee. If near-vertical launch is not achievable, positive lift (and thrust vector control) are used during the early phase of the climb out to bring the vehicle close to the $\theta = f(z)$ characteristic of the optimal vertical-launch case.

The types of trajectories that must be examined to obtain optimal range of a high thrust-to-weight (F/W) ratio vehicle (e.g., the rocket) are typified by

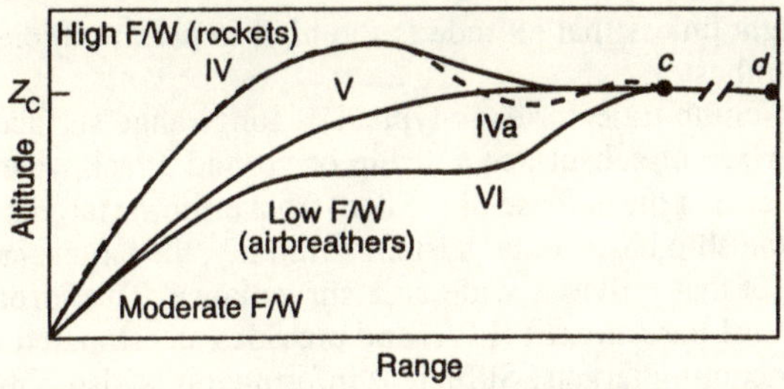

Fig. 13 Effect of thrust-to-weight ratio on ascent trajectories for missiles designed for near-level flight at an altitude Z_c.

curves IV, IVa, and V in Fig. 13. A high F/W vehicle has the accelerative capability to overshoot Z_c by a number of possible paths with the introduction of positive lift. Curve IV corresponds to a case with a lift-to-weight ratio $(L/W) > 1$ as Z_c is approached. With $L/W = 1$ at Z_c, the vehicle will oscillate about Z_c with a decreasing excursion as time progresses, as typified by curve IVa. With $L/W < 1$, the vehicle can approach Z_c at $\theta = 0$ asymptotically from below (not shown). The permissible altitude at apogee is limited to the aforementioned requirement to maintain control; however, in general, optimal range is obtained on trajectories having lower apogees. These constraints significantly restrict the region of permissible climb-out trajectories to a narrow corridor in ψ_a , where ψ_a is the elevation angle of the apogee with respect to the launch point. For example, given that the flow rate is 80 lb/s and that the rocket is constrained to an apogee of 100,000–160,000 ft, then the range in elevation angle is from 11 to 15 deg.

Curve V of Fig. 13 schematically represents the climb out of a rocket with a moderate thrust-to-weight ratio or of a ramjet having a relatively large ratio of inlet areas to body cross-sectional area. For the same Z_c, however, the optimal climb out for an air breather would lie below that of a rocket. That is a consequence of the need to optimize thrust and fuel flow rate simultaneously with the minimization of the integrand in Eq. (3) in the climb out of an air breather. Typically, those trajectories require modulation of the lateral force that produces turn down following launch, followed by positive lift at low and intermediate altitudes and negative lift as Z_c is approached.

Curve VI is representative of the climb out of a low thrust-to-weight ratio engine. The accelerative capability is inadequate to climb to Z_c and accelerate to the desired u_c simultaneously. Instead, the vehicle must climb to a lower altitude, use lift to maintain this altitude, accelerate to u_c, then turn up with $L > W$, and finally approach Z_c with $L < W$. Usually this type of climb out results in poorer performance than a case V climb out, which is one of the factors that influences the choice of the proper thrust-to-weight ratio propulsion system.

Optimal cruise range for the air breathers or glide range for the rockets usually results when the vehicle flies from c to d at an angle of attack corresponding to $(L/D)_{max}$. The maximum lift-to-drag ratio varies with the aerodynamic configuration, velocity, and altitude, but for tactical missiles in the range of 4000–7000 fps, $(L/D)_{max} \approx 3$ and the corresponding angle of attack α is between 8 and 12 deg. During cruise, the rocket is lighter than the air breather, and the optimum altitude at the same velocity is somewhat higher. As the rocket slows down during the glide, the optimum altitude decreases. On the other hand, as the air breather burns fuel to maintain constant velocity, the optimum altitude increases with the decreasing weight unless that altitude is too high to permit efficient operation of the inlet and combustor.

This class of climb trajectories is typical of long-range surface-to-air defense missiles and surface or subsurface antiship or ground-attack weapons. Figure 14 depicts the scenario of the defense of the fleet from airborne targets located beyond the horizon of the ship based radar systems. Initially, the targets are detected by a space-based asset that provides wide-area surveillance. This information is used to position forward-pass aircraft (FPA) and provides an estimated range and "raid count" of the incoming targets. Sufficient information is also obtained to permit launching of the surface-to-air missile (SAM) in the approximate azimuthal plane of the target aircraft. During the climb-out and/or cruise phase of the SAM, the target aircraft crosses the radar horizon of the FPA which then provides a real-time target track for the SAM. When the SAM reaches the general vicinity of the target aircraft, an onboard terminal seeker provides the final targeting information. Point d from Fig. 13 would represent the intercept of the SAM with the target aircraft.

Figure 15 shows the altitude-range envelope for points d for the same tactical missiles (i.e., SAMs) used for the flight envelope of Fig. 11. For the rocket, boundaries are shown for slowdown to $u_d = 4000$ and 6000 fps. For the air breathers, three boundaries are shown. The area above the dashed line is shaded to show that angles of attack greater than 20 deg would be required to maintain level flight of a coasting rocket. Indeed, at altitudes above about 130,000 ft, level flight is not possible for this vehicle. The $u_d = 6000$ fps is for a missile designed to cruise at 6000 fps. Two 4000-fps boundaries are shown; the one having the greatest extent is for a missile designed to cruise at 4000 fps and the other is for the 6000-fps design, coasting to 4000 fps following fuel exhaustion.

As in Fig. 11, the range that can be reached by the air-breathing missile is considerably greater than that reached by the rocket. On the other hand, for ranges less than that corresponding to equal velocities of the rocket or air breather (i.e., 6000 or 4000 fps) for this example, the rocket could be the preferred propulsion system for a tactical missile, the argument being that the mean velocity would be greater, hence, the time to target would be less for the rocket. This, in turn, could perhaps permit more time for assessment prior to firing or time for more salvos in a shoot-look-shoot scenario.

The optimum cruise altitude for the rocket, assuming $C_L = L/q_0 A_R \approx 1$ at $(L/D)_{\max}$, where $A_R = \pi D^2/4$, would be about 111,000 ft at $u = 6000$ fps and 93,000 ft at $u = 4000$ fps. Maximum range, however, occurs at lower altitudes because cruise range for the coasting rocket is small compared to the ranges of the optimal paths to points c. For the air breathers, the corresponding altitudes for flying at $(L/D)_{\max}$ would be 2000–4000 ft lower at the same velocity, but better performance would be achieved at slightly lower altitudes due to the aforementioned problems of low inlet and combustor efficiency at low-pressure levels. Thus, the cruise phase of the air breather occurs at a somewhat lower angle of attack than that which yields $(L/D)_{\max}$. For this example, optimum cruise altitude is about 95,000 ft for $u = 6000$ fps and 80,000 ft for $u = 4000$ fps.

The scenario shown in Fig. 14 can also serve to explain the benefits that could be derived by higher speeds of the interceptor. If it is assumed that the target aircraft are carrying antiship missiles (ASM) and have not been intercepted prior to ASM release, then the ASM become targets of the type shown in Fig. 10, with intercept capabilities typified by Fig. 11. In reality, there is but a finite probability of successful intercept for a given SAM. Thus, more than one SAM needs to be

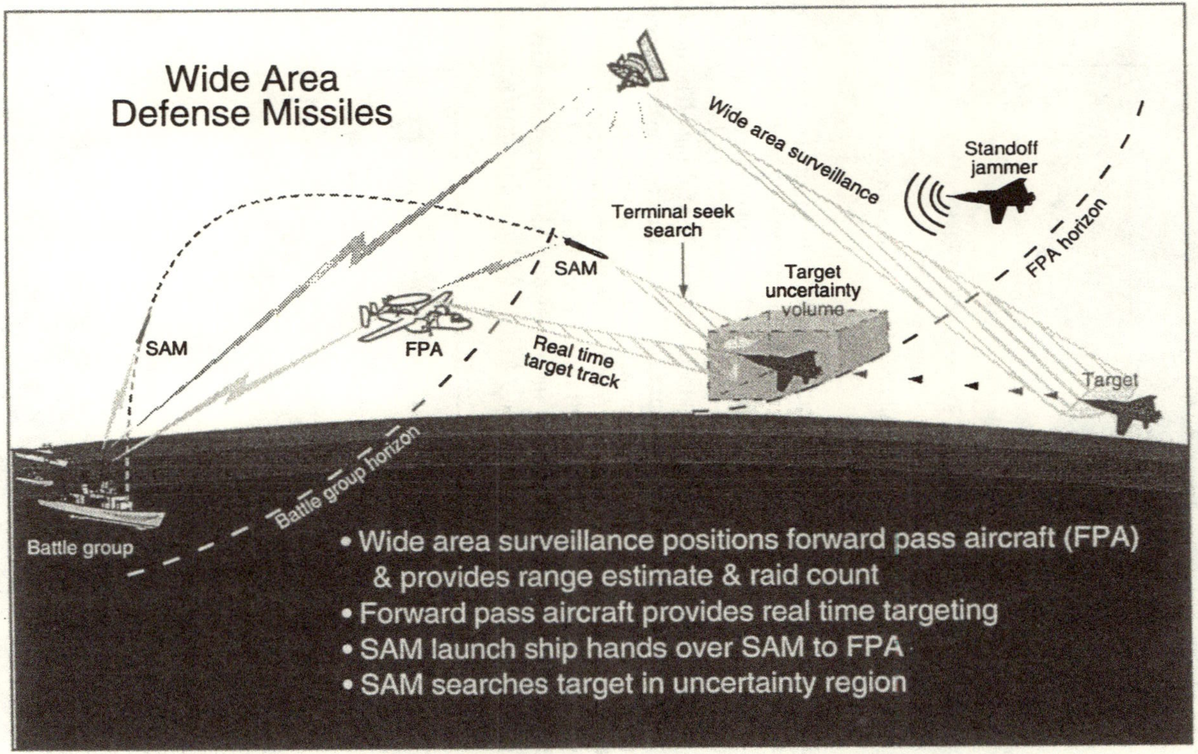

Fig. 14 Scenario of outer air-battle cooperative engagement.

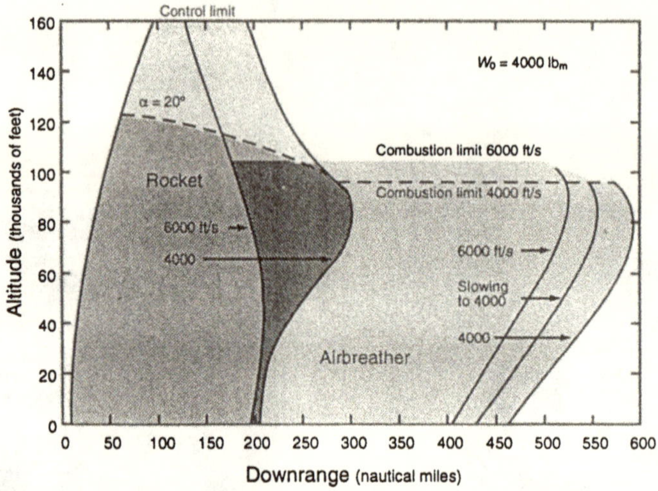

Fig. 15 Comparison of rockets and ramjets for long-range cruise missions.

fired at each ASM to increase the overall kill probability. Figure 16 shows the depth of fire, i.e., the number of SAMs that could be targeted at a SAM which is launched at a range of 150 n mile and must be intercepted at a range no less than 10 n mile. Results are shown for Mach 3 and Mach 4 ASMs. In this shoot-look-shoot firing doctrine, 5 s of kill assessment is assumed between SAM firings. The minimum speed of the SAM is about that of the speed of the ASM. Increasing the block speed of the SAM by about one Mach number increases the depth of fire by one unit. This increase in depth of fire results in a corresponding decrease in the probability of the ASM penetrating the postulated 10-n mile boundary. Figure 17 shows results for an assumed kill probability of 0.5 for any given SAM. The probability of the ASM penetrating decreases dramatically as the SAM speed increases.

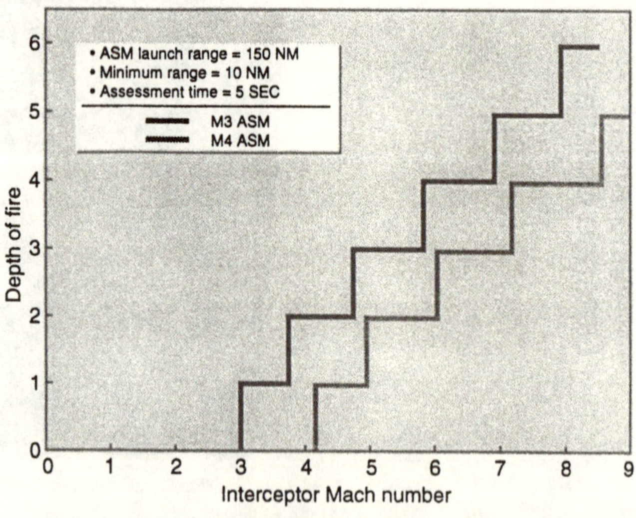

Fig. 16 Fire power against ASMs screened by jammers.

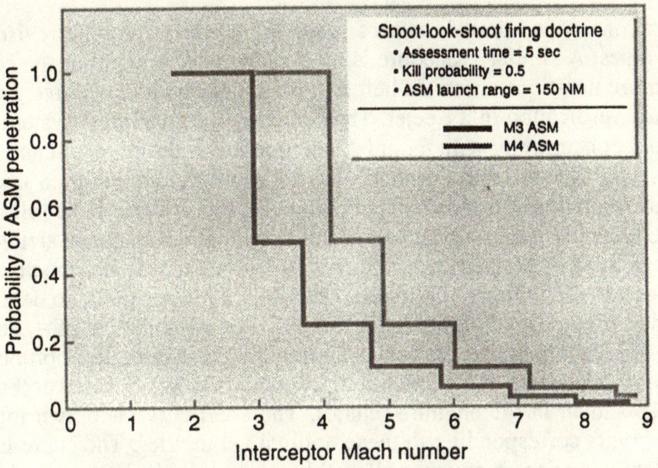

Fig. 17 Probability of ASM penetration to 10 n mile.

A. Design Procedures for an Air-Breathing Missile

Figure 18 is a flow diagram that in general terms shows the design and development stages of a ramjet powered missile. Initially, engine performance is determined using assumed efficiencies for the engine components, weights, and aerodynamics based on preliminary composite designs and constraints imposed by the guidance system, ordnance, stowage, etc. Parametric variations are made in each of the categories and trajectories are flown to determine which candidate-engine concepts can perform the proposed mission. Based on these results, computational techniques are used to obtain preliminary designs for candidate-inlet

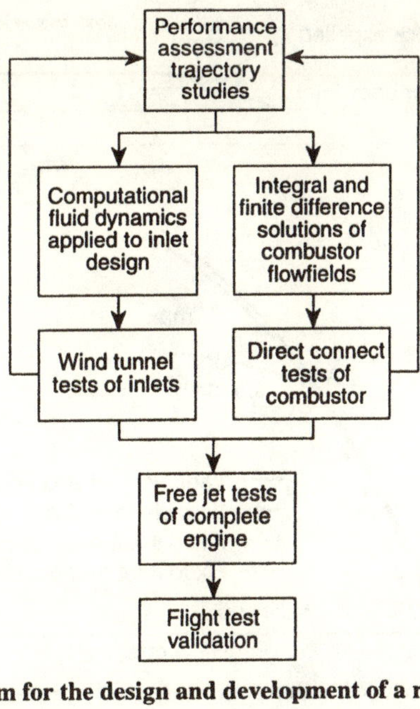

Fig. 18 Flow diagram for the design and development of a ramjet powered missile.

configurations. Wind-tunnel tests of inlets are made to provide revised performance values. A similar procedure is used in the development of the combustor. One or more iterations are made before a principal candidate-engine geometry is chosen for verification in a freejet. The fuel-supply and control-system design is carried out concurrently with the inlet and combustor development and generally is used in the freejet engine test. If all goes well, the engine then undergoes a flight-test validation. A cursory examination of each of these steps follows.

Figure 19 shows results from one of the steps that take place in sizing the inlet and nozzle throat. Parametric variations of the inlet area A_i and the nozzle throat area A^* at various flight conditions and fuel-flow rates are made to determine the thrust level. For a given fuel-flow rate, there exists a combination of A_i and A^* that yields the maximum thrust. For a subsonic combustion engine, this optimum occurs when the normal shock in the inlet-compression process is located at the cowl-lip station, as shown in the engine sketches. The solid curve is drawn through the locus of points corresponding to these optimal geometries. The curve terminates at the point where the maximum allowable A^* and the fuel/air equivalence ratio equals 1. This curve gives the performance of a variable geometry engine.

Take the A_i and A^* that correspond to any point on the curve, e.g., point A permits comparison of partially variable and fixed geometry engines. If only the inlet can be varied, optimum performance is obtained with the shock on the cowl lip, but more fuel is required to produce the same thrust, and an equivalence ratio equal to 1 occurs at a lower thrust level. Likewise, with a fixed inlet and a variable exit, the normal shock is on the cowl lip, more fuel is required than for the variable geometry engine, and the maximum thrust is lower. With a fixed inlet and exit, the normal shock will be located on the cowl lip only at point a.

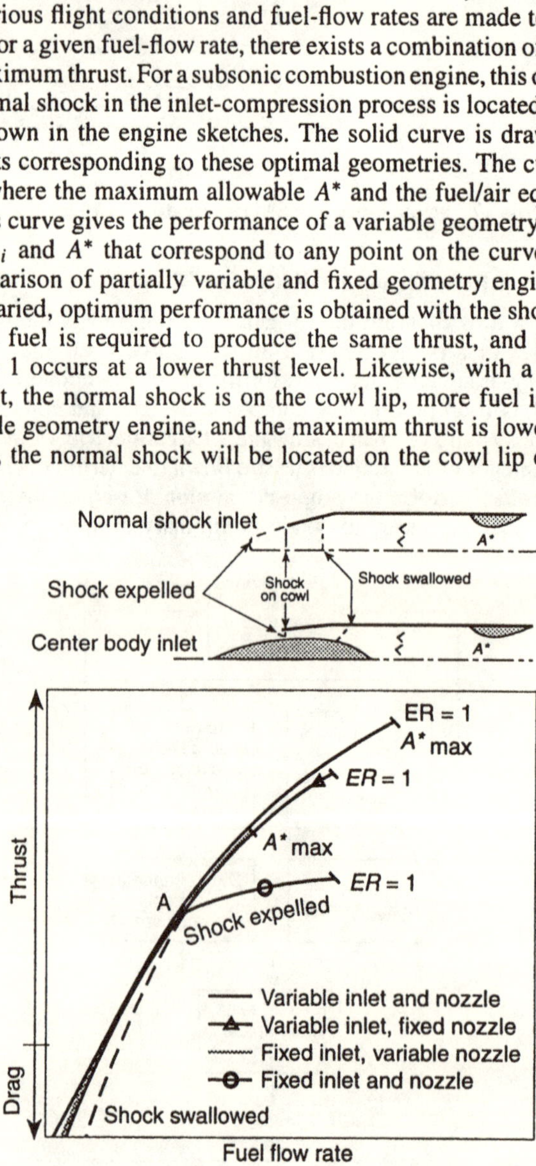

Fig. 19 Procedure used to determine inlet and nozzle-throat dimensions.

For lower fuel rates, this shock is swallowed, as shown in the sketches, and the performance is correspondingly degraded. For higher fuel-flow rates, the normal shock must be expelled which may or may not be acceptable. In normal shock inlets that are used at relatively low-flight speeds (i.e., less than Mach 2.2), no problems are encountered; however, as shown in Fig. 19, the gains in thrust accrue only with large increases in fuel-flow rates. With centerbody inlets that are used at higher speeds, expelling the shock much beyond the cowl lip generally results in the generation of an unacceptable instability, sometimes referred to as buzz. In a buzz condition, the airflow into the ramjet pulsates with a frequency equal to the fundamental or to a harmonic of the equivalent Helmholtz resonator.

The choice of point A in this discussion was arbitrary and, if more thrust were needed in a fixed geometry engine, a larger inlet would be dictated, albeit with an increased fuel flow and less efficiency at lower thrust levels.

B. Inlet Testing

Typical inlet test performance data are illustrated in Fig. 20. The flowfield of an inlet in flight can be simulated in a wind-tunnel test as long as the Mach number and Reynolds number [(velocity diameter)/kinematic viscosity] are duplicated. If the tests are performed at pressures higher than in flight or at temperatures lower than flight or both, the ratio ud/v can be several times larger and the diameter can thus be smaller. Typically, wind-tunnel tests are carried out at one-fifth to one-third of the flight size. Tests are run over the entire range of Mach numbers and Reynolds numbers at all angles of attack expected in flight. For asymmetric designs, various roll orientations must also be examined.

The principal parameters that are measured in the test are pressure recovery, air capture, and cowl and additive drag. The details of both the internal and external flowfields and the inlet-starting characteristics may also be of interest, especially if boundary-layer separations or large gradients in the internal flow are present or if problems are expected in starting the engine. In Fig. 20, pressure recovery as a function of air-capture ratio A_o/A_i is shown for Mach 2–5 for a centerbody inlet designed for full capture at $M = M_D = 4$, angle of attack = 0 (M_D is the design

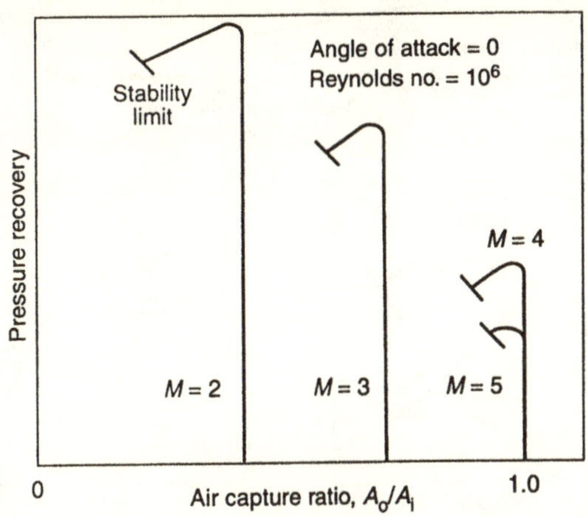

Fig. 20 Typical air-capture pressure-recovery characteristics of an air-breathing missile inlet.

Mach number). These data are obtained by closing a calibrated throttle attached to the inlet which simulates the pressure rise that would be caused if combustion were present. As the throttle closes, the normal shock moves toward the cowl lip and the pressure recovery increases. When the normal shock passes upstream of the cowl, air capture decreases. With further closing of the throttle, the inlet reaches the aforementioned stability limit. The ratio of the cross-sectional area of the steamtube captured by the inlet to A_i is 1 for $M \geq M_D$. Pressure recovery corresponding to normal shock at the cowl lip increases with decreasing M.

Systems integration issues often have considerable impact on the design of missile inlets. Figure 21 lends some insight to the problem of the integration of guidance-system components with the inlet-compression surfaces. In the design that is depicted, both an infrared (IR) and an rf seeker must be integrated in the missile forebody which also serves as the principal compression surface of the air inlet. Propulsion and guidance present competing demands. The IR seeker desires a hemispherical window; bluntness decreases the pressure recovery and air capture of the inlet. The rf seeker would prefer a low-fineness ratio to minimize boresight-error slope and to position the cowl lip aft to maximize seeker-look angle. The inlet desires a high-fineness ratio to maximize pressure recovery and needs to position the cowl forward to maximize air capture and optimize wave cancellation. An overall system design including trajectory studies is required to obtain the optimum-system designs as qualitatively shown in the lower portion of the figure. Other integration issues, e.g., warhead shape, can also impact the inlet design but generally can be accommodated without as great an impact.

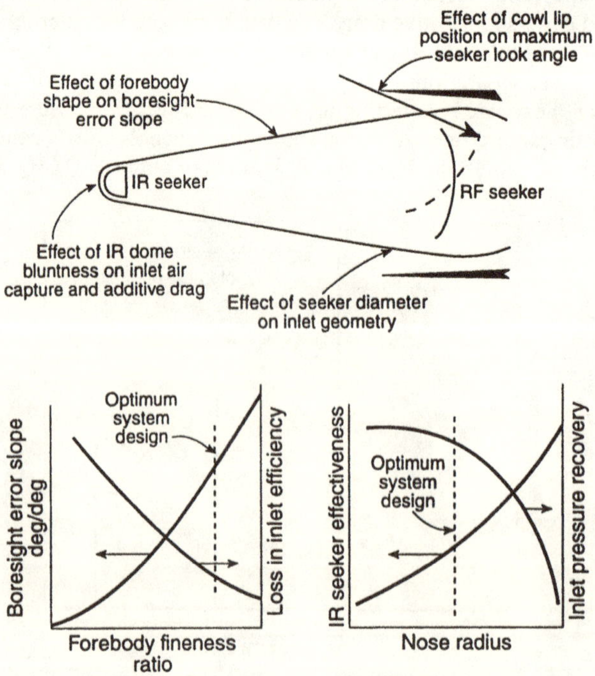

Fig. 21 Integration of guidance-system seekers with inlet-compression system.

C. Combustor Development

Combustor development involves the direct-connect testing of the combustor chamber to an air ducting system that provides air at conditions simulating those provided by the inlet in flight. Since there is no unique similarity parameter for combustion, the tests must be made in full-scale hardware. In the course of the development of the combustor, numerous fuel injector and flame-holder geometries are investigated. Other parameters that may be studied include the nozzle throat area, combustor length, port-to-dump area ratio in sudden expansion burners, and the effects of swirl. The standard measures of performance are combustion efficiency η_c and total (stagnation) pressure loss. The former is the ratio of the heat release actually realized to what is theoretically available in the fuel. The latter is a measure of all of the nonkinetic losses that are present.

Typical direct-connect testing data are shown in Fig. 22. Combustion efficiency varies with equivalence ratio, air-total temperature, and pressure or weight flow. Both rich and lean blowout limits can be present, and regions where the combustion is unstable are often encountered. These instabilities can be either longitudinal or transverse. The former, relatively low-frequency oscillations, can cause unfavorable interactions with the inlet; the latter, relatively high-frequency oscillations, can produce detrimental effects on insulation materials, electronics components, etc.

Finally, in freejet testing, the entire engine is placed in a large ground-test facility. It is installed on a thrust stand to measure the net force on the engine. The entire test chamber is enclosed so that tests can be made that simulate high-altitude flight. Figure 23 shows typical test data. Successful completion of the freejet engine tests provides the opportunity to proceed into a flight-test program with reduced risk.

The development of other types of propulsion systems follows a similar approach to that diagrammed in Fig. 18, but the steps may differ depending on the components that must be tested. For example, rocket-motor development obviously does not require inlet testing but, frequently, attention must be given to tailoring the propellant composition to meet specific flight requirements. Because the coefficient a and exponent n in Eq. (4) are only weakly dependent on scale, most of the combustion characteristics of propellants can be obtained in a small-scale apparatus. Then, when a candidate propellant has been selected, full-scale motor firings are made to verify the small-scale results and to examine behaviors such as combustion instability that are scale dependent.

Much of the emphasis in the full-scale tests of rocket motors, and indeed in the testing of other propulsion cycles, addresses issues other than propulsion per

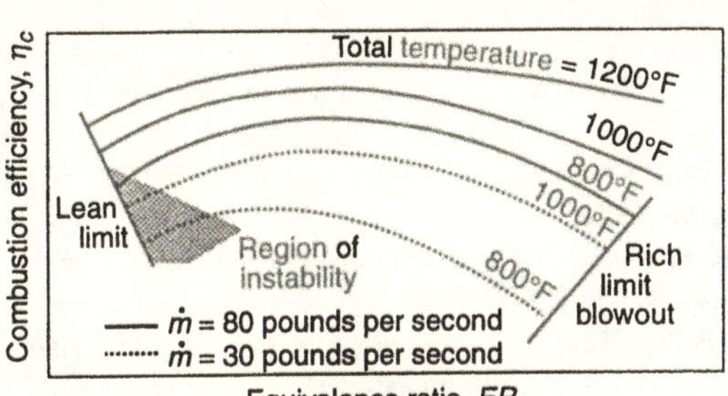

Fig. 22 Effects of temperature, pressure, and equivalence ratio on combustion efficiency in direct-connect testing of the combustion chamber.

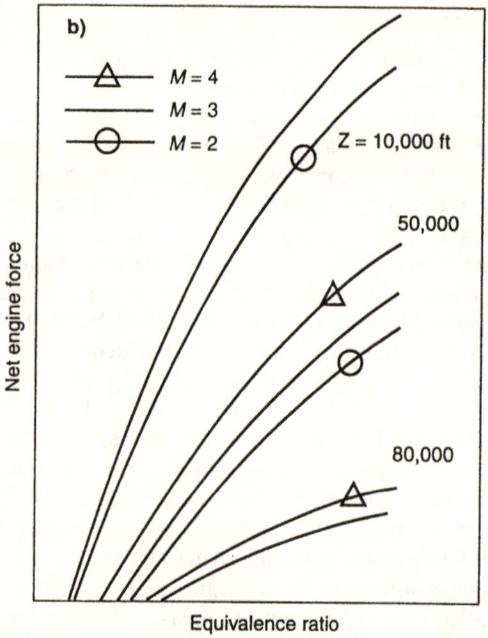

Fig. 23 Freejet engine testing, typical measurements of engine force at various fuel-flow rates and altitudes.

se, e.g., the structural integrity of the propellant, rocket-motor case, and nozzle. When the nozzle must provide thrust vector control, the operating characteristics of the actuating system, the structural adequacy of the gimbal or vanes, and the determination of the thrust degradation as a function of the lateral force generated must be examined.

The development of integral rocket ramjets requires not only all of the steps for both the rocket and the ramjet individually but also a ground-test demonstration of the transition from operation as a rocket to that as a ramjet. Even for cycles of this complexity, experience has shown that when meticulous attention is given to the resolution of all problems encountered in ground tests, near-flawless performance in flight-test programs can be expected, as exemplified by the tests of the Advanced Strategic Air-Launched Missile for the U.S. Air Force.[5]

References

[1]Bryson, A. E., Jr., and Ho, Y. C., *Applied Optimal Control,* Halsted, Washington, DC, 1975.

[2]Miele, A., "Recent Advances in Gradient Algorithms for Optimal Control Problems," *Journal of Optimization Theory Application,* Vol. 17, Dec. 1975, pp. 361–430.

[3]Hill, P. G., and Peterson, C. R., *Mechanics and Thermodynamics of Propulsion,* 2nd ed., Addison-Wesley, Reading, MA, 1992.

[4]Waltrup, P. J., Billig, F. S., and Stockbridge, R. D., "Engine Sizing and Integration Requirements for Hypersonic Missile Applications," NATO/AGARD/PEP 58th Symposium, CP 307, Oct. 1981.

[5]Cushman, J., "Ramjet? Roger, Says DeLauer," *Defense Week,* Vol. 6–7, Sept. 27, 1982.

Chapter 3

Liquid Rockets

Ben F. Wilson
U.S. Army Missile Command, Redstone Arsenal, Alabama 35809

I. Introduction

A HISTORY of 50 years of liquid rocket propulsion technology is available to the propulsion designer. The development has included all types of engines from the very small thrusters used for satellite maneuvering to the large propulsion systems used for space launches. Each component of the propulsion system has been analyzed, designed, fabricated, tested, and optimized for many individual missions. The technology is well documented, it has been reported in detail, and it has been abstracted. Generally, the development had the goal of achieving "maximum performance per unit mass of propulsion system" because the mass of the system was proportional to launch cost. Included in this technology development is a limited number of liquid rocket systems for tactical applications. The final designs of these tactical systems are well documented and many of the development problems are available through historical searches. However, the reasons for and considerations in selecting a specific design for a tactical application are not readily available.

Based on the quantity of development that has gone into propulsion systems in the last 40 years, one would expect a comprehensive model available for the optimum design of a tactical liquid propulsion system to meet any specific requirements. A general model for tactical liquid propulsion design is not available. Perhaps the limited number of fielded systems has not warranted its development. Some general trajectory models are available that can establish an optimized thrust profile for the projected mission. The propulsion companies provide limited models to aid the design, however, these models are based on the specific engine design for that company and are not optimized for the industry. Additionally, these models are based on experimental data and cover a narrow range of sizes and capabilities. Generalized models have been solicited but funding for their development has been unavailable.

The purpose of this chapter is to provide sufficient insight into system requirements, the source of these requirements, and the general technology available for tactical liquid propulsion systems to permit the designer to achieve the optimum propulsion system for a specific weapon system. The specific design technology for system components has been well documented and will not be repeated, and it is assumed the reader is well versed in the fundamentals of liquid propulsion.

II. Design Requirements and Restraints

Each tactical missile system has a specific mission and the propulsion system is a major contributor to the accomplishment of this mission. The propulsion must be integral with the system design and not be considered an "add on" module. A liquid propulsion system can add significantly to the versatility of the missile system, and its attributes of thrust management and control along with its signature characteristics should be used to complement and simplify the other missile components. There are some missions for which the liquid propulsion system is not suited and into which it should not be forced.

The requirements that follow are general requirements for tactical systems and are designed primarily to ensure their consideration in developing specific propulsion requirements.

1) The initial missile performance parameters guide the missile design, but often the initial requirements are arbitrarily established and can be modified through negotiations. The range limits, both maximum and minimum, normally are derived from the battlefield scenario and almost always are critical parameters. Missile velocity during end-game maneuvers may be based on the characteristics of the target as well as the type of warhead. The "fly out" velocity is normally negotiable and is the area in which the propulsion designer may make the greatest contribution.

The flight trajectory and the velocity profile are factors that influence aerodynamic drag and, therefore, the impulse requirement. The drag, a function of the velocity squared, can be reduced by decreasing velocity in the dense atmosphere. The drag also can be reduced by maximizing the flight range in the lower density upper atmosphere. Simple trajectory models can be used to optimize the trajectory and velocity to achieve minimum required impulse. The thrust management capabilities of the liquid system make it ideal for this optimization. Both thrust level modulation and pulse width modulation should be utilized in the optimization to achieve the full potential of the liquid system.

2) The propulsion system envelope is the next consideration. The volume available for the propulsion system in a tactical missile typically is restricted. The missile length is restricted by the launch platform; the missile length cannot exceed the available space for launch or storage. The missile diameter is normally restricted to maximize the number of missiles on each launcher or transporter.

A section of the missile normally is set aside for propulsion. The propulsion designer must determine if the allotted space is adequate, if the exhaust is correctly located to minimize interference with the sensor or communications, if the propellant is located to minimize the effect of the center of gravity (c.g.) shift, and if the required impulse can be achieved. The liquid propulsion system offers some flexibility in locating components because the nozzle and reactor can be separated from the propellant. The variety of liquid propellants offers choices both in density packaging and in performance. The possible integration of the liquid components with the other modules (warhead, guidance, controls, and sensor) provides an opportunity to increase the density packing of the missile.

3) The operational requirements of the missile typically are functions of the mission and launch platform. The military must be prepared to operate in any climate from the Arctic to the desert. For surface-launched weapons this usually translates to a temperature range from 230 to 335 K. For air-launched systems the temperature range is often wider, because the missile must withstand the cold temperatures at altitude and aerodynamic heating if mounted outside the airplane.

Vibration must be considered in the design of any tactical system. The launch platform can impose high-vibration loads on the missile. The missile also is exposed to high vibration and shock loads during transportation. These loads normally are determined from experimental data, and the propulsion system and its shipping and launch canister must combine to survive these conditions.

The wide variation in temperature imposes many restrictions on the propulsion system design. The ullage (unfilled volume in liquid tank) in the propellant tanks must be adequate for thermal expansion of the liquid at the highest temperature limit. As the temperature increases the density of liquid decreases and affects the mass flow rate. To assure adequate pressurant the gas pressurization system must be sized to provide the required pressures at the lowest operating temperature. At high temperatures the gas system is considerably oversized and may require a system for dumping the excess gas.

Design of a system to operate over wide temperature limits decreases performance. Therefore, the limits should be as tight as possible. Sometimes the other missile components will have tighter temperature limits than those initially proposed for the propulsion system. In this case the temperature limits for propulsion operation should be no larger than those imposed for other components.

4) Typical storage requirements are straightforward; the system is to be stored for 10–20 years without any preventative maintenance and be ready to perform at any moment. This philosophy of operation has become known as the "wooden round" concept. Temperature limits during storage usually are wider than operational limits to permit unprotected storage in all climatic conditions. This increases the ullage requirement and decreases the packing density. Vibration and shock requirements during storage also are greater than operational requirements. Experience has shown that the most severe shock, vibration, and acceleration loads are imposed during storage and transportation.

5) Restrictions on missile signature are imposed to prevent detection of the missile, prevent detection of the launch site or launcher, and to permit communications with the missile during flight. Any emission that will enhance detection by visual, infrared, ultraviolet, or radar sensors or will interfere with radar communication with the missile is objectionable.

A theoretical analysis of the propellant systems will indicate any undesirable particles in the exhaust and guide propellant selection. Additionally, the propellant and propulsion system should be evaluated in a signature characterization facility to determine any signature problems. These evaluations should be performed throughout the operation temperature range and under various humidity conditions to thoroughly characterize the signature effects.

6) The next requirement is safety. The Department of Defense has established an insensitive munitions (IM) policy for all weapon systems, and criteria and tests to ensure compliance with this policy are being established. Details of the policy requirements are documented elsewhere in this book. Liquid propellants are nondetonable; the energy release takes place either through combustion or

decomposition alone, in combination with another propellant, or in combination with the surrounding materials. In all instances, the liquid propulsion system can be designed to meet all the IM requirements.

The major safety concern of the user of liquid propulsion is toxicity, and the propulsion system designer must assuage this concern. For most liquid systems the exhaust products are nontoxic. The use of interhalogen oxidizers will produce some quantities of hydrogen chloride and hydrogen fluoride. These products pose little toxicity hazards because the hot gases rise and quickly disperse. The greater concern with these oxidizers is the pollution of the environment through acid rain.

The liquid fuels and oxidizers are toxic, and their toxicity mandates zero leakage of the propellants during the storage life of the system. This is a standard requirement for the prepackaged liquid propulsion system, and the capability to meet this requirement has been demonstrated in the Lance Missile System.

The safety characteristics of gelled propellants have prompted designers to consider their use in the tactical missile propulsion. The active ingredients of the gelled propellants are the same as in liquid propellants, but the flow properties have been modified to slow the flow of the oxidizer and fuel and decrease the intensity of the combustion in a spill.

7) Some other requirements of the propulsion system may not be well defined and require searching to identify. Examples of the items to search out that may sell the liquid propulsion system and its inherent thrust management capabilities are listed as follows. Will thrust management provide sufficient flexibility to permit the system to be used for two missions? Can thrust management increase the probability of kill for the system? Has the customer been presented sufficient data to overcome doubts about the liquid system? Are biases built into the requirements that preclude the use of a liquid system, and if so can the requirements be modified?

III. System Selection

The system designer must consider many factors in selecting the correct liquid propulsion system for the mission and use engineering judgment to arrive at an optimal solution. In narrowing the choice to a liquid system, many factors already will have been examined. The following sections describe the characteristics of the three types of systems to permit the selection of the best system to meet the requirements. One should remember that thrust management is the primary characteristic of all liquid systems.

Hydrazine monopropellant systems are the typical choice for monopropellant use in tactical systems and their characteristics will be used in describing the advantages and disadvantages of monopropellant systems. As its name shows, a monopropellant system requires only one liquid and permits the use of a single flow system and a simple control system.

The low temperature (1150–1300 K) of the gas products permits the use of conventional materials of construction even for long duration missions. The use of a spontaneous catalyst or a thermal initiator provides rapid thrust response with a simple on/off fuel control. A major advantage of the monopropellant is its lack of signature. The low-temperature exhaust products are invisible in the visible region and provide high transmission in the infrared spectrum. The monopropellant also offers a safety advantage because it eliminates the need for a highly reactive oxidizer.

Table 1 Typical performance for tactical liquids/gels

Parameter	Monopropellant	Bipropellant	Gel propellant
Specific impulse, N/kg/s	2000	2500	2450
Density, g/cm^3	1.0	1.4	1.5
Volume impulse, N/L/s	2000	3500	3675

The disadvantage of the monopropellant system is its relatively low specific impulse. The 2000 N/kg/s is no match for the 2500–3000 N/kg/s specific impulse provided by the bipropellants.

The bipropellants deliver the highest specific impulse of any tactical propulsion system, giving them an advantage in limited weight systems. The delivered performance at sea level is in the range of 2500–3000 N/kg/s, depending on the propellant system selected. The higher impulse is associated with highly reactive oxidizers and high-combustion gas temperature. Continued improvements in high-temperature materials for propulsion system components lead to the possibility of increased usage of the bipropellant liquid propulsion system. The many possible combinations of fuel and oxidizer along with the adjustment in oxidizer to fuel (O/F) ratio provides a means to optimize the system for specific requirements.

The gelled propellants offer two potential advantages for the tactical propulsion system, improved safety and increased volumetric impulse. Improvement in safety is attributed to the thixotropic nature of the gels. In a spill involving the fuel and oxidizer, the flow of gelled materials is greatly diminished over the flow which would occur with neat propellants, thereby reducing the magnitude of the fire. The improved volumetric impulse (the impulse per unit volume) is achieved by loading the gelled propellant with solid fuel. This increase in volumetric impulse is normally associated with a slight decrease in specific impulse.

The loaded gelled propellants currently are limited to the bipropellant system. Gelled propellants that contain solid fuels are generally unsuitable for the monopropellant system because the solid fuels currently in use poison the catalyst in the monopropellant reactor and render it unusable.

Because of the improved safety of the gels, the gelled propellant system appears most favorable to the end user, which may hasten its use in a tactical system.

A model to select the optimum system for any application would be impossible. Many factors, including the biases of the user, cannot be quantified. Most critical is selection of a system that will be fielded and will perform its mission. Usually the cost and volume of a tactical system are more important than its weight. Packing the propulsion in the missile envelope may also be a major factor in the selection process. Most of the future tactical missile systems will be designed to fit an existing launcher rather than develop a complete new weapon system.

As a guide in selecting a propulsion system, some typical performance values of each liquid/gel system are shown in Table 1.

A. Monopropellant Propulsion System

The liquid monopropellant system is very attractive because the simple hardware and controls provide an economical tactical system with thrust control and no signature. Research on monopropellant systems has produced a limited number of good monopropellants. A material that produces high energy during decomposition is not stable during long duration storage. Hydrazine, hydrazine derivatives, and hydrazine blends are currently the best monopropellant systems being considered for tactical propulsion applications.

The hydrazine monopropellants were developed primarily for space applications and have proven to be extremely reliable. Many of the thrusters on satellites have seen 10–20 years of service with millions of cycles. The spontaneous catalyst used with hydrazine was developed for long life and millions of cycles.

Monopropellant systems for tactical systems require only one-time operation after a storage period of at least 10 years. The major difference in space and tactical applications is the system cost. An acceptable tactical propulsion system must be cost effective. Additionally, it must meet the requirements of the operational and storage environments dictated by the battlefield.

1. Propellant

Over 30 years of technology development has provided an enormous amount of data on the use of hydrazine as a monopropellant. The major part of this technology was directed to propulsion for space applications and emphasized low thrust, low chamber pressure, long duty cycles, and multiple restarts. Schmidt in Ref. 1 has an excellent compilation of this effort. Here the requirements of tactical missile propulsion will be discussed and related to the available monopropellant technology.

Tactical systems are typically required to operate at 230 K and be capable of storage temperatures to 220 K. Hydrazine freezes at 275 K, rendering it unsuitable for tactical applications. Virtually all tactical liquid monopropellant propulsion systems require a hydrazine propellant blend to achieve the required operating and storage temperature. Water is the best freezing point depressant for hydrazine but it contributes little to thrust generation. Hydrazine/water blends containing 70% hydrazine have been used for gas generation but the thrust is only about 70% of that achieved with hydrazine alone. To partially compensate for the performance lost by the addition of water, hydrazinium salts have been added to the hydrazine/water blends. Hydrazinium nitrate (HN) is the most frequently used additive. A hydrazine propellant blend (HPB-2012) containing 20% HN and 12% water will freeze at about 241 K, is about 9% more dense than hydrazine (1.00 g/cm^3) and will achieve about 95% of the specific impulse of hydrazine (2000 N/kg/s). HPB-2012, due to the presence of the nitrate, is not compatible with most stainless steels or aluminum alloys for long-term storage but is compatible with titanium alloys.

As with any energetic material, safety precautions must be observed during operations with hydrazine and hydrazine blends. Hydrazine and its decomposition products will burn in air when ignited. Hydrazine is not detonable but some of the hydrazinium salts are. Mixtures of hydrazine and HN that contain more than 25% HN will detonate. In specifying the hydrazine/HN/water mixture one must exercise care to avoid the detonable compositions.

Hydrazine is known to be toxic and ongoing studies continue to update the allowable limits. Most of the data are based on animal research, as few major incidents with humans have been reported. The studies also indicate hydrazine is a potential carcinogen. For tactical missile applications the hydrazine blend will be prepackaged for long-term storage, and the toxicity risks are expected to be minimal.

The exhaust products of hydrazine decomposition are ammonia, hydrogen, and nitrogen. The exhaust products produce no toxicity problems, and no burning of the products with air has been reported. The absence of any afterburning, the relatively low-temperature exhaust products, and the low concentration of condensable gases produces one of the cleanest exhaust plumes available.

2. Decomposition

Tactical propulsion requirements permit more options for the decomposition mechanism than is available for space propulsion. Whereas space propulsion requires many restarts over a period of several years, tactical systems have no cold restarts and a limited number of starts while the reactor is hot. Thermal decomposition, one-shot catalyst, thermal initiator plus catalyst, as well as spontaneous catalyst are available to the tactical reactor designer.

The Shell-405® catalyst, which has become the standard of the industry for hydrazine decomposition, would perform in an excellent manner for tactical applications, but the cost of the catalyst alone could increase the propulsion system cost as much as 100%. In the development of the system, where repeated tests of the reactor and system are required, the use of Shell-405 may be cost effective. The use of Shell-405 as an initiator in combination with a less reactive, and less expensive, catalyst may prove economical for the tactical system.

The active ingredient in Shell-405 catalyst is iridium metal which provides spontaneous initiation even at cold temperature. Many metals will catalyze the decomposition at elevated temperature. Nickel has been used to decompose hydrazine after the catalyst bed was heated by reacting hydrazine with iodine pentoxide.

Hydrazine will decompose thermally, and beds of firebrick substrate and metal spheres have been used to transfer the heat from the decomposition products to the liquid. A solid oxidizer has been deposited on the substrate to react with the first flow of hydrazine and provide heat to initiate decomposition. The bed loading (flow rate per unit area) may be lower for a thermal bed reactor than for the catalytic reactor because the heat transfer from the gas to the liquid controls the decomposition rate.

The system designer has many options in developing the optimal reactor for a specific mission. Although much technology development has been documented, a tactical monopropellant propulsion system has not been fielded.

3. Storage and Expulsion of Propellants

Many options for the storage and expulsion of liquid propellants are available to the designer. The final selection generally will be the least expensive system that meets all other requirements of the missile system. These requirements include storage life, weight, volume, materials compatibility, center of gravity control, sloshing, permeability, and packaging. The method used to feed the propellant to the engine will also be a major consideration in selecting the expulsion device. The expulsion systems generally considered for tactical systems and their limitations will be discussed as follows.

Direct expulsion. Direct expulsion generally refers to direct contact between the pressurizing gas and the liquid propellant. Direct expulsion is probably the simplest and cheapest system for tactical propulsion but may not meet the system requirements. First, there is no method to control the center of gravity (c.g.) of the missile as the propellant is consumed. Second, the sloshing of the liquid may interact with the guidance system, causing unstable missile operation. Third, the choice of pressurization gases is limited to those compatible with the liquid propellant. Last, the use of warm pressurization gases may not be an advantage because the heat will be transferred from the gas to the propellant. Care also must be exercised to prevent localized heating of the liquid.

The c.g. control for each application must be addressed and a simulation of the missile dynamics for each application will normally indicate any problem areas. If problems are indicated, baffles in the propellant tanks can be used to control liquid sloshing. For small tactical missiles, c.g. control may not be a major problem. The mass of the liquid compared to the total mass of the missile normally is small, and depletion of the liquid and movement of the liquid in the tank do not cause a major shift in missile c.g. During boost propulsion the missile is under positive acceleration, and the liquid position is predictable.

When the missile is under positive acceleration, locating the propellant outlet to the engine is simple, but during other periods the liquid may not be located over the outlet. To ensure adequate flow of liquid to the engine to initiate thrust and provide a positive acceleration, propellant traps can be used. These are normally simple geometric barriers to ensure the initial flow of liquid to the engine when a downstream valve is opened. The use of surface tension screens to trap some of the liquid have been demonstrated, but their use in long-term storage of tactical propellants has not been proven. One area that must be examined is the high surface area of the screen material exposed to liquid and the compatibility of this material with the liquid for long-term storage.

The gases used to pressurize the liquid in direct expulsion are limited to those that are compatible with the propellant. High-pressure, stored inert gas presents no problem with most tactical propellants. In using heated gases one must exercise care to prevent premature decomposition of the propellant from localized overheating. The heat transfer between the gas and liquid and between the gas and the fuel tank must be considered for each application. For the short duration normally associated with tactical booster applications the effect of heat transfer is small, but in sustain applications the effect may be very significant. In all cases a heat transfer analysis should be used to aid in selecting the type of expulsion and pressurization systems.

Piston expulsion. Piston expulsion offers several advantages for tactical expulsion, and its use in long term storage has been well demonstrated in the Lance Missile System. The piston provides a separation between the liquid and gas presenting more options for pressurization. The piston also provides positive c.g. control and eliminates sloshing. Piston expulsion is limited to cylindrical tanks, and the area behind the piston is wasted ullage unless it can be used for packing other components (e.g., the pressurization system).

The major problem associated with piston expulsion is the dynamic seal between the tank wall and the piston. The seal is required to maintain a static seal for 10 or more years and then provide a dynamic seal as the piston moves along the tank wall for the duration of missile operation. The requirement is difficult for any elastomer; additionally, the elastomer must exhibit long term compatibility with the liquid propellant. In the Lance Missile System a breakaway metallic seal was used to provide the long-term static seal between the liquid and the dynamic piston seal. The metallic seal joined the leading edge of the piston with the tank wall until the piston started to move down the tank. This seal protected the elastomeric dynamic seal from the propellant until the piston moved. Additionally, the elastomeric seals were designed for storage under no compression and were compressed upon first movement of the piston. This prevented the permanent distortion of the elastomeric seal by long-term compression.

Piston expulsion requires a smooth surface on the internal tank wall to provide a good dynamic seal with the O ring. This precludes a simple roll and weld tank construction and dictates some machining of the internal surface. One other problem occurs with the use of overwrapped composite fibers to provide the structural properties with a reduced weight system; under tension the composite fibers elongate considerably more than metal, increasing the diameter of the propellant tank wall more than normally found with metal. This increased elongation produces a greater challenge for sealing with elastomeric seals.

Piston expulsion has been demonstrated in a bipropellant liquid system with good results and should be considered for application in any tactical liquid propulsion system.

Elastomeric bladders. Elastomeric bladders have been successfully used for long term storage and expulsion for space applications and appear to offer cost and volumetric advantages for tactical propulsion. The bladder provides a positive separation of the liquid and pressurizing gas, ensures propellant availability to the engine at all times, and provides some thermal insulation between the liquid and the gas. The elastomeric bladder does not provide positive c.g. control and does not prevent sloshing. The stored propellant will permeate the elastomer and contaminate the area between the bladder and the pressure vessel. This requires the pressure vessel to be compatible with the propellant.

An elastomeric bladder using an ethylene propylene dimer (EPDM) as the base material was considered for a tactical monopropellant application using HPB-2012 propellant. The AF-E-332 compound successfully used for long-term space propulsion with hydrazine fuel was proposed for this application. A cylindrical tank configuration bladder was successfully fabricated for the tactical propulsion application, but the AF-E-332 was not compatible with the hydrazine blend for long-term storage. The HPB-2012 removes the metallic oxides used as a filler in the elastomer and produces a solid powder in the liquid. Current technology indicates an elastomeric bladder can be successfully formulated and fabricated for tactical missile propulsion. The elastomeric bladder would permit the use of an aluminum alloy for the pressure vessel. The propellant would still permeate the bladder and would cause some corrosion, but the liquid would not be contaminated with solids and the corrosion of the aluminum is so slight that the structural integrity of the tank would not be a concern.

One area shows potential for application of the elastomeric bladder. The conformal tank and its associated geometry does not lend itself to any expulsion scheme other than direct expulsion or elastomeric bladders. These conformal tanks normally conform to the missile geometry and may be a truncated cone or of unsymmetrical geometry.

Bonded rolling diaphragm. The bonded rolling diaphragm (BRD) is another option for storage and expulsion of liquid propellants from a cylindrical pressure vessel. This system provides all metal containment for the stored liquid during storage and operation. This system can be designed to provide double wall containment for the more reactive propellants. The BRD provides one wall, the second is the pressure vessel itself. The BRD is bonded to the wall of the vessel using an adhesive such as room temperature vulcanizing (RTV) rubber or Teflon. When gas pressure is applied to the back side of the diaphragm head, the metal diaphragm rolls through 180 deg and peels the diaphragm from the case wall.

The BRD provides excellent positive expulsion for control of sloshing and missile c.g. The expulsion efficiency has been demonstrated to exceed 98%, and both the diaphragm and the pressure vessel wall can be insulated for solid propellant gas generator pressurization. Composite structures can be used for the pressure vessel to provide a lightweight structure. Both aluminum and stainless steel have been used to fabricate the diaphragms and both have proven successful in prototype demonstrations. Efforts are underway to demonstrate a BRD system using titanium BRD to significantly reduce the weight of the system.

The most common problem experienced in development demonstrations is the quality of bond between the diaphragm and the case wall. The bond system is not readily inspected by nondestructive testing, and this mandates a good quality assurance program. Another problem associated with using the BRD for tactical missiles is the cost. The components can be mass produced, but the assembly and bonding process is very labor intensive.

Metal bladders and diaphragms. Metal bladders and diaphragms have been designed in many shapes and sizes for liquid expulsion demonstrations, but have not been used in any major production programs for tactical propulsion. The ring-stabilized diaphragm provides positive c.g. control but its use has been limited to near-spherical tank designs. The collapsing lobe, sometimes called the bathtub, provides a barrier between the propellant and pressurization gases but offers no c.g. control. The metal bellows provide an alternative approach but are inefficient from the volumetric packaging standpoint. Each of these concepts should be investigated for application to new propulsion requirements, but their general use in tactical systems is limited.

4. *Pressurization*

Several options for pressurizing the propellant expulsion system are available to the designer. The attributes of the primary systems will be discussed, but one must be careful to select the type of pressurization system based on the total integrated propulsion system rather than the merits of the subsystem alone. The principal systems are: pressurized cold gas, solid propellant gas generator, and liquid propellant gas generator. Combinations and modifications of the primary systems offer some advantages, and typical examples of these will be mentioned.

Stored cold gas. High-pressure stored gas provides dependable pressurization with excellent control. A simple gas regulator supplies the gas at a programmed pressure to the expulsion system. The inert stored gas permits direct expulsion of the liquid propellant, simplifying the expulsion system.

Gaseous nitrogen can be effectively stored at pressures to 41 MN/m^2. At higher pressure the compressibility factor for nitrogen increases sharply and prevents the efficient storage of the gas. If the system dictates gas storage at pressures above 6000 MN/m^2, helium is normally used due to its lower compressibility factor at these pressures.

The solubility of the pressurization gas in the propellants has presented some problems in space propulsion but has not been a problem for tactical systems. The gas is not stored in contact with the liquid, and the short exposure time during engine operation causes no problem.

For long term storage the gas is normally stored in metal or metal lined pressure vessels to prevent loss through permeability of nonmetallic materials. The composite structures for high-pressure storage of gas are normally lined with a thin light metal to reduce system weight.

To reduce missile system weight the trend is toward the use of composite structures to store the pressurized gas. Some questions on the degradation of the composite structures under high-tensile loading during long-term storage have yet to be answered. Currently higher design safety factors are required for composite pressure vessels than metallic vessels.

The design safety factor for the pressure is normally dictated by its intended use and handling during storage. If the system containing the pressurized gas is to be transported by commercial transportation, the ratio of design burst pressure to ambient fill pressure established by the U.S. Department of Transportation is 2.75 to 1.

Gas flow from the pressure vessel is normally initiated by penetrating the metallic seal with an electroexplosive device (squib valve). The squib must meet all the safety requirements normally imposed on initiators for solid propulsion systems. The basic requirements are for no squib actuation at 1 A and 1 W and always actuation at 5 W.

Pressure control to the propellant tanks is normally controlled with a simple spring regulator. The requirement for complete shutoff of the regulator is again system dependent. For short "off-times" of the engine a complete shutoff of gas flow to the propellant tanks is not required.

Interaction of the propulsion requirements with the total missile system design may prove very productive in reducing the weight and volume of the pressurization system. If the thrust profile can be regressive during the last part of the mission rather than requiring a fixed thrust level, the quantity of stored gas can be significantly reduced. This reduction in gas may permit an increase in propellant with a corresponding increase in total impulse per unit volume.

Solid propellant gas generator. The solid propellant gas generator (SPGG) is the most compact of the systems normally used for propellant pressurization. The solid grain produces a fuel-rich gas about 1300–1600 K. The grain can be tailored to produce the required quantity of gas at a given temperature and pressure for any ambient temperature. Because the burning rate of the grain varies with grain temperature, the mass of produced gas varies with the ambient temperature. To maintain a given pressure level for liquid pressurization, the SPGG is designed to provide the desired pressure at the lowest ambient temperature limit, and a hot-gas relief valve is installed to vent any excess gas produced at higher ambient temperatures.

The SPGG is best suited for short duration expulsions with little or no thrust management requirements. Heat transfer to the liquid propellant and the expulsion system can present a problem. The system must be designed to prevent hot gases from heating the propellant above its decomposition temperature. Localized heating of the propellant must be avoided as this may trigger decomposition. The system design also must consider heat transfer to the pressure vessel. The lightweight composite materials used for the structure of some expulsion systems experience a severe reduction in strength above 550 K. For all applications of the SPGG, a thorough evaluation of the heat transfer is a necessity.

The on-demand requirement for liquid flow may present some problems in the design of a system using the SPGG. The generator is normally designed to provide gases for maximum demand at the lowest operating temperature. For periods when the demand is zero the gases produced must be dumped overboard. For some applications the vented gases may cause a severe contamination or signature problem. If the vented gases comprise a large fraction of the total gas produced, the SPGG becomes less competitive in weight and volume with other pressurization options.

Liquid propellant gas generator—monopropellant. The liquid propellant gas generator (LPGG) offers many options for pressurizing the liquid expulsion system. It can provide a simple pressurization such as provided by the SPGG or on-demand pressurization to meet any flow requirements. The LPGG is ideally suited to a pulse-width modulation or proportional throttling. The LPGG can use a separate propellant feed system or can utilize the onboard propellant. If the main propellant supply is used to feed the gas generator, a pressure intensifier is needed to boost the feed pressure to the gas generator. The total system requirements dictate the type of gas generator, and the options to be incorporated. Generally, the least expensive system that meets the basic requirements will be the choice for tactical propulsion systems. The possibility of altering the system requirements to reduce the cost and size of the propulsion system should also be investigated. For instance, the use of a regressive thrust profile instead of constant thrust throughout the mission can significantly reduce the complexity of the gas generator.

Both the monopropellant and the bipropellant are viable options for the gas generator. Generally, the monopropellant provides lower temperature gases and is simpler to produce than the bipropellant. The bipropellant is more efficient and may be desirable in a bipropellant system if the onboard propellants can be used for the gas generator. The primary attributes of the two options are discussed next and more details of the characteristics of each can be obtained from the descriptions of the propulsion systems.

Hydrazine and hydrazine blend fuels using catalytic decomposition reactors are the state-of-the-art technology for monopropellant gas generators. The gas temperature can be varied from 1300 to about 1000 K by controlling the amount of ammonia decomposition in the gas produced. The ammonia decomposition is regulated by the length of the catalyst bed; the longer bed decomposes more ammonia and the endothermic reaction produces lower gas temperatures.

The gas temperatures from the bipropellant LPGG are regulated by varying the fuel to oxidizer (*O/F*) ratio of the system. Generally, the LPGG is operated on the fuel-rich side of the stoichiometric ratio to provide fuel-rich gases. The minimum gas temperature attainable on the fuel-rich gas generator is about 1350 K depending on the specific propellant combination. The lower temperature is limited on the fuel-rich side because the fuel decomposes and provides energy to the system. Theoretically, lower (800 K) temperature gases can be obtained by operating in the oxidizer-rich region of the *O/F* ratio; however, the temperature is very sensitive to ratio changes and control is difficult. Care must be exercised in the use of oxidizer-rich gases to expel fuel as the two are hypergolic.

Other pressurization systems. Combinations of the described pressurization systems provide the propulsion system designer many options. Some of the combinations proposed will be briefly mentioned to illustrate possible uses.

The use of a solid propellant grain to heat high-pressure stored gas is one possible combination. The gas pressure can still be regulated with a simple spring operated regulator to provide close control of pressure. The heat from the grain is very effective in overcoming the temperature loss due to adiabatic expansion. One problem with trying to heat the gas above ambient temperature is the heat loss to the propellant tank and liquid propellant. For short duration, heat addition to stored gas can be effective but for long offtimes the heat transfer to the ambient surroundings tends to nullify any significant gains. A heat transfer analysis on the total expulsion and pressurization system will provide insight to the feasibility of heating stored pressurized gas.

The use of liquids to cool the higher temperature gases from a SPGG or a LPGG has been proposed but never implemented on a tactical system. In some instances this application may be advantageous, but it contains several disadvantages. One possibility is the condensation of the coolant liquid in the propellant tank with a corresponding decrease in expulsion efficiency. Again, a good heat transfer analysis is recommended prior to implementation of any design.

5. *Reactor*

The reactor is designed to decompose the liquid fuel to produce high-temperature gases. For maximum thruster performance it is desirable to maintain the gas temperature as high as possible. To accomplish this the decomposition of ammonia is minimized by optimization of the dwell time in the catalyst bed. The optimum dwell time is the minimum required to decompose all fuel and is usually determined through experimental evaluation. For a given flow rate through the bed, the optimum dwell time is obtained by reducing the length of bed until maximum gas temperature is achieved.

The flow rate of fuel per unit area of catalyst bed is the bed loading for a reactor. The permissible bed loading varies with reactor operating pressure, type of catalyst, length of catalyst bed, fuel and catalyst temperatures, and type of injector. To minimize the quantity of catalyst and the packing volume, one should maximize the bed loading for each system. For space applications, for which the monopropellant was primarily developed, the nominal bed loading is less than 7.0 g/cm^2. For tactical systems operating at 3.5 Mpa., bed loadings of 20–35 g/cm^2 are typical. For this type of system, bed lengths of 1.8 cm are common.

The injector is designed to provide an even flow of liquid through the catalyst bed. Injector types are many and varied including a showerhead, platelet, and fused screen (Rigimesh®). For tactical applications the selection from those that will provide an adequate flow distribution is normally based on production cost.

Heat flow in the reactor must be addressed to prevent heating the liquid in the injector or flow passages. The major heat source is the hot-gas products of the reactor which transfer from the nozzle back to the lower end of the catalyst bed. Depending on the design and length of duty cycle, insulation may be required to prevent heat flow to the injection area. Silicone phenolic insulation has proven satisfactory for repeated test of the reactor.

Materials for construction of the reactor may depend somewhat on the specific design. Any material that will contain the pressure at the gas temperature is adequate. For long duration tests, a high-temperature nickel alloy such as Haynes® 25, Haynes® 188, Haynes® 320, or Hastalloy® X is very satisfactory. Stainless-steel injectors have proven very satisfactory and easily can be joined to the reactor by welding. Compatibility with the fuel is not a problem because the dwell time of the fuel in the reactor is very limited.

6. *Nozzle*

The nozzle design will be dictated primarily by the missile system packaging. The nozzle is usually fabricated as part of the reactor body and the material of construction is the same as the reactor. Either the conical or contoured nozzle is readily cast and should be very inexpensive in production. The added cost of expansion deflection or force deflection nozzles usually does not warrant their use.

7. *Flow Control*

Flow control of the liquid system is critical to its performance. Start transients, thrust level during operation, and the thrust termination can all be influenced by the liquid flow control devices. During storage the liquid propellant is isolated in its compatible storage vessel with pressure rupture disks. These disks, used to prevent contamination of the pressurization system as well as the downstream flow components, must possess long-term compatibility with the propellant. Additionally the rupture disk plays an important role in shaping the start transient for the system. Selecting the design rupture pressure of the disk should be a part of the overall dynamic analysis of the system.

In addition to serving a distribution and mixing function, the injector plays a major role in flow control. The pressure drop across the injector is often used to control the liquid flow and maintain thrust control during throttling.

Cavitating venturies are an excellent means of flow control in any liquid system. The venturi creates a vapor pocket downstream of the throat which separates the feed system from the reactor. Any pressure fluctuations in the reactor will not be transmitted to the feed system. If the downstream pressure is no more than 85% of the upstream pressure, flow is a function of the difference in pressure between the upstream pressure and the vapor pressure of the liquid (a constant for a given temperature). The cavitating venturi permits excellent flow control of the system by regulating only the upstream pressure.

Valves perform a major function in providing thrust management for a missile system. They can be used to provide an on–off function only, or they can be used to throttle the flow to provide thrust modulation. A simple on–off solenoid valve can be used to provide pulse width modulation and achieve the same results as a throttling valve. One should remember that the valve must perform the duty cycle for the specific mission, which is very mild compared to the millions of cycles normally required for aerospace applications. Additionally, the cost of a solenoid valve is dependent on the specified response characteristics. Required response times of less than 5 ms will significantly increase the valve cost. For some applications an electroexplosive (squib) valve may meet the specific requirements at a lower cost than the solenoid valve.

8. *Packaging*

The available space and the configuration of that space has a major impact on the design and selection of the propulsion system and also the components of the system. The tactical missile is an integration of many functions into a package that will perform its specific mission. Sharing a nitrogen supply with the controls function may significantly reduce missile cost and volume. An electrical supply may fit into a void space around the propulsion nozzle to provide a reduction in overall missile volume. The liquid propulsion system with its multiple components is readily adaptable to flexible packaging, and the utilization of this characteristic may provide a liquid system that is competitive with other propulsion systems on both a volumetric basis and a weight basis.

B. Bipropellant Propulsion System

The bipropellant propulsion system can meet all of the basic requirements for a tactical propulsion system, and the propulsion system designer must provide the proper combinations to meet the specific requirement. The prepackaged bipropellant propulsion system provides all attributes of the wooden round concept: long-term storage under all temperature extremes, no field maintenance, rugged for handling and operational conditions, and instant readiness. The propellants for tactical systems are nondetonable, but will burn. The toxicity of the propellant ingredients, even though the propellant is always contained, is one objection of the user to liquid systems. The other misconception concerning liquids is the unit cost being too high for tactical systems that may require hundreds of thousands of missile systems. One major design problem is presenting a system that satisfies the user's concern in these areas. Presented next are the major characteristics of the bipropellant propulsion system that must be integrated to optimize a system for a specific application.

1. Propellants

Selection of a propellant system for a specific missile application is based on all characteristics of the propellants rather than the energy release alone. Some of the characteristics of the ingredients are now discussed.

Fuels that are storable, deliver high specific impulse, have high volumetric impulse, and are nondetonable, whose exhaust products do not pollute the environment and do not produce a visible or infrared signature, are limited. The hydrazine fuels (hydrazine, hydrazine derivatives, and fuel blends of the hydrazines) are the top choice for any propulsion system. This family of fuels, in combination with appropriate oxidizers, will provide specific impulses (at 6.89 Mpa and sea level) of 2500–3100 N/kg/s. Although hydrazine freezes at 274 K, some of the derivatives have freezing points below 220 K. The hydrazine fuels are hypergolic with most oxidizers and do not require a separate ignition system. Hydrazine (density of 1.0 g/cm^3) is compatible with most common metals and some elastomers, insuring long-term storage for tactical applications.

Hydrocarbon fuels have been used in combination with liquid oxygen for launch vehicles but have never seen application in tactical liquid rocket propulsion systems. Unlike the hydrazine-based fuels, the hydrocarbons are not hypergolic with most tactical oxidizers. In addition, the specific impulse of the hydrocarbons in conjunction with the storable liquid oxidizers is about 10% lower than that achieved using hydrazine-based fuels.

The types of liquid oxidizers that are suitable for tactical liquid propulsion are limited. The three basic types are nitrogen tetroxide and mixed oxides of nitrogen, nitric acid and mixtures of nitric acid and nitrogen tetroxide, and interhalogen oxidizers. The characteristics of each of these are discussed next.

Nitrogen tetroxide is a nonflammable, toxic, strongly oxidizing, corrosive material. Its vapor pressure is about one atmosphere at 298 K. Its freezing point is 264 K, and the density at 293 K is 1.44 gr/cc. To lower the freezing point, nitric oxide is added to form the mixed oxides of nitrogen (MON) fuels. The freezing point of 70% nitrogen tetroxide and 30% nitric oxide is below 230 K. The addition of nitric oxide will decrease the theoretical performance of the mixture, and so the minimum quantity to achieve the required freezing point should be used. Also, the addition of nitric oxide has caused combustion instability in some combustion chambers. This instability is attributed to lower kinetic reaction rate of nitric oxide compared to nitrogen tetroxide.

Nitric acid is a strong oxidizing agent, is hypergolic with the hydrazine fuels, and will react with most metals. Its density at room temperature is about 1.55 g/cm^3, and its freezing point is about 231 K. The addition of nitrogen tetroxide will decrease the freezing point and increase the density of the oxidizer.

Red fuming nitric acid (RFNA) is a mixture of 84.6% nitric acid, 13.4% nitrogen tetroxide, and 2% water. The mixture has a freezing point less than 222 K. To increase its compatibility with metals, a small percentage of hydrofluoric acid (HF) is added to RFNA to form inhibited red fuming nitric acid (IRFNA). The HF provides a fluoride coating on aluminum and stainless steel to decrease the corrosion rate and increase the storage life of the oxidizer.

Aluminum and stainless-steel alloys are normally used to store nitric acid and red fuming nitric acid. Titanium and titanium alloys are not recommended for use with these oxidizers. For long-term storage, a program to monitor the corrosion of the storage vessels should be initiated. The major problem associated with IRFNA is the formation of solid particles in the liquid which interfere with the engine injector. The HF reacts with the metal to form a fluoride coating and prevent further corrosion, but over long periods of time the coating increases in thickness and finally separates from the container walls to contaminate the liquid. IRFNA has been successfully stored for more than 10 years in a tactical system using aluminum alloy tanks.

The interhalogen oxidizers, chlorine trifluoride (CTF) and chlorine pentafluoride (CPF), are extremely reactive oxidizers that are generally more energetic than nitric acid and nitrogen tetroxide. CPF is of interest as the theoretical specific impulse with the hydrazine fuels is about 10% higher than with nitrogen tetroxide. Additionally CPF has a density greater than 1.8 g/cm^3 compared to 1.45 g/cm^3 for nitrogen tetroxide. This increased density coupled with the higher theoretical impulse makes CPF attractive for volume limited tactical systems.

One objection to the interhalogen oxidizers for some applications is the composition of the exhaust products formed from combustion. Significant amounts of both hydrogen chloride (HCL) and hydrogen fluoride (HF) are emitted from the nozzle. Also, the interhalogens are very reactive with organic compounds, restricting the materials for use in storage and combustion of these materials. An all-metal bonded rolling diaphragm has proven very satisfactory for this application.

The performance of the propellant system is a major factor in selecting a propulsion system but it is only one of the factors to be considered. Cost, exhaust products, packaging, volume restraints, and storability are some of the other factors that must be considered. Thermodynamic equilibrium calculations are normally used to determine the theoretical performance of a propellant system. The NASA Lewis Performance Equilibrium Program is considered the standard for the industry, and most of the required thermodynamic data is contained within the program. Several versions of the equilibrium program are available for use on personal computers, but care must be exercised to ensure that the correct thermodynamic data is used. For example, an erroneous value for the heat of formation of an ingredient can significantly change the output data.

The oxidizer to fuel (*O/F*) ratio of the propellant combination can significantly change the performance. The maximum specific impulse is achieved at the stoichiometric ratio, but this may not be the optimum *O/F* for the propulsion system. For a limited-volume system it may be desirable to optimize the *O/F* for maximum density impulse (specific impulse times density). Other factors, such as combustion gas temperature and exhaust products, may also enter into the selection of the *O/F* ratio.

Ignition of the propellants may be a major factor in the selection of a propellant system. Most of the propellants just described are hypergolic and require no external ignition system. If the fuel and oxidizer are not hypergolic over the required operating range, an ignition system must be added to the propulsion system. The addition of an ignition system adds significantly to the complexity and cost of the propulsion system.

Compatibility of the fuel and oxidizer with the materials of construction and with the pressurizing gases is a major consideration in the system design. Both long term compatibility with the storage vessel and short term compatibility with the flow system, the injector, the combustion chamber, any pumps, and the pressurization medium are required. To prevent degradation of the propellants, the entire propulsion system that comes in contact with the liquids should be free of all contamination.

2. *Storage, Pressurization, and Expulsion of Bipropellants*

The major components available for storage, pressurization, and expulsion of liquids have been discussed earlier in this chapter, and this discussion will be limited to characteristics associated with the bipropellants.

The tactical missile system requirements have a major influence on the selection of the propulsion system components and the propulsion system designer should thoroughly understand the mission and operation of the complete system. The design of the propulsion system is an iterative process, and the missile design should be optimized to produce the most cost effective tactical missile system.

The major requirements that influence the selection of propulsion components are the center of gravity control, thrust management, duty cycle, thrust level, total impulse, response time, and storage life. Each of these requirements should be examined to ascertain the degree of control required.

For maximum c.g. control, the piston or the metal rolling diaphragm provide maximum control and the c.g. location is a function of the quantity of propellant remaining in the tanks. If required, the oxidizer and fuel tanks can be positioned to provide compensation for propellant usage and minimize the total c.g. shift in the missile. Often the propellant weight is small in comparison to the total missile weight and c.g. control is not a factor in missile performance. In these instances elastomeric bladders, metal diaphragms, liquid traps, or direct impingement expulsion devices offer cost effective solutions.

The thrust level during the duty cycle has a major influence on propulsion component selection. Increasing the thrust level increases the size and capacity of the combustor, valves, flow system, pressurization system, and flow control components. The use of a regressive thrust profile during the duty cycle permits the use of a blowdown pressurization system and significant reductions in pressurization requirements.

The duty cycle plays a major role in selecting the pressurization and expulsion components. A solid propellant gas generator may prove effective for pressurizing a simple boost system, but for long offtimes during the duty cycle, heat loss to the hardware and propellant may negate any advantage of a warm gas system. For long offtimes, a cold gas pressurant may gain heat from the hardware and propellant and increase the normal efficiency of the system. Additionally, the warm gas system for long duty cycles may require thermal insulation to prevent overheating the propellant and structural material.

3. Bipropellant Engine

The bipropellant rocket engine has been the focal point of extensive development, and volumes have been devoted to reporting this effort during the past 40 years. A large portion of this development effort has been devoted to space applications for which maximum performance per unit weight was the ultimate goal. Tools are available to determine the performance in relation to the maximum theoretical performance available and to assign the causes for this difference within fractions of a percent.

The requirements for a bipropellant engine for tactical systems do not dictate maximum performance per unit weight. For a tactical application, in which the engine is used for one short duty cycle, the production cost becomes a major factor.

The thrust duty cycle for tactical missiles is relatively short (normally measured in seconds) and the materials of construction are required only to survive for this short duration. The missile is normally volume limited rather than weight limited, which leads to requirements for smaller engines. Performance trades based on volume limitations may dictate smaller volume engines even though performance is degraded.

For tactical applications the regenerative cooled engine is too expensive to be cost effective. Ablative liners for thrust chambers are acceptable for most applications. Ablative nozzle throat inserts are normally acceptable even though there is a small degrading in nozzle performance.

The size of the combustion chamber has decreased with increased technology. Increased mixing of the propellants by the injector has permitted a reduction in combustor size by a factor of four. An L^* (volume of chamber divided by nozzle area) of 25 cm is the current state of the art. This has been made possible by use of vortex injection and platelet technology for improved mixing of liquids by the injector.

For tactical engines, conical nozzles should prove very effective. The contoured nozzle may reduce the length slightly, and it appears to be readily producible by casting. For operating near sea level the expansion deflection nozzles increase the nozzle performance less than 1% which normally does not justify the extra cost caused by the closer tolerances at the throat.

C. Gelled Propellant

Thixotropic gel bipropellant propulsion systems offer two potential advantages over the liquid bipropellant systems. The gels provide the mechanism to suspend solids in the propellants and significantly increase the density of the system. The gels, having low flow rate under low shear conditions, offer some safety advantages over the liquid propellant systems during a spill or leak. The characteristics of the gels and the unique requirements for their use are presented next.

1. Characteristics of Gelled Propellants

The thixotropic gel propellant is formed from the neat liquid propellants, so that the gel has essentially the same performance as the liquid and is used in the same combinations as the liquid propellants. The gelling agent is an inert material that will not react with the liquid and permits long-term storage of the gel propellant. Being inert, the gelling agent does not add to the energy of the system, and the specific impulse of a given propellant system is slightly lower than the corresponding liquid propellants. The gelling agent is normally a very fine particulate. The small quantity of fine particulate does little to change the characteristic of the exhaust products, and the propellant should meet the requirements for a minimum signature propulsion system; however, each system must be evaluated for signature characteristics.

At rest the thixotropic gel emulates a solid, and flow is slow or nonexistent. When pressure is applied to produce a high shear rate, the gel possesses the properties of a liquid and can be injected into a combustion chamber in the same manner as a liquid. A solid can be dispersed in a gel and stored for years in all temperature conditions without any separation of the liquid and solid. This property allows the addition of solid particulate to the gel to increase the density impulse of the system.

2. *Gel Fuels*

Hydrazine, hydrazine derivatives, and hydrazine blends are the primary choice for liquid propellants and have received the greatest attention in preparation of the gel fuels. Monomethylhydrazine (MMH) gel, because of its low freezing point, has been the primary choice for development and characterization. The gelling agent is a cellulose compound that is inert and causes no degradation of the propellant. The amount of gelling agent, normally less than 5%, can be varied to achieve specific rheological properties.

Aluminum powder of the type used in solid propellant fuels has been the material used to increase the density impulse of the propellant system. Up to 60% aluminum by weight has been successfully suspended in the MMH gel without long-term settling or separation. The addition of aluminum to the propellant significantly increases the volumetric impulse (impulse per unit volume) of the propulsion system, however, it does alter the exhaust products and the system is no longer considered a minimum signature propellant. For applications in which the plume signature is not a major concern, the aluminum loaded MMH gel is an excellent method of achieving high-density loading required for the volume-limited tactical system.

For the minimum signature gel propellants requiring high density, other solid materials are being used to develop a solids-loaded gel fuel. Materials such as carbon that enter into the combustion process and exit the nozzle as gaseous products are being developed. These gels are a compromise between minimum signature and high-density propellant.

There is current interest in gelling hydrocarbon fuels such as RP-1 and JP-10 and suspending solids to provide a high-density fuel. The gelling agents being considered are the cellulose compounds and copolymers of polyethylene. Again, the choice of a suspended solid will be determined by the specific application. The hydrocarbon gel offers additional options; however, its role in tactical propulsion is yet to be determined because it is not hypergolic with most oxidizers.

3. *Gel Oxidizers*

Gelled inhibited red fuming nitric acid (IRFNA) is the only tactical oxidizer that has been extensively developed. IRFNA has a low freezing point, has a relatively high density, and has demonstrated field storage in excess of 10 years. The IRFNA gelling agent is silicon dioxide. Currently, there are no known energetic materials that can be added to the oxidizer and stored in excess of 10 years. Inert solids have been added to the oxidizer to match the rheological properties of the loaded fuel throughout the operating temperature range. A match of flow properties permits flow *O/F* (oxidizer to fuel ratio) control throughout the temperature range and maximizes propellant utilization. Lithium nitrate, an inert solid particulate, has been added to IRFNA gel in quantities to 30 wt% to achieve the desired rheological properties and optimize performance.

Because chlorine pentafluoride in combination with MMH has a higher specific impulse and also has a higher density (1.82 g/cm^3) than IRFNA, there is current interest in gelling this oxidizer. To date little development of this gel has been accomplished and due to its reactivity the task will not be easy. One problem with the use of chlorine pentafluoride is the hydrogen chloride and hydrogen fluoride produced as combustion products. These materials provide a visual signature and rule out its use as a minimum signature propellant. The most probable gelling agent for chlorine pentafluoride is the inert silicon dioxide.

4. *Performance*

The theoretical performance of the gel propellants is determined using an equilibrium performance program in the same manner as liquid propellants. The specific impulse of the gel propellants is normally 2–5% less than the neat liquids because the gellants are inert and do not provide energy to the system. For the metal-loaded gels, the specific impulse is lower than the neat liquids but the volumetric impulse is significantly larger than the liquids because the density is much greater. This large density impulse is a major advantage for use in the volume-limited tactical systems.

The delivered specific impulse of the gel systems investigated to date is generally lower than the liquid propellant system. The combination of MMH gel with IRFNA gel generally produced delivered performance which is 4–5% lower than that achieved using liquid propellants under similar conditions.

The mixing and combustion of the gel propellants is more difficult than liquid propellants. The MMH fuel that is gelled with a cellulose containing polymer such as Klucel exhibits much greater cohesive forces that make it difficult to diffuse and mix with the oxidizer. The addition of carbon to the fuel gel to increase density creates an additional problem in the combustor. The kinetic reaction of carbon is slower than MMH and requires additional dwell time in the reactor. Combustion chambers with an L^* (combustor volume/throat area) of less than 25 cm normally produce very good efficiency with liquid propellants, however an L^* greater than 60 cm is currently required to achieve good performance with a fuel containing 50-wt% carbon in MMH gel. Innovative mixing techniques, as well as some modifications to the gelling agents, may be required to achieve high performance in a volume limited combustor currently achieved with liquid propellants.

The lower performance achieved with the gel propellant systems using hardware designed for liquid propellants does not mean that high performance cannot be achieved. To achieve acceptable performance from the gel propellant system the injector, the combustor, the nozzle, and the expulsion system must be tailored for the specific gel propellant system. The gel propellant systems show great promise of meeting the propulsion requirements for the limited volume tactical propulsion systems.

5. *Flow Characteristics*

The flow characteristics of both the gel fuel and the gel oxidizer are critical parameters in the design of a tactical propulsion system. The tactical system normally is required to operate over a wide temperature range (typically 230–335 K). The system designer desires a gel propellant system in which the flow properties of the fuel and oxidizer are matched over this temperature range so that a constant *O/F* ratio can be maintained. A shift in *O/F* ratio between the low- and high-temperature limits will result in reduced performance and under utilization of one of the propellant ingredients. The gel systems can normally be formulated to achieve a fair match in flow rate over the desired temperature range, but it may result in system designs that operate with higher pressure drops than desired. The high-pressure drops in the flow system and injector increase the weight and volume of the pressurization and expulsion subsystems.

The flow characteristics of the gel propellants are a major factor in the design of the system, and the total propulsion system must be considered in the formulation of the gels.

6. *Plume Characteristics*

The plume characteristics of the gel propellants have not been determined in a signature characterization facility or under actual flight conditions so that no definitive data can be cited. As the gel systems are considered for system applications, experimental evaluation of the exhaust products will be required. Based on the theoretical combustion products the gel propellants are expected to be the same as the propellant system without gelling agents. The MMH fuel and IRFNA oxidizer combination should provide a minimum-signature propellant even with the addition of the gelling agent. Addition of metals to the gel fuel and oxidizer in order to increase the density impulse will increase the plume signature and change the plume classification from minimum signature to smoky propellants.

7. *Storage and Expulsion of Gel Propellants*

With a few minor exceptions a gel propellant will store the same as its parent liquid and the expulsion techniques will be the same. The known and postulated exceptions are discussed next.

In the area of long-term compatibility of the IRFNA gel with its storage tank, one possible problem has surfaced. IRFNA contains a small percentage of hydrogen fluoride (HF) to inhibit the exposed surface of the storage vessel. The HF normally reacts with the metal to form a fluoride coating on the metal and prevent the acid from reaching and reacting with the metal. The HF in the acid keeps the fluoride coating replenished over the life of the system. In the oxidizer gel, some concern exists that the HF near the tank wall will be depleted with time, and the HF in the bulk of the oxidizer will not transfer to that area and maintain a concentration of HF near the metal to form the fluoride coating. This potential problem must be investigated prior to fielding a gelled system.

The use of direct pressurization with a compatible gas does not appear to be feasible in expelling the liquid gels. The direct pressurization of liquids in many cases simplifies the expulsion system for tactical systems. The use of gravity or acceleration forces to position the liquid over the liquid pickup point works well for neat liquids. For gel propellants the acceleration forces will not impart sufficient energy to move the gel to the pickup point. This indicates that the gel propellants must use a positive expulsion device such as a metal diaphragm, an elastomeric bladder, or a piston to ensure expulsion of the gel.

Pressurization of the gel system using a positive expulsion device should be very similar to that of neat liquids. One area that deserves some attention is the heat dissipation into the liquid. With neat liquids any heat transferred to the liquid is rapidly transferred to the total mass of liquid, and localized heating is not a major concern. Any heat transferred to the static gel system may not be dissipated as rapidly as that into a neat liquid, and localized heating may occur. This localized heat may initiate decomposition of the gel and result in uncontrolled gas generation. No major problem is anticipated, but the effect of reduced heat transfer in the gel system should be considered for any design.

8. *Combustion Chamber and Nozzle Design*

The function and design of the injector, combustion chamber, and nozzle for gel propellant systems is essentially the same as for neat propellants. Some precautions should be observed for the tactical systems that are required to operate over the wide temperature and environmental conditions. The major considerations are the mixing of the fuel and oxidizer, the cooling of the combustion chamber, and nozzle materials to accommodate the metal-loaded gels.

The flow rates of the gelled propellants are generally more sensitive to temperature variations than the neat propellants. To maintain a constant O/F mixture ratio throughout the temperature range it is desirable that the flow properties of the oxidizer and fuel remain relatively constant. In formulating a propellant system it is necessary to vary the solids content of each ingredient to ensure compatibility of the flow properties of each ingredient at all temperatures of interest. Another method of ensuring a compatible system is to vary the amount of gelling agent in each liquid. To ensure proper operation of a gelled propellant system, the flow properties of each component over the temperature range of interest must be a part of the specification.

9. *Safety of Gel Propellants*

One of the major advantages of the gelled propellant system over the liquid bipropellants is the added safety advantage. The flow properties of the gelled liquids under low shear rates provide this feature. To initiate and sustain combustion of the bipropellant gels, the oxidizer and fuel must come together. Damage to both storage tanks in which liquids are released normally results in a rapid combustion of the total ingredients. In the gelled system, the propellant flows will be much slower resulting in a less intense flame than experienced with the neat propellants.

For spills in which the oxidizer and fuel do not get together, there is still some safety advantage. When exposed to the atmosphere the liquid propellant normally evaporates from the gel forming a crust of gelling agent and solids filler. This crust formation retards further evaporation of the liquid and decreases the concentration of toxic materials in the atmosphere.

Currently, there is no fielded tactical system that has utilized gel propellants. Some development effort has been performed, and system designs have been postulated. Component development, especially in the injector and combustion chamber areas, has been promising but also indicates problem areas that must be solved. The current trend is toward higher injection pressure drops, higher chamber pressures, and higher injection velocities. Techniques to break up the fuels and improve mixing are needed. The use of gel propellant systems opens an exciting and challenging field for the propulsion system designer.

IV. Summary

The liquid propulsion systems can be used for tactical missile systems and offer some unique advantages. Thrust management capability is an obvious characteristic, but its minimum signature characteristics and the ability to meet the required insensitive munitions requirements are major assets.

The monopropellant propulsion systems are normally considered for applications with low total impulse requirements because the sea level specific impulse of the monopropellants are generally around 2000 N/(L/s). For smaller systems the simplicity of the monopropellant offers both weight, volume, and cost advantages. The low temperature (less than 1350 K.) of the exhaust gases permits use of conventional materials and lowers the signature of the missile.

The bipropellant liquid propulsion system offers high performance [specific impulse in excess of 2500 N/(kg/s)] in addition to thrust management, low signature exhaust, and the capability to meet the insensitive munitions requirement. Generally the volume requirements exceed that of solid propulsion systems.

The gel propellant system offers a means of achieving high-density impulse while maintaining the capabilities of thrust management and meeting the insensitive munitions requirement. If the gels are loaded with metals to achieve high-density impulse, they may not meet the requirements for a minimum-signature propellant system. Loading the gel fuel with carbon to achieve high-volumetric impulse and minimum signature is currently underway.

Liquid propulsion systems have been perceived as expensive because their primary application has been for space applications in which high performance, high reliability, and long operating life were required. For tactical applications the production cost of the system must be a major design factor along with weight and volume requirements. For many tactical applications the liquid/gel propulsion system can be a very cost effective system.

References

[1]Schmidt, E. W., *Hydrazine and Its Derivatives,* Wiley, New York, 1981.

[2]Sutton, G. P., and Ross, D. M., *Rocket Propulsion Elements,* Wiley, New York, 4th ed., 1976.

Chapter 4

Solid Rocket Motor Design

Alain Davenas*
SNPE Défense Espace, 75004 Paris, France

Nomenclature

A = area
a = constant in burning rate law for solid propellants
C_D = discharge coefficient
C_F = nozzle thrust coefficient
c_p = specific heat at constant pressure
c_R = volumetric loading fraction
c_v = specific heat at constant volume
c^* = characteristic velocity
d = diameter
E = Young's modulus
e = propellant web thickness
F = thrust
g = acceleration of gravity
I_{sp} = specific impulse
I_T = total impulse
K = A_p/A_t
L^* = characteristic length, v_c/A_t
l = length
M = molecular mass
m = mass
$\dot{m}$ = mass flow rate
n = exponent in burning rate expression for solid propellants
p = pressure
R = propellant acoustic response
$\mathbf{R}$ = universal gas constant
$\dot{r}$ = burning rate
T = temperature
t = time

*Research and Technology Director, 12 quai Henri IV.

V = velocity
v = volume
α = damping or gain factor
Γ = Eulerian gamma function
γ = specific heat ratio
π_k = temperature sensitivity of motor pressure
ρ = density
σ_p = temperature sensitivity of propellant burning rate

Subscripts
a = ambient
c = chamber conditions
e = exit
i = inert
p = propellant
t = throat

I. Introduction

WHEN compared to other types of motors, the advantages of the solid rocket motor for tactical use probably outweigh its disadvantages since it is by far the most widely used in missile propulsion today.

A major advantage of solid propellants compared to liquid propellants is high density; this is an asset in volume limited designs. Solid propellant is also chemically stable for many years and ages well; today the service life of many tactical motors is close to 15 years. Another advantage compared to either air-breathing or liquid propulsion systems is simplicity, and this is an important point for tactical systems with high production rates. Because of this simplicity the cost of development and production is the lowest (except for subsonic propulsion of medium to long range missiles in which the turbojet today can be the lower cost solution) and the cost of maintenance is low. When compared to other propulsion systems the two main disadvantages are inferior specific impulse and fundamental difficulty in varying thrust on demand, at least for high-performance motors. Some important progress, however, has been made in this area with pulse motors that will be described subsequently. Finally, solid propellant motors provide the easiest way to achieve the very high mass flow rates and thrusts required for the initial propulsion phase of a missile; all air-breathing sustainers will use a solid propellant booster.

For all of these reasons it can be expected that the solid rocket propellant motor will retain a predominant position in the future for tactical missiles propulsion except, perhaps for medium to long range mission sustainers where turbojets and ramjets will be considered more and more.

II. Overall Design Approach

A. General Requirements

The mission of a propulsion unit is to bring a missile to a given point in space with a given velocity vector after all of the propellant has been burned. The mission of the motor can, in some cases, be completed before burnout, when interception of the target occurs during the propulsion phase. In other cases, the missile continues its flight after the powered phase, with or without guidance (in the case of tactical missiles or ballistic rockets). To be able to use various trajectories to intercept a target under a variety of possible conditions, the tactical missile must remain controllable during and after the powered phase, within a certain range. The mechanical and aerodynamic loads are the most important factors, and they have great impact on the design of the various elements of the missile and, specifically, on the motor case.

Different types of motors, depending on the general architecture of the missile, are used with accordingly different types of loads and environment effects: that is, integral motors, which are part of the load bearing structure, the case being usually part of the outer skin; discrete motors, which are housed within the missile without being part of the load bearing structure; and jettisonable motors, attached to the rear of the missile or mounted on the side and jettisoned when burning is complete.

Most current motors are integral, some are jettisonable. In any case, although the type of motor will influence the environment to which it is subjected (vibrations, for instance) the design principles will be the same.

The various missions expected from a motor and the conditions under which these missions are carried out lead to the identification of requirements. These must not only include the energetic performances but also the requirements for the various operational phases of the motor life cycle, e.g., transport, storage, operational use, firing, etc.

The main requirements can be broken down into performance requirements and other types of requirements. The latter include weight and inertia, operational constraints, interface requirements, and in some cases specific constraints, including test requirements.

B. Functional Requirements

Taking into account the weight and the trajectory pursued, specified propulsion conditions are usually as follows: total impulse supplied by the motor, together with a required thrust-time profile (this is the stage at which there can be a requirement for a single thrust level, two successive thrust levels, or multiple thrust levels); maximum thrust or acceleration levels, for instance, at burnout; a profile of pressure or thrust as a function of time for the ignition phase; minimum or maximum speed at a given time during the propulsion phase (integral of the acceleration at a specific time); acceptable variations for the length of the propelling phase; and thrust vector control.

Weight and inertia characteristics include total weight of the motor, total weight of the inert parts, location of the center of gravity and limits during the flight, and inertia under rolling, pitching, and yawing.

Structural integrity and critical case constraints include stiffness, bow and ovality which are very critical for high length/diameter tactical missiles. Safety factors are generally specified by the missile designer.

Operational requirements include environmental conditions in operation and firing, i.e., temperatures, thermal cycles, atmospheric conditions, and vibration levels; signature of the combustion products released by the motor; ease of maintenance; reliability; safety; and ergonomy.

Even if they are part of the operational requirements the growing importance of the vulnerability requirements needs additional discussion. Insensitive munitions (IM) requirements are detailed in DOD-STD-2105B and are discussed in detail in Chapter 9. This Department of Defense regulation outlines the specific tests and criteria to meet the requirements. The main points deal with the response to bullet impact, sympathetic detonation, cookoff, and fragment impact.

These tests will have to be increasingly run on development motors with their typical launch systems. To assure that the requirements will be satisfied, the designer will have to use all possible data and models accumulated on the response of similar motors or subscale motors and specimens. For example, he should already have data on the effect of the case materials and grain architectures on the response to bullet impact, or the response of the propellant with various confinements to cookoff, etc.

C. Interface Requirements

These are mostly design constraints including, in particular, the maximum length and diameters, launch hook features, etc. These requirements must result in the determination of the borders between the subassemblies of the missile. Table 1 gives a description of the main types of specifications for a missile and how they induce the specifications of the different subsystems.

D. Description of a Typical Solid Propellant Rocket Motor

A solid propellant rocket motor (Fig. 1) is designed to ensure the combustion under pressure of the propellant grain it contains. The resulting gases are expanded through the nozzle whose function is to convert this internal pressure into a supersonic exhaust velocity.

As a rule, such a rocket motor has the following five major components.

1) The typical solid rocket motor case, the first component, is basically a double-domed right circular cylinder with openings at both ends and cylindrical extensions called skirts. The aft opening interfaces with the nozzle, and the forward opening accomodates the safe and arm device as well as the igniter. Additionally, numerous conduits, ducts, fin attachments, and carry features may be required on the case.

Table 1 Main specifications for a missile system and subsystems

Missile system	Missile	Propulsion	Propellant	Case and hardware
Mission	Trajectories	Thrust, Impulse Combustion time Mechanical loads	Propellant Grain geometry	Materials Shape Dimensions
	Temperature envelope	Motor temperature envelope	Propellant temperature envelope	Case temperature envelope
	Guidance	Optical (visible, EM, IR) transmission through the plume	Propellant	Insulation and nozzle materials
Launch platform	Mass Size volume	Mass Size volume	Mass Size volume	Mass Size volume
Reliability	Reliability	Reliability	Reliability	Reliability
Operational environment	Thermomechanical environment	Thermomechanical environment	Thermomechanical environment	Thermomechanical environment
Ergonomy		Noise-toxicity	Propellant Grain geometry	Nozzle
Battlefield	Signature	IR, uv, EM signature Afterburning	Propellant Propellant	TVC system Materials
	Vulnerability	IM requirements	Propellant	Materials
Cost	Cost	Cost	Cost	Cost

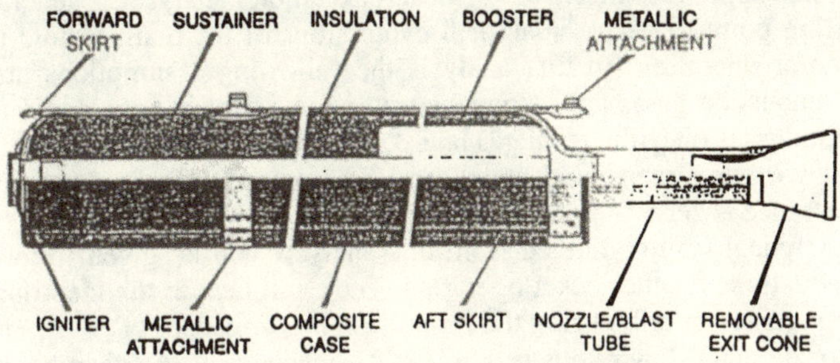

Fig. 1 Solid rocket motor for tactical missile.

The solid rocket motor case is essentially a minimum-weight pressure vessel whose design, as we have seen, is complicated by the presence of significant levels of thrust and bending and locally concentrated loads from the skirts and attachments. Tactical motors typically operate from 3 to 25 MPa. Aerodynamic heating profiles may also have to be considered since temperatures as high as 800 K are possible in some cases.

2) The propellant grain is the next component. Grains are made of solid propellant shaped according to a given geometry during manufacture. To prevent combustion, their surface may be locally restricted or may be inhibited with a flame resistant adhesive material, an inhibitor. The regression of the free propellant surface, once it has begun to burn, is normal to this surface and dictates the evolution of the pressure and thrust. When the propellant grain is case bonded, a chemical bonding system, generally a liner, attaches the grain to the case wall interior and insulator.

The burning time of the propellant in tactical systems will generally vary between a few seconds and 30 s. Using end-burning grains, sustainers with a few minutes burning times are used for some long-range applications.

3) The thermal insulation is the next component. The motor case interior is covered with an insulator which provides thermal protection and flow erosion resistance to gases whose temperature ranges from 1500 K to more than 3500 K, in regions where propellant burns to the wall before the entire grain is consumed.

4) The general shape of the nozzle, the fourth component, includes three major parts: the convergent zone, which channels the gas flow; the throat, whose dimensions determine the operating pressure of the rocket motor; and the exit cone, which increases the exhaust velocity of the gases in their expansion phase, consequently improving the propulsive effect.

In tactical systems, the nozzle is sometimes placed at the end of a tube, called a blast tube, in order to provide space for different devices such as those that activate the steering controls of the missile.

5) The last component, the ignition system, brings the energy necessary to the surface of the propellant to start burning.

E. Propulsion Performance of a Solid Rocket Motor

This subsection gives the basic definitions and some equations used to analyze a motor operation. For more detailed information, the reader should refer to general textbooks, such as Ref. 1.

A standard approach, known as the ideal rocket motor analysis, is generally used as a starting point to derive analytical expressions of the main factors involved in the motor operation. In this analysis the following assumptions are made. 1) The combustion gases are homogeneous and invariant in composition. 2) The combustion gases obey the ideal gas law. 3) The average specific heat is constant. 4) The flow is one dimensional, steady and isentropic. 5) The flow velocity at the nozzle entrance is zero.

The analytical results derived from this analysis will be given following the general equations for the rocket operation. They will help in the identification of the main parameters influencing the propulsion performance.

The nozzle mass flow rate is $\dot{m} = C_D p_c A_t$, where p_c is the chamber pressure and A_t the area of the nozzle throat. The proportionality coefficient C_D is called the discharge coefficient. In the ideal rocket motor analysis it can be shown that C_D is affected only by the nature and temperature of the gases

$$C_D = \Gamma(\gamma)/\sqrt{\gamma R T_c}$$

where T_c is the chamber temperature. For solid propellant gases $R = 300$ J/kg · K and $\gamma = 1.2$.

T_c and γ are not very susceptible to pressure variations, therefore, in many cases the independance of C_D from pressure is accepted. A typical value of C_D is in the range of $6.5 \cdot 10^{-4}$ s/m.

The parameter $c^* = 1/C_D$, called the characteristic velocity, is sometimes also used.

The combustion of a solid propellant is characterized by the way its surface regresses once it has begun to burn. The burning rate is the distance traveled by the flame per unit of time measured normally to the burning surface. This front is assumed to be regular and, in most cases, progresses in a direction normal to itself. This has been experimentally verified by interrupting the combustion and examining the surface.

The flow rate resulting from the combustion of the propellant is

$$\dot{m} = \rho_p A_p \dot{r}$$

where ρ_p is the density of the propellant, A_p the area of the propellant burning surface, and $\dot{r}$ the burning rate of the propellant.

The steady-state rocket motor operation point corresponds to the equality of gas flow rates generated by the propellant and ejected by the nozzle. The burning rate and the operating pressure are then related by

$$p_c = \rho_p A_p \dot{r} / C_D A_t \qquad (1)$$

This equation neglects the volume of gas filling the space resulting from the combustion of the propellant.

The parameter $K = A_p/A_t$, the burning surface area to throat area ratio (Klemmung in German, serrage in French) is an important parameter for the internal ballistics (flowfield) of the motor.

The burning rate of a propellant is a function of pressure

$$\dot{r} = f(p) \qquad (2)$$

Generally, in a limited pressure range, it can be given by a power law (de Saint-Robert's law)

$$\dot{r} = ap^n$$

so that

$$p_c^{n-1} = \rho_p A_p a / C_D A_t \tag{3}$$

which shows that once the evolution of the burning surface is known the evolution of internal pressure is determined.

Propellant temperature T_p affects the burning rate and, thus, the pressure and thrust. Two coefficients are often used to quantify this sensitivity.

$$\Pi_K = \frac{1}{p_c}\left(\frac{\partial p_c}{\partial T_p}\right)_K$$

characterizes the sensitivity of the motor (constant Klemmung) and

$$\sigma_p = \frac{1}{\dot{r}}\left(\frac{\partial \dot{r}}{\partial T_p}\right)_p$$

characterizes the sensitivity of the propellant (constant pressure).

Thc rocket motor operation is stable when $n < 1$, and the effect of burning surface area variations on the pressure curve will be enhanced when n is higher. This effect is generally undesirable and low-pressure exponents are generally required, but high exponents can be of help for special applications, for instance, thrust modulation.

It is easy to demonstrate the relation

$$\Pi_K = \sigma_p/(1-n)$$

between the temperature coefficients. Here also, high values of n will have a detrimental effect since, generally, variations on burning time are required to be very limited.

We will see later that high values of n lead to higher values of the acoustic response of the propellant which will aggravate the sensitivity to acoustic instabilities.

To express the thrust F, for practical reasons it is convenient to use the coefficient C_F

$$F = C_F p_c A_t \tag{4}$$

With the assumptions of the ideal rocket analysis it is possible to show for the ideal gas hypothesis that

$$C_F = \sqrt{2\frac{\gamma^2}{\gamma-1}\left(\frac{2}{\gamma+1}\right)^{(\gamma+1)/\gamma}\left[1-\left(\frac{p_e}{p_c}\right)^{(\gamma-1)/\gamma}\right]} + \frac{p_e - p_a}{p_c}\left(\frac{A_e}{A_t}\right)$$

C_F is a nondimensional parameter that depends only on the combustion gases γ, on the expansion ratio of the nozzle ($\epsilon = A_e/A_t$, the ratio between area of the nozzle exit and area of the throat), and on the ratio p_c/p_a between chamber pressure and ambient pressure (since p_c/p_e is only a function of γ and ϵ). C_F gives the efficiency of a nozzle for a given propellant and nozzle geometry.

Figure 2 gives the evolution of C_F as a function of ϵ for various values of p_c/p_a for $\gamma = 1.2$. The same results can be displayed in the form of a table based on the values of γ.

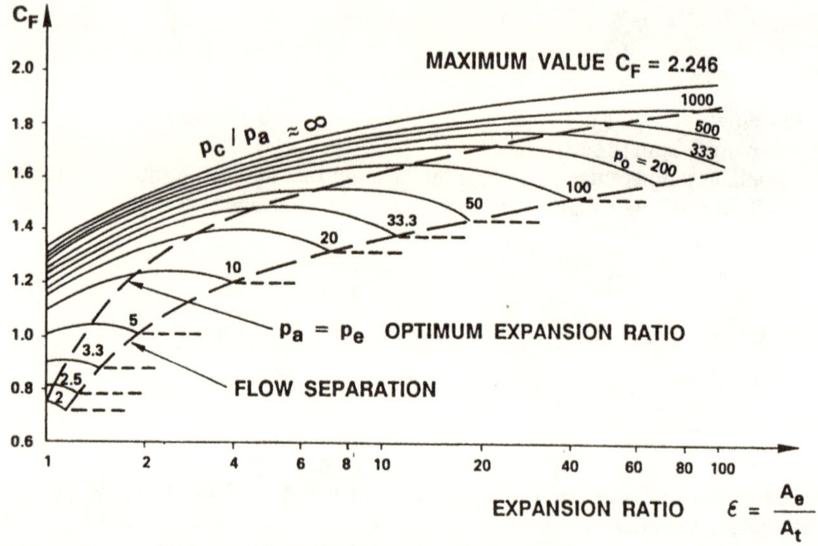

Fig. 2 C_F **diagram for** $\gamma = 1.20$.

It is easy to show that the thrust is maximum at a given ambient pressure when $p_e = p_a$. The expansion of the combustion gases is then optimum, and the nozzle is said to be adapted. Since p_e is a function of the chamber pressure and nozzle geometry and p_a a function of altitude, different situations will arise. When $p_e > p_a$ the jet is underexpanded and expands at the exit section. When $p_e < p_a$ the jet is overexpanded and a shock forms at the nozzle exit. When $p_e \ll p_a$ the jet separates from the wall of the nozzle. A classical criterion for the occurence of this undesirable phenomenon, which is valid for half-angles of the exit cone smaller than 15 deg, indicates that separation occurs in an area where pressure is close to 0.4 Pa.

The propulsion performance of a given motor is generally characterized by its specific impulse I_{sp}

$$I_{sp} = F/(g_0 \dot{m})$$

where $g_0 = 9.81$ ms^{-2}, since the velocity increment of a missile using this motor under gravity-free, drag-free conditions is

$$\Delta V = g_0 I_{sp} \log(1 + m_p/m_i)$$

where m_p is the propellant mass and m_i the inert mass, that is, total mass of the missile minus the propellant mass.

The value of the delivered I_{sp} is related to a given motor and to its operating conditions. To use I_{sp} as a performance index to compare different propellants, one then has to define a standard motor and standard conditions with which a standard specific impulse is measured. These standard motors are also used as ballistic evaluation motors for the propellant.

Computer programs of increasing degrees of complexity have been developed in order to calculate the theoretical (or maximum) specific impulse of the propellant. Based on thermodynamics and some simplifying assumptions the program of Ref. 2 gives the composition and thermochemical values of the combustion gases and a theoretical specific impulse of an ideal rocket motor having no losses for the operating conditions. Programs have also been developed which calculate the losses due to flow conditions in the nozzle and which estimate the losses due to combustion chamber conditions. The deviations from ideal performance are generally calculated by using a number of independant models for the different losses: two-dimensional flow (divergence), two-phase flow, combustion efficiency, chemical kinetics, and boundary-layer losses.[3,4]

These codes provide guidance to the chemist for the research of new energetic formulations, and they enable the designer to predict the delivered specific impulse of a new motor. In the case of the preliminary design of tactical motors semiempirical approaches are often applied. They use the data obtained through firing of similar motors or ballistic evaluation motors of approximately the same size loaded with the same propellants and the equations for variations of C_D and C_F previously presented.

F. Structural Performance

The equation for the velocity increase shows that the two main parameters that characterize the performance of a motor are the specific impulse and the mass fraction, i.e., the ratio of propellant mass to inert mass (case, nozzle, insulation, igniter). From a structural point of view the coefficient often used is the ratio of the propellant mass to the total mass of the motor; Fig. 3 gives an idea of the evolution of this ratio with history for tactical motors having approximately the same size. The best performance is, of course, obtained by the use of high-specific-strength composite materials for the case and a minimum-weight design for the nozzle.

The higher the operating pressure, the higher is the specific impulse of the motor, but also the higher is the thickness and mass of the case. There must be a tradeoff and optimization of the motor design. In many cases for integral motors with high length to diameter ratios, the most important design criteria will be bending stiffness after combustion of the propellant.

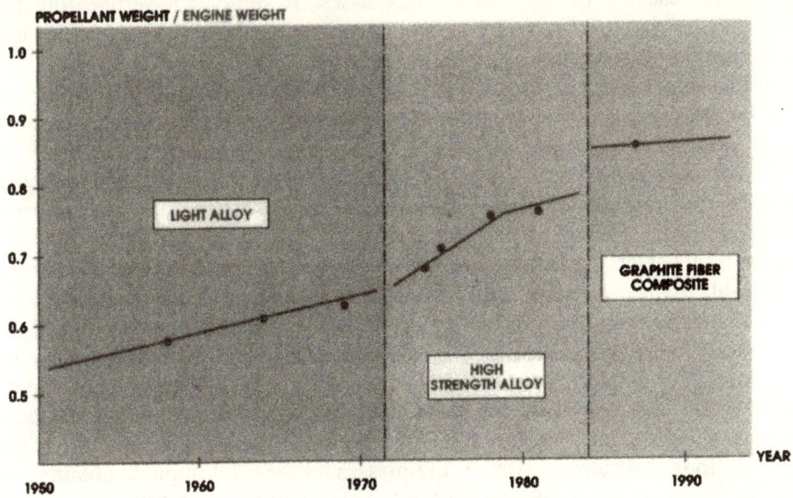

Fig. 3 Structural ratio evolution.

G. General Method for the Design and Optimization of a Solid Rocket Motor

The general procedure summarized next is a typical procedure applied to a typical tactical motor. There is, of course, no such thing as a typical rocket motor, especially in the field of tactical systems, because of the great variety of designs (the range of combustion times can vary, for instance, from a few milliseconds to a few minutes) and the great variety of missions. In addition there are preferences of the company or organization developing the propulsion unit based on past experience or specialization in some types of materials for cases, nozzles, propellants, etc. Therefore, the development must be considered as a didactic description that should be amended in every specific case.

As already stated the solid rocket motor case is essentially a minimum-weight pressure vessel on which bending and locally concentrated mechanical loads are applied, also sometimes local deformations have to be considered. The design of the motor must take into account these loads from the beginning.

The preliminary design will generally start with some general technology decisions, such as type of case (composite, metallic, etc.), type of propellant (ability to case bond or not, signature characteristics, hazard classification, etc.), type of thrust vector control if any, characteristics related to vulnerability to be incorporated, characteristics related to production (for instance, type of assembly of the aft closure) and maintenance, and characteristics related to cost, etc.

The overall envelope (maximum diameter and length) is already dictated by the missile designer. The initial design will generally begin by the choice of an operating pressure (based on past experience, extreme operating temperatures, rough optimization between impulse and weight, specific considerations, etc.), and a first calculation of the case will be done with a thickness of the case material adapted to the internal pressure and to the bending moments or local deformations according to a given coefficient of safety. Coefficients of safety are often supplied to the designer and may vary from customer to customer and depend on the application. They will generally be higher for air-launch applications or man portable systems. The range is generally from 1.25 to 1.6. Local reinforcements can be represented by equivalent thicknesses. If significant aeroheating takes place the design must be checked according to the case material properties at the corresponding temperature.

At this stage analytical calculations or simple codes may be used. The result will be a first drawing of the case with a determination of the deformation properties under pressure.

The second step will be the choice of a propellant and a grain geometry able to satisfy the propulsion requirements. The choice of the main parameters of the nozzle (throat diameter and expansion ratio) is done at the same time. This preliminary design of the propellant grain can be made with the help of computer codes able to make an optimization through mechanical and ballistics calculations according to the main propulsion requirements. The margin of safety level for the operational thermal cycles taken for the propellant is an important parameter at this point. The internal thermal insulation will be then defined as a consequence of the grain geometry. The results will be thrust and pressure performance as close as possible to the requirements. These results and, especially, the value of the maximum operating pressure, the mass, and the total impulse obtained may lead to iterations on the design by modifications of the case followed by a new adaptation of the propellant grain and insulation designs or even to a modification of the case volume or type of propellant.

The next step will generally be a review of the various design solutions and their compliance to the requirements with the missile designer. Also, production costs and capability of performance improvements will be compared. At this stage simulations of the missile performance with the various possible solutions will be performed. Functional and value analysis are efficient tools to help the program managers to make the right decisions.

When an agreement is reached a more detailed design will be undertaken. It involves more detailed calculations of the case and propellant grain, optimization of the nozzle, introduction of a possible throat erosion law for the nozzle, etc. Pressure and thrust laws at ambient and extreme temperatures are determined, and structural integrity safety margins for the various thermomechanical or aerodynamic loads are checked.

Two-dimensional codes, and today more and more three-dimensional codes, may be used. The results of these analyses may lead to the definition of future testing to be performed for some critical loads.

III. Propellant Properties and Selection

A. Case-Bonded Grains and Free-Standing Grains

There are two main types of grain architecture: free-standing grains and case-bonded grains. Grains of the first type are introduced into rocket motor cases (cartridge loaded) after manufacture. Grains of the second type are bonded to the motor case during the casting (or injection) and curing steps of the propellant grain manufacturing process (Fig. 4).

There is not a single, well-defined procedure for selecting a free-standing grain architecture or a case-bonded grain architecture for a given rocket motor, except when one of these two architectures is obviously most appropriate for a specific reason. Nevertheless, case-bonded grains generally give higher performances than free-standing grains for equal available volumes (the volume occupied by the inhibitor, the case insulation, and the space between them can be filled with propellant). On the other hand, the mechanical loads will be very different in the two cases.

The service life of a propellant grain before firing consists in a succession of transport and storage phases, under conditions that are contingent to the mission of the missile and, particularly, involve thermal cycling (−54 to 65°C is typical for air-launched missiles) which induces stress/strain in the propellant, especially when it is bonded to a rigid structure having a lower thermal expansion coefficient. This is why the first step of the preliminary design analysis is a rough structural analysis, particularly for case-bonded grains.

The second most important mechanical load for the grain is a consequence of the pressure rise at firing. In a case-bonded grain an intense stress field will develop in the propellant due to the case deformation (the propellant is incompressible). In a free-standing grain, if the pressure can equilibrate properly between the port and the interstice between the motor and the grain, the level of stresses will be limited. In any case these stresses will be maximum at ignition when the volumetric loading coefficient is maximum. Also they will be generally maximum at the lowest operating temperature because of the modulus of the propellant even if the

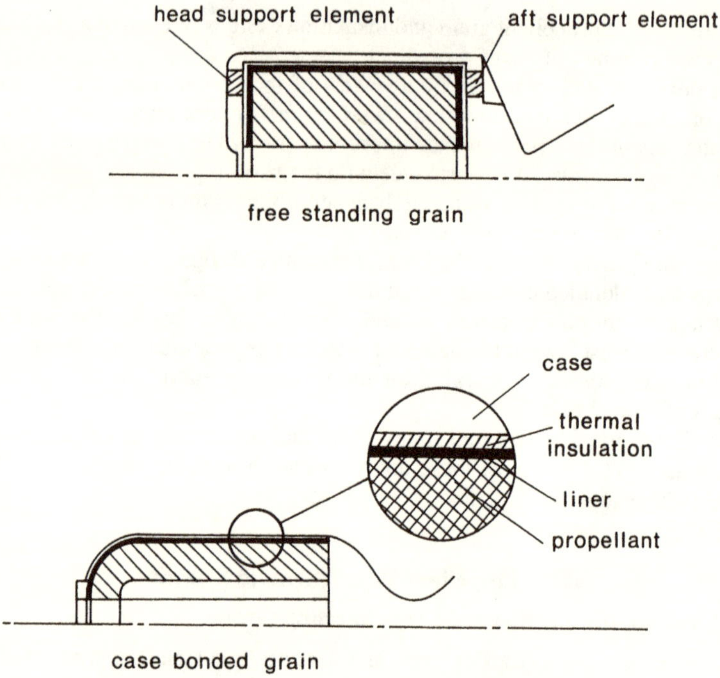

Fig. 4 Case-bonded and free-standing grains.

operating pressure is the lowest. This enables the first analysis of criticality to be done at ignition at the lowest operating temperature.

In conclusion it is easy to guess that case-bonded grains require the use of propellants having very good (elastomeric type) mechanical properties in a wide temperature range.

B. Grain Configurations and Geometries

Since the pressure and thrust curves will be dictated by the burning surface area evolution, the initial geometry of the propellant grain is a key factor for the grain design. In tactical motors with generally high length over diameter ratios (the greatest part of the motor being cylindrical and the greatest part of the grain also being cylindrical or nearly cylindrical) the burning surface area evolution is generally directly related to the evolution of the cross section of the grain port. This evolution will be characterized as a function of the web thickness; the web thickness is generally defined as the minimum-propellant thickness obtained between the beginning of combustion and the time when the flame front becomes tangent to itself or to the outside contour of the grain.

1. Case-Bonded Grains

When propellant grains have an outer diameter larger than 500 mm or a weight of more than 300 kg, they are almost always case bonded. High-performance, middle-sized (outer diameter between 100 and 500 mm, weight between 10 and 300 kg) are generally case bonded, but free-standing smaller sized grains are still common.

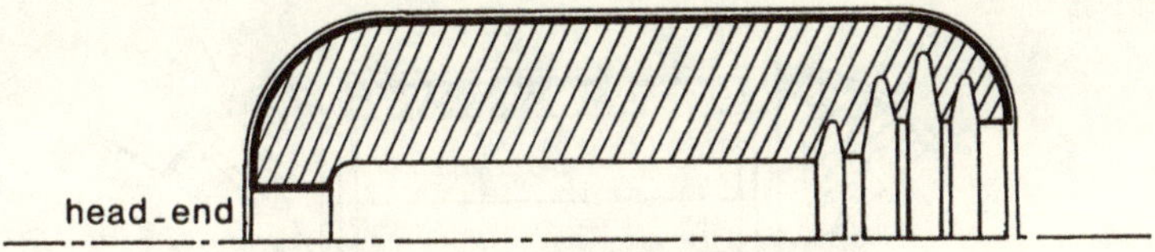

Fig. 5 Axisymmetric configuration.

Case-bonded grains generally have a central port; the outer surface of the grain is bonded by a liner (and sometimes a thermal insulation) to the motor case. During firing combustion of the propellant is initiated on the internal surface of the central port and proceeds radially toward the case (and to a certain extent longitudinally, depending on the exact geometry).

Exact grain geometry is generally obtained during manufacture of the grain by direct casting of the propellant in the case around the mandrel and curing. In some cases the port can be obtained by machining after curing. The most common configurations are discussed next. The axisymmetric with annular slot configuration has slots that are circular; their axis is the same as the grain axis. They are located along the central port (Fig. 5). Cylindrical configurations are either star shaped or wagon wheel. In the star, the cross section of the central port has the shape of an *n* point star. The contour of the star is constant along the axis [in some cases it may be slightly evolving for manufacture practicality (Fig. 6) or for internal ballistics reasons]. In the wagon wheel, the cross section of the central port looks like a wagon wheel (Fig. 7). Numerous parent configurations exist like dendrite, anchor, and dogbone configurations.

Other configurations may be obtained by combination of some of the described configurations; for instance, the bipropellant star configuration (to eliminate sliver). Full head-end web grains are also used. Simpler configurations such as the internal-burning tube are commonly used; the ends may be partially or totally inhibited.

Three-dimensional geometries. The described configurations are considered as one or two dimensional, although, of course, they are actually three dimensional. They are either axisymmetric or cylindrical, often with an order-*n* symmetry. It is, therefore, not too difficult to calculate burning area vs web burned or the stress-strain field. Today, three-dimensional configurations are getting more and more popular in the designer community; three-dimensional configurations are also much more difficult to design. Most of these configurations are referred to as finocyl, which is a contraction of fin and cylinder. The fins may be located either at the head end or at the aft end of the grain (and sometimes at both ends); they merge into a central cylindrical port. They may have the shape of slots, which simplifies the geometry (Fig. 8) and the production processes. The slot-and-tube

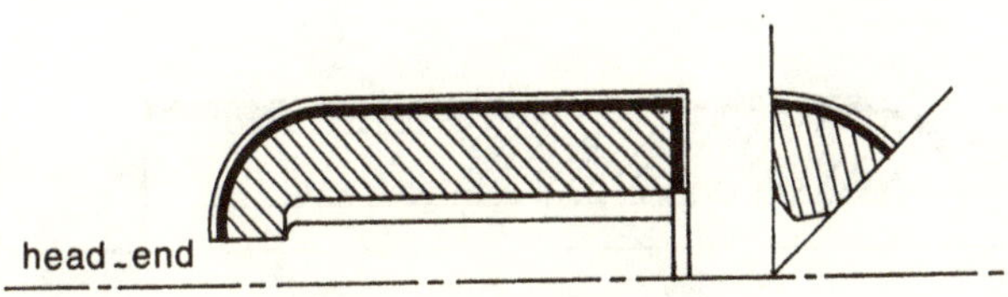

Fig. 6 Star configuration.

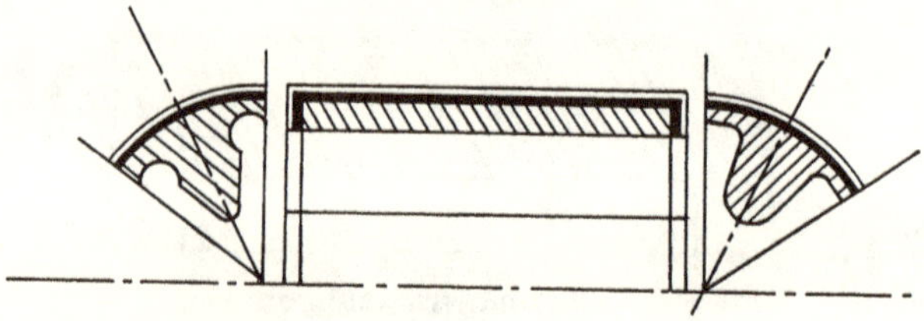

Fig. 7 Wagon wheel configurations.

configuration is today probably the most widely used for tactical systems. Often, for mechanical stress relieving, there are also radial slots.

These configurations require three-dimensional analysis for calculating burning area vs web burned, as well as for the stress-strain field or the gas flow inside the central port. The flexibility of this design is illustrated in Fig. 9, which shows how the pressure (and thrust) curve can be modified easily using, for instance, the length of the slots as the parameter.

End burning grains. The end-burning configuration is not well adapted to case-bonded architecture because of structural integrity if the propellant is directly bonded to the case. However, it is possible to design such case-bonded grains using stress-relieving grain support and retention systems. These allow thermal shrinkage due to propellant cooling after curing and to thermal cycles and enable pressure to equilibrate like in a free-standing grain configuration during firing.

2. *Free-standing grains*

Because they are not bonded to the case wall, except sometimes locally, for instance, at the forward end, they allow configurations which cannot be obtained with case-bonded grains (for instance, internal-external burning tube).

Cylindrical configurations. Star, wagon wheel, and tube configurations similar to those already described may be found for free-standing grains. Grain ends are generally simpler: they may be plane.

Configurations with evolving port cross section (trumpet). This configuration was developed to reach high-volume loading fractions for free-standing grains: the cross section of the central port is circular in the forward section and becomes progressively star shaped in the aft section of the grain (Fig. 10).

End-burning grains. The orientation of burning is totally in the longitudinal direction. This configuration has some advantages: gas generation rate is almost constant, volumetric loading fraction is high, and grain manufacture is easy. Side and forward end faces are inhibited. Burning times are long and thrust levels are low or moderate. Thermal insulations and inhibitors play an important role to protect the chamber walls from the long exposure to hot gas and to restrict the combustion to the desired area, respectively. They also generate a high level of pyrolysis gaseous products during firing which must be taken into account in the

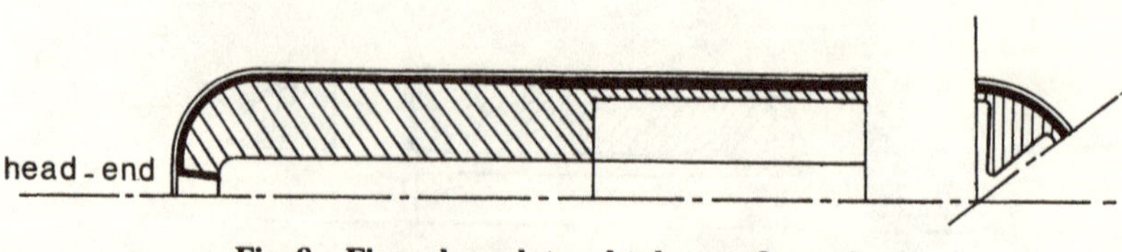

Fig. 8 Finocyl, or slot and tube, configuration.

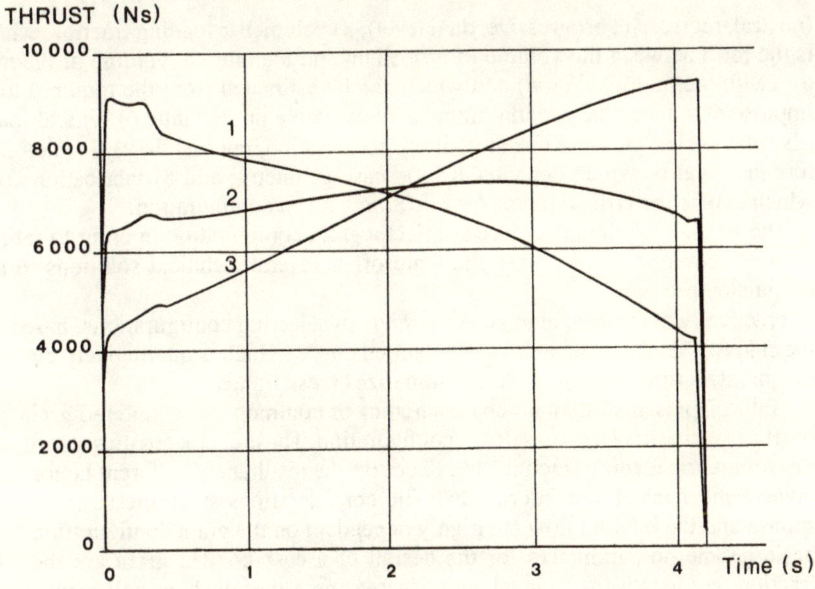

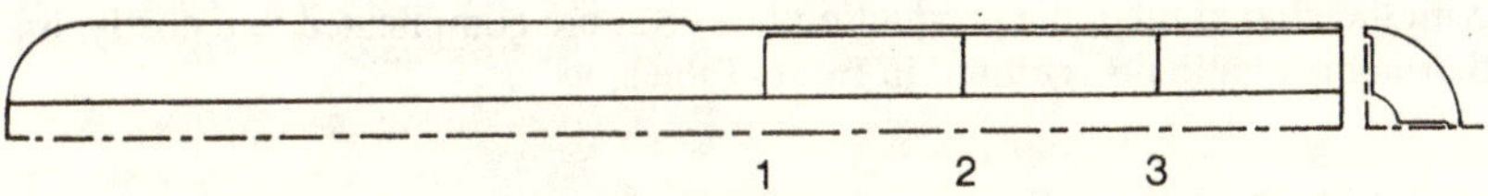

Fig. 9 Influence of slot length on the thrust curve.

total amount of gas generated by the grain. They are used mainly for the sustaining phase of the flight of some missiles.

Other type of geometries can be found in some cases: rod and shell, cruciform, wrapped spirals, etc.

C. General Principles for Selection of Grain Configuration

For selection of grain configuration, the main factors which are taken into account are 1) volume available for the propellant grain; 2) grain length-to-diameter ratio (l/d); 3) grain diameter-to-web thickness ratio (d/e); 4) thrust vs time curve, which gives a good idea of what should be the burning area vs web burned curve

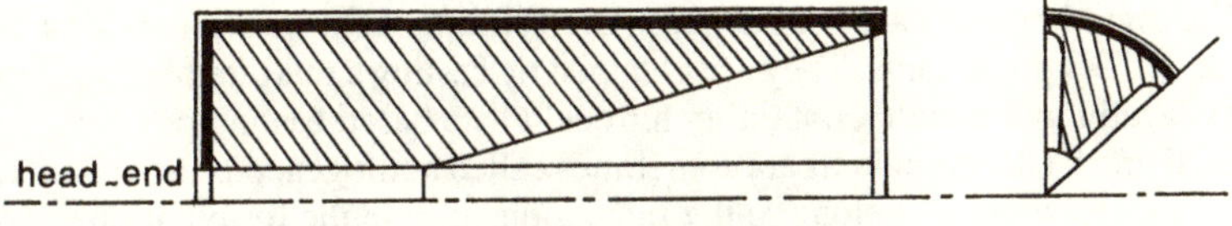

Fig. 10 Trumpet configuration.

(neutral, regressive, progressive, dual level); 5) volumetric loading fraction, which is the ratio between the volume of propellant and the internal volume of the motor (with thermal insulation) and which can be estimated from the required total impulse and the actual specific impulse of available propellants; 6) critical loads (thermal cycles, pressure rise at ignition, acceleration, internal flow); 7) manufacture practicality, which depends also on case geometry; and 8) fabrication cost, which can be the critical factor for selecting a given configuration.

There is no definite procedure to select a grain configuration in order to satisfy a set of requirements, because there are often several technical solutions to the propulsion problem.

Practically, there are some general trends in selecting configurations, based on the shape of the burning area vs web burned curve (which is qualitatively close to the thrust vs time curve). Table 2 summarizes these trends.

Table 3 presents the main characteristics of commonly encountered grain geometries which may help to select a configuration. These characteristics, especially the volumetric loading fraction, are, of course the resultant of different factors that involve mechanical and internal ballistic considerations since the structural response and the internal flow are highly dependant on the grain configuration. The basic geometric parameters for the design of a case-bonded grain are the web fraction and length-to-diameter ratio. Increasing either of them will result in an increase of the volumetric loading fraction and a simultaneous increase in bond stress and bore strain.

The two most used configurations in practical applications for tactical motors are the star and the slotted tube designs. If some better loading can be obtained with the slotted tube, the production is a bit more complicated and costly since a thermal insulation is required in front of the slots.

D. Propellant Families

Six families of propellants are today commonly manufactured and used. They are described more specifically in specialized references.[5]

1. Extruded Double-Base Propellants

The main ingredients of extruded double-base (EDB) propellants are nitrocellulose and nitroglycerine. After impregnation of nitrocellulose with nitroglycerine, generally in a water medium to get a paste, the composition is obtained through kneading with additives and carpet rolling. Configuration is obtained by extrusion through a die having the desired shape. The outer diameter is limited to about 300 mm. Additional grain machining may be performed. More modern production processes involve continuous screw extrusion.

2. Cast Double-Base Propellants

The ingredients of cast double-base (CDB) propellants are similar or parents to those of EDB propellants. They are obtained by casting a mixture of nitroglycerine and triacetin into a mold containing nitrocellulose based casting powder.

EDB and CDB propellants are sometimes called homogeneous propellants since both oxidizer and fuel belong to the same molecules in the formulation.

Table 2 Burning area neutrality vs propellant grain configuration

Burning area evolution	Grain configuration	Remarks and comments
Good neutrality[a] (less than 15% relative change in burning area)	Wagon wheel, dendrite	Short burning times ($\leq$3 s), low volumetric loading fraction
	Trumpet, tube and slots	Constant burning area vs web burned obtained by progressive cylinder slots and regressive slots or trumpet zone
	Star	Significant sliver
	Tube (end faces may be restricted)	Volumetric loading fraction less than 0.2
	Bipropellant star	Two different propellants. Internal propellant has higher burning rate than external propellant by a factor of about 2. Costly configuration requiring long manufacture cycles (two curing stages), neutrality kept, even for high volumetric fraction, and sliver is eliminated.
	End burning	Long burning times, low or moderate thrust.
Dual level	Trumpet, slotted tube	The ratio for the two levels can be adjusted by varying the geometry of the aft-end section. Volumetric fraction may reach 0.88. l/d may reach 10.
		Adjusted by number and geometry of the annular slots.
	Bipropellant star	Adjusted by the geometry of the star and propellant burning rates.
	end-burning (with annular slots in the aft-end face)	Adjusted by the geometry of the annular slot(s) (boost with radial burning, sustain with end-burning).
Progressive	Tube (right circular port section)	Most common is the grain with restricted faces.
	Star	With restricted end faces; slivers induce long tail-off.
Regressive	Tube with internal external burning	
	Star with internal external burning	Unrestricted end faces and low to moderate l/d ratio.
	End-burning with tapered aft-end	Long burning time, low to moderate thrust.

[a]The lower the volumetric loading fraction, the better the neutrality. A constant burning area can be accurately obtained by adding axisymmetric slots in the central port or by restricting specific propellants surfaces. It raises grain cost because it requires additional phases during grain manufacture.

Table 3 Main characteristics of common grain configurations

Configuration	Volumetric loading fraction	Web thickness	Burning area	Burning area neutrality	Sliver fraction, %	Web fraction, d/e	Comment
Star	0.75–0.84	Intermediate	Intermediate	Good	5–10	3.5–5.5	Case-bonded and free standing grains
Trumpet	>0.88	Large	Intermediate	Excellent	0	>2	Free-standing grains
Slotted tube	0.75–0.85	Large	Large	Good	0	3	Case-bonded grains
Wagon wheel	0.5–0.7	Small	Very large	Excellent	5–10	6–12	Free-standing and case-bonded grains
Star with full head end web	0.75–0.85	Intermediate	Intermediate	Good	5–10	3	Case-bonded grains
Dual propellant	0.9	Large	Intermediate	Excellent	< 5	2.5–3	Case-bonded grains
End-burning grains	0.98–1	Very large	Small	Excellent	0	1	Low thrust

3. *Composite Modified Cast Double-Base Propellants*

Composite modified cast double-base (CMDB) are derived from CDB propellants by addition of RDX, HMX and, possibly, nitroglycerine in the casting powder.

4. *Elastomeric Modified Cast Double-Base Propellants (EMCDB)*

Elastomeric modified cast double-base (EMCDB) propellants are produced by the same process as CDB and CMDB with the same type of basic formulations. A hydroxyl prepolymer (polyester, polycaprolactone) and an isocyanate crosslinking agent are introduced in the liquid casting solvent.

These four types of propellant families may be filled with energetic solids (RDX, HMX) but are nowadays generally used without any solid mineral oxidizers or fuels as fillers. In this way their formulation, based on C, H, O, N only, gives them "minimum smoke," sometimes called smokeless, characteristics since there are very few condensed species in the nozzle exhausts for the most generally encountered atmospheric conditions.

5. *Composite Propellants*

These are based on a nonenergetic polymeric binder (generally polybutadiene or polyurethane) and on ammonium perchlorate (AP). They may or may not contain aluminum powder as a fuel. They are obtained by mixing the solids and the liquid ingredients of the binder, casting, and curing. The propellants without aluminum are reduced-smoke propellants, i.e., there is only secondary smoke formation, by condensation of water with hydrochloric gas resulting from the combustion of AP, in certain conditions of ambient atmosphere temperature and humidity.

6. *High-Energy Propellants*

These are based on a binder highly plasticized by a liquid nitric ester or a mixture of nitric esters, on a nitramine (RDX or HMX). They also may contain ammonium perchlorate and aluminum. These are called cross-linked double base (XLDB) even if there is very little or no nitrocellulose in the binder and a very high level of energetic solids in the formulation which give them a physical behavior of the same type as composite propellants. Their production process is roughly the same as composite propellants. According to the precise formulation they can fall in the minimum-smoke (binder + nitramine), reduced-smoke (binder + AP) or smoky (binder + AP + HMX + Al) category.

Besides the main ingredients, propellants may contain several other ingredients, generally at low contents, used as stabilizers, afterburning suppressants, combustion instabilities suppressants, and burning rate modifiers. One of the important tasks of propellant designers is to find a practical way (fillers particle size, burning rate modifier, etc.) to control the burning rate and pressure exponent, which are key factors in designing solid propellant grains.

E. Mechanical Properties

EDB, CDB, and CMDB propellants are generally stiff with high modulus and low-elongation capability at low temperatures so their use is more or less limited to free-standing grains. EMCDB, composite, and high-energy propellants exhibit a viscoelastic behavior. They are very well suited for case-bonded grain applications because of their mechanical behavior: low modulus and high-elongation capability in a wide temperature range.

F. Energetic Characteristics

The standard delivered specific impulses and densities of these families are summarized in Fig. 11.

Because sometimes different standard motors and conditions of operations are used to measure the standard delivered specific impulse it may be useful to give the precise values in the conditions of Fig. 11 for a few typical propellants: for a composite propellant (HTPB binder, 68% AP, 20% Al), 245 s; for a reduced-smoke propellant (HTPB binder, 85% AP), 236 s; and for a high-energy propellant (high-energy binder, 10% AP, 46% HMX, 19% Al), 254 s.

G. Burning Rate Characteristics

Limitations of propellant burning rates available for a specific project is one of the most frustrating difficulties for the designer who would prefer the use of the greatest possible range of burning rates.

Current practice is to determine them over specified operational temperature ranges utilizing standard or ballistic evaluation motors which are designed to provide neutral pressure traces with sharp tailoffs in order to obtain combustion characteristics with good precision. More economically efficient methods are

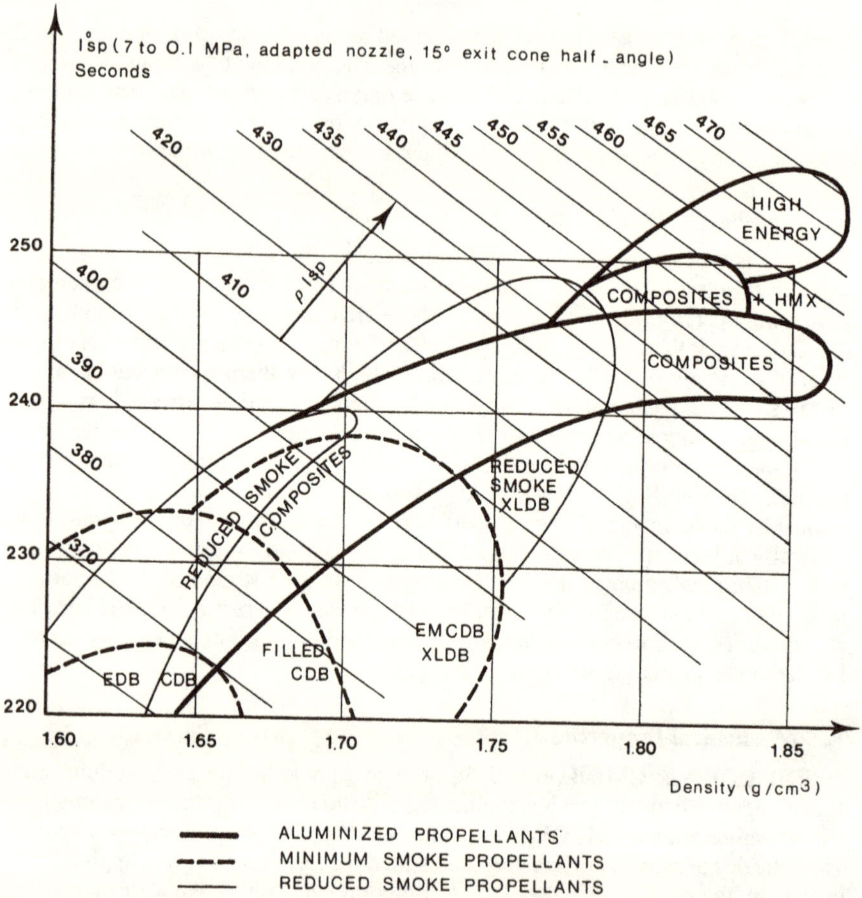

Fig. 11 Standard specific impulse and density of the main propellant families.

being developed to obtain the rate of combustion law over a whole range of pressure by using devices with variable K. The regression rate of the propellant can even be obtained directly by ultrasonic measurements of the thickness of unburned propellant during firing.

The basic combustion characteristics and mechanisms for these propellants have been described by Kuo and Summerfield.[6] The more practical aspects will be described now for the three most used families: EDB, CDB, and composite propellants. As previously stated, it is customary to use an empirical expression of the $\dot{r} = ap_c^n$ type for the linear burning rate.

Some propellants show regions where the pressure exponent is markedly reduced (plateau effect) or even negative (mesa effect). These very desirable effects, since they reduce the temperature coefficient sensitivity of the motor (at constant K), are typical of double-base propellants using adequate (lead and copper salts, generally) ballistic modifiers.

The burning rate of composite AP propellants is controlled by the AP particle size and the level of AP in the formulation, fine particle size leading to higher burning rates (and also to higher pressure exponents). Solid (copper chromite, iron oxide) or liquid (ferrocenic derivatives) additives are commonly used in order to get high rates of combustion. Good pressure exponents (0.5 or lower) can generally be obtained for pressures lower than 20 MPa. It may become a difficult problem for higher pressures. The effect of aluminum when used is generally second order.

With the help of the plateau effect double-base propellants can show very low-temperature coefficients. This is generally more difficult to obtain with the most energetic formulations. Values for composite propellants are generally of the order of $0.2 \cdot 10^{-2}{}^{\circ}C^{-1}$.

Figure 12 shows the burning rate range, at a given pressure of 7 MPa or at the middle of the plateau for double-base propellants, for the most commonly used propellants. When seeking the highest burning rates, the answer lies with composite propellants for the usual burning times for missile propulsion stages (several seconds to several tens of seconds). However, for very short burning times, on the order of a few tens of milliseconds, which are often used for light antitank missiles where a low signature is generally required, EDB propellants (solventless) are a prime solution because of their physical properties lending themselves particularly well to designs with thin webs (fraction of a millimeter).

For very long burning times, solutions can be found in every family, although they are always accompanied by a significant decrease in the energy characteristics.

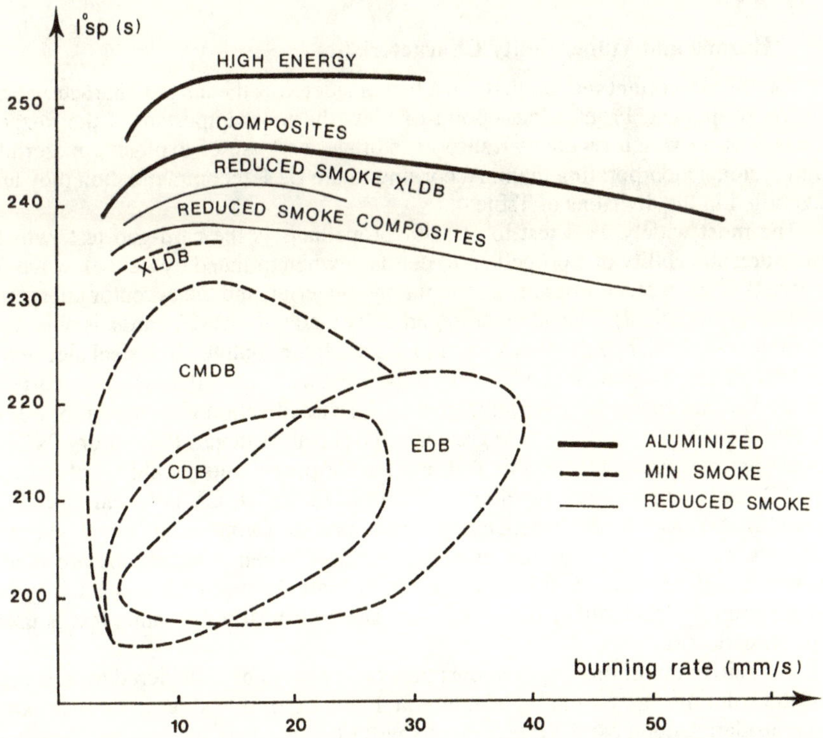

Fig. 12 Standard specific impulse and rate of combustion of the main propellant families.

"Average" (6–16 mm/s at 7 MPa) burning rates can be obtained with every type of propellant. It is useful to know, however, that EDB propellants offer greater burning rates than the CDBs for an equal level of energy.

This is a characteristic inherent in the product. In applications where a large range of operating temperatures is required, the lower temperature coefficient of the homogenous propellants may compensate for low specific impulse per unit volume. With an identical burning rate at 20°C, the propellant with the highest temperature coefficient results in a higher maximum working pressure of the grain at high temperatures, necessitating a thicker structure for the motor, resulting in a weight increase. Similarly, the decreased flow rate at low temperatures can have negative effects because of the resultant reduction in thrust.

Ballistic evaluation motor burning rates must frequently be scaled up by a few percents to correlate with full-scale motor rates. Also, some effects related to the casting process of the grain can sligtly modify the burning rate characteristics of the propellant. All these factors have to be taken into account for a good motor operation prediction. Their influence is generally estimated with the help of data gathered through past experience or specifically designed research programs.

The possible error on the real burning rate of a new propellant grain design is probably the most important factor influencing the precision of the thrust or pressure curve prediction of a new motor. Because past experience is important, the relevant data is often considered proprietary by the organizations involved in the design.

H. Hazard and Vulnerability Characteristics

Another important feature that must be considered is the hazard characteristics of the propellant. From a legal point of view the most important is the hazard classification which regulates transport, storage, and use of explosive materials and systems incorporating them. According to the U.N. recommendation they are classified in the divisions of Table 4.

The most widely used test to classify propellants is the card-gap test, which measures the ability of a propellant to detonate when initiated by a shock wave. It consists of an explosive donor, an attenuating material, and the acceptor energetic material (propellant). The attenuator, generally made of PMMA cards, is adjusted in thickness to change the shock level to the acceptor confined in a steel sleeve. A mild steel witness plate is used to determine whether or not a detonation occurred. In the U.S. an energetic material is given a 1.1 classification if the acceptor detonates when the gap is 70 cards or greater at the Naval Ordnance Laboratory (NOL) card-gap test. According to this test, composite propellants would be classified as 1.3, and high-energy propellants as 1.1; EDB, CDB, CMDB, and EMCDB are classified as 1.3 or 1.1 depending on the precise formulation (nitroglycerine and nitramine levels). Caution must be exercised when comparisons are made between different countries because different limits between 1.3 and 1.1 may be considered by local authorities even if the same, or nearly the same, test is used for classification.

Insensitive munitions requirements may have also to be considered even if part of the solutions are on the architecture and general design side. If slow cookoff is considered, composite AP propellants will often not pass the test, but they will easily pass the sympathetic detonation test when this will be much more difficult for EDB, CDB, CMDB, and EMCB.

Table 4 Classification of hazards of explosive substances and objects

Class number	Division number	Characteristics of the substances or articles in this section
1	1	Substances or articles essentially involving a mass explosion hazard, i.e., affecting almost the entire load virtually instantaneously.
1	2	Substances or articles involving a projection hazard but no mass explosion hazard.
1	3	Substances or articles involving a fire hazard with a minor blast or projection hazard, and no mass explosion hazard. Subdivision 3a consisting of substances or articles whose combustion gives rise to considerable radiant heat; Subdivision 3b consisting of substances or articles that burn fairly slowly, or burning one after the other with minor blast and projection effects.
1	4	Substances or articles involving no significant hazard, designed or packed so as to exhibit only relatively minor hazard or whose effects, in case of ignition or initiation, do not give rise to the projection of fragments of appreciable size, and remain in any case small enough not to significantly hinder fire-fighting operations and the application of emergency measures.
1	5	Substances that are, when they explode, as those of Section 1, but that are relatively insensitive. These substances display a very low probability of initiation and transition from combustion to detonation except when in large amounts in a small and confined space. They may not explode when exposed to outside fire.

I. Production and Cost Characteristics

EDB grains obtained by extrusion are clearly limited in their mass and size. These limitations are mainly due to the size and the performance of the presses that can be used. Diameters are generally limited to 250 mm in the western world. Similarly, only cylindrical grains simple to manufacture and, therefore, inexpensive can be produced.

The EDB process, on the other hand, lends itself particularly well to industrial production as well as to high-production rates. It is, therefore, particularly suited for the production of small ammunitions. The homogeneity of this product and its high degree of rigidity, allowing the production of very thin webs as well as the possibility of machining within very precise limits (a few hundredths of a millimeter), makes it very useful when seeking high-precision impulses. In addition, the combination of rigidity and very thin webs is attractive for grains combining short combustion times with high accelerations, useful for light antitank missiles.

Finally, processes such as continuous screw extrusion and stamping may well give new impetus to this product by contributing to a decrease in costs or improvements in working conditions, production rates, or possible geometries.

The CDB and CMDB propellants have the advantage of allowing the production of free-standing grains in any type of shape with performances that are comparable to that of the EDBs.

The length of the propellant production cycle is of critical importance for the client. It is, of course, determined not only by the type of propellant but also by the type of production capability available and by the number of specific tools and equipments available for production.

All things being equal, a classification of the length of production cycles would be in the following increasing order: EDB, composites or XLDB, CDB (or CMDB).

The production of EDB grains with selected raw materials is almost instantaneous as they are thermoplastics shaped directly. The length of the cycle is controlled by finishing and quality control operations.

At the other end of the scale, the production of CDB is very slow, requiring various intermediate production steps, as well as ballistic adjustment of the casting powder, requiring firing test controls on specially cast specimens.

Composite propellants are easy to produce and can be used for grains of nearly any size, mass, and geometry. This is probably one of the reasons why they can be found in most of the practical applications today. A lot of research and developement work is devoted today to continuous production of composite propellants and to efficient high rate of casting processes for tactical motors.

The cost of a propellant grain is linked to four main parameters: 1) the cost of the raw material, 2) the manufacturing process, 3) the quantities required and delivery schedules stipulated, and 4) the technical specifications, including conditions for acceptance and control.

Of course, there is an interaction between the selection of the propellant grain (and its manufacturing process) and the specifications requested by the client. A constructive dialogue is necessary to avoid certain specifications unnecessarily increasing the complexity and, thereby, the cost of the propellant grain.

When the volume and the duration of the manufacturing process allow the organization of a specific production facility the cost may decrease significantly.

The cost of raw materials may also be an essential factor, e.g., XLDB using HMX of very specific particle size, or special nitrate esters (BTTN) are intrinsically more expensive than a simple composite propellant.

J. Propellant Selection

Selection of a propellant for designing a given grain is based on numerous criteria and, here again, there is no strict procedure for selecting a given composition. The type of architecture (case bonded or free standing), energy and burning rate criteria, structural integrity considerations, smokelessness and safety considerations may lead toward a given propellant family. Each of the propellant families covers a certain range of properties, and it is necessary that the properties of the selected propellant allow the design and manufacture of a grain satisfying all of the requirements.

Table 5 Main characteristics of common propellants

Propellant	Delivered specific impulse in standard conditions, s	Maximum density, kg/dm^3	Range of burning rates at 7 Mpa (or at the plateau) mm/s	Pressure exponent	Temprature coefficient	Architecture	Smoke	Manufacturing cost
Extruded double base	225	1.65	5–40 (plateau)	0	Very low	Free standing	No	Low
Cast double base	215	1.60	4–22 (plateau)	0	Very low	Free standing	No	Low
Cast composite modified double base	230	1.70	3–28 (plateau)	0–0.2	Low	Free standing and case bonded	No	Moderate
EMCDB	235	1.69	3–40		Moderate	case bonded	No	Moderate
Non aluminized composite polybutadiène propellant	240	1.73	4–60	0.3–0.5	Low to moderate	Free standing and case bonded	Secondary	Moderate
Aluminized composite polybutadiène propellant	240	1.86	5.5–80	0.2–0.5	Low to moderate	Free standing and case bonded	Primary and Secondary	Moderate
Non aluminized cross-linked double base	245 (with AP)	1.79	10–25	0.4–0.6	Moderate	Case bonded	No	Moderate
	235 (without AP)	1.75	5–15					
Aluminized cross-linked double base	254	1.86	9–25	0.5–0.7	Moderate	Case bonded	Primary	Moderate

Table 5 summarizes some properties of the main propellant families. It allows, in combination with Tables 2 and 3, a first approach in the selection of the configuration/propellant which is detailed in the next section.

K. Liners and Inhibitors

Liners are used to bond the propellant to the case or to the thermal insulations of the case. They are usually deposited in very thin thickness, 0.4 mm or less, by spraying. The main requirements are the mechanical strength of the liner and of the bond. Since thermal insulations have higher mechanical strength than propellants, the optimum bond strength may be considered to be obtained when the rupture of the specimen in a bond test happens in the propellant.

Combustion inhibitors must also have a very good bonding to the propellant, but they also require good thermal and ablation resistance properties. Since their ablation products will be part of the exhaust, these products must not condense into smoke for low-signature motor designs. This can be achieved either by using inhibitors whose pyrolysis products remain in place, like silicones with refractory fillers, or which are purely gaseous without condensed compounds; this can be achieved with formulations of the C, H, O, N type with enough oxygen, for instance, a combination of a polyurethane binder and an organic oxygenated solid. Inhibitors can be cast or injected on or around the propellant grain, starting with liquid mixes, or the propellant can be cast into preformed inhibitor boots (free-standing grains). Some typical inhibitor characteristics are given in Table 6.

L. Aging and Service Life

This is one of the most difficult questions that will face the individuals in charge of the development of a new motor or a new grain. The prime contractor responsible for the missile system wants to know the estimated service life or—a direct consequence—the replacement cycle of the motors.

Unfortunately, that question can only be answered with great caution at the beginning of a new design. It is, therefore, advisable to use a maximum of relevant information obtained during propellant development. Similarly, it is necessary to use the greatest possible amount of information that can be provided by results from the operational behavior of existing motors and grains, whose designs are as similar as possible to the one being considered.

During the development of the motor, although at a time when the designs are sufficiently determined, it is possible to start what is commonly known as accelerated aging, which will provide the opportunity of estimating the potential life at the end of the development phase. The principle of these accelerated agings usually consists in using an increase in temperature to increase the rate of the chemical phenomena responsible for the evolution of the materials that make up the grain. Extreme caution is recommended, however, because the failure processes can be of different natures, and the temperature can impact differently on each of them, so that, in the worst case, it is possible to create a failure that would not occur in the real application. There have been such instances.

General recommendations in this regard are as follows:

1) The temperature increase and accelerated aging must remain within modest ranges to be representative.

Table 6 Main characteristics of some cast inhibitors used for double-base or CMDB grain, minimum smoke propellant

	Mechanical properties			Physicochemical characteristics				Behavior		
Inhibitors[a]	m, MPa	E, MPa	%	Temperature,[a] °C	L/E[b]	C_p^c		Burning rate enhancement	Ablation,[d] %	Transmission,[e] %
				−40	0.44	1.13				
EPE 01	9.0	540	10	−20	1.07	1.58	1.50	yes	15	82
				−60	1.67	1.76				
				−40	0.55	1.17				
EPE 02	4.5	530	17	−20	1.20	1.53	1.50	yes	20	85
				−60	1.44	1.71				
				−40	4.1	1.54				
SI 111	4.5	1.90	45	−20	3.2	1.36	1.32	no	5	88
				−60	4.4	1.43				
				−40	1.0	1.45				
SI 113	5.5	2.35	37	−20	1.0	1.33	1.50	no	5	90
				−60	0.9	1.42				
				−40	1.0	1.42				
SI 120 A	3.8	3.70	15	−20	0.9	1.31	1.43	yes(in some cases)	2	90
				−60	0.9	1.39				
				−40	0.40	1.10				
PU 003	7.3	596	30	−20	1.30	1.68	1.30	yes	40	90–92
				−60	1.82	1.96				
				−40	0.54	1.26				
PU 004	8.5	523	35	−20	1.25	1.76	1.33	yes	30	90–92
				−60	1.40	1.83				
				−40	0.46	1.44				
PU 005	7.5	450	50	−20	1.30	1.73	1.35	yes	35	90–92
				−60	1.10	2.15				
				−40	0.44	1.59				
PU 11 C	7.5	517	60	−20	1.0	1.87	1.36	no	30	90
				−60	1.58	2.29				

[a]EPE: Polyester, — SI: Sillicon — PU: Polyurethane.
[b]Linear expansion coefficient, $10^{-4} \cdot K^{-1}$.
[c]C_p: Specific heat value, $j \cdot g^{-1} \cdot K^{-1}$.
[d]Weight loss in firing an end-burning standard motor.
[e]Transverse transmission in the plume in a standard test.

2) When elementary propellant or propellant to liner bonding samples are subjected to aging, they must be as representative as possible of the propellant that will be industrially manufactured and cast into motors.

3) Not only nondestructive and firing tests must be performed on the aged motors but also detailed and analytic evaluations. It is these evaluations alone that provide the capability of creating models of the evolution of the margin of safety throughout the service life, as various properties of the propellant and bonds change.

Experience has shown that the evolution of the ballistic properties of the propellant through aging is generally not very significant except in cases where migration of mobile species (liquid plasticizers or burning rate modifiers) may occur. Most of the failures observed in current practice are related to debonding of the propellant or oxidative aging if proper care is not taken.

IV. Ballistic Analysis and Grain Design

The grain design will use the propellants and grains characteristics previously described to elaborate the best possible solution to the requirements through a ballistic analysis based on the equations presented in Sec. II.E.

A. Requirements on Solid Propellant Grains

The general requirements just described must be transformed into requirements for the propellant grain in order to conduct its design.

1. Requirements Related to Functional Specifications

The main internal ballistics requirements are generally of the following types: 1) average, minimum, and maximum values of chamber pressure; 2) thrust, total impulse, and burning times specified within the full operating temperature range; and 3) envelopes of thrust time or mass flow rate vs time curves.

Other requirements are necessary to the designer in order to define a satisfactory propellant grain: maximum weight of propellant grain, maximum axial and transverse acceleration undergone by the propellant grain during operation of the rocket motor, rocket spin rate (for instance for unguided rockets), and dispersions on pressure, thrust, total impulse, burning time have to be specified. Depending on the corresponding requirements, manufacturing process and control operations may be stongly affected and thus the cost of the grain as well. Plume characteristics (emission and transmission in the visible, infrared, electromagnetic wavelength range) are also requirements necessary to define a satisfactory grain.

2. Requirements Related to Operational Specifications

Depending on environmental conditions, definition of the propellant grain may be significantly affected. Such conditions must, therefore, be well defined in order to be correctly taken into account during the grain structural design phase.

Long term storage. Desired shelf life, related temperature cycles, and storage conditions must be defined. Particular conditions (e.g., relative humidity, atmosphere in the motor) which could directly affect propellant behavior must be specified.

Thermal environmental conditions. The nature and number of thermal cycles undergone by missiles (for instance, during operational flights for airborne missiles) must be defined. Generally, they are the limiting factors for structural grain design because very low temperatures may be encountered.

Acceleration, handling, and transportation. Acceleration before and during rocket motor operation, i.e., longitudinal acceleration undergone by the rocket motor, must be specified, as well as radial acceleration due to rocket spin which can influence the burning rate of the propellant. Additionally, handling and transportation, i.e., dynamic loads such as shocks and vibrations encountered during handling (drops) and transportation, must also be specified.

Reliability. A level of reliability is more and more commonly required. It is essential to define in which conditions it has to be satisfied and what part of the quantitative reliability objective is allocated to the propellant.

Maintenance. The content and the planning of missile surveillance, inspection, and maintenance must be defined, in so far as they may have an effect on the rocket motor environmental conditions.

Safety and vulnerability. These requirements may induce an a priori selection of a type of propellant (e.g., a nondetonable propellant or a propellant having a large critical detonation diameter) or, during engineering development, the performance of safety and vulnerability tests.

3. *Interface Specifications*

Case geometry and properties. A blueprint of the case, or at least, its geometry (length, diameter, configuration of head and aft ends) is mandatory in order to perform grain preliminary design analysis. Physical and mechanical characteristics of the case have a direct effect on structural and ballistic design: type of case, thermal expansion, hoops and longitudinal strains as function of internal pressure, maximum allowable peak pressure (depending on ultimate elastic elongation of case wall), maximum temperature allowable at case wall at the end of motor firing etc.

Thermal insulations. The nature and geometry of thermal insulations must be determined in order to settle the grain definition, either from a ballistic point of view or from a structural point of view. Compatibility between the propellant and the insulation materials must be checked.

Nozzle. The characteristics of the nozzle dictate the practical ballistic performance of a rocket motor. The following characteristics are of particular interest to the grain designer: number and orientation of the nozzles (the angle between nozzle center line and rocket motor center line must be known), degree of nozzle submergence, erosion of the nozzle (diameter evolution) vs operation time at throat and exit planes, angle of the exit cone (or a dimensioned sketch, in the case of a contoured nozzle), failure pressure of the frangible closure disk (it allows the definition of the ignition system and the control of pressurization at ignition), and dimensions of the blast pipe (between chamber and nozzle), when existing; it affects rocket motor efficiency.

Ignition system. The conditions of propellant grain ignition depend on its configuration (location, volume, design) including pressure at the end of ignition and pressurization rate (which affects structural integrity during firing).

Minimum and maximum values of delivered pressure and pressurization rate must be accurately known because they are important factors governing grain structural integrity. An envelope of ignition pressure vs time is of interest for this task.

B. Ballistic Modeling and Grain Design Methods

1. Input

To design a propellant grain, two types of data are needed.

Section IV.A gives an almost complete list of technical specifications, the first type of data needed. The list of requirements must be clear, complete, and consistent. Past experience and databases already existing will enable the designer to address only as a first step some of them which are more critical for the design. Usually it will involve length, volume, total impulse, thrust profile, maximum operating pressure, nozzle geometry and signature, and hazard classification for the propellant.

The second type is a database on propellants, liners, inhibitors, and thermal insulations. It allows the designer to have at his disposal, quickly and with a low probability of error, chemical, physical, kinetic, mechanical, thermodynamical, etc. characteristics of the various candidate materials which may be used in a rocket motor. The values of these characteristics will be used as input data in analytical and computational design tools.

2. Procedure

When performing a solid propellant grain design analysis, two levels of design accuracy have to be distinguished.

First level. It is the level of preliminary design analysis. The tools used at this level must be simple and friendly. They usually are small computer codes based on analytical models, or even graphs which give very simply the first results.

In any case, the method involves four main stages: selection of a couple of propellants/configurations, definition of a grain geometry satisfying internal ballistic and structural integrity (vs temperature cycles related loads) requirements, approximate assessment of erosive burning and potential combustion instabilities, and assessment of grain structural integrity during pressure rise at ignition.

The method is iterative. Depending on the results obtained at the third or fourth stage, it allows starting again at the second stage or even at the first stage if it appears that the first definition needs strong modifications.

Today, grain designers are requested to quickly provide fairly precise preliminary design analysis for a given project. To satisfy this request computer-aided grain preliminary design analysis methods have been developed. An example is presented in the next section.

Second level. It is the level of grain final design. The tools required for this task are more sophisticated. They are operated by grain design experts. They are mainly finite difference or finite element computer codes based on two- or three-dimensional models of physical phenomena related to internal ballistics, fluid dynamics, continuous media structural analysis, etc. They allow accurate calculations and, therefore, optimization of the grain final definition.

The principle of the method is parent to the one developed for preliminary design analysis, but it starts from the final result of this analysis, that is to say, the geometry and the propellant selected at the end of the preliminary design analysis.

Starting from this geometry, the evolution of the burning surface area vs web burned is accurately calculated. Taking into account propellant properties, one obtains the evolution of chamber pressure vs time $p_c(t)$ and thrust vs time $F(t)$. If necessary, the effect of erosive burning has to be taken into consideration at this stage. The results must then be compared with corresponding requirements (maximum pressure, combustion time, total impulse, etc.). Afterwards the structural factor of safety (related to thermal cycles and pressure rise loads) has to be assessed with the aid of advanced structural analysis computer codes.

If the results are satisfactory, the design is correct and the propellant grain definition is accepted for starting engineering development. If they are not, the grain definition must be modified so as to increase safety factors in grain critical areas. Additional structural analysis must be performed in order to check the benefits of geometry modification. Evolution of burning area vs web burned, pressure, and thrust vs time must also be checked so that the ballistic requirements remain satisfied. It may happen that, after these modifications, some of the requirements no longer be satisfied. In this event, the propellant and/or geometry have to be changed, or, if there is no other possibility, modification of some requirements has to be considered.

3. *Ballistic Analysis*

The basic equations used have been presented Sec. II.E [Eqs. (1), (2), and (4)]. A quick examination of these equations indicate the effects of various parameters on motor operation and, therefore, on motor and propellant grains design. Evolution of burning area vs web burned is directly connected to pressure evolution vs time. Sensitivity of burning rate to factors like propellant initial temperature, rocket motor acceleration, chamber pressure, and gas flow will have an effect on motor operation. Density ρ_p and C_D, which are specific of a propellant, may be considered for propellant selection. Initial values and possible evolutions during firing of A_t and C_F, which are directly related to nozzle definition (and also regarding C_F to propellant nature), must be accurately known.

In the following sections, the series of stages encountered in ballistic design analysis are described.

4. *Selection of a Geometry Associated with a Propellant*

This important part of design work has been approached in Sec. III.C but only through a semiquantitative analysis. With experience the designer can generally directly guess the best solution. However, we will describe a simple method which can guide the selection.

This selection is performed with the aid of charts and graphs like the one presented in Fig. 13. The example of this figure illustrates the logical method used, which permanently takes into account technical requirements, properties of actual propellants, and characteristics of widespread actual grain configurations.

The steps are as follows.

First, calculate propellant mass m_p, given total impulse I_T, and standard delivered specific impulse I_{sp} (for an expansion ratio of 70/1 and an optimum expansion ratio nozzle) measured for the propellant likely to be selected. This first calculation is iterative since the value of I_{sp} has to be corrected so as to be representative of the average conditions of motor operation. The average chamber pressure p_c is estimated from the specified maximum pressure, and the required nozzle expansion ratio is found depending on the maximum allowable nozzle exit cone diameter $\epsilon = A_e/A_t$. A_e is limited by the specification on maximum diameter of

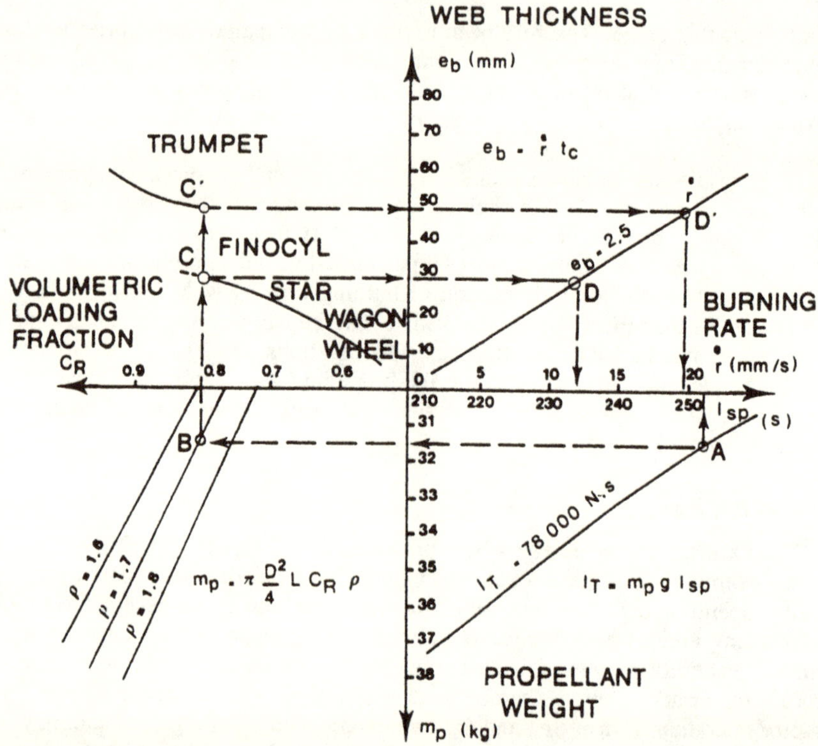

Fig. 13 Example of a graph for aiding the initial selection of geometry and propellant.

the nozzle exit cone. From Eq. (1), A_t equals $(m_p/p_c C_D t_c)$, where t_c, burning time, is specified.

Second, assess the volumetric loading fraction (C_R) required to obtain the specified total impulse, given the mass density of the propellant likely to be selected and the volume available for the propellant grain.

Third, select grain configurations. For each family of grain configuration, an empirical maximum volumetric loading fraction has been determined. Thus, given the volumetric fraction required, one or several configurations can be selected. Other criteria, like processing practicality, difficulty of structural analysis, and propellant web thickness have also to be taken into consideration.

Next define propellant burning rate $\dot{r} = e_b/t_c$.

Last, verify consistency betweeen specific impulse, density, and burning rate (at the average chamber pressure).

This approach must be completed by an accurate calculation of nozzle throat diameter generating a maximum pressure lower than that required by the specifications. This step requires a precise definition of grain geometry in order to calculate the burning area evolution that is needed for the determination of A_t.

$$A_t = \rho_p A_p \dot{r} / C_D p_{max}$$

In Fig. 13, the various steps of the method can be represented by the path from A to B, then to C and D, or to C′ and D′.

5. *Calculation of Propellant Grain Burning Area*

Accurate prediction of chamber pressure evolution vs time depends on accurate calculation of propellant burning area vs web burned. Computational tools which are commonly used belong to two families: one for two-dimensional configurations and the other for three-dimensional configurations. Actual grain configurations are three dimensional, but in numerous cases their geometry is defined by only two coordinates (r, Θ) or (r, z); in that case, configurations are said to be two dimensional (Fig. 14).

Two-dimensional geometry computer codes. These programs calculate the evolution of burning area of the following propellant grain configurations: cylindrical grains (constant port area section); axisymmetric grains, presenting a symmetry of revolution with respect to the motor centerline; and end-burning grains with axisymmetric slots on the aft-end face.

These various codes require the description (expressed in plane coordinates) of the initial burning area and of every section of the propellant grain. The computing

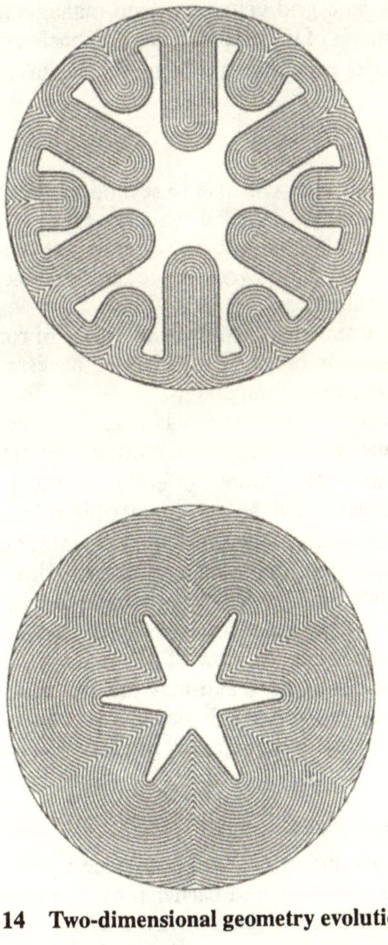

Fig. 14 Two-dimensional geometry evolutions.

time can be adjusted according to the level of accuracy desired. Because of the rapidity at which the computations can be done, a visual display of the computed burning areas is possible. As a rule, the level of accuracy is excellent. In a more complex case where the local burning rate of the propellant is not assumed to be independent of the curvilinear abscissa, the evolution of the burning area as a function of time can be computed with the help of a specially designed computerized numerical model.

Three-dimensional geometry computer codes. These codes (Fig. 15) allow the calculation of burning area of complex configurations, for example, finocyl grains having one or several slots. A first method, limited to the assumption of a constant burning rate throughout the whole grain, uses for each computation step the principle of generation of a surface at a constant distance from the initial surface. This software requires the generation of an initial volumetric grid with a density suitable for the level of accuracy required in the highly three-dimensional zones. A grid generation can only be performed by an expert, and the analysis of the evolution of the grain burning area, in spite of the existence of grid generation preprocessed data, represents a significant amount of work.

Another method, which allows the burning rate to be a variable function of time and space, uses automatic grid generation and management of the burning area evolution. The assessment of the perpendicular for each triangle of the grid is done using a numerical model using hyperbolic nonlinear differential equations.[7]

6. *Burning Rates*

Propellant burning rates are sensitive to several factors.

Pressure. In the pressure range in which rocket motors operate, the power law is generally used. It is also possible to directly use plots of actually measured burning rates vs pressure. The lower the pressure exponent, the more stable the rocket motor internal ballistics.

Temperature. Environmental and use conditions of rocket motors may correspond to a wide temperature range. It is, therefore, necessary to know burning rate sensitivity to initial propellant temperature.

Acceleration. Propellant burning rate is sensitive to acceleration directed normal and into the surface, but it is taken into account only when it is more than 10 g.

Manufacture process. The "hump" effect[8] is the result of a change in burning rate as a function of web burned (enhancement of burning rate in radially burning grains in the zone between the central port and the motor case walls). It is related to the manufacturing process. Empirical correlations, drawn from experience, are generally applied to take into account this phenomenon in ballistic design. This effect and other causes of ballistic variations will be detailed later.

Internal flow. In the motor the flow of combustion products interact with the propellant combustion phenomena and may locally change the burning rate law which is no longer the one expected. Because of the significant effect of this phenomenon, it is discussed in more detail in the following section.

7. *Pressure and Thrust Laws*

Burning rate laws, evolution of burning surface vs web burned, and basic internal ballistics equations discussed earlier provide pressure vs time and thrust vs time evolutions. In the simple case where internal flow does not significantly

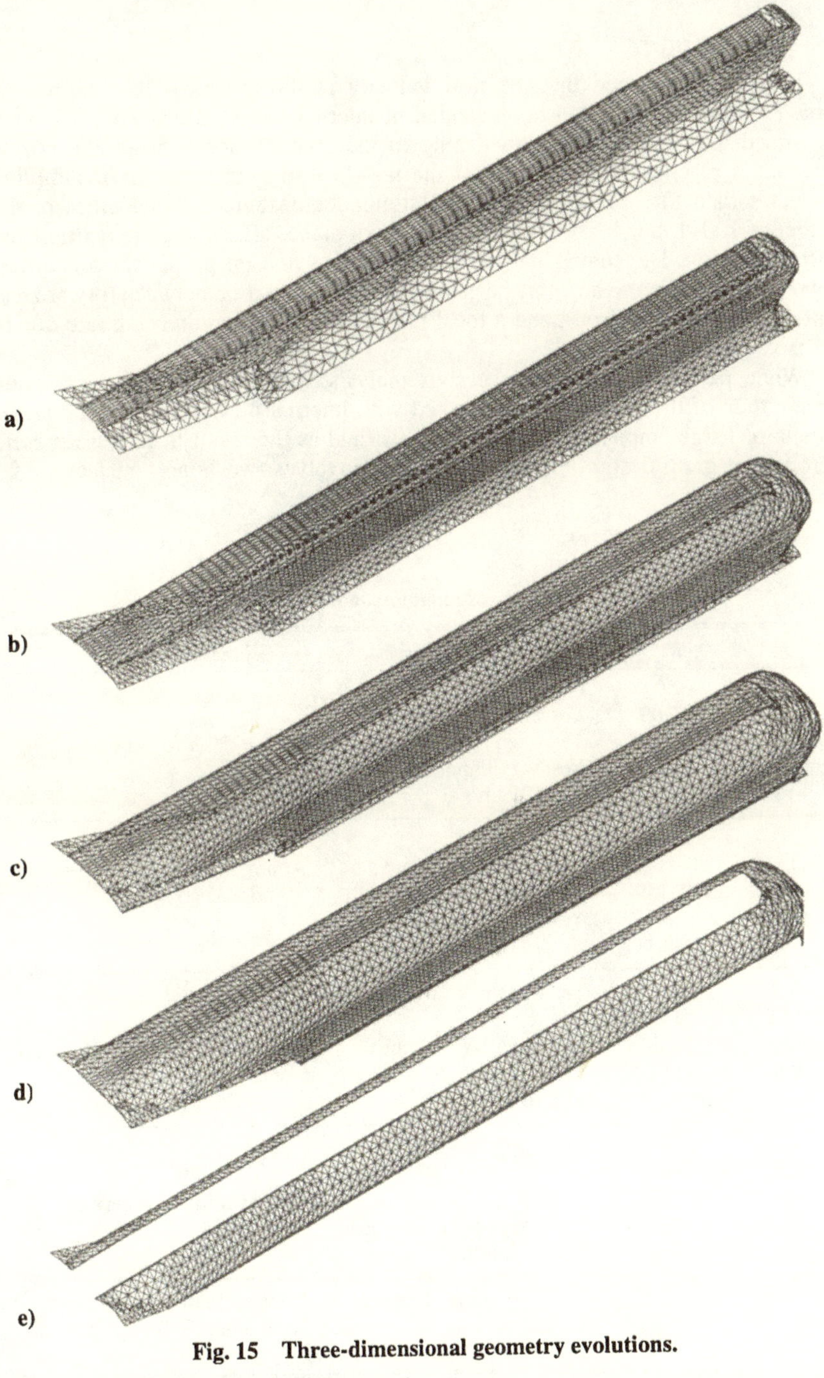

Fig. 15 Three-dimensional geometry evolutions.

interact with burning rate, Eqs. (1) and (2) combined with $\dot{r} = de/dt$, lead to a differential equation that is numerically solved and that provides web vs time $e(t)$, burning area vs time $A(t)$, pressure vs time $p(t)$, and thrust vs time $F(t)$.

8. *Effect of Internal Flow*

It is often assumed that the flow velocity in the central port exit plane is low enough so that it can be neglected in internal ballistics analysis. It is then assumed that flow is accelerated only in the convergence zone of the nozzle so that it reaches sonic velocity at the nozzle throat. In fact, this assumption is not satisfactory because flow calculations demonstrate that velocities of the order of 100–150 m/s are observed in port exit planes after complete ignition and pressurization. Depending on grain configuration and on propellant properties, two types of phenomenon may be generated: a pressure drop between forward and aft-end of the central port and a local increase of propellant burning rate due to erosive burning.

When performing a ballistic design analysis, one has to quickly assess the magnitude of the phenomena connected with internal flow. Table 7 summarizes the knowledge empirically acquired in this field as the result of numerous solid propellant grain design analyses. This table involves two important factors K_p and J:

$$J = K_p/K \qquad K_p = A/A_c \qquad K = A_p/A_t$$

Table 7 Intensity of phenomena due to internal flow

J	K	Erosive burning	Pressure drop
<0.2	<50 50–100 100–150 >150	No Yes when $r < 10$ mm/s Yes when $r < 20$ mm/s Yes, very important when $r < 10$ mm/s	Low < 5% p forward end
0.2 to 0.35	<50 50–100 100–150 >150	No Yes when $r < 10$ mm/s Yes when $r < 20$ mm/s Yes, very important when $r < 10$ mm/s	Approximately 10% p forward end when $J = 0.3$
0.35 to 0.5	<50 50–150 >150	Yes when $r < 10$ mm/s Yes when $r < 20$ mm/s Yes, very important when $r < 10$ mm/s	Approximately 10% p forward end when $J = 0.4$
0.5 to 0.8	<50 and 50–150 >150	Yes very important when $r < 20$ mm/s Yes, very important when $r < 10$ mm/s	40% of p forward may be observed
1	Any value	Yes, very important when $r < 20$ mm/s Low when $r < 39$ mm/s	Pressure in the sonic section is $p \cong 0.56p$ forward

where A_c is the area of a given cross section of the central port, A the propellant burning area upstream of the cross section, A_p the propellant grain burning area, and A_t the nozzle throat area. K_p is directly related to the mass flow rate per unit area G.

$$\rho_p \dot{r} K_p = \rho_p \dot{r} A / A_c = G$$

J gives an idea of the maximum gas velocity (as a rough approximation $J = M$, the Mach number in the aft end port section) and of the difficulties of the internal ballistic design.

High values, even $J = 1$ (nozzleless motors) can be used if the geometry of the grain is adapted to high-pressure drops along this axis, highly erosive burning, and a critical ignition phase.

Pressure drop is related to a decrease of pressure from grain head end to grain aft end. It induces an increase of head-end pressure at the first phase of motor firing and, therefore, the maximum pressure generally increases. Pressure drops are generally due to energy losses inside the flow and to phenomena occurring at the interface of flow and propellant surface or to sharp changes of port section or of flow direction, and to side injections from burning propellant walls.

One of the critical steps in rocket motor operation, therefore, occurs just after ignition when port sections (through which combustion gas must flow) are minimum. Average pressure drops values encountered are of the order of 0.1 MPa between head and aft end. In some cases, for special configurations, pressure drops of more than 1 MPa have been observed.

A gaseous flow is fully characterized by the knowledge of local velocities, pressures, and temperatures. Computer codes have been developed in order to know such characteristics. They are used to describe unsteady phases during pressure rise at ignition and to know steady flow just after ignition, in the whole cavity and in the nozzle convergence section.

To use these codes, one has to generate a grid of the combustion chamber. Order-n symmetry (when existing) is taken into account so as to reduce the analysis to a sector of $\epsilon\pi/n$ (n = symmetry number). Results may be presented either as gas velocity or pressure contour line in various cavity sections (Fig. 16) or as curves representing for instance gas velocity as a function of radial distance from the central axis in the central port cross section.

9. *Erosive Burning*

Enhancement of propellant burning rate due to tangential gas flow (compared to propellant burning rate without tangential flow) is known as erosive burning. It occurs when the propellant burning surface is subjected to a high-velocity combustion gas flow parallel to it. The value of K_p of the chosen design gives

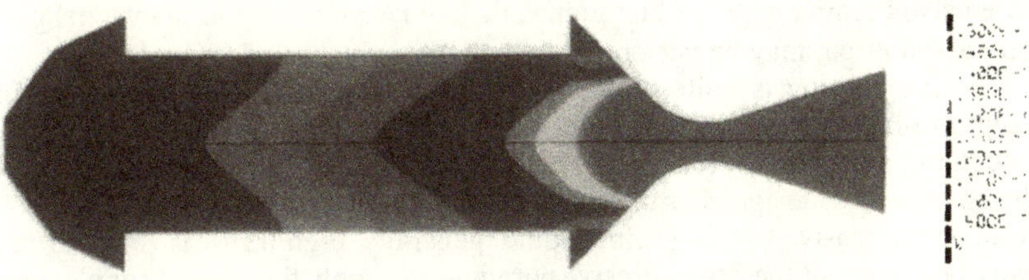

Fig. 16 Gas velocity field, Mach number representation.

a rough idea of the presence of erosive burning. The phenomenon is due to an increase of heat transfer from the flame zone to the propellant surface. There are numerous physical models to explain and to quantify this phenomenon.[9,10]

One-dimensional models are generally well suited to compute internal ballistic of large l/d tactical motors with smooth changes in cross section along the axis. The accuracy of such models (within 5% on peak pressure at ignition) allows the design of grains with high J factors and high level of erosive burning in the aft-end region.

Predictive and accurate erosive burning models are available today for composite propellants, but a lack still remains for nitramine based (CMDB, XLDB) propellants which seem to be experimentally far less sensitive (Fig. 17). On the other hand, higher pressure exponents which may be observed with this family of propellants impose on the designers a higher accuracy in pressure drop predictions.

Two-dimensional models need more sophisticated flow solvers which cannot yet take into account erosive burning in complex geometries. Fortunately, axisymetrical configurations, with high-internal-flow velocities, are not frequently met or can often be treated with one-dimensional computer codes.

In the case of three-dimensional geometries, the results of flow calculations can only be analyzed to point at zones with erosive burning occurence risk but still fail in taking into account coupling between flow and combustion. Designers, in that case, must evaluate mass flow rate increase and generally take a higher safety margin for predicted ignition pressure.

As an example, a simple computer code, based on a one-dimensional flow model, allows the calculation head-end and aft-end pressure evolutions in a grain experiencing erosive burning. The erosive burning law selected for the model is

$$\dot{r}_e = \dot{r}_0[1 + (G - G_0)]$$

where $\dot{r}_e$ is the burning rate with erosive burning, $\dot{r}_0$ the burning rate without erosive burning, G the mass flow rate per unit area in the given port cross section and G_0 is the mass flow rate per unit area threshold (beyond which erosive burning occurs).

This law may be experimentally established for a precise case or may be the result of modeling work for some well-known propellants like AP composites.[11]

This type of code is used systematically in preliminary ballistic design analysis. It implies that grain geometry be described by the cross section contour perimeter evolution along the grain axis. Erosive burning is calculated in several cross sections of the central port according to local flow characteristics (static pressure p and local mass flow rate G) and to the preceeding erosive burning law. The ignition phase is simulated as an unsteady phenomenon; time steps range from 1 to 5 ms. A complete motor firing may be simulated, using a steady-state model and time steps generally ranging from 0.05 to 0.1 s (Fig. 18). A more comprehensive investigation of erosive burning in propellant grains, though keeping a one-dimensional geometry assumption, may be performed with more sophisticated codes.[12]

Erosive burning is quite often a problem in tactical motors because of high length-to-diameter ratios and high-volumetric loadings. Predictive models are becoming more and more reliable, but some surprises can still happen at the first firing of a new design. On the other hand, more and more designs incorporate voluntarily erosive burning. Since, quite generally, high thrust is required at the very beginning of the firing, erosive burning can supply the needed supplement of thrust (and pressure) that the burning surface evolution does not offer alone.

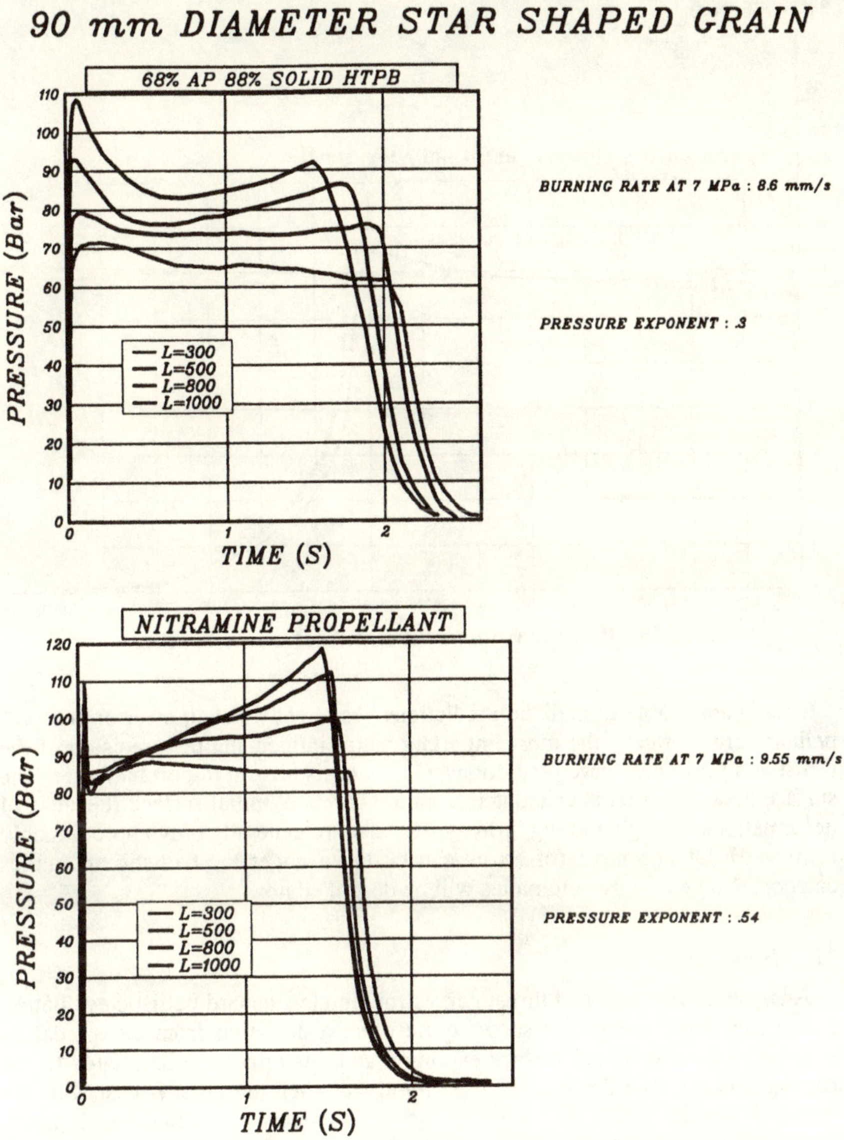

Fig. 17 Erosive burning response for HTPB and nitramine propellant.

10. *Motor Ballistic Characteristic Predictions, Ballistic Anomalies*

A systematic analysis of burn time variations in solid rocket motors has been recently presented.[13] Propellant weight, propellant density, and characteristic velocity variations have a small effect (0.1–0.2%) on the burn time variations. Throat area variations have a slightly more pronounced effect, especially for small motors with high throat erosion. The main influence factors are the burn surface area variations (due to variations in the geometry of the propellant grain with production, for instance, port ovality or port misalignment) and the burning rate variations.

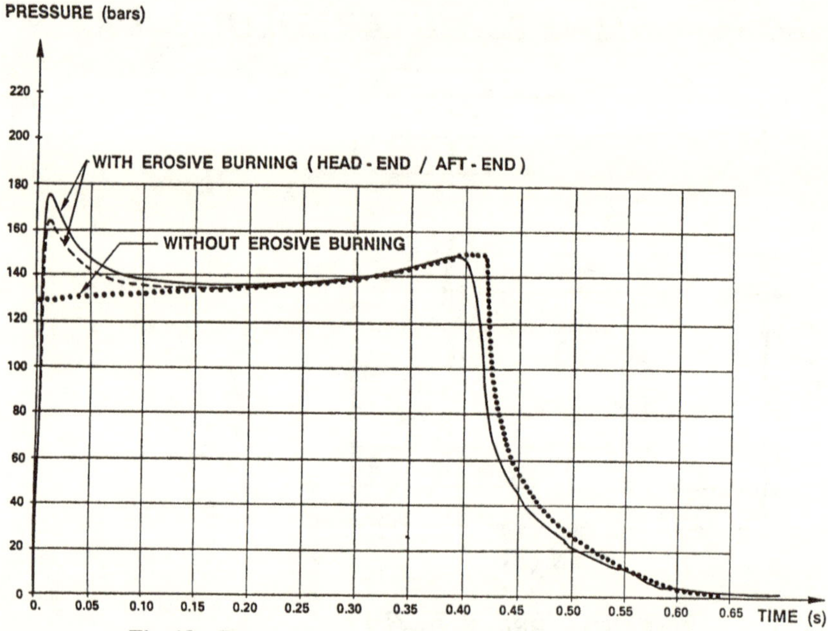

Fig. 18 Pressure vs time with or without erosive burning.

This is applicable to ballistic predictions. As already stated, error on the propellant burning rate is the most important factor influencing the precision of the thrust and presssure curve predictions. Geometry is also an important factor. The surface area evolution is calculated from a theoretical initial surface (because of deformations through cooling, gravity, etc. that are generally only approximately estimated). Other sources of errors can be listed under the ballistic anomalies category. Some of these anomalies will be described now.

11. Hump Effect

Analysis of pressure and thrust curves for simple standard ballistic evaluation composite propellant grains shows quite often a deviation from curves calculated from the theoretical surface evolution and this effect can be related to the casting process. The deviation can be observed when the grain is cast with the mandrel already placed in the mold and it disappears if the mandrel is introduced after casting. The pressure curve shows a characteristic hump occurring approximately halfway through the web burned. Calculations demonstrate that the burning rate is higher than predicted by 3–7%, and the size of the hump is not a function of the burning rate of the propellant. The intensity of the effect is dependent on AP, its ratio and size distributions in coarse and fine particles, and on the presence of aluminum. Analysis of results obtained after extinction shows that it is caused by a burning rate law change as a function of propellant thickness. It is probably related to the distribution of the fillers, induced by the manufacturing process. Taking into account the hump effect for ballistic predictions is only possible through accumulating data from firings to obtain empirical laws of burning rate deviation $\dot{r}/\dot{r}_0$ as a function of web burned. With tubular or star grain shapes with mostly radial burning, this law exhibits a hump, with a finocyl geometry with combustion in every direction, the curve has the shape of sinusoïd.

12. Mechanical Stresses

In case-bonded composite propellant grains, in regions where high stresses may be present at firing, the burning rate law can be modified because of dewetting between the energetic fillers and the binder causing some internal porosity in the propellant.

13. Burning Rate Modification at the Propellant–Insulator Interface

This effect is particularly important for laterally inhibited end-burning grains, and its causes will be detailed later. It can, of course, occur on every type of geometry, depending on the propellant and liner or insulator used and modifies the theoretical surface evolution.

To summarize all of these considerations (if we had to make a statement on the precision of an a priori prediction of the pressure curve, for instance, of a new motor), based on our experience, we would consider two extreme cases, in the absence of erosive burning.

If the propellant and the geometry are well known and also the inert materials used in the motor have been tested in motors of approximately the same size, the difference between the prediction and the result of the first firing may be less than 2%.

If a very new geometry with new propellants and associated materials are used, the difference may be as big or bigger than 5%.

In case of erosive burning one may add an error of 10–20% on the overpressure due to erosive burning.

14. Combustion Instabilities

Combustion instabilities during a rocket motor operation are detected as oscillations superimposed to the average pressure. Their amplitude depends on the motor definition and on the operating conditions. They are always undesirable, even if they do not all have catastrophic consequences. Average pressure shifts modify the motor performance, including exceeding specified limits; they sometimes lead to motor failure. Heat transfers are intensified by pressure oscillations and induce such a rapid degradation of inert parts (thermal insulation, nozzle) that in some cases they are destroyed before burnout. Pressure oscillations create vibrations which are transmitted to the whole motor case and to the missile.

Two sets of phenomena are generally referred to as instabilities. The first are irregular combustion phenomena. Pressure oscillations, even extinction and reignition are observed. These oscillations of a few hertz, in phase throughout the whole combustion chamber are called "chuffing" or L^* instabilities. They occur under very specific operating conditions, at low pressure and small characteristic length L^* (combustion chamber volume to nozzle throat area ratio). The second are pressure oscillations matching cavity acoustic modes. This is by far the most common case.

Grain design must then incorporate an assessment of combustion stability during motor firing. The phenomenon of combustion instability may occur when perturbations excite acoustic oscillation modes of the chamber cavity. Damping or amplification interactions with combustion, flow, particles, nozzle, etc. may induce either an increase or a decrease of the phenomenon. To assess combustion stability, a linear analysis procedure is usually followed first (linear acoustic balance). It will try to predict the instability risk.

The pressure inside the combustion chamber cavity is assumed to be

$$\frac{p'}{p_0} = \sum_{i-1}^{n} e^{\alpha_i t} e^{j\omega_i t} \psi_i(M) \qquad (j^2 = -1)$$

where p_0 is the average chamber pressure, p' the instantaneous pressure at point M, ω_i the pulsation of mode of rank i and of frequency f_i, ψ_i the spatial form of mode of rank i, M the point in grain cavity, and α_i the damping coefficient (when $\alpha_i < 0$), or gain factor (when $\alpha_i > 0$) of the mode of rank i.

The first step of the analysis consists of calculating the various acoustical modes which are specific to the grain cavity. Finite element methods are generally used. Examples of grids and corresponding results are presented in Figs. 19 and 20. The second step of the analysis consists of calculating the value of α (equal to the algebric sum of the various gain and damping factors). These calculations need

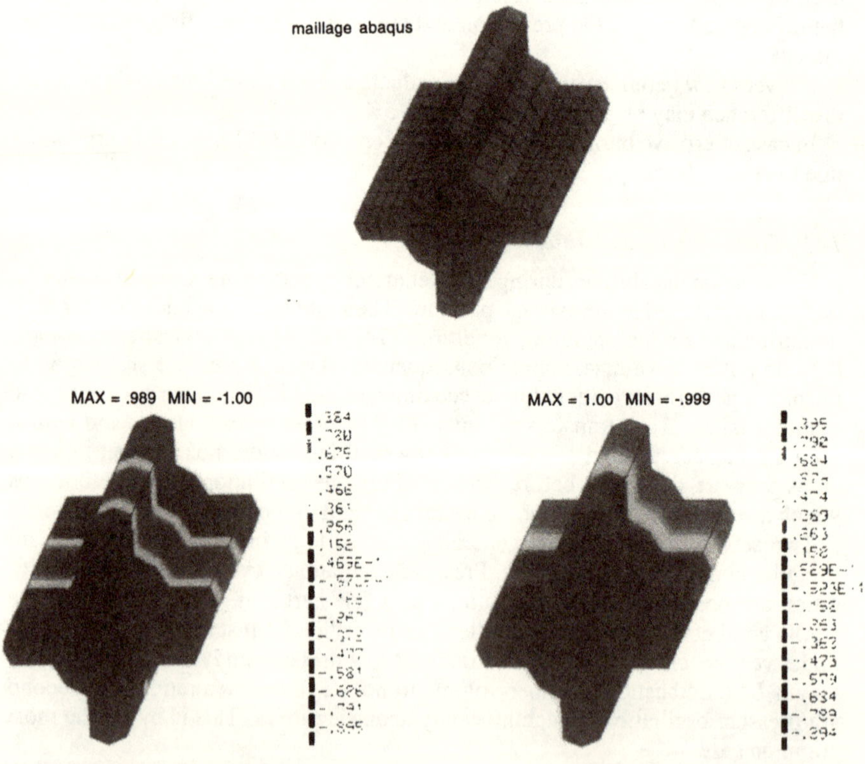

Fig. 19 Acoustic instabilities, grid of the cavity used for calculation and representation of the first two modes.

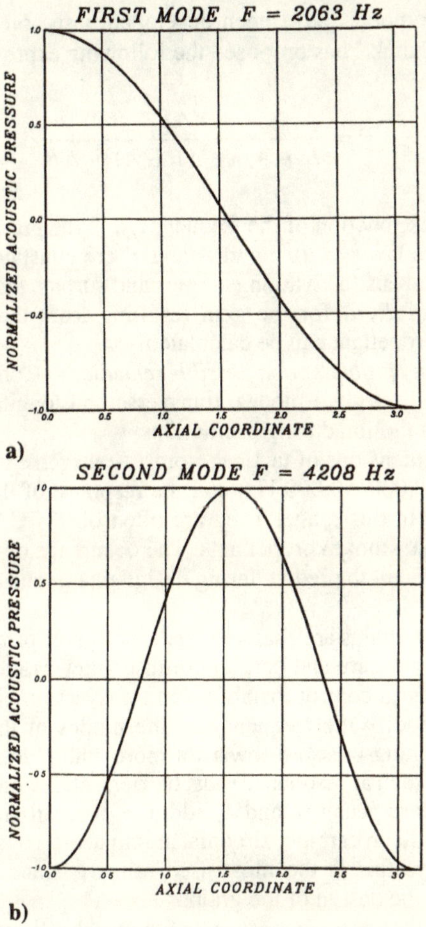

Fig. 20 Acoustic pressures for the first two modes.

not only the results of the first step but also data describing propellant response to pressure, effect of condensed particles, etc.

Depending on the value of α it is possible to conclude on the grain propensity to experience combustion instabilities; for a mode of frequency f_i, a value of α_i, larger than $0.1 f_i$ indicates that there is a significant probability that combustion instability may occur. Grain configuration (or propellant) then has to be modified.

A more refined analysis would then involve the computation of limiting cycle amplitude as well as the magnitude of average pressure shift for each unstable mode (nonlinear analysis).

Propellant acoustic response. The gain factor related to the acoustic response of the propellant is often determined with the help of measurents of the (pressure coupled) acoustic response of the propellant R as a function of frequency. R gives the mass flow rate response to pressure fluctuations. Devices like the T-burner or the modulated exhaust motor are used. These devices are severely limited in frequency. Some models have been developed based on a description of the combustion zone. Culick[14] has proposed the following expression.

$$R = \frac{nAB}{\lambda + A/\lambda - (1 + A) + AB}$$

where λ is a complex function of the frequency, n is the pressure exponent of the steady-state pressure law $\dot{r} = ap^n$, and A and B are constants, worked out from physicochemical analysis (activation energies and surface temperature).

From the value of R, α_i for the gain resulting from the (pressure coupled) combustion of the propellant can be calculated.

Practical instability problems and possible remedies. Combustion instabilities are usually classified into two modes, transverse and longitudinal, according to the nature of the gas motion during oscillations.

Because of the dimensions of tactical motors, transverse oscillations will have high frequencies (10,000–50,000 Hz) and the response of the propellant is very difficult to measure in this range. They are often observed with nonaluminized (reduced or minimum smoke) propellants. The occurence of transverse modes is, in any case, very difficult to predict during design phase, and only some empirical criteria can be applied.

Generally, stable domains are characterized, for a given propellant, by reference to average chamber pressure and port length-to-diameter ratio.

The main methods to control instability in transverse modes are 1) geometry changes that will modify the frequency of the modes of the cavity, 2) change of the average operating pressure toward a more stable domain, 3) mechanical damping structures, such as resonance rods, barriers, etc. (which can cause an inert mass or propellant loss penalty), and 4) addition of small amounts of refractory damping particles (silicon carbide, zirconia, alumina, etc.) optimized for the given frequency. Nowadays this is the most used remedy, since it has generally no significant effect on the design of the grain.

Examples of each type of treatment can be found in the literature. Nugeyre et al.[15] have described an interesting case of rupture of the propellant grain followed by explosion of the motor at ignition at low temperature in which the failure was caused by a coupling between vibrations related to acoustic instabilities of the cavity coupled with a mode of resonance of the solid (brittle) CMDB grain. A change in star-shaped geometry of the port modifying the acoustic modes could solve the problem. Brownlee and Marble[16] have described the influence of pressure, port-to-throat area ratio, and l/d on the stable and unstable domains of transverse instabilities. Kaczmarek et al.[17] have described a typical treatment of transverse instabilities by the use of resonance rods in double-base grains. The basic theory and practical examples of instability suppression in CDB propellants by the use of refractory particles of optimized size has been presented by Evans and Smith.[18]

Longitudinal oscillations will have lower frequencies (200–2000 Hz). The response of the propellant is easier to measure, and the linear analysis easy to perform. On the other hand, remedies are more difficult to find. One way to deal with the problem is, of course, to modify the response of the propellant by tailoring the formulation. AP and aluminum particle sizes are possible parameters. If this is not sufficient, modifications of the grain geometry must be considered.

C. Structural Design Considerations

Various loads are imposed to the propellant all along the motor's lifetime. These loads depend not only on the rocket motor environment, i.e., type and number of thermal cycles undergone during storage and transportation (for instance, captive flights for airborne missiles), acceleration during boost phase, and pressurization during grain ignition, but also on manufacturing conditions, e.g., curing temperature, gravity, etc.

The goal of the detailed structural analysis is to calculate a factor of safety defined as $K = C/S$ or a safety margin $(C - S)/S$. C is the propellant (or bond) structural capability (allowable) and S is a functional related to stress/strain induced in the propellant grain region undergoing the more severe loads. To compare them, C and S must be of the same physical nature.

The factor of safety must remain, until the end of motor firing, higher than 1 during the rocket motor lifetime. According to this definition, it is assumed that grain cracking or propellant/liner debonding induce significant modifications to rocket motor internal ballistics having consequences ranging from failure of missile mission to rocket motor explosion. It is assumed that failure at the most stressed (strained) point does not depend on stress (strain) gradient in the surrounding region.

If the factor of safety calculated for a given propellant grain and given imposed loads is lower than a required value, the grain has to be redesigned until a satisfactory factor of safety is obtained and this, of course, can modify the ballistic properties of the design.

The most typical types of design adjustments deal with modifications of geometric discontinuity zones where the highest stresses usually occur, i.e., bondline, termination, and discontinuities in the bore surface area, like star termination in a forward dome or forward fin or slot termination for a finocyl or slotted tube design.

There are two ways to treat the bond stress problem at the ends of the grain. One option is to incorporate a stress relief boot or flap, which has the effect of reducing the maximum stress by creating an unconfined volume in the vicinity of the critical point. Since both faces of the stress relief boot will be inhibited, there will be no change in the burning area evolution but a small loss of propellant. Another option is to keep a completely bonded propellant dome and create a radial slot in the same area. This radial slot can even be combined with a star-shaped bore termination. In this case, there can be a strong influence on the surface evolution of the grain.

For plane grain end terminations that are common at the aft end in tactical motors, the same general types of consequences occur on the ballistic design.

V. General Procedure for a Propellant Grain-Design Optimization and Computer-Aided Preliminary Design

The procedure involving the different steps already described to make a total design can be summarized by the presentation of a computer-aided preliminary design code. All of the phases of the optimization procedure would be the same if the task was performed manually.

A. General Description

To optimize the motor and missile design, there is an increasing pressure to have, as early as the preliminary design phase, quick and rather accurate results defining

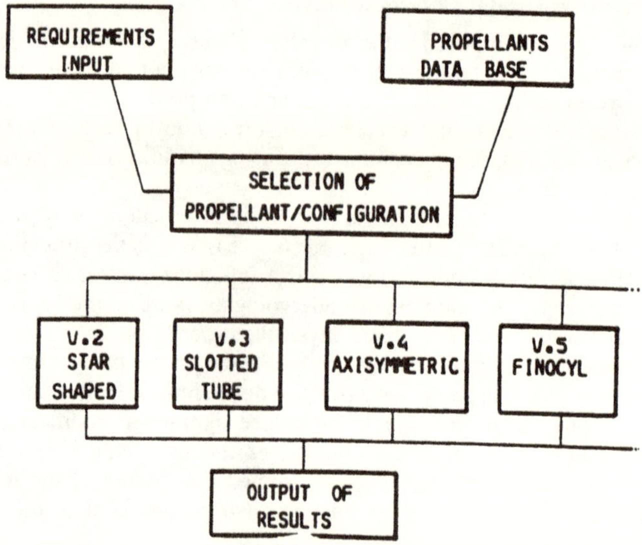

Fig. 21 General structure for the first two modes.

the propellant grain. As an example, a computer code allowing satisfaction of these needs will be presented. Figure 21 presents the general architecture of the code. Each type of grain configuration (star-shaped, slotted tube, axisymmetric, finocyl, etc.) is individually treated inside the code. The procedure for any of these configurations is the one which is generally followed to perform propellant grain preliminary design analysis. The architecture of the code is modular so that any addition of new modules or any improvement of an existing module may be very simply worked out.

Runs are controlled by the user from the graphic terminal. CPU time is negligible as compared to time spent by the user performing the design analysis.

The process is iterative and, besides the input of technical specifications, the user has only to answer yes or no to the option which is proposed on the screen. Results are presented either as tables or as curves. The diagram presented in Fig. 21 emphasizes the role of propellant/configuration selection, which allows one to provide several possibilities, ranked according to a given set of criteria. The selection of propellant/configuration depends on the one hand, on technical requirements (total impulse, burning time, etc.), or on the other hand, on semiquantitative requirements related, for instance, to manufacturing process praticality, industrial and economical aspects, etc.

Because of the dual nature of the criteria, Expert Systems has been selected and implemented for this critical stage of preliminary design analysis.

B. Description of the Code

All of the branches of the code have the same basic structure. The slotted tube branch is detailed hereafter. On the flow chart of Fig. 22, the main stages of the analysis appear. Two possibilities are provided to the user: design of a case-bonded (or free-standing) grain meeting technical requirements or for a given rocket motor, calculation of motor operation (pressure vs time, thrust vs time, etc.).

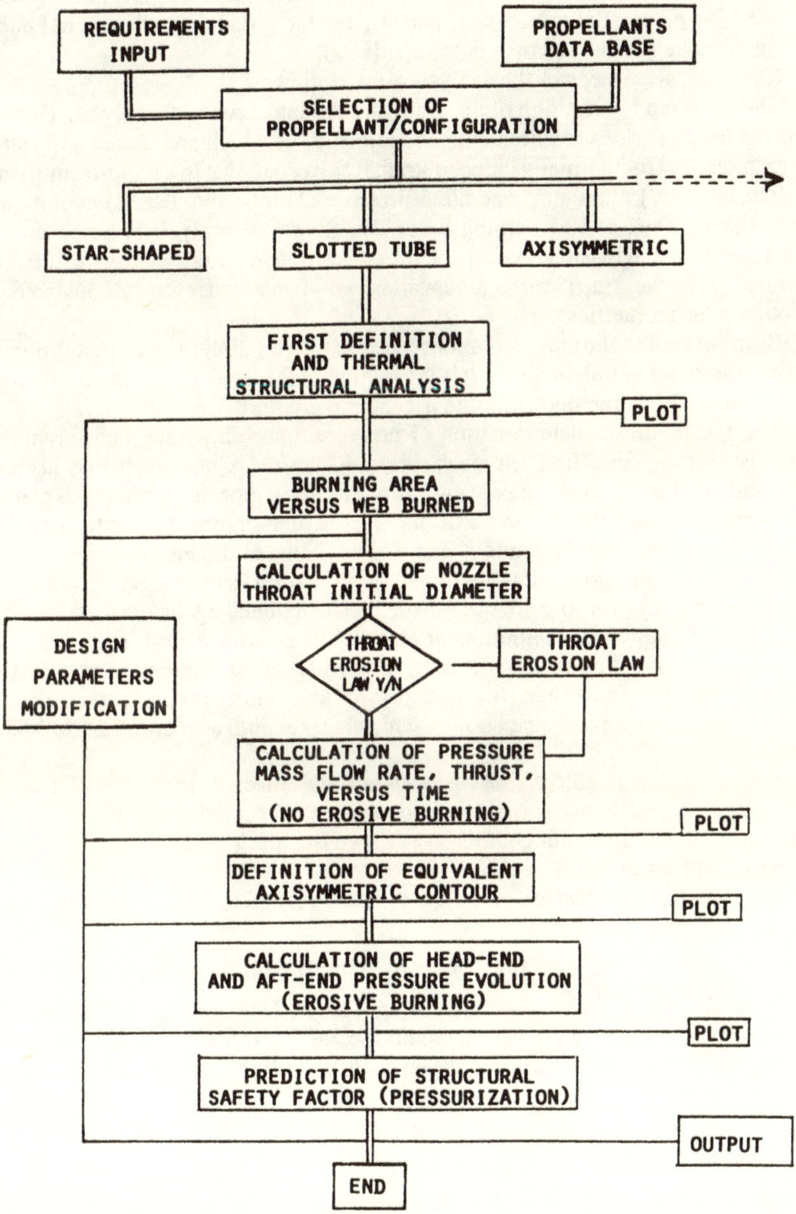

Fig. 22 Detailed flow chart of a slotted tube propellant grain design computation.

In the case of slotted tube configurations, geometrical characteristics which are taken into account for design analysis are the cylindrical motor case, presenting possibly thermal insulation over-thickness in the aft-end zone (slotted zone); the plane grain aft-end zone (slotted zone); the cylindrical central port; the slot walls, which may be parallel or not (star shaped); and the possibility of a tapered port in grain aft zone (to limit erosive burning effects).

The successive steps of the analysis are as follows.

The first step is definition of the grain and thermal structural analysis. This definition meets various requirements, the priorities of which are ranked as 1) structural integrity (for thermally induced strains) corresponding to a required minimum safety factor, 2) maximum operating pressure, 3) total impulse, 4) evolution of pressure vs time, and 5) burning time. The structural analysis is based on data obtained from regression analysis of stress/strain field determination results, performed with the aid of three-dimensional computer codes on various selected slotted tube geometries.

Burning area evolution calculation is the second step, followed by determination of nozzle throat initial diameter. It is the minimum throat diameter value consistent with the specification on maximum acceptable pressure.

The fourth step is determination of pressure, mass flow rate, and thrust evolutions vs time. In a first approach, erosive burning is not taken into account. The calculations provide expansion ratio and, thus, nozzle optimum expansion ratio and exit diameter. Nozzle exit diameter is then compared to corresponding requirement. Afterwards, ratios K and J are calculated. If needed, a tapered zone is designed in the grain slots region so as to meet a criterion on J (maximum permitted value). Burning area vs time is then calculated again.

The next step is the definition of equivalent axisymmetric longitudinal port contour. It is based on equal flow rates in any port cross section for actual (three dimensional) and equivalent (two-dimensional axisymmetric) contours. It allows a simplified analysis of erosive burning which is taken into account at the following stage.

The sixth step is calculation of head-end and aft-end pressure evolution. At this stage erosive burning is taken into account. The module provides pressure evolution inside the grain central port, as well as peak pressure at ignition and pressure rise time.

The last step is prediction of the safety factor related to ignition pressurization.

VI. Some Specific Cases

Probably 90% of the cases practically encountered are included in the preceding developments. There are, however, some special applications that require special configurations or designs, such as, nozzleless motors, end burners, and pulse motors.

A. Nozzleless Motors

The nozzleless motor is an early design that resulted from an analysis of the conventional rocket motor, demonstrating that the nozzle accounted for a significant portion of the cost, the weight, and the size of a rocket motor.

The propellant grain of a nozzleless motor is usually case bonded with a generally cylindrical central port, connected at the aft end to an exit cone molded in the propellant (Fig. 23). The combustion pressure will decrease with time because of

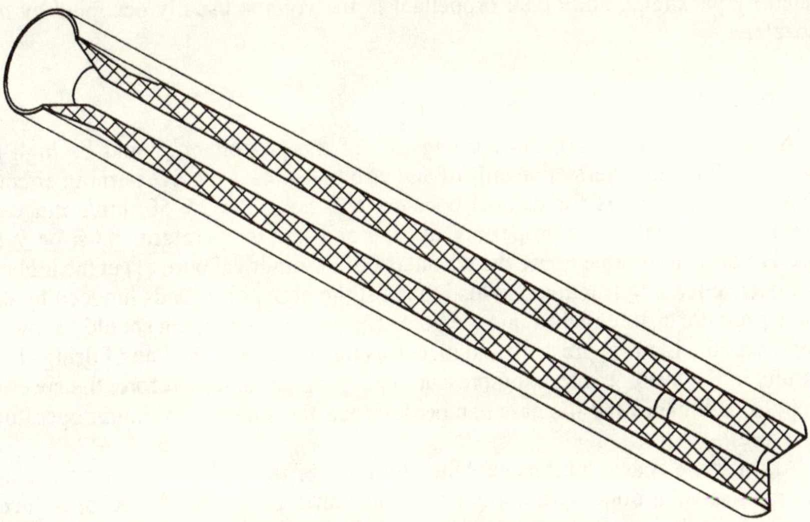

Fig. 23 Nozzleless configuration.

the absence of a throat. In the absence of nozzle throat, the operating conditions of the nozzleless rocket motor are governed by an aerodynamic constriction of the gas flow. The exit cone section at the aft end of the grain is designed to allow gas expansion, thereby improving performance.

The development of nozzleless rocket motors is linked to composite propellants with high-burning rates, because this type of grain requires a minimum flow rate to ensure a stable performance, in turn, requiring prohibitive length-to-diameter ratios when other propellant compositions are used. In addition, the propellant pressure exponent should be less than 0.5 in order to obtain designs which are competitive with nozzled motor configurations. Length over diameter ratios over five are generally used.

The main advantages of nozzleless rocket motors are simplicity of the propellant grain geometry, greater loading ratio than in a conventional rocket motor, weight reduction due to absence of mechanical parts at the aft end and absence of nozzle, significant decrease of the cost of the rocket motor, and elimination of a nozzle which otherwise would have to be ejected at the end of the acceleration phase, making this concept very interesting for use as an integral booster for rocket-ramjet or ducted-rocket systems. In fact the first real applications (advanced development programs) until today were conducted on ducted rockets.

There are some drawbacks as the price to pay for these advantages. There is a loss of approximately 15–20% of specific impulse in comparison to a conventional system depending on the geometry and optimization of the grain. There is a danger of combustion instabilities occuring at low pressure. There is a need to have a thorough knowledge of the normal burning rate and erosive combustion for a very large range of pressures, to enable performance predictions. Furthermore, the propellant grain mechanical deformation is directly reflected in the ballistics of the motor.

Although nozzleless motors have much lower specific impulse values than conventional motors using the same propellant, this loss can be largely compensated by packaging additional propellant in the volume usually occupied by the nozzle.

1. Propellants for Nozzleless Designs

As already mentioned, the burning rate of the propellant should be high to prevent a long and ineffective tail-off and also to decrease erosive burning effects. Typical burning rates for tactical boosters will be in the 25–50 mm/s range at 7 MPa. The mechanical properties, at high and low temperatures must be very good in order to be able to use the advantage of a cylindrical bore to get the highest volumetric loading fraction and also to resist the high shear loads induced by the large pressure differentials along the bore. The pressure exponent should be low: it decreases the ignition pressure and increases the pressure at the end of firing. This results in the average operating pressure being higher and, therefore, the specific impulse is higher (and the case can be designed for a lower maximum operating pressure).

All of these considerations explain why most of the real testing on nozzleless motors has been done with composite propellants, either metallized or reduced smoke, with burning rate accelerators, mostly ferrocene derivatives. Minimum-smoke high-energy propellants because of their ballistic limitations (rate of combustion and pressure range, with low-pressure exponent) have to be improved before being considered.

2. Performances

Specific impulses of the order of 85% of the specific impulse values obtained with an optimum designed nozzled motor using the same propellant can be obtained with aluminized or reduced smoke composites. This is also true for zirconium–AP–HTPB propellants that are sometimes considered for integral ducted rockets boosters because of their high density and high-volumetric specific impulse.

3. Temperature Sensitivity

The temperature sensitivity of a nozzleless motor using the same propellant is generally smaller than with the nozzled motor. The deformation under pressure of the propellant that forms the nozzle is significant at high temperatures and partially compensates for the higher burning rate. At lower temperatures the increased modulus of the propellant limits this phenomena. Typically, nozzleless motors exhibit only 50–65% of the operating pressure temperature sensitivity of nozzled motors.

4. Design Analysis

A highly regressive combustion pressure curve is a characteristic of a nozzleless rocket motor operation, reaching very low pressure (Fig. 24). Because of these particular operational characteristics, the design analysis of a nozzleless rocket motor can be performed only with very specific models and computer codes.

All of these models are one-dimensional programs, a typical code using a single-phase, quasistationary description of the flow. The precision of the calculations depends on a very exact knowledge of propellant grain combustion laws as well

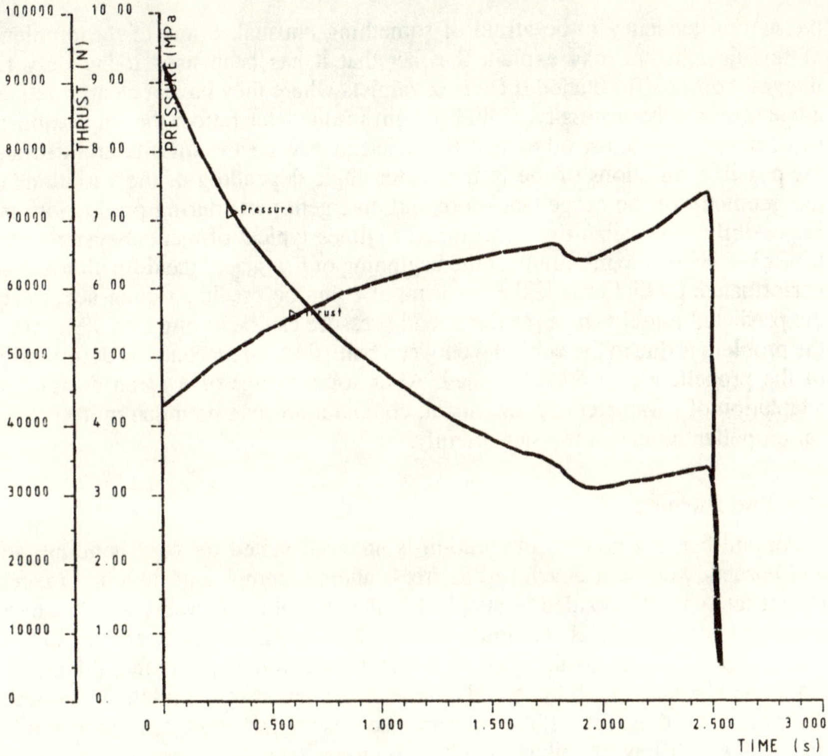

Fig. 24 Pressure and thrust curve for a nozzleless motor.

as an exact description of the geometry of the case and the grain's central port, including structural strains.

Before undertaking the computer analysis and in order to limit computer time, the following method may be used to perform a preliminary design analysis. Limit the minimum central port diameter for structural integrity reasons. In the case of a given diameter, determine the required length of the propellant grain by using an I_{sp} of approximately 215 s for an aluminized–AP–HTB composite and 200 for a reduced-smoke composite propellant. Determine the diameter of the central port, for a given maximum pressure, using the following simplified formula.

$$d = \frac{4\rho_p l c^*}{0.8} \quad \frac{\dot{r}(p)}{p}$$

where p is the maximum pressure, $\dot{r}(p)$ the burning rate at maximum pressure, ρ_p the density, c^* the characteristic velocity, and l the estimated length of the propellant grain.

These preliminary values for the diameter and the length are used in the computer analysis to optimize the propellant grain.

5. *Design Problems*

Considering all of the potential advantages of the nozzleless motor one may wonder why there have not been more applications today. There is, of course, the natural tendency to be afraid of something unusual. Some of the problems of this design that may explain the fact that it has been used today only for integral boosters for ducted rockets or ramjets where they have a clearly defined advantage are the necessity of high length to diameter ratios; the impossibility, of course, to use blast tubes and the consequence on the missile architecture; the possible variations of the thrust vector angle depending of the variations of the geometry of the propellant (bore and divergent part) during production; the impossibility to realize two level thrust regimes typical of tactical systems; the thrust levels that are minimum at the beginning of firing, and the difficult accurate performance predictions. If the total impulse can be predicted quite accurately, the predicted initial values for thrust and pressure can be wrong by 10%. Part of the problem is due to the coupling between ballistics and mechanical deformation of the propellant already mentioned. After some firings of a given design and adaptation of parameters of the model, computations can be much more precise for propellant grains of the same family.

B. End Burners

An end-burning grain configuration is not well suited for case bonding, and end burners were first developed as free-standing, completely inhibited (except the aft face) grains, bonded or attached to the case at the forward end. A careful structural analysis of this area must be done. It is also generally necessary to have some devices to support the grain at the aft end to allow for thermal dilatations. Care must be taken with the aerodynamic alimentation of the interstice between the inhibitor and the case. If the pressure rise was not fast enough, there would be, especially with low modulus propellant (composites), compression of the grain in the forward end of the case due to the pressure on the aft face and possible ruptures of the inhibitor or the propellant in that zone.

These types of problems can be solved by using thermal stress releasing thermal insulation boots made of longitudinal alveolus able to open under thermal cycles and bonded to the propellant on one face and to the case on the other face.The same concept can be used more simply, but with may be less reliability, by bonding the grain inhibitor to the case on a few generatrices.

1. Burning Rate Enhancement Along the Inhibitor

Most of the problems encountered with the design of end-burning grains deal with the observation of progressive pressure and thrust curves observed at firing when a neutral curve was expected, due to numerous burning rate enhancement phenomena possible at the propellant-inhibitor interface. The consequence is a burning surface area increase or "coning" phenomenon.

Six or seven different mechanisms have been described. Some of them are general and can be observed with any type of propellant and inhibitors. Some of them are specific of composites, others of nitrate-esters plasticized propellants. The burning rate increase along the inhibitor may be the result of heating of the propellant under the inhibitor due to the circulation of hot gases in the interstice during a long burning time. It may result from segregation of the fine oxidizer particles in the vicinity of the inhibitor. This possibility exists only in the case of composite propellants cast in their inhibitors. It may result from the existence in the inhibitor of solids that are burning rate accelerators (iron oxide for instance) or the existence of porosities or microscopic cracks, increasing the burning rate (for instance, if the stress relief boot is not carefully designed and the local strain of the propellant is very high). It may result from alteration of the local composition of the propellant through the migration of mobile molecules toward the inhibitor, and its reciprocal (active or inactive plasticizers, cross linking agents, liquid combustion catalysts, and others) or from localized variations in the radiation density of the flame in the vicinity of the inhibitor.

For instance, if the plasticizer of a propellant can migrate to the inhibitor, the propellant close to the inhibitor appears, locally, richer in oxidizing species. This results in an increase in the burning rate as well as in the pressure exponent, so that the burning rate enhancement grows with the temperature and operating pressure of the motor.

To minimize this phenomenon, the grain designer selects materials that are impermeable to the mobile components of the propellants, and balances, a priori, the chemical potentials on each side of the bonding plane. When equilibrium cannot be obtained, he may, on the other hand, introduce into the inhibitor a plasticizer moderating the burning rate of the propellant.

2. *Design Issues*

As already seen, most design issues deal with the design of the inhibitor and the interstice between the grain inhibitor and the case (or the thermal insulation of the case).

Aerodynamic problems in the interstice. The first of the two mains problems is related to the alimentation by hot gases that cool down progressively. An example is given at Fig. 25. This temperature elevation can be the cause of burning rate enhancement and the inhibitor thickness must be designed accordingly. The second is related to a gas circulation that can lead to high ablation rates of the inhibitor in the aft part of the grain when the space between case and grain is high, and the grain is supported by its lower generatrix. This can be solved by the longitudinal partial case bonding already described or an increase of the thickness of the inhibitor in the aft part. Since it has been shown that gases circulate in a length approximately

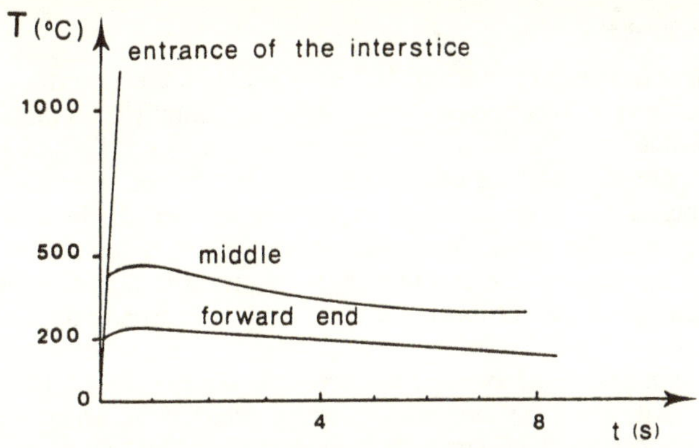

Fig. 25 Temperature evolution in the interstice between the case and the propellant grain.

equal to the diameter of the grain, one can consider a protection for a time equal to the diameter-to-burning rate ratio and calculate the necessary inhibitor thickness according to its pyrolysis law.

Burning rate enhancement due to physicochemical factors. Previous testing on subscale specimens with the same inhibitor-propellant couple must be performed in order to quantify the phenomena at the different operating temperatures. If none is observed, some representative accelerated aging must be done again with the same testing. If the phenomenon is observed, it is suggested to find some physicochemical remedies before accepting to live with it.

The burning rate enhancement phenomenon can be taken in account to calculate the burning surface area by firing specimens and extinguishing them in order to measure the angle of the coning of the burning surface and calculating the burning surface of the real grain considering $\dot{r} = \dot{r}_s \sin\alpha$ ($\dot{r}_s$ = burning rate along the inhibitor).

Blocking of the nozzle throat by condensation of combustion products. End burners operate generally with small sized nozzle throats that may generate intermittent or continuous reduction of the throat with generally two origins. The first involves pieces of the inhibitor. This problem can be solved by a design of the inhibitor that will stand as a complete boot until the end of firing or by a design of the inhibitor which will regress regularly at more or less the same rate as the burning surface.

The second is caused by particles condensing in the nozzle. This type of generally intermittent blocking can be solved by using propellants and inhibitors giving gases that will not condense (highly aluminized propellants will generate problems) and by designing the nozzle with smooth shapes and a high surface temperature.

Contribution of inert materials to performances. For long burning times (100 s or more) inert materials (insulations and inhibitors) can provide an important fraction of the gas flow, with a proportion changing with time which can influence the thermodynamical and ballistics phenomena. It has been demonstrated[19] that temperature can decrease by 25% and thrust by 10% during firing.

C. Wired End Burners

For typical tactical motor length/diameter ratios, the end-burning configurations tend to lead to long burning times, whereas centrally perforated configurations yield much shorter times. The designer sometimes finds himself in situations where the burning time of an end burner is too long whereas the burning time of a perforated grain is too short, even if he used the highest propellant burning rate available in the first case and the lowest in the second case. One way around the problem for end burners is to embed in the propellant metallic wires, parallel to the axis of the motor, to increase effective, or apparent, burning rate (Fig. 26).

The metallic wires lead to amplification of the mass burning rate by providing heat conductivity along the wires with a corresponding temperature increase of the propellant adjacent to them. The local burning rate increase generates cones in the propellant with half-angles equal to the arc sine of the base propellant burning rate divided by burning rate along the wires. The propellant not immediatly adjacent to the wires burns with its normal rate, perpendicular to the cone surface, but because

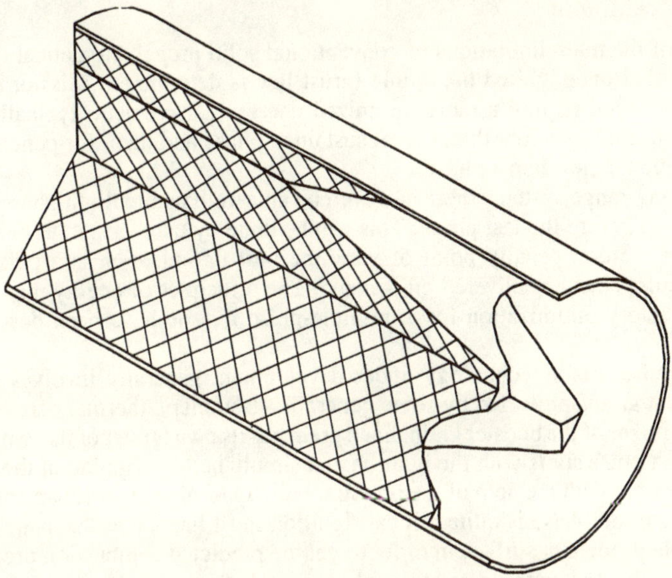

Fig. 26 Combustion of a wired end burner.

of the increased surface area, the mass generation rate and effective burning rate are increased.

For heat conductivity considerations, silver, copper and aluminum, and, sometimes, tungsten wires are generally considered. Typical burning rate augmentation ratios (effective burn rate to normal burn rate ratio) for various 5-mil-diam wires at 6.97 MPa will be 5.3 for silver, 4.6 for copper, and 2.3 for aluminum.

The preliminary analysis of the performance of this type of propellant grain involves two distinct steps that cannot be handled with traditional means. The first is the determination of the overburning rate coefficient of a given wire/propellant combination. Currently, this determination is mostly experimental using propellant strands with an embedded wire. Theoretical approaches are being developed, requiring the generation of a complete database for thermal data on wires and on acceptable propellant, e.g., diffusivity, conductibility, thermal capacity, etc.[20]

The second is the determination of the evolution of the grain burning surface.When dealing with end-burning propellant grains, the computational programs are relatively easy to generate, since the evolution of the grain's burning surface involves nothing but cones.The difficulties increase with the number of wires embedded in the propellant grain.

The main problems encountered with wired end burners are related to the production of these grains: the wires are thin and delicate and must be perfectly rectilinear and unbroken in the final grain. To achieve this, the propellant viscosity when casting must not be too high. In any case, the cost will be higher than with conventional designs. Some problems may also arise with high length-to-diameter ratio end burners after thermal cycling. The difference of thermal expansion coefficients between the propellant and the wire may result in unbonds and gaps between them. This will cause reductions and variations in burning rate and operating pressure.

D. Pulse Motors

One of the main limitations of conventional solid propellant tactical motors is the fact that once ignited the whole thrust law is determined. Mission trajectories quite often require a more optimized energy management: typically, a first boost followed by a low-thrust level sustained phase and then, for penetration or interception, a new boost phase.

For long-range systems there can even be an aerodynamic flight phase without propulsion before the last phase. This can be done by using separate motors, but from a structural penalty point of view the best optimization is a pulse motor where pulses can be delivered on demand. The principles of energy management and trajectory optimization for a multiple-pulse missile have been described in Ref. 21.

The pulse motor technology under development generally involves separate propellant grains protected by some generally elastomeric thermal barriers of the thermal fluxes of the booster grain. Each grain has its own igniter or its own igniting system. A pulse starts with the ignition and simultaneous stripping of the thermal barrier, either with the help of the pressure buildup or of some mechanical device.

The ignition phase is quite critical. Ignition must happen at the same time on the whole propellant surface in order to get the predicted combustion profile. It is important that the parts ejected through the nozzle throat be very small, otherwise some pressure burst and thrust perturbation could occur and instabilities could be triggered, especially with reduced- or minimum-smoke propellants. The case must be designed according to the production requirements of the casting of different propellant grains. This design also raises new issues on safety during the production process since ignition systems in close contact to the propellant may be involved in production and handling phases of the motor.

Acknowledgments

The author wants to express his gratitude to Bernard Zeller for the possibility to use his presentation in AGARD Lecture Series 150 as a starting material and also J. M. Laurent of Aerospatiale and G. Frut of SEP for useful discussions on the principles of design and optimization of a solid rocket motor.

References

[1]Sutton, G. P., "Nozzle Theory and Thermodynamic Relations," *Rocket Propulsion Elements,* 6th ed., Wiley, New York, 1992, pp. 41–88.

[2]Gordon, S., and MacBride, B. J., "Computer Program for Calculation of Complex Chemical Equilibrium Compositions, Rocket Performance," NASA SP 273, 1971.

[3]Coats, D. E., et al., "A Critical Review of the SPP Loss Mechanisms," AIAA Paper 89-2781, July 1989.

[4]Anon., "Performance of Rocket Motors with Metallized Propellant," AGARD Propulsion and Energetics Panel, Working Group 17, AR 230, 1986.

[5]Davenas, A., *Solid Rocket Propulsion Technology,* Ist ed., Pergamon Press, Oxford, England, UK, 1993.

[6]Kuo, K. K., and Summerfield M. (eds.), *Fundamentals of Solid-Propellant Combustion,* Progress in Astronautics and Aeronautics, Vol. 90, AIAA, New York, 1984.

[7]Saintoux, E., et al., "ELEA tool for 3D Surface Regression Analysis in Propellant Grains," AIAA Paper 89-2782, July 1989.

[8]Friedlander, M., and Jordan, F. W., "Radial Variation of Burning Rate in Center Perforated Grains," AIAA Paper 84-1442, July 1984.

[9]Razdan, M. K., and Kuo, K. K., "Erosive Burning of Solid Propellants," Progress in Astronautics and Aeronautics, Vol. 90, AIAA, New York, 1984, pp. 515–598.

[10]King, M. J., "A Model of the Effects of Pressure and Cross Flow Velocity on Composite Propellant Burning Rate," AIAA Paper 79-1171, July 1979.

[11]King, M. J., "Erosive Burning of Composite Solid Propellants: Experimental and Modelling Studies," AIAA Paper 78-979, July 1978.

[12]Delannoy, G., "Prediction of Antitank Solid Propellant Rockets Internal Ballistics," AIAA Paper 84-1355, July 1984.

[13]Eister, S. D., and Davis, R. J., "Predicting Burn Time Variations in Solid Rocket Motor," AIAA Paper 90-2736, July 1990.

[14]Culick, F. E. C., "Stability of High Frequence Pressure Oscillations in Rocket Combustion Chambers," *AIAA Journal,* Vol. 1, No. 5, 1963, pp. 1097–1104.

[15]Nugeyre J. C., et al., "An Example of Failure by Acoustic Coupling of a Free Standing Double Base Grain," AIAA Paper 81-1525, July 1981.

[16]Brownlee, N. G., and Marble, P. B., "An Experimental Investigation in Solid Propellant Rocket Motors," Solid Propellant Rocket Research Conf., American Rocket Society, ARS Paper 1067, 1960.

[17]Kaczmarek, R. E., et al., "Tangential Mode Instability in Double-Base Solid Rocket Motors, A Case Study," AIAA Paper 82-1063, July 1982.

[18]Evans, G. I., and Smith, P. K., "The Suppression of Combustion Instability by Particulate Damping in Smokeless Solid Propellant Motors," *Solid Rocket Motor Technology,* AGARD CP 259, 1979, pp. 27-1–27-11.

[19]Banon, S., and Astier, J., "The Contribution of Inert Material to End Burning Propellant Grain Performance," AIAA Paper 86-1421, Huntsville, AL, June 1986.

[20]King, M. J., "Analytical Modeling of Effects of Wires on Solid Motor Ballistics," AIAA Paper 89-2784, July 1989.

[21]Phillips, C. A., "Energy Management for a Multiple Pulse Missile," AIAA Paper 88-0334, Jan. 1988.

Chapter 5

Solid Propellant Grain Structural Behavior and Service Life Prediction

Thomas H. Duerr*

Stone Engineering, Huntsville, Alabama 35805

and

Barbara P. Marsh†

U.S. Army Missile Command, Redstone Arsenal, Alabama 35898-5940

Nomenclature

A_i = amplitude of daily temperature variation
a_T = time–temperature shift factor
B_i = amplitude of yearly temperature variation
C_{V1} = strength coefficient of variation
C_{V2} = stress coefficient of variation
D = stress cumulative damage
F_{age} = age degradation factor for strength
$F_{\mathrm{m/ax}}$ = multiaxial factor for strength
$F_1(S)$ = probability density function for induced stress
$F_2(S)$ = probability density function for allowable strength
N_1 = strength statistical factor
N_2 = stress statistical factor
n = number of harmonics
P = probability of failure
P_f = probability of propellant grain failure under multiple cycles
P_{fn} = probability of propellant grain failure under nth cycle
P_i = period
R = reliability
S_{all} = allowable strength standard deviation
S_{app} = applied stress standard deviation
T_D = mean daily temperature
$T(t)$ = temperature at time t

*Senior Aerospace Engineer, Huntsville Division.

†Supervisory Aerospace Engineer, Propulsion Directorate.

T_{sf} = strain-free temperature
T_Y = mean yearly temperature
t = time
V_0 = volume at time = 0
z = safety index
α = coefficient of thermal expansion of propellant
α_i = daily phase angle
β = damage exponent
β_i = yearly phase angle
ΔT = temperature difference
ΔV = change in volume
ρ = stress–strength correlation coefficient, 0 = no correlation and 1 = perfect correlation
σ_{all} = allowable uniaxial strength, $\bar{\sigma}_{all}(1 \pm N_1 C_{V1}) F_{age} F_{m/ax}$
σ_{app} = applied maximum principle stress, $\bar{\sigma}_{app}(1 \pm N_2 C_{V2})$
$\bar{\sigma}_{all}$ = mean allowable strength
$\bar{\sigma}_{app}$ = mean applied stress
σ_i = applied stress at t_i
σ_0 = allowable stress at $\log t = 0$

I. Background

THE advent of stringent low- and high-temperature specifications for solid propellant tactical motors during storage and operation has required increasing attention to propellant grain structural issues. The trend to larger sized and more expensive motor inventories also drives this requirement in addition to the need for more accurate and reliable service life prediction.

A. Case Bonding and Grain Structural Requirements

Typical modern tactical solid rocket motors are case bonded such that the propellant grain is chemically bonded to a very thin layer of liner material that in turn is adhesively bonded to the motor case or to insulation (Fig. 1). This latter

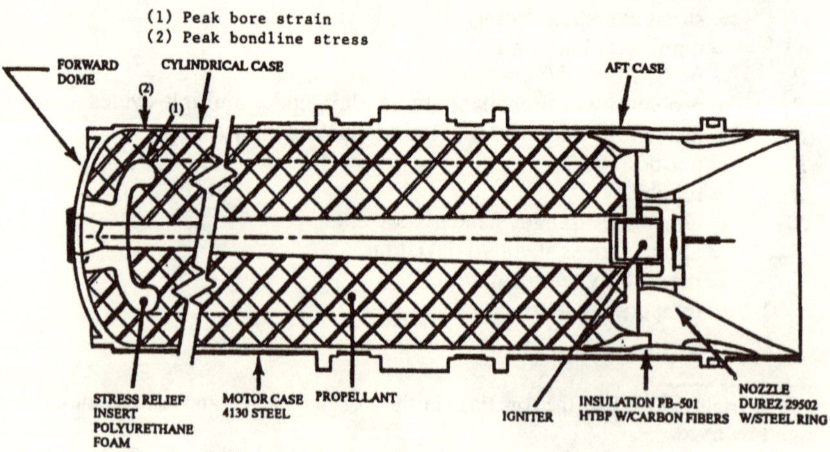

Fig. 1 Typical case-bonded rocket motor.

material is adhesively bonded to the motor case at selected critical locations likely to be exposed to hot gases during the early part of motor operation.

The concept of bonding the propellant grain to the motor case was exploited in the 1950s and permitted the development of larger motors with very high mass fractions. With case-bonded grains, the propellant mass is substantially immobilized against drag from internal gas flow, gravity storage loads, and flight acceleration loads; cartridge-loaded grains do not possess any of these advantages. Furthermore, case bonding generally results in very low average bondline stresses to resist the aforementioned loads.

However, case bonding does result in structural conditions that are of significant concern in the design and analysis of tactical motor propellant grains. After propellant casting, tactical motors are typically cured at 120–140°F for several days to a week to polymerize the propellant mass and create the desired mechanical strength. This chemical polymerization always causes a slight volume shrinkage of the propellant that is accommodated by polymer chain adjustments during the early portion of the cure process and by small internal stresses during the later portion of the cure process. Because the outer boundary of the propellant mass is now bonded to the much more rigid motor case, this volume contraction can only occur by a slight expansion of the grain's free surface, that is, the inner bore or cavity formed by the casting mandrel. By raising the motor temperature 10–20°F above the cure temperature, these cure shrinkage stresses and bore deformations essentially disappear. The temperature at which this occurs is the stress- or strain-free temperature that is typically then used for subsequent grain thermal stress or strain calculations. Equation (1) can be used to estimate the temperature increase above the cure temperature in a long case-bonded grain with a circular bore:

$$\Delta T = \frac{(\Delta V / V_0)_c}{4.5\alpha} \tag{1}$$

Because of thermal expansion differences between case and propellant, stresses and stains develop in any case-bonded grain at any temperature below the stress-free temperature. This difference is approximately a factor of 10 between metallic motor cases and composite propellants. The tactical motor requirement to withstand these thermal stresses and strains at low storage temperatures and long storage times has been a dominant factor in the development of five or six generations of propellants over the past 40 years.

The thermal stresses and strains in a typical case-bonded grain configuration generally peak at the grain bore and are therefore considered most critical relative to failure. Although bondline stresses are usually much lower than bore stresses, they do increase rapidly near grain termination points or other geometric discontinuities. Figure 2 shows results from a finite element analysis of a 3-in.-diam × 11-in.-length case-bonded grain configuration subject to a 110°F temperature cooldown or change from its stress-free temperature. The analysis assumes the propellant to behave elastically with a modulus of 660 psi and indicates that the peak stress and strain occurs at the grain bore nearer the bonded head or dome end. Also indicated in Fig. 2 is the steep strain gradient across the propellant web and the 2/1 ratio of the bore hoop stress (125 psi) to the bondline radial stress (58 psi) in the plane through the peak hoop stress axial location. If the head end was assumed to be unbonded, as in many actual motor designs, these peak values would shift significantly toward the center of the configuration. Additionally, the bondline stresses, near the bonded to unbonded discontinuity, all exhibit very

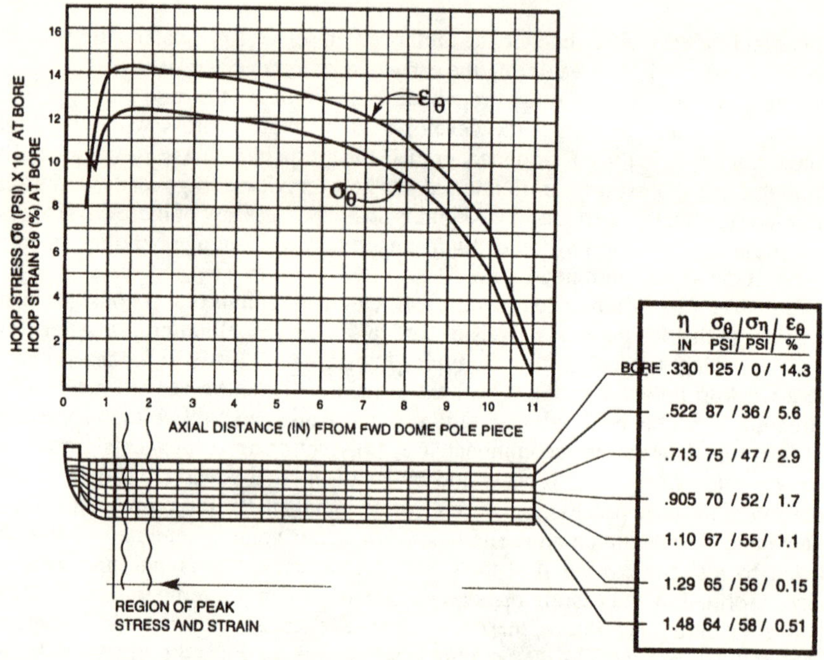

Fig. 2 Finite element thermal analysis results.

significant gradients. The peak values are as much as four to five times the value near the lengthwise center of the motor. Controlling these peak values so that they do not become the primary structural failure mode for the grain is an important objective during the design phase of any case-bonded propellant grain.

B. Micromechanical Origins of Solid Propellants as a Structural Material

Materials like solid propellants, with room temperature strengths of 100 psi, are not considered typical structural materials, even though they must demonstrate a definite level of structural integrity as well as ballistic performance to meet tactical motor system requirements. Case-bondable composite propellants contain 10–20% by weight of rubber-based binder material that exhibits strong time- and temperature-dependent response and failure behavior caused by polymeric network entanglement, slippage, extension, and cleavage. This small percentage of material acts as the coating and matrix for the remaining large percentage of solid ammonium perchlorate and aluminum particles that vary widely in size and shape. An important component of composite propellant thermoviscoelastic behavior results from the interaction of these rigid oxidizer and fuel particles and the soft binder material. This interaction manifests itself under stress or deformation as dewetting or binder unbonding from the particle surfaces and binder cohesive failure at peak stress locations slightly displaced from the particle surfaces. These microscopic stress risers are the origins of propellant mechanical failure, as they develop into microflaws and propagate or coalesce into observable cracks. A uniaxial tensile strain dilatation measurement is often used to indicate the extent of this microscopic void formation as a function of both tensile stress and strain. Although the test does not distinguish between oxidizer to binder dewetting

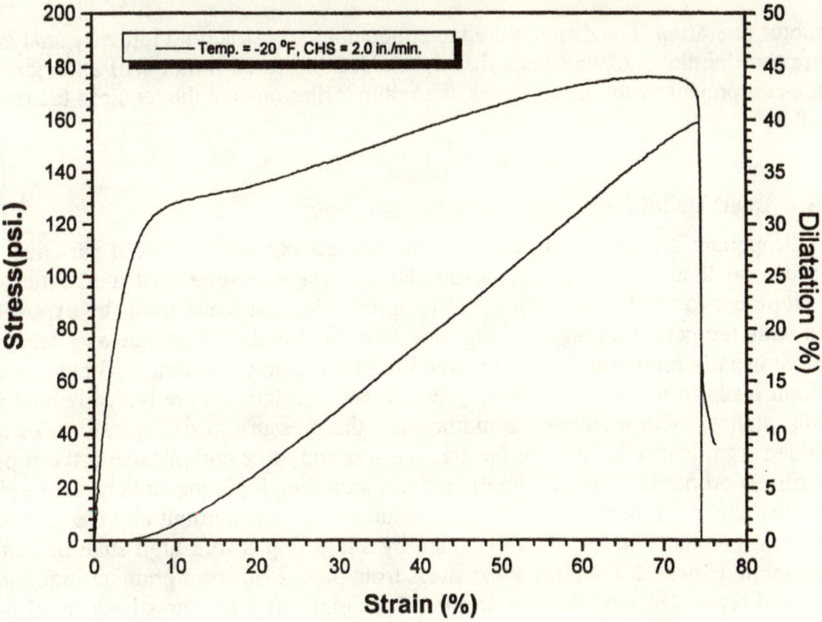

Fig. 3 Dilatation behavior for a composite propellant.

and binder cohesive void formation or isolate the specific location of the initial failure, it clearly shows critical stress or strain levels where the rate of change of void volume increases dramatically. The data beyond this so-called threshold become a measure of the propellant's departure from classical continuum behavior and its approach toward becoming a particulate-filled structural foam. Figure 3 indicates typical results from a low-temperature uniaxial strain dilatation test on a composite propellant specimen. For the particular test temperature and crosshead rate used here, the stress and strain thresholds for onset of volume change (dV/V_0) are approximately 110 psi and 5%. Also, the volume dilatation at large stress or strain levels near failure is indicated to be approximately 40%.

As shown in Fig. 3, the dilatation at large stress levels near failure can easily reach 15–20% volume increase during a uniaxial test. This level of internal void increase during loading is a clear indication of departure from continuum material behavior even at moderate strain levels. The consequences of this microstructural behavior, combined with the binder's thermoviscoelastic effects, are significant difficulties in modeling both constitutive and failure behavior for structural and margin of safety analyses.[1,2]

II. Structural Design Considerations

A grain structural integrity analysis is an evaluation of the capability of the solid rocket motor to meet mission requirements under a specified environment. The grain structural analysis determines the stresses, strains, and deformations that the solid propellant grain will experience under the specified environment. The structural analysis is then coupled with a failure analysis to define the environment in which the motor will operate successfully. Since the operational environment is usually defined, the analysis is used to determine a minimum factor of safety for motor operation. The data required to conduct a grain structural integrity analysis are the definition of the loads that the fielded or stored motor will experience, the component material properties, the failure criterion, and the required factor of safety.

A. Load Definition

Propellant grains in tactical rocket motors are exposed to several varieties of structural loads that require analysis during design, testing, and hardware development to verify grain structural integrity. Thermal loads from the exposure to wide temperature ranges during long-term field and fleet storage and deployment usually represent the most taxing long-term load environment. Single event flight loads from motor operating pressure and acceleration are typically next in importance. With a composite motor case, the pressure loads typically take on added significance because of the case's lower and more complicated orthotropic stiffness properties. The acceleration loads, including flight maneuvers, are rarely critical since the peak values usually occur near motor burnout and the internal pressure tends to immobilize the grain by subjecting it to a high state of compression. However, bondline shear stress from these loads near grain terminations should always be analyzed by detailed finite element procedures because of the previously discussed large and very localized stress gradients.

Slump, shock, and vibration loads are continuously or repeatedly imposed upon tactical motors during a typical lifetime from storage, transportation, deployment, training, and combat activities. Estimates of these g-loads are usually consolidated into a peak value for a given time span or at a given frequency for the purpose of grain design analysis. Because the propellant is a highly damping viscoelastic material, shock and vibration amplitudes are typically quite low, and the shock and vibration peak induced stresses and strains are generally very low.[3] Although the slump load from vertical or horizontal storage is continuously applied, the 1.0 g-load level and the relatively small size of tactical motors keep the resulting stresses and strains at a very low level, even in the largest tactical motors. However, significant propellant softening from long-term aging or high-temperature storage can cause a significant slump problem at these very low stresses.

Tactical propellant grains are sometimes prestressed by internal pressure during cure to lower the bore strain-free temperature to approximately room temperature. This procedure is most effective with flexible composite motor cases made of fiberglass or Kevlar epoxy materials. The cure pressure applied to the open end of the uncured grain expands the case and propellant grain during the cure process. After completion of the cure process, the pressure is released, allowing the case to contract and to apply compressive loading to the propellant grain that is sufficient to offset the bore strain caused by the cure shrinkage and the bore strain associated with cooling the motor from the cure temperature to room temperature.

Table 1 shows a typical design load environment and the resulting stresses or strains at two structurally important locations in a tactical motor grain. Firing the motor at low temperatures generally produces the highest bore strains and case-bonded stresses and the lowest safety margins. Also indicated are the slump and vibration loads resulting in very low stresses and strains and very high safety margins. However, for this particular motor the specified shock loads, and especially the one associated with the lateral mode, do result in relatively high case-bonded stresses and relatively low but still very adequate safety margins.

Table 1 Typical load environment for design analyses of a high-performance tactical motor propellant grain

	Propellant centerport[a]		Case bond	
Loading conditions	Maximum strain, in./in.	Margin of safety	Maximum deviatoric stress, psi	Margin of safety
Thermal storage at −35°F	0.10	2.5	4.0	4.0
Thermal cycling, 25°F–125°F	0.08	3.8	8.0	4.6
Slump—vertical at 135°F	0.005	69.0	1.1	17.0
Slump—horizontal at 135°F	0.003	116.0	2.0	9.0
Mechanical shock—axial, 20 *g* at 135°F	0.008	60.0	21.5	3.2
Mechanical shock—lateral, 30 *g* at 135°F	0.005	97.0	40.0	1.6
Vibration—axial, 2.5 *g* at 135°F	0.004	104.0	5.3	10.3
Vibration—lateral, 2.5 *g* at 135°F	0.003	139.0	10.0	5.0
Static firing at −35°F	0.18	0.8	—	—
Static firing at 70°F	0.15	2.4	65.0	0.5
Static firing at 135°F	0.12	2.8	27.0	1.2
Flight at −35°F	0.18	0.8	—	—
Flight at 70°F	0.15	2.4	68.0	0.4
Flight at 135°F	0.12	2.8	30.0	1.0

[a] Grain configuration—circular port with forward and aft end stress relief; motor $L/D = 2.6$ and grain OD/ID = 5.3.

1. *Thermal Loads*

Thermal loads induce stresses and strains within the solid rocket motor grain by temperature gradients and the differences in thermal expansion coefficients between case materials and insulation/liner and propellant materials. The thermal expansion coefficient is generally an order of magnitude greater for propellants, liners, and insulation materials than that of motor case materials. Thermal loading occurs from environmental temperature cycling, with the most severe loading usually resulting from low-temperature cycling. Motor specifications usually indicate the number of cycles and temperature limits that the motor must be able to withstand. Aerodynamic heating is also a source of thermal loading. As the temperature of the case is raised from flight, there is an appreciable difference from the temperature of the insulation/liner and propellant materials. The structural problem that generally arises from aerodynamic heating is increased stress on the bondline as a result of case expansion with a decrease in the strength capability of the bondline system during the rapidly rising temperature.

2. *Acceleration Loads*

Acceleration loading conditions are generally classed as slump during storage and dynamic acceleration loads as a result of missile launch and maneuvering. Storage slump from vertical storage can result in grain deformations because of the gravitational force exerted on the grain. Horizontal storage can also produce propellant deformation in which star points and/or longitudinal slots may close together. For flight loads, the maximum shear stresses may occur near the time of maximum pressure, generally immediately after ignition rise time. Shear stresses are usually highest at the bondline interface with acceleration at high temperatures being the most critical since bondline strength is reduced as temperature is increased. In general, in-flight maneuvering does not induce significant loads in the propellant grain.

3. *Transportation and Handling*

Transportation and handling loads encompass loads imposed by vibration during transport, vibration under the wing of the airplane for air-launched missiles, and shock loads imposed by dropping missiles during transport. The vibration loads, which are often of the bending mode, generally induce very small stresses compared with thermal, acceleration, and pressure loads, but they should be considered in combined loads failure analysis.[3]

4. *Cure Shrinkage*

As one would expect, cure shrinkage stresses and strains are induced during the propellant curing process as the propellant volume contracts during cure. In composite propellants this volume shrinkage is proportional to the volume of binder present. This shrinkage results in a stress-free state at temperatures usually 10–20° higher than the cure temperature.

5. *Pressurization Loads*

Pressurization loads occur during ignition and continue throughout the burning of the propellant grain during flight. These loads impose essentially hydrostatic compressive stresses on the grain ends. When ignition takes place, the pressure that typically builds rapidly to the maximum level induces the critical load. Propellant properties, grain design, and motor case stiffness determine the imposed pressurization stresses and strains.

B. Propellant Mechanical Behavior

The stress fields that may be present in a solid rocket motor grain vary according to the life phase. During cure/cooldown, biaxial and triaxial stresses may be present at the bondline, and biaxial stresses at the bore surface may be present because of shrinkage and thermal contraction. Simple shear stresses as a result of gravity loading may be present in the grain during storage, as well as biaxial tension because of thermal contraction caused by thermal cycling. Simple shear stresses may also be present during transportation because of vibration. During ignition, tensile strains as a result of pressurization are present. Flight loads produce combined tension and shear because of burning and acceleration. For computation of stress fields under the various types of loadings, the characteristic material properties of the propellant must be determined experimentally and then used in equations of state. The propellant behavior is characterized by failure properties and response to deformation. The response is a function of the interdependent relationships of stress, strain, time, and temperature.

1. *Failure and Response Behavior Tests*

Solid propellants are thermoviscoelastic materials, which means that their response and stress/strain capability are time and temperature dependent. Master

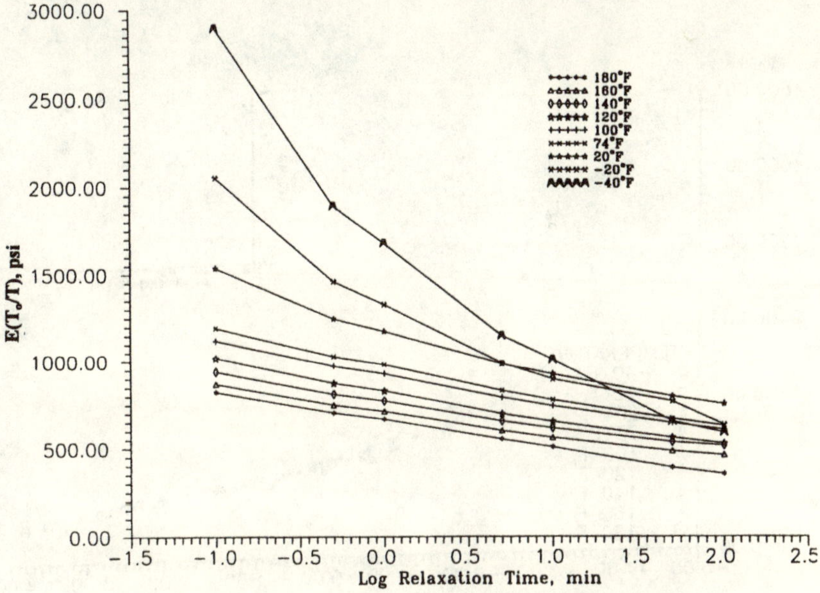

Fig. 4 Isothermal response curves for a composite propellant.

modulus curves must be experimentally determined for propellant response properties by conducting stress relaxation tests over a wide range of temperatures. Typical temperature ranges are from −62.2°C (−80°F) to 71.1°C (160°F). The specimen is rapidly loaded to a fixed strain level considered to be in the linear response region, usually 1–5%, and the load decay over time is recorded. The individual curves at each temperature are shifted along the time axis to form a master relaxation modulus curve. The time–temperature superposition principle is applied to materials whose response and failure behavior show a dependence on temperature and rate (time). The principle is based on the assumption that a shift in temperature is equal to some shift in time. Figure 4 shows modulus curves from several isothermal tests. One chooses a reference temperature, usually 24°C (75°F), and shifts the curves along the x axis to determine shift factors and to form a master curve (Fig. 5). The response curve is used in design and analysis to define the response behavior of the propellant under operational loading and storage conditions. The stress capability master curve (Fig. 6) is used similarly to determine stress capabilities under various operational and storage environments. For failure analysis, failure properties, as well as response properties, are required. Uniaxial tensile tests[4] are also conducted over a wide range of temperatures and crosshead rates. From these tests, master curves of the stress and strain are constructed, as well as the Smith stress–strain failure plot. Data from the master curves are used in propellant grain stress analyses. Once the analysis is completed, analysis results can be superimposed on the failure curves to determine if those conditions fall within the safe operating range (Fig. 7).

It is desirable to obtain properties under the same conditions as will be present in the grain, although this is not always feasible. Some of the conditions that should be considered are ignition pressurization, simultaneous straining, and thermal cycling, cooling, and acceleration.

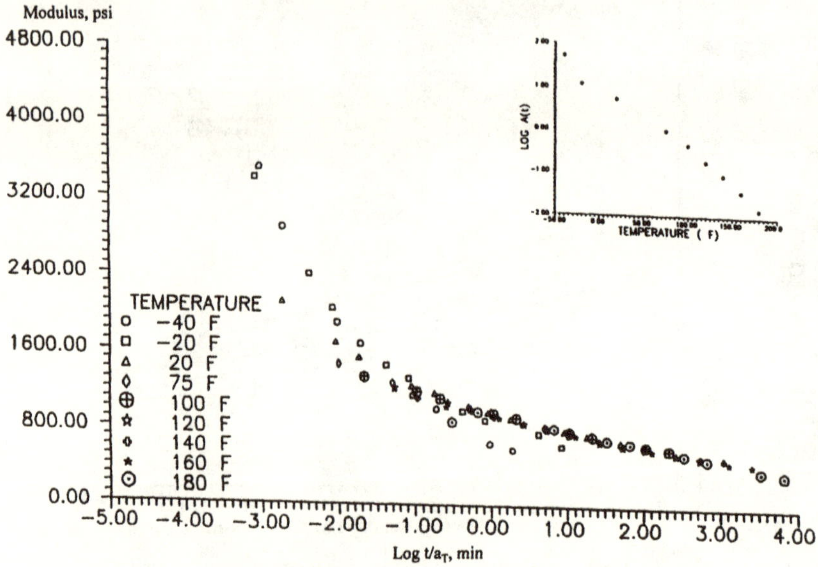

Fig. 5 Master stress relaxation curve.

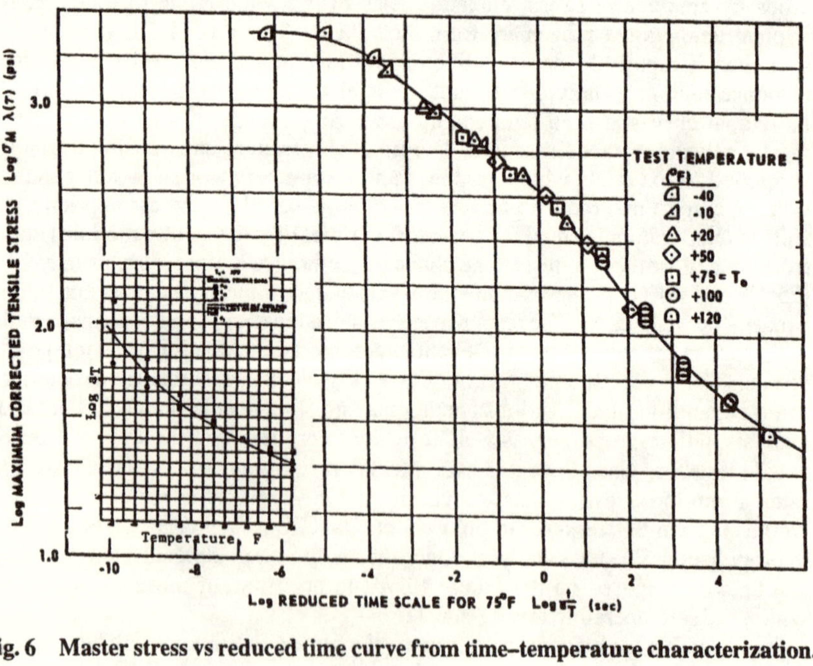

Fig. 6 Master stress vs reduced time curve from time–temperature characterization.

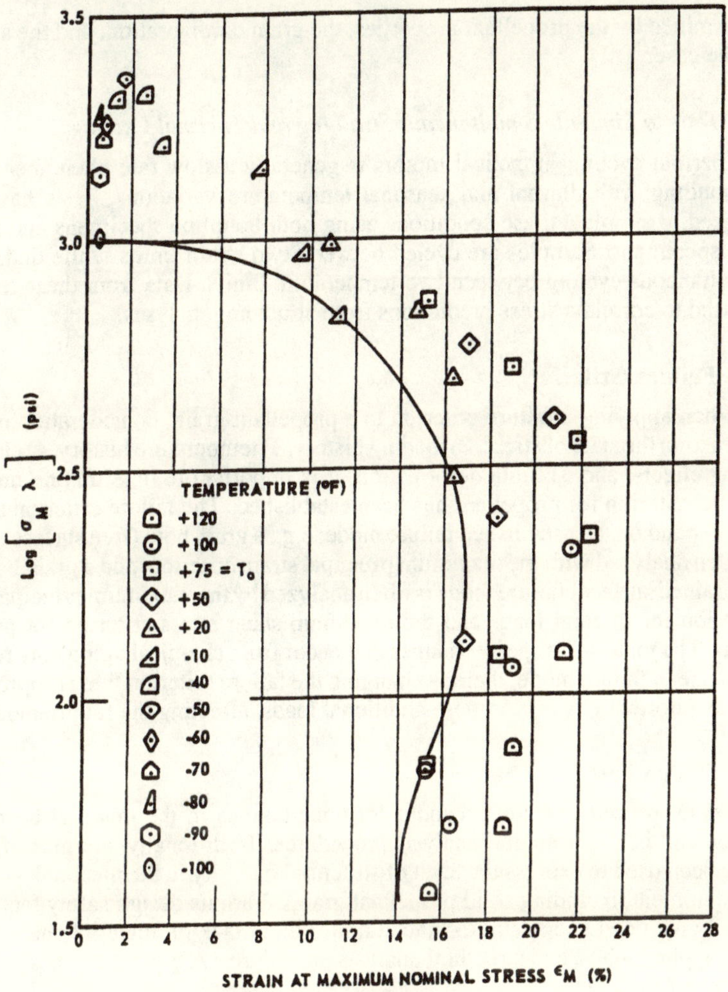

Fig. 7 Smith failure plot.

2. *Test to Simulate Ignition Pressurization*

Tests have been conducted in an attempt to simulate ignition pressurization by preloading uniaxial or biaxial specimens to strain levels equivalent to thermal loading, then pressurizing and straining at a high rate of failure.[5] When pressurized tensile data are available, these tests provide the appropriate modulus and failure properties for structural integrity investigations of the propellant grain under ignition loads. Generally, when no pressurized tensile data are available, the value for the modulus corresponding to the ignition rise time is chosen from the master stress relaxation curve from unpressurized tests. If pressurization test data are not avaliable, failure properties are also taken from the master strength curves (time–temperature characterization). These values need to be modified to account for the effects of pressure and propellant aging. Pressurization loads are ballistically determined. The stresses and strains resulting from the pressurization loads are determined by the propellant properties, the grain configuration, and the stiffness of the case.

3. *Tests to Simulate Simultaneous Straining and Thermal Cycling*

Thermal cycling in tactical motors is generally a slow rate phenomenon corresponding with diurnal and seasonal temperature variations. Tests have been devised to simulate these conditions using both bondline specimens and propellant specimens. Samples are cycled between two strain limits while undergoing simultaneous cycling between two temperature limits. Data from these tests can be used to correlate stress predictions from structural analyses.

C. Failure Criteria

When applying a failure criterion to a propellant grain, consideration must be given to 1) the state of stress, 2) loading history, 3) temperature history, 4) chemical aging effects, and 5) finite deformation. It is important to note that no universal failure criterion for propellant has been established. The failure criterion applied will depend on the established failure mode; e.g., a grain bore (free surface) failure is often analyzed with the maximum principal strain criterion, and a grain bondline (restrained surface) failure mode is often analyzed by the maximum principal stress criterion for thermal loads and the maximum shear stress criterion for pressure loads. The material property changes that occur from chemical aging may result in a change in failure mode, thereby changing the failure criterion that is appropriate. Finite deformations may impose additional loads, affecting the failure mode.

D. Design Factors of Safety

Design procedures must account for uncertainties in the material characterization and in the structural analysis procedures. Traditionally, margins of safety have been used to express the level of structural integrity in the mechanical design of equipment, including solid propellant grains. Various design safety factors are usually included to account for uncertainties in, or lack of, material characterization, applied loads, and structural analysis procedures. A typical expression of the margin of safety is

$$MS = \frac{\sigma_{\text{all}}}{\sigma_{\text{app}} \times \text{Safety Factor}} - 1 \tag{2}$$

Although this expression may include estimates of statistical variability, it is essentially deterministic since the final result gives no indication of reliability. In the design phase of propellant grains, the production property mean values and variabilities for a new formulation can only be estimated from previous experience. This is also true of the aging and multiaxial factors.

To estimate propellant grain structural reliability, a model that couples the stress–strength means and their variabilities is necessary. Such a relation is expressed in the safety index or coupling coefficient,[6] i.e.,

$$Z = \frac{\bar{\sigma}_{\text{all}} - \bar{\sigma}_{\text{app}}}{\sqrt{s_{\text{all}}^2 + s_{\text{app}}^2 - 2\rho s_{\text{all}} s_{\text{app}}}} \tag{3}$$

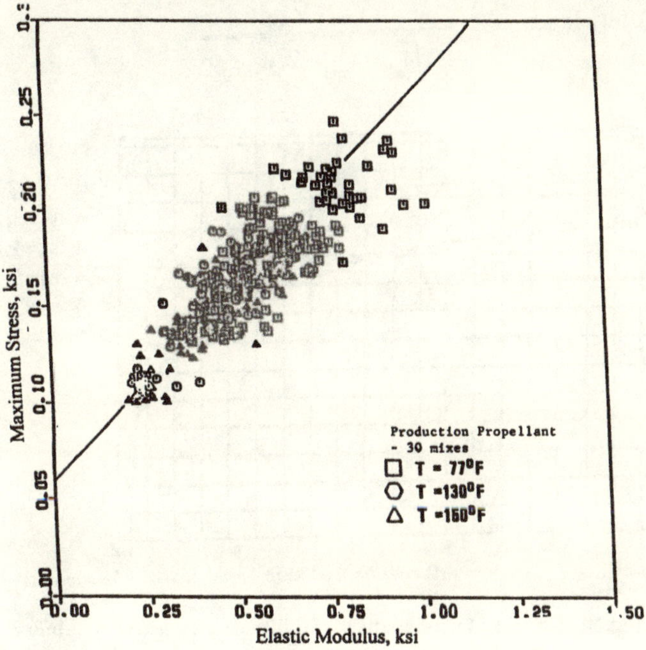

Fig. 8 TP-H8208 production propellant—30 mixes (maximum stress vs elastic modulus).

The reliability for normally distributed stress and strength can be directly estimated from a cumulative normal distribution table. Equation (3) also permits an estimate of the effect for the stress–strength correlation indicated in any grain stress margin analyses because of the direct relationship of modulus on applied stress and the observed strength modulus correlation from test results. Typical experimental data from production mixes of a modern composite propellant are shown in Fig. 8 and indicate a strength–modulus correlation coefficient in the range of 0.70–0.85. Figure 9 indicates the significant impact of a typical range of propellant variabilities on structural reliability for a constant mean margin of safety. The effect of a typical stress–strength correlation coefficient is also shown.

Even when all parameters in Eqs. (2) and (3) based on experience or previous experimental data, Eq. (3) provides a more informed basis for design decisions because it relates the grain structural integrity in a fully compatible way to a critical system requirement reliability. Use of the safety index has also been applied to estimating the probabilistic service life of rocket motors.[7]

III. Surveillance and Service Life Prediction

Many tactical missiles, with solid rocket motors, spend much of their useful life in storage depots in various geographical locations throughout the world where the thermal environments are changing continuously with time. Since the stress-free temperatures of most solid rocket motors are considerably higher than the storage temperature, the motors are subjected to continuously varying thermal stresses that can produce structural failures. Case-bonded double-base and minimum-smoke nitrimine propellants are also susceptible to structural failure in this manner as

$$\text{Safety Index} \equiv Z_o = \frac{\sigma_{all} - \sigma_{app}}{\sqrt{s_{all}^2 - s_{app}^2 - 2\rho s_{all} s_{app}}}$$

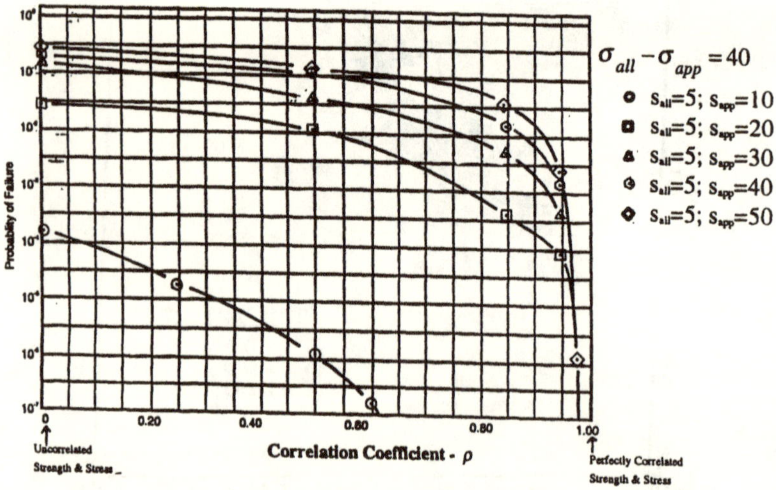

Fig. 9 Impact of propellant variability on structural reliability.

are composite propellants. These double-base-type propellants also generate gases internally when exposed to high temperatures, and the gas generation may cause the propellant grain to crack. Loss of stabilizer in double-base-type propellants is also a life-limiting factor that is a function of time and temperature.

A. Failure Mode Identification

In general, a solid propellant rocket motor has reached the end of its useful service life when it is no longer capable of performing the mission for which it was designed. Determining what condition constitutes motor failure is then the first step in a service life analysis. Failure modes can be the following: 1) grain bore cracking that may be caused by thermally induced stresses, pressurization, vibration, shock, internal gas generation (double-base propellants), or aging; 2) case-liner-propellant bond separations, i.e., debonds because of mechanical or thermal loads, acceleration, aging, or improper processing; 3) ignition (igniter) failure; and 4) ballistic failure. Determining probable failure modes at the beginning of a service life program will provide the guidelines used in developing an individualized analysis and testing program for that system.

Methods for determining the failure mode(s) of a system are testing of full-scale or analog motors and prediction by finite element analysis. The best method at present for determining structural failure modes is to thermally overtest a full-scale motor. Although this provides the method and location of failure, it is often impractical to use this approach because of expense or limited availability of large rocket motors. Failure information may be available from the development phase of a system, or occasionally, an in-service motor may fail and its examination will give valuable insight in determining failure modes.

Another approach is to overtest analog motors. Successfully scaling all of the significant structural and ballistic parameters pertaining to the full-scale motor in one analog, or subscale, is generally not possible. Furthermore, the analog may produce a different failure mode than the full-scale motor if not properly designed.

Finite element analysis is used most often to predict failure mode and location. This technique has the advantage that it does not require expenditure of motors. However, the accuracy of this structural analysis is limited by the accuracy of input data, failure data, and the mathematical models.

B. Real-Time Assessment

To conduct a service life program, the most reliable source of data for predictive purposes is from full-scale motors that have seen the deployment environment. Since it is difficult to monitor deployed tactical rocket motors, the most acceptable method to determine service life is to store rocket motors at climatic locations that represent extremes of temperatures and sample from this controlled storage population to conduct ballistic and structural testing over a period of at least 10 years and, if possible, 20 years. The service life program should be structured for periodic sampling, adjusting time intervals if necessary as data are collected. Baseline data should be taken from motors fresh from production with a full range of mechanical, physical, chemical, and ballistic tests performed. Types and quantities of tests performed may also be adjusted to concentrate on suspected failure modes.

As motors are withdrawn from storage and properties are determined, comparison can be made to ascertain changes to the properties and to determine if margins of safety or failure probabilities are improving or degrading.

Although this method is currently used on many programs, often the service life prediction is based on successful static firing, i.e., no failure or substantial change in ballistic performance. This limits the service life to the maximum test period plus one or two years.

C. Predictive Assessment

Although comparison of real-time data is informative, it is desirable to be able to predict motor failure in advance. This allows for more efficient planning for replacement and for a higher degree of safety. Three basic methods for predicting the location of failure and when it is likely to happen are discussed next.[8]

1. Deterministic Approach

The simplest approach to service life analysis is the deterministic approach, which involves applying failure prediction methods to changing properties and loads. The first step is to identify the most likely failure mode or modes, as discussed previously. After the failure mode is identified, then a failure analysis must be conducted using properties of the components to determine if failure will occur under a given set of loading conditions, usually a "worst case" condition. When the properties degrade to a point where failure is predicted, the motor has reached the end of service life. For instance, with a bondline failure mode, when the strength of the bondline degrades below the level at which failure is predicted, the end of service life has been reached. If the failure mode is determined to be

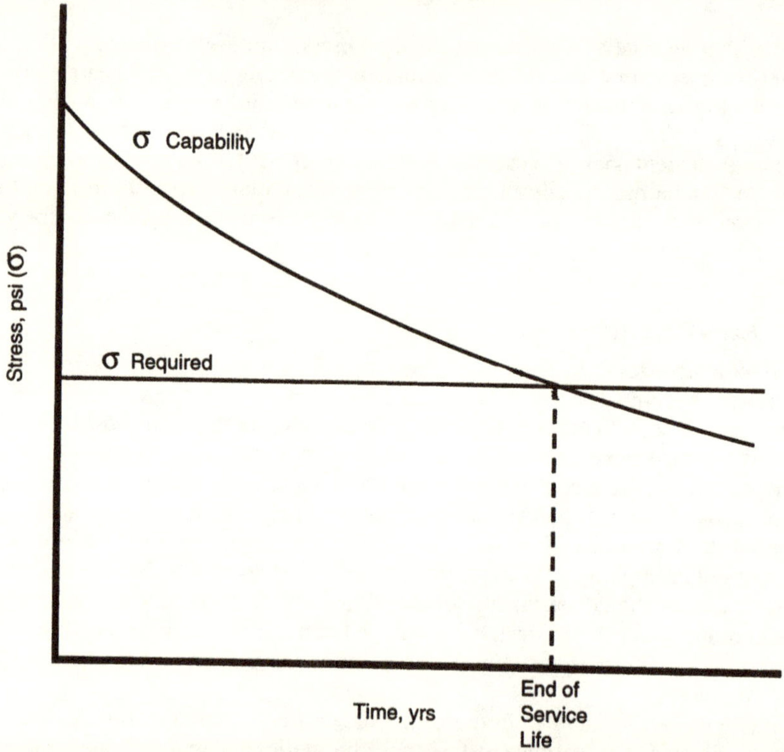

Fig. 10 Deterministic service life prediction.

grain cracking at the inner bore, then when the stress or strain allowable is equal to or less than the predicted induced value, the service life has ended (Fig. 10).

One limitation of this approach is that it is often difficult to determine how much of the change in properties is caused by aging or by other factors such as raw material variations, sample preparation variations, etc.

2. *Probabilistic Approach*

Procedures have been developed for conducting service life analysis of rocket motors using the probabilistic approach in conjunction with various failure criteria based on maximum stress, strain at maximum stress, or stress linear cumulative damage. In the use of each failure criterion, the induced quantity is compared with the corresponding material capability. The induced value and the material capability are both treated as random variables with associated statistical models in the form of probability density functions. The most used approaches involve the stress–strength interference method or the use of the stress linear cumulative damage theory. Both of these approaches require the determination of the most probable failure mode and a statistical description of the induced value and the material capability.

The first step in using the stress–strength interference method is to determine the most probable failure mode, expressed in terms of a stress functional. The failure criteria can be maximum stress, shear stress, or any stress or strain functional. For the purpose of this discussion, it is assumed that the failure criterion will be the

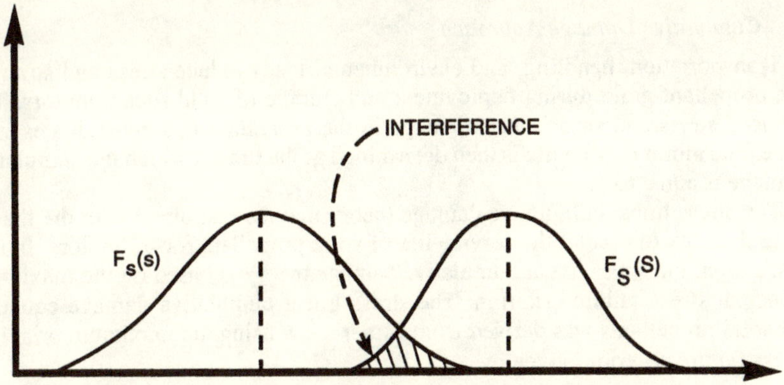

Fig. 11 Stress–strength probability interference curve.

maximum stress. Once the limiting failure mode has been determined, the induced value of maximum stress is then described by a probability density function $f_1(s)$, which accounts for the variations in the material properties and loading conditions. The allowable value of the material capability or strength is then described by a probability density function $f_2(S)$. The probability of failure occurring under the prescribed loading conditions can be calculated by the following relationship:

$$\rho = \int_{s_1}^{s_2} F_1(s)\left(\int_{s}^{s_2} F_2(S)\mathrm{d}S\right)\mathrm{d}s \tag{4}$$

This equation is related to the interference between the induced density function $F_1(s)$ and the allowable density function $F_2(S)$, as illustrated by the darkened region in Fig. 11. When changes occur in the material properties as a result of age and applied loads, the probability density functions $F_1(s)$ and $F_2(S)$ will also change. The effect of age-related material changes normally result in increased probability of motor failure and must be considered in a probabilistic service life analysis.

For cyclic loading, the probability of motor failure is determined for each cycle of loading. The motor reliability is defined as one minus the probability of failure. The probability of motor failure under multiple cycles is then determined as one minus the product of the motor reliability for the total number of cycles. Therefore, the probability of motor failure under multiple cycles is

$$P_f = 1.0 - (1.0 - p_{F1})(1.0 - p_{F2})\cdots(1.0 - p_{Fn}) \tag{5}$$

where p_{Fn} is the probability of motor failure during the nth cycle. The probability of motor failure is increased with each additional cycle of load. The motor service life is determined as the time or cycle at which the calculated probability of failure reaches the allowable probability of failure for the motor. The allowable probability of failure is normally set by the project office controlling the motor manufacture and deployment.

The main requirement of the application of probabilistic service life predictions is that all variables should be statistically described. Adequate statistical descriptions require a large database, and the accuracy of the analysis depends on the accuracy of the statistical distributions used.

3. Cumulative Damage Approach

Transportation, handling, and environmental loads induce stress and strain in the propellant grain during deployment and storage of solid rocket motors. The induced stress and strain cause damage in the propellant that accumulates with time. The motor service life is then determined as the time at which the cumulative damage is equal to 1.

The stress linear cumulative damage theory may be combined with the theory of probability to predict the service life of solid propellant rocket motors. In this discussion, the stress linear cumulative damage theory is based on the maximum principal stress failure criterion. The stress linear cumulative damage equation for solid propellants was derived from Miner's law using the maximum principal stress failure criterion[9] given by

$$D = \int_{i=0}^{t} \left(\frac{\sigma_i}{\sigma_0}\right)^{\beta} \mathrm{d}t \tag{6}$$

This equation incorporates the assumption that the critical stress below which no damage occurs is zero.

The damage parameters σ_0 and β are determined from constant stress vs time-to-failure data. The constant stress vs time-to-failure data may be obtained from either constant load or constant strain rate tests. A plot of the log of the constant stress σ vs the log of the time to failure ($\log t/a_T$) constructed based on the corrected stress is presented in Fig. 12. The parameter β is determined as the negative reciprocal of the slope of the straight line. The parameter σ_0 is the intercept of $\log \sigma$ at a reduced time of $\log t/a_T = 0$.

Linear viscoelastic analyses are used to determine the stress and strain histories at deployment conditions. The stress linear cumulative damage theory is then used

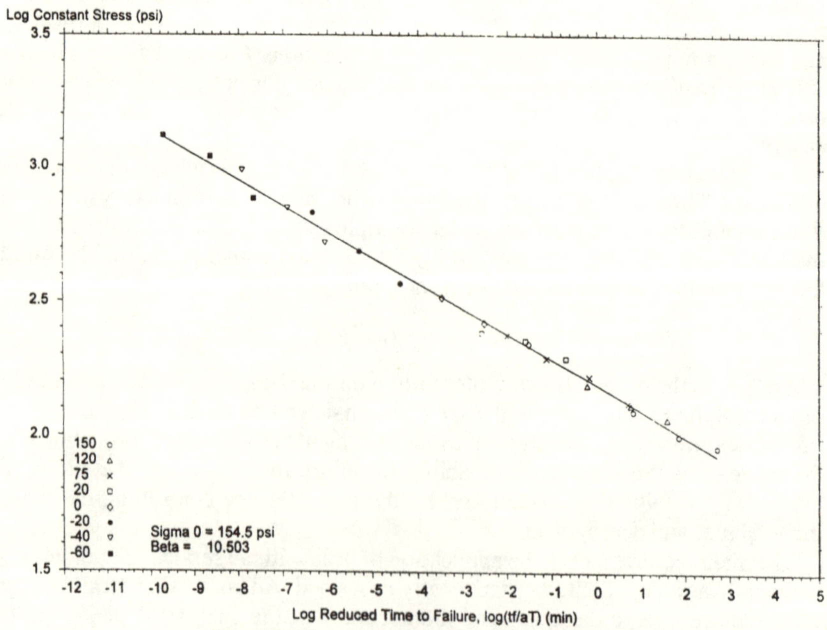

Fig. 12 Constant stress vs reduced time.

to predict the cumulative damage for one year at deployment conditions (damage index). The damage index is used with material property statistical distributions in a Monte Carlo analysis to determine the distribution of the yearly damage index as a result of the material property variations. The damage index at the desired reliability level is modified to include propellant aging. The predicted service life is then determined as the number of years required for the cumulative damage to reach the critical value.

The reliability of the propellant grain to survive a given number of years is determined by first modifying the damage index to include propellant aging. The cumulative damage at the desired number of years is then used with the material property distribution models in a Monte Carlo analysis to determine the distribution of damage expected as a result of the material property variations. The motor reliability is then defined as the probability that the cumulative damage is less than the critical value.

D. Statistical Load Models

Service life prediction that incorporates a probabilistic approach requires a statistical description of the critical loading function. For tactical motors, this appears to be the long-term cyclic temperature environment to which the motors are subjected during storage and field deployment. A commonly used method to describe the daily and seasonal temperature variations on an hourly basis is a Fourier series with six or more terms:

$$T(t) = \frac{1}{2}\left[\bar{T}_D + \sum_{i=1}^{n} A_i \cos\left(\frac{it_i}{P_i - \alpha_i}\right)\right] \sin\left(\frac{t+17}{12\pi}\right) + \bar{T}_Y + \sum_{i=1}^{n} B_i \cos\left(\frac{it_i}{P_i - \beta_i}\right) \tag{7}$$

The drawback to the use of the Fourier series model is that the temperature extremes are eliminated by the averaging process of Fourier analysis. Use of actual hourly temperature records for the average year at a given deployment location would be preferable; however, this requires large blocks of storage in computer memory and is not always feasible.

The following model has been developed for igloo storage[10] since the dai.· yearly temperature variations are not wide:

$$T = 64 + 20\cos\left[\left(\frac{t}{4380} - 1\right)\pi\right] \tag{8}$$

As stated, hourly temperature files are not always feasible. Mathematical models are useful for generating temperature profiles for various locations to be used in calculating stresses induced in the propellant grain during thermal storage and deployment.

E. Reliability Alloca..on

To determine the end of service life, the necessary reliability of the missile system must be determined. One method of determining the missile system reliability is to determine or to assign reliabilities for each of the critical components, i.e.,

Table 2 Rocket reliability allocations

System element	Reliability allocation
Rocket motor	0.985
Warhead structure	0.998
Explosive core assembly	0.993
Electrical	0.998
Fuse	0.995
Fin assembly	0.995
Fin release system	0.995

case, grain, bondline, nozzle, electronics, etc., and then to multiply the reliabilities to give the total system reliability, as shown in the example in Table 2.

The total reliability is determined by multiplying the individual reliabilities of each of the components as follows:

$$R_{\text{Total}} = R_{rm} R_{whs} R_{eca} R_e R_f R_{fa} R_{frs}$$

In the example shown in Table 2 the minimum reliability for the rocket design would be 0.959, using the preceding equation. This method assumes that each component is independent of the others but that if one component fails, the entire system will fail.

IV. Summary

In tactical solid rocket motors in which the propellant grain, liner, insulation, and case are chemically bonded, the propellant grain becomes a load-bearing structure. As a result, the structural integrity of the propellant grain and the bond systems becomes critical to the successful operation of the motor. In order to conduct a structural analysis of a solid rocket motor propellant grain, a definition of the loads the motor will experience during its life cycle, the material properties of the components, the failure criterion, and the required factor of safety are required. Results of structural analyses on unaged components should be used to design surveillance programs to monitor changes in the properties of the components and provide the necessary data to conduct structural analysis on the aged motor and predict service life. The structural analysis conducted with the aged properties can indicate the changes in the margin of safety as a function of time. Real-time and predictive assessments using appropriate failure criterion can be used to predict the service life of the rocket motor. Generally, application of predictive methodologies using real-time assessment data is most effective in service life prediction.

References

[1]Anon., *JANNAF Solid Propellant Structural Integrity Handbook*, Chemical Propulsion Information Agency, Columbia, MD, CPIA Publication 230, Sept. 1972.

[2]Anon., *Predictive Techniques for Failure Mechanisms in Solid Rocket Motors*, Vols. 1–6, Chemical Systems Division, United Technologies, San Diego, CA, Air Force Rocket Propulsion Laboratory-TR-79-87, Jan. 1980.

[3]Heller, R. A., and Singh, M. P., "Random Probability Techniques for Rocket Motor Service Life Prediction, Part I: Tansportation Stresses in Rocket Motors," Final Rept. VPR-E-81.25, Virginia Polytechnic Institute, Blacksburg, VA, Contract DAAK-40-79-C-0231, Sept. 1981.

[4]Anon., *Solid Propellant Mechanical Behavior Manual,* Chemical Propulsion Information Agency, CPIA Pub. 21.

[5]Nelson, J. M., "Influence of Pressure upon the Mechanical Properties of Solid Propellant," *JANNAF Structures and Mechanical Behavior Subcommittee Proceedings,* Chemical Propulsion Information Agency, Columbia, MD, CPIA Pub. 388, Nov. 1983, pp. 197–215.

[6]Kapur, K. C., and Lamberson, L. R., *Reliability in Engineering Design,* Wiley, New York, 1977.

[7]Heller, R. A., Thanjitham, S., and Janajreh, I., "The Safety Index Approach to Storage Life Calculations for Rocket Motors," U.S. Army Missile Command, Redstone Arsenal, AL, TR CR-RD-PR-91-1, Sept. 1991.

[8]Anon., *Tools Required for a Meaningful Service Life Prediction,* Chemical Propulsion Information Agency, Columbia, MD, CPIA Pub. 506, March 1989.

[9]Bills, K. W., Sampson, R. C., Steele, R. D., Peterson, F. E., Briar, H. P., and Plant, R. W., "Solid Propellant Cumulative Damage Program," Aerojet General Corp., Sacramento, CA, Final Rept. Air Force Rocket Propulsion Laboratory-TR-68-131, Oct. 1968.

[10]Martin, D. L., Jr., "Service Life Prediction of the MLRS Propellant Grain," U.S. Army Missile Command, Redstone Arsenal, AL, TR RK-81-1, Oct. 1981.

Chapter 6

Nozzle Design

B. H. Prescott*
Thiokol Corporation
and
Michael Macocha†
Morton International

I. Overview

MOST tactical solid rocket motors are cost-performance driven requiring maximum performance from the nozzle and other components within very restrictive geometric envelopes and at the lowest costs possible.

Maximizing nozzle performance requires optimizing motor ballistics (i.e., pressures and burn times), exit cone thrust, and weight to meet mission requirements for the entire missile system. Fulfilling mission requirements may require a simple fixed nozzle configuration or multiple (2–4) highly complex movable nozzles employing noneroding throats and thrust vector control (TVC) concepts. In addition to providing motor ballistic and mission performance capabilities, the nozzle design must also meet environmental and system life requirements with a highly reliable design.

Because other monographs have been written discussing solid rocket motor nozzles in general,[1] this portion of Chapter 6 will address only those design issues unique to tactical nozzles, specifically, basic nozzle function, key design features and operating conditions, nozzle types, design considerations, materials, TVC systems, and design process. Examples of typical production or development systems will be incorporated wherever possible.

A. Basic Nozzle Function

The basic function of the nozzle is to 1) direct subsonic motor gases, 2) provide a ballistic throat, and 3) provide additional motor thrust. The nozzle throat diameter (i.e., area) inversely affects motor operating pressures as noted by the

---

*Supervisor, Nozzle/TVC Section, Structures Design/Analysis Organization, Science and Engineering.

†Project Manager, Chrysler Programs; formerly Thiokol Corporation, Utah Tactical Division.

simplified steady-state mass balance equation involving the propellant physical grain geometry and burning characteristics

$$P_c = \left[\frac{\rho_p C^* A_b r_b}{A_t P^n}\right]^{(1/1-n)}$$

where

P = chamber pressure
ρ_p = propellant density
A_b = propellant burn surface area
r_b = propellant burn rate
A_t = nozzle throat area, $\pi D_T^2/4$
n = propellant burn rate exponent
C^* = characteristic velocity
g_c = acceleration due to gravity

Nozzle throat area as a function of time is related to the initial diameter, the material selected, and its erosion rate with burn time. The minimum throat area usually occurs at motor ignition. Maximum chamber pressure (Pmax) may occur after motor ignition due to propellant mass flow rate (i.e., burn rate) and burn surface area. It directly affects the design of critical nozzle structural components and secondarily effects erosion-char rates of ablatives and insulators.

The nozzle exit diameter is a function of overall motor performance as noted in the overall theoretical thrust coefficient equation,

$$c_F = \left\{\left(\frac{2\gamma^2}{\gamma-1}\right)\left(\frac{2}{\gamma+1}\right)^{(\gamma+1)/(\gamma-1)}\left[1-\left(\frac{P_e}{P_c}\right)^{\frac{\gamma-1}{\gamma}}\right]\right\}^{\frac{1}{2}} + \varepsilon\left(\frac{P_e - P_a}{P_c}\right)$$

where

γ = ratio of specific heats
P_e = nozzle exit pressure
P_c = chamber pressure
P_a = atmospheric pressure
ε = nozzle expansion ratio, exit area (A_e)/throat area (A_t)

Thus, one can see a relationship among the nozzle throat diameter, throat material and erosion rate, and the exit diameter and its effect on expansion ratio and/or thrust.

B. Design Features and Operation Conditions

A nozzle consists of three distinct regions: 1) the inlet or entrance where subsonic motor gases are directed into the throat, 2) the throat where choked flow maintains motor chamber pressure, and 3) the exit cone where supersonic gases are efficiently expanded and controlled to obtain additional thrust. Typical external and submerged nozzle schematics are shown in Fig. 1.

Operating temperatures, pressures, and gas velocities (Mach numbers) vary widely from region to region, as shown in Fig. 2. Pressures equivalent to those in the motor chamber can be as high as 3500 psi in the inlet region decreasing to

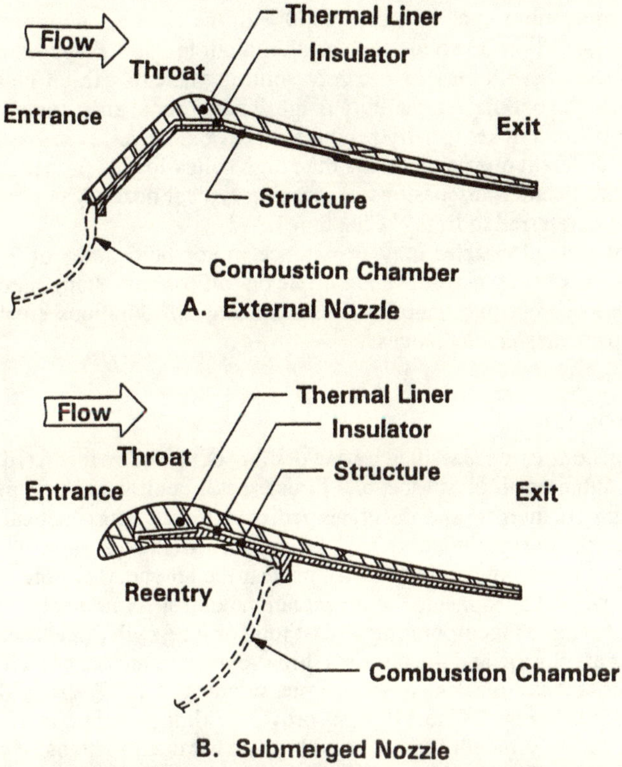

Fig. 1 External and submerged nozzle regions.

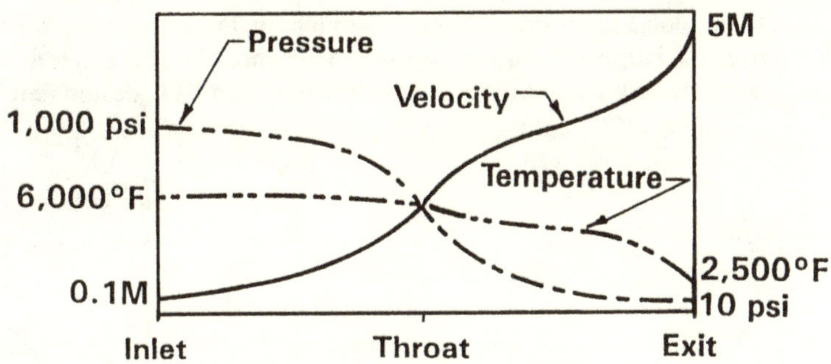

Fig. 2 Typical nozzle conditions.

ambient (i.e., typically 10–12 psi) at the exit plane. Gas flow velocities range from nearly stagnant Mach numbers less than 0.1 to Mach 1.0 (i.e., sonic) at the throat to supersonic Mach numbers greater than 5 at the exit plane.

Gas temperatures can vary from 6000°F in the chamber with propellants employing high (18–21%) solids such as aluminum to 2500°F approximately at the exit. Propellants with very low to zero solid ingredients exhibit higher gas temperatures in the 6500°F temperature range. These propellants, commonly referred to as minimum smoke, usually produce more "oxidative" exhaust environments requiring different nozzle materials than high solids-loaded propellants.

For more detailed discussions concerning typical nozzle gas flow conditions, the reader is referred to Ref. 2, Chapters 2.3–4.

Typical tactical nozzles may experience motor burn times of 2–200 s, dual operating pressure levels, or even multiple on–off (or start–stop) operating pulses. High-pressure, multiple on–off pulses with long off durations create additional unique nozzle design challenges.

C. Nozzle Types

Tactical nozzles are classified by one or more of the following: fixed comprising, external, submerged, blast tube, and thrust vector control and/or movable (TVC) comprising submerged and desubmerged or external. Most tactical nozzles are fixed, external designs with no TVC requirements. Tactical missiles often incorporate external aerodynamic control fins near the aft end, therefore, requiring the nozzle to be packaged inside the fin actuation control components. This envelope restriction requires incorporating a blast tube or long cylindrical section between the motor aft closure and the nozzle's throat-exit cone region, as shown in Fig. 3.

Smaller tactical nozzles are sometimes submerged 100% inside the motor aft end, as shown in Fig. 4. These designs provide additional propellant loading in the motor in severely length- and aft-envelope-constrained systems with the nozzle and exit cone packaged within the motor diameter and length. Most tactical motors operate at low altitudes resulting in low expansion ratio (approximately 10 : 1) exit cones designs.

Some larger tactical motors greater than 10 in. in diameter utilize partially submerged nozzles where the nozzle inlet, throat, and a portion of the exit cone are inside the motor aft end, as shown earlier in Fig. 1B. These systems have been optimized to provide extra propellant in a given length, but submerging the nozzle does partially restrict propellant loading in the aft end. Nozzles can augment or replace external aerodynamic fin steering with a movable TVC design employing

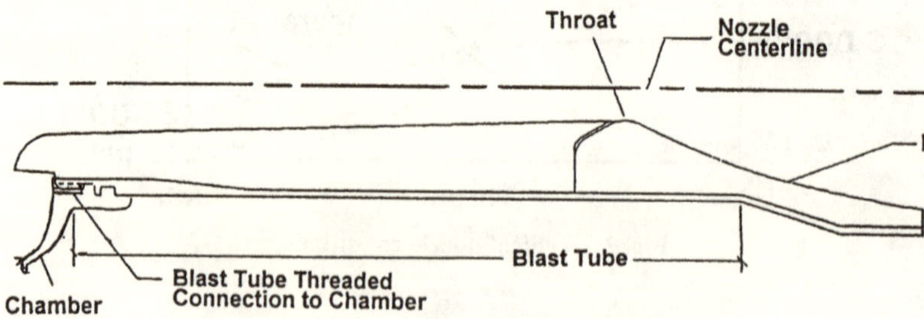

Fig. 3 Fixed, external nozzle with blast tube.

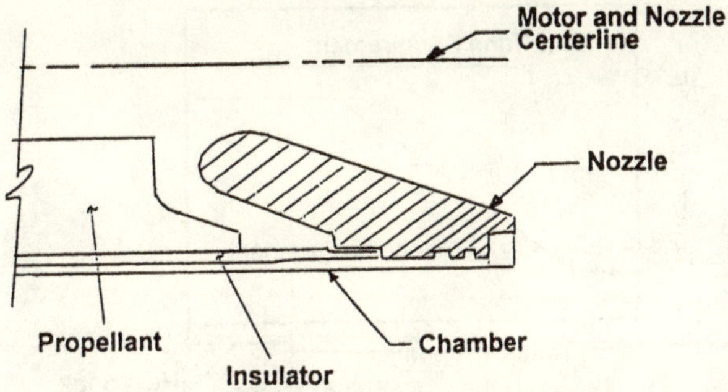

Fig. 4 Fully submerged nozzle design.

either fixed or movable nozzles. A more detailed discussion of TVC systems is presented later in Sec. V.

II. Design Requirements

Many tactical nozzle design requirements are direct systems-related items defined by systems contractors in interface control documents or motor specifications. Others are derived from motor and performance optimizations conducted to meet overall mission requirements as summarized in Table 1.

A. Systems Requirements

Typical systems requirements provided to the nozzle designer are discussed in the following paragraphs.

Table 1 Tactical nozzle design requirements

Explicit requirements (customer specification)	Derived Requirements (from motor design process)
Environmental	Ballistics/performance
Temperature	Pressure
Shock and vibration	Thrust
Humidity, etc	Action time
	Throat diameter
Interfaces	Expansion ratio
Functional	
Mechanical envelope	Propellant
Loads	Gas temperature
	Erosivity
Design	Combustion products
Safety factors	Slag accumulation
Materials	
	TVA constraints
Performance	Packaging/envelope
Thrust versus time	Actuation loads
Acceleration	Thermal environment
Thrust alignment	
TVA requirements	
Vector angle	
Slew rate	
Frequency response	

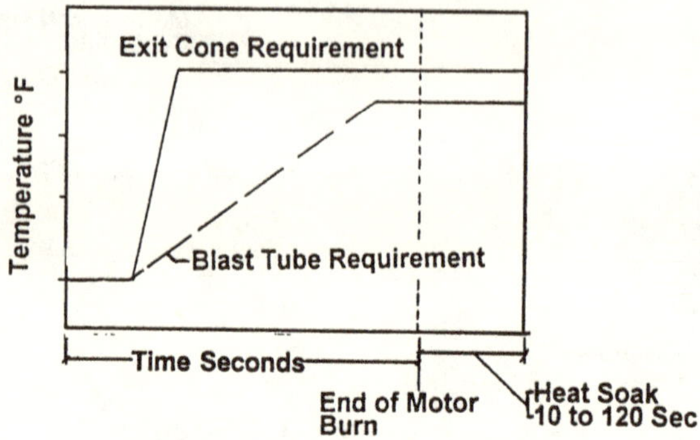

Fig. 5 Typical blast tube, exit cone temperature requirements.

1. Envelope and Interface Constraints

The nozzle external envelope is determined from the missile's aerodynamic profile constraints, often including a restricted region to package a missile's external aerodynamic fin actuation control system, as noted earlier. To fit within the control system envelope, a long cylindrical blast tube concept is used. The nozzle designer must fit not only the structural components but also internal insulators inside of the structure to protect it from the high-temperature motor gases as well. The inner diameter of the blast tube insulator must be slightly larger than the throat diameter to prevent nearly sonic flow and subsequent high erosion and performance losses. Blast tubes have been static-tested with Mach numbers as high as 0.8. In addition to these two design challenges, additional structural outer surface temperatures may be restricted to prevent radiant heating of the fin control actuation components. A typical blast tube–exit cone outer diameter temperature requirements profile is shown in Fig. 5.

Occasionally the blast tube–exit cone maximum temperature requirement extends beyond motor burnout as indicated in Fig. 5. This is an additional design challenge since now the nozzle designer must not only provide enough insulation on the inside diameter of the blast tube and exit cone for motor burn conditions, as well as additional insulation to be able to absorb the energy generated in the insulators during motor operation. The postmotor burn requirements are often referred to as heat-soak requirements.

During missile deployed life, access to the fin actuation system requires the nozzle exit cone be removable and replaceable. Such a configuration is shown in Fig. 6. This design illustrates a threaded joint between the blast tube and exit cone structures since this concept requires less radial space than a typical bolted configuration. If the fin actuation system envelope is too restrictive to thread the exit cone onto the aft end of the blast tube, sometimes the threaded interface is moved to the forward end of the blast tube, as illustrated earlier in Fig. 3. This design is not as desirable as the previous concept since it places the threaded joint and o-ring seal in more of a blind area.

2. Thrust Vector Control

Incorporation of TVC onto a system involves another set of operational and nonoperational requirements. The TVC system may either employ a fixed nozzle with movable components, such as jet vanes or tabs to deflect the exhaust gases, or a movable flex bearing or trapped ball (TB) nozzle system.

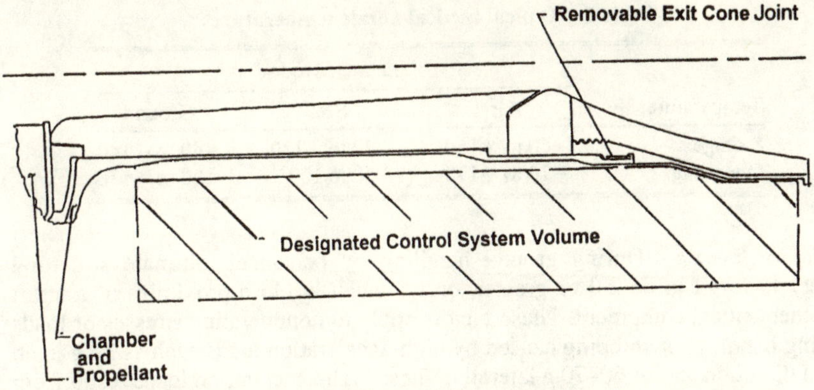

Fig. 6 Blast tube nozzle with removable-replaceable exit cone.

3. *Performance*

The only nozzle-related performance parameter sometimes specified in systems requirements is a maximum nozzle weight or, more often, natural frequency for movable TVC systems. Otherwise, most other nozzle design requirements are determined from optimizing the overall motor to meet system requirements.

4. *Reliability Factor*

Reliability is usually allocated from overall motor requirements and typically results in a factor like 0.999+ for the nozzle subsystem, either fixed or movable. A reliability factor is attributed to most tactical systems operating in close proximity to personnel or equipment, such as launching from the deck of a ship or from mounts on an aircraft.

5. *Nonoperating Requirements*

Commonly referred to as shipping and handling, these requirements do not typically drive nozzle designs but may require interfacing features to accommodate them. They may include requirements such as electrical grounding, safety factors, temperatures, motor sealings and environmental requirements. Each of these is briefly discussed as follows.

Electrical grounding. Requirements are usually limited to those defined in specification MIL-B-5087, Class S, which require conductive components to be able to bleed off static charge buildup during storage or handling. Most nozzle materials or components that generate or collect static electricity are naturally grounded through physical interfaces such as bolted or threaded joints or physical material interfaces, and so forth. Usually maximum resistance between two components is limited to 1 ohm or less. Typically, nozzles will measure less than 0.5 ohms of resistance between components, excluding some movable nozzle TVC concepts, such as flex bearings or trapped ball designs, which usually are grounded via a motor interface by a grounding strap.

Electrical grounding of nozzle and motor components recently has become a much more serious safety concern to prevent, for example, accidental ignition of the motor's solid propellant during handling. This phenomenon has been referred to as electrostatic discharge or ESD.

Table 2 Typical tactical nozzle temperatures

	Launch Modes		
Temperature, °F	Air	Sea	Ground
Storage	−60 to +150	−20 to +130	−40 to +140
Operating	−70 to +170	50 to 100	−30 to +130

Safety factors. During ground handling by personnel, ultimate structural safety factors can be 2.0 or greater, or 1.5 when in close proximity of aircraft or other critical equipment. These factors apply to nonoperating stresses or loads during handling or shipping caused by high-acceleration loads such as 30 g in an axial direction and/or 60–70 g laterally. These high-acceleration loads result from a motor being dropped or from in-flight conditions on an aircraft.

Temperatures. During shipping, handling, or flight preoperational conditions, nozzle components can be exposed to temperatures from −60°F to as high as +170°F approximately as noted in Table 2.

Motor sealing. This is usually required for most tactical motors and generally installed somewhere in the nozzle exit cone region. Three types of motor seals are illustrated in Fig. 7.

The first example is a urethane foam barrier with a thin aluminum foil moisture barrier on the forward face and an environmental sealant coating on the aft face. This seal is bonded into the forward exit cone region. Valves allowing the motor to breathe can be incorporated if required.

The second example is a simple thin PTFE or polypropylene film barrier trapped between two nozzle components during exit cone installation.

The third example is a more complex structural design consisting of a thin stainless-steel cup with a valve located in the center to allow the motor to breathe under certain delta pressure conditions. This seal is bonded to the aft end of the nozzle exit cone structural shell. All seals must be expelled from the nozzle on ignition without contributing to motor pressure pertubations causing any damage to the nozzle, or leaving any debris in the exit cone creating adverse thrust misalignment.

Environmental. In addition to the temperature extremes discussed earlier, tactical systems are exposed to harsh environments during handling, shipping, and preoperational deployment. Service environments may involve humidity effects from 50 to 100% combined with temperatures of 40–100°F often including salt spray.

6. *Operating Requirements*

During motor operation times, the severe internal pressures, temperatures, and gas species of the propellant combustion process usually overrides any other system induced loads or conditions on the nozzle and its components.

Safety factors. For structural components, safety factor requirements of 1.25–1.50 are typically applied to ultimate conditions, whereas factors of 1.0–1.2 are applied to yield parameters. Occasionally, a nozzle closure or adapter is routinely hydroproofed during manufacturing acceptance. Usually a safety factor of 1.0 is applied to thermal requirements. But, as noted earlier, additional insulator thickness may be required to maintain structural component temperatures below acceptable limits after motor burnout and during subsequent heat soak.

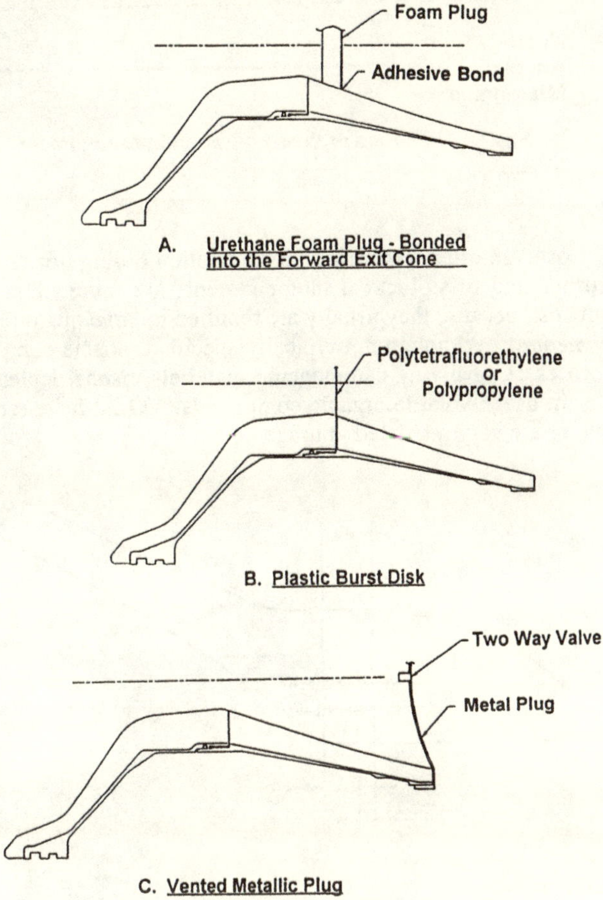

Fig. 7 Examples of nozzle mounted motor seals.

Thrust misalignment. Thrust misalignment is defined as the maximum allowable offset and angular difference between the motor centerline (defined by centerlines of the motor attach features) and the nozzle centerline (defined as the centerline connecting the center of the throat and the exit cone, exit plane), as shown schematically in Fig. 8. Typically maximum allowable offset requirements are less than 0.020 in. at the throat, whereas maximum allowable angular misalignment values typically range from 1.5 to 3.5 mrad (0.085–0.20 deg). These requirements are usually verified by geometric evaluation of the nozzle and loaded motor design drawings.

If more than two or three interfaces are involved in calculating offset and misalignment, a root-sum-square (RSS) approach may be employed that results in significantly smaller effects on dimensional tolerances.[3] An example of this approach is presented in Fig. 9.

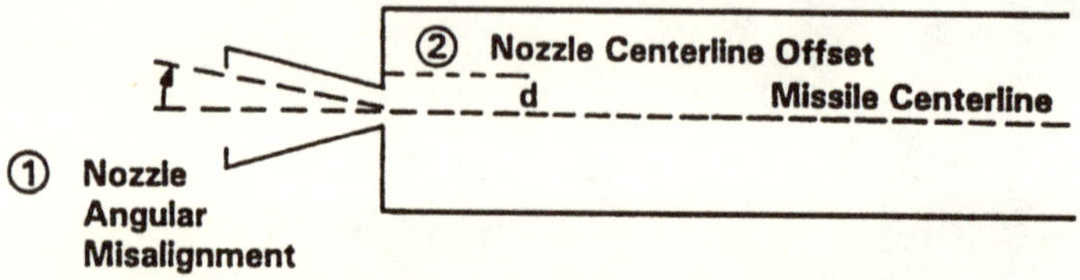

Fig. 8 Definition of thrust offset and misalignment.

7. *Others*

Cost. Costs are often a dominant consideration during nozzle design and/or manufacturing tradeoffs. Tactical motor systems are more often driven to the lowest unit cost, because they usually are required in large quantities.

Environmental. Although not typically specified, nozzle components, materials, or processes involving carcinogenic materials, ozone depleting chemicals (ODCs), or in usage volatile organic compounds (VOCs) have recently become unavailable or are very restricted in usage.

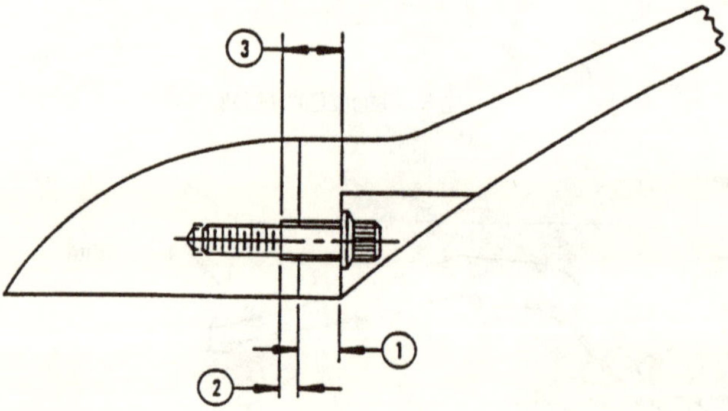

FEATURE DIMENSION

FEATURE	SIZE (IN.)	TOLERANCE (IN.)
1. FLANGE THICKNESS	0.410	±0.010
2. COUNTER-BORE DEPTH	0.150	±0.010
3. BOLT SHANK LENGTH	0.430	±0.010

CLEARANCE CALCULATIONS

I. WORST-ON-WORST (WOW) CALCULATION:

$$(0.410 \pm 0.010) + (0.150 \pm 0.010) - (0.430 \pm 0.010) = 0.130 \pm 0.03 \quad \text{WOW}$$

II. ROOT-SUM-SQUARE (RSS) CALCULATION:

$$(0.410 \pm (0.010)^2) + (0.150 \pm (0.010)^2) - (0.430 \pm (0.010)^2) = 0.130 \pm \sqrt{0.0003} \quad \text{RSS}$$
$$= 0.130 \pm 0.017 \quad \text{RSS}$$

Fig. 9 Example of RSS vs WOW tolerance methods.

Of prime concern since passage of Public Law 102-484 (June 1993), and President George Bush's mandate February 1992, the solid propulsion industry has been fully aware of the phase-out of certain ODCs and VOCs that were involved in tactical nozzle fabrication. Compounds involved are certain chlorofluorocarbons such as Freon®; 1,1,1-trichloroethanes such as methyl chloroform, and trichlorethane (TCA). Carbon tetrachloride also was banned in systems after 1994–1995.

Known carcinogenic compounds used in tactical nozzle materials manufacture such as di-methyl formamide (DMF) will not be allowed on future systems. All asbestos-containing materials have been phased out since the mid-1980s and some slightly radioactive fillers such as Thoria in tungsten alloys will soon be unavailable.*

Producibility. Producibility involves incorporating concurrent engineering, or total quality management (TQM) practices to involve materials suppliers, manufacturing, quality, and prime contractor personnel early in the design phase to define a producible tactical nozzle. Involving the prime system contractor early ensures that the designer is provided with customer requirements not often spelled out in a systems requirements document.

III. Materials

Nozzle materials are separated into structural and ablative/insulator components. Current design practices for tactical nozzle structural components include metals, glass, carbon or graphite reinforced composites, or a simple overwrapped ablative insulator with the same or other material at a different ply angle.

Selection of an insulator or ablative material is determined by the gas pressure, temperature, and the presence or lack of reactive gas species, or solid particles in the exhaust gases. An entire volume by itself would be required to cover all nozzle materials, processing, properties, and design techniques so this section will only briefly address those materials as state of the art (SOTA) for tactical nozzles. Tactical nozzles are generally manufactured in high production rates and large lot quantities utilizing low-cost materials and/or processes.

A brief discussion concerning materials tradeoffs and industry standard quality control follows at the end of the materials usage, advantages/disadvantages, and brief manufacturing description.

A. Structural Materials

Structural components maintain nozzle integrity by supporting the insulators and ablative liners, as well as reacting motor chamber pressure loads and external loads during flight and TVC systems.

Steels are commonly used for structural components because of their relatively low cost, high strength, and modulus. They are used where minimal structural deflections and high load reactions are required, but require corrosion and oxidation protection. High-strength stainless steels, such as 17-4 Ph and 17-7 Ph, are used due to their corrosion resistance, but their costs are higher than carbon steels.

High-strength (75–80 ksi) corrosion resistant aluminum alloys, such as 7075 or 7175, are available that can be used when weight needs to be minimized.

*More details of the aforementioned efforts and updated status can be obtained from the U.S. Environmental Protection Agency's Industrial Toxics Program (33/50 Program).

Table 3 Typical properties of nozzle structural metals (room temperature properties)

Property	Aluminum (7075-T6)	Titanium (6 A1-4V)	4130 Steel Normalized	4130 Steel Heat treated	18% Ni steel (200 class)
Density, gm/cc	2.80	4.47	7.83	7.83	8.00
Ultimate tensile strength, psi in thousands	82	178	90	180	210
Yield strength, 0.2% offset,	72	160	70	179	200
Yield strength/density	25.7	35.8	8.9	22.9	25.0
Modulus of elasticity, psi $\times 10^{-6}$	10.4	16.0	29.0	29.0	27.5
Shear strength, psi in thousands	46	90	55	109	—
Elongation, % in 2 in.	11.0	8.0	10–20	15	9–10
Poisson's ratio	0.33	0.31	0.30	0.30	0.31

Additional envelope is required for the slightly thicker aluminum cross sections than steel counterparts for equivalent stiffness or deflections.

High-strength D6AC or 300M steels are often used for nozzle aft closures or structures subjected to the high chamber pressure environment, whereas lower grades of carbon steels can be used for exit cone or attach structures. Military handbook MIL-STD-5 is often used for metals properties and recommended usages. Fracture toughness should be a prime consideration when selecting stainless or carbon steels for nozzle usage. Typical properties for 7075-T6 aluminum, 6AL-4V titanium, A1S1 4130 carbon steel, and a low-strength 200 series stainless steel are summarized on Table 3. Proper protection against stress or galvanic corrosion should also be considered when combining different metals or alloys in structural components.

Titanium alloys are not used very often for tactical nozzles because of cost and availability. Composites are not generally used either because of their higher costs, as well as low interlaminer shear capabilities in high bending or rotational operating modes. Some weight critical designs, however, have used glass epoxy or silica or glass phenolic structures for the exit cones.

Most metallic structures are machined plate, tubing, or bar stock. Forgings or rolled ring forgings offer higher strengths and specifically oriented properties capabilities. Castings are sometimes used on high production rate components where strength and grain orientations are not design drivers. However, castings require more expensive process controls or nondestructive evaluation such as x rays.

B. Insulators/Ablatives

Nozzle insulator or ablative materials generally consist of glass, silica, carbon, or graphite fabric reinforced phenolics or fiber reinforced elastomers. Often special erosion-resistant materials such as polycrystalline graphites (bulk graphite), pyrolytic graphite, tungsten (pure or reinforced) or carbon–carbons are used in the throat region.

Carbon and graphite phenolics are made from wood pulp extracts, as shown in Fig. 10. Rayon fibers are woven, carbonized, or graphitized and then impregnated

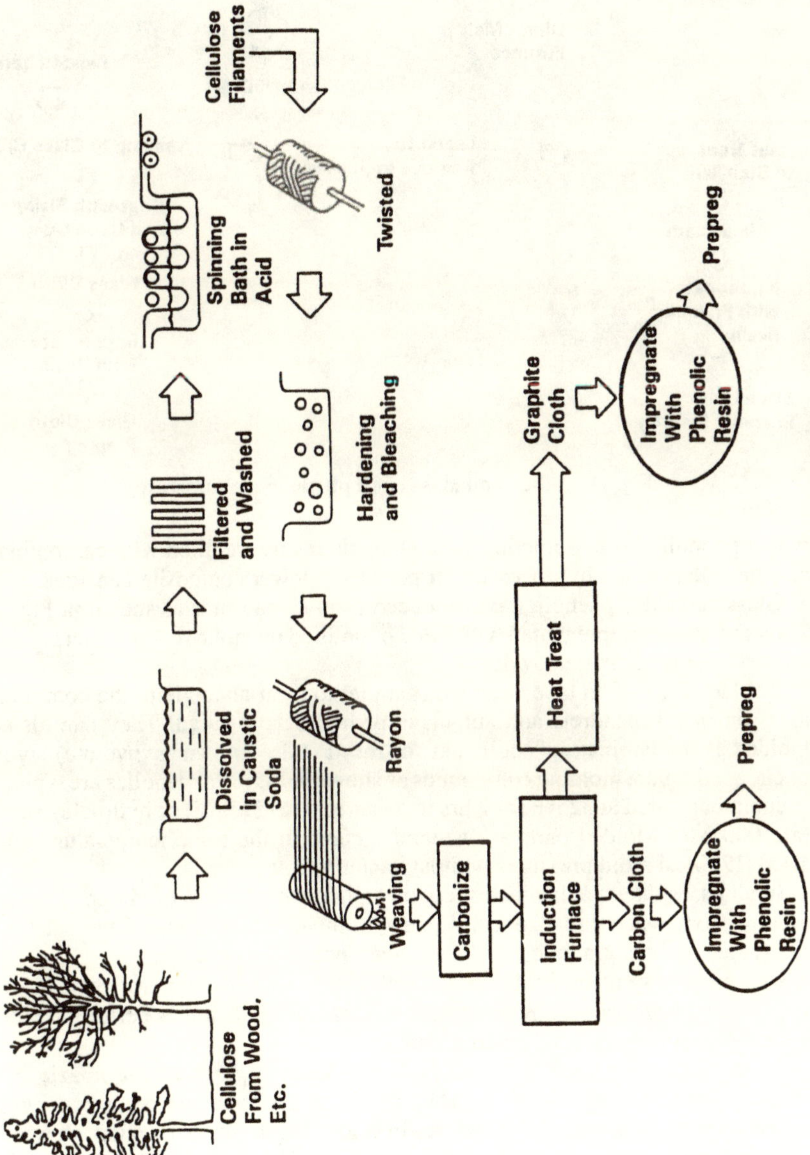

Fig. 10 Carbon and graphite cloth phenolic manufacturing.

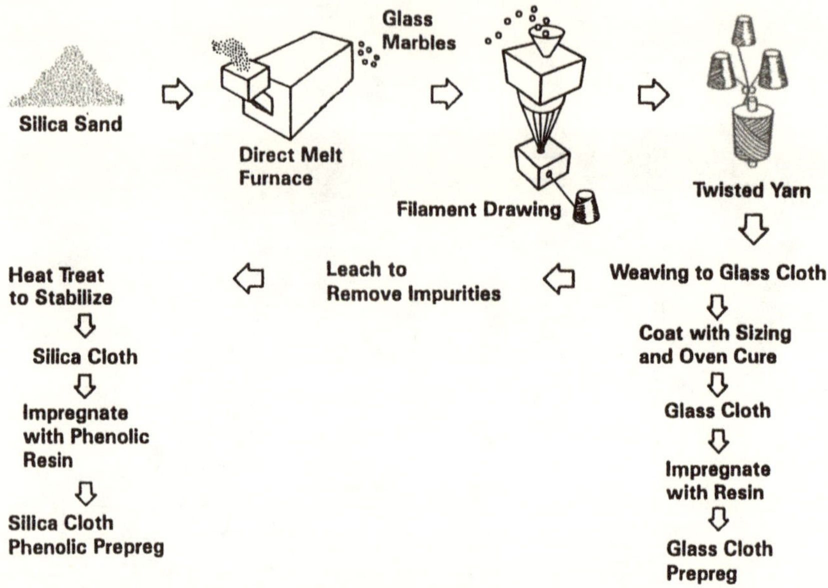

Fig. 11 Silica and glass cloth phenolic manufacture.

with a phenolic resin. Sometimes the phenolic resins are filled with carbon black or other fillers to aid in processing or producing lower composite densities.

Glass and silica phenolics are produced from silica sands as shown in Fig. 11. They can also be impregnated with filled or unfilled phenolic resins. Fillers usually consist of silica dioxide powders.

Insulators are often tape wrapped (parallel to or at an angle to the component axis) on metal mandrels and autoclave or hydroclave cured. They can also be molded as flat laminates, "coolie hat" or rosette (also known as involute) layups, or chopped square molding compounds as shown in Fig. 12. Phenolics are typically vacuum bag cured at 325°F for 2 hrs in an autoclave (225 psi) or hydroclave cured (at 1000 psi). Molded parts are generally cured at the same temperature under 1000–1250 psi mold pressures without vacuum requirements.

In all cases the tape, laminates, coolie hat/rosette patterns, or chopped squares are cut from wider broadgoods. Table 4 compares properties of typical fabric reinforced carbon, graphite, silica, and glass phenolic composites.

Carbon or graphite phenolics are usually used for maximum erosion resistance in aluminized propellant environments, whereas silica phenolics perform better as ablatives in nonaluminized environments.

Elastomers are sometimes used in low-velocity regions of the nozzle inlet, such as fixed housing insulators. They are fairly low cost and readily processed by casting directly onto the structure. In a low Mach number region, they offer sufficient insulation and excellent elongation capabilities allowing them to deflect along with the nozzle structure without cracking or becoming unbonded, even at low temperature. Formulations that have been used include nitrile butadiene rubber (NBR), carbon or silica filled EPDM, and silicone rubber.

Other applications for elastomers involve their use as adhesives/sealants. Room temperature vulcanized (RTV) silicone elastomers have been used to bond the phenolic closure insulation components to the shell. RTV's pliability allows large

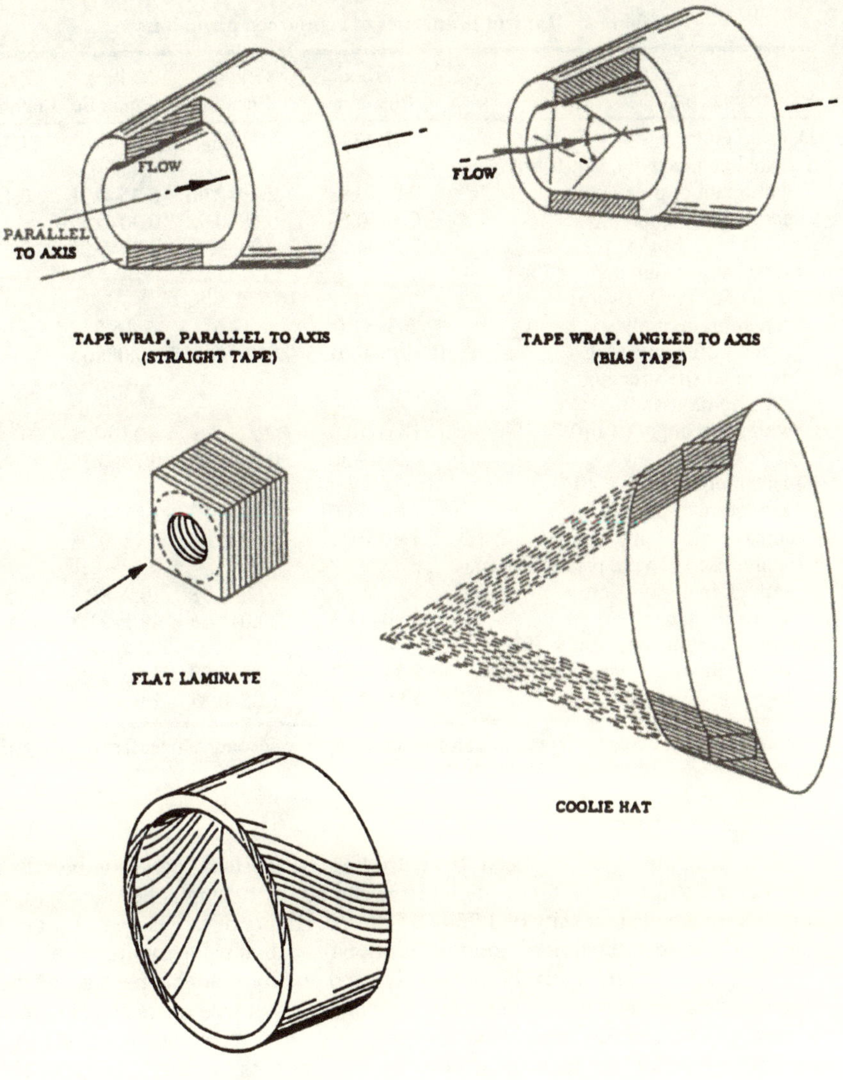

Fig. 12 Nozzle ablative/insulator component manufacturing methods.

shell deflections without opening up a gas path or inducing large stresses in the adjoining ablative insulation components.

C. Throat Materials

Sometimes tactical motors require maximum delivered performance, hence lower eroding (or noneroding) throat materials than silica or carbon phenolics described earlier. These materials are typically polycrystalline or bulk graphites, pyrolytic graphites, three or four directionally reinforced carbon–carbons, or tungsten, as will be described. Table 5 compares properties of typical throat materials and Fig. 13 shows their manufacturing methods.

Table 4 Typical properties of reinforced phenolics[a]

Property	Carbon Phenolic	Graphite Phenolic	Silica Phenolic	Glass Phenolic
Density, gm/cc	1.43	1.45	1.75	1.94
Thermal conductivity, Btu-ft/h-ft^2-°F				
with grain or ply (warp)	0.83–0.93	2.29–2.90	0.35–0.38	0.16
across grain or ply	0.48–0.58	0.69–0.92	0.30–0.32	
Specific heat, Btu/lb °F	0.20–0.36	0.24–0.39	0.24–0.30	0.22
Thermal expansion, in./in. °F × 10^6				
with grain or ply (warp)	3.8	5.3	3.9	—
across grain or ply	5.3–31.0	17.6	16.5	
Thermal diffusivity, ft^2/h	0.0108–0.0125	0.0126–0.018	0.0080	—
Ultimate tensile strength, psi in thousands				
with grain or ply (warp)	18.0–10.5	10.5–7.6	12.0–7.6	60
across grain or ply	0.90–0.30	0.74–0.33	0.72–0.39	
Tensile modulus, psi × 10^{-6}				
with grain or ply (warp)	2.64–1.60	1.57–1.23	2.62–1.99	4.6
across grain or ply	1.80-0.05	0.44–0.08	0.48–0.06	
Compressive strength, psi in thousands				
with grain or ply (warp)	36.1–13.5	13.0–3.98	16.2–8.13	50.6
across grain or ply	62.9–42.5	33.0–21.6	49.1–21.3	
Compressive modulus, psi × 10^6				
with grain or ply (warp)	2.34–1.73	1.50–0.89	3.50–1.95	3.7
across grain or ply	1.85–0.75	1.05–0.37	2.07–0.78	

[a]Values shown are room temperature properties except the second value refers to properties at 750°F.

1. *Bulk Graphites*

Bulk graphites are produced by extruding or isostatic pressing blends of petroleum coke and coal tar or pitch resins and finally baking at elevated temperatures with a typical density of 1.75–1.85 gm/cc. Higher densities up to 1.9 gm/cc can be achieved by chemical vapor infiltration of carbon using methane gas.

Bulk graphites are relatively inexpensive and readily available because of their large process lots and widespread commerical use as furnace liners and electrodes in steel production furnaces. However, they are thermal shock and notch sensitive and, therefore, do not lend themselves to unusual cross-sectional shapes. They have been replaced recently with much tougher multidirectionally (three or four) reinforced carbon–carbon described later.

2. *Pyrolytic Graphites*

Pyrolytic graphites (PG) are produced by depositing carbonaceous material from a methane gas source under vacuum and a temperature of approximately 2000°C. This process produces growth cones with their apex toward the initial deposition surface. Pyrolytic graphites are very erosion resistant and must be used as thin washers, 0.38–0.50 in. thick in the throat, with the growth cones or strata oriented parallel to the gas flow. Pyrolytic graphite also tends to be very laminate oriented with the laminates formed perpendicular to the growth cone direction. Pyrolytic graphite throats are also much more thermal shock sensitive compared to carbon–carbon composites and usually employ thermal expansion spacer materials

Table 5 Typical properties of throat insert materials[a]

Property	Tungsten: Forged or extruded	Tungsten: Pressed and sintered[b]	Pyrolytic Graphite	Polycrystalline Graphite (bulk graphite)
Density, gm/c	19.0	17.4	2.2	1.75
Sublimation or melting temp., °F	6170	6170	6600	6600
Thermal conductivity, Btu-ft/h-ft^2-°F	96–60	54–33		
with grain or ply (warp)			200–40	70–16
across grain or ply			1.2–0.03	40–15
Specific heat, Btu/lb °F	0.033–0.047	0.033–0.047	0.22–0.5	0.25–0.6
Thermal Expansion, in./in. °F × 10^6	2.5	2.3		
with grain or ply (warp)			1.33	1.50
across grain or ply			20.0	2.2
Ultimate tensile strength, psi in thousands	160–10	55–10		
with grain or ply (warp)			10–15	4.5–7.0
across grain or ply			0.4	3.0–5.0
Tensile modulus, psi in thousands	59	40	4–2.5	
with grain or ply (warp)				0.75–0.80
across grain or ply				0.90–1.25
Compressive strength, psi in thousands				
with grain or ply (warp)			10	9.0–11.0
across grain or ply			45	10.0–12.0
Compressive modulus, psi × 10^{-6}				
with grain or ply (warp)				0.9–1.1
across grain or ply				0.8–1.0
Thermal shock parameter (Ks/aE) (Ref. 20)	104	32	564	159

[a]Values shown are room temperature properties except the second value which refers to properties at 4000°F.
[b]Properties for 90% dense unalloyed pressed and sintered tungsten.

to allow for growth in the axial (with growth cone) and radial (with laminate or strata) directions.

3. *Carbon–Carbons*

Carbon–carbons (C–Cs) are very thermal shock resistant, with densities the same or higher than bulk graphites and can be made into very stable geometries. Carbon–carbons typically erode slightly higher than graphites and can cost more because of materials and processing costs.

Carbon–carbon preforms can be reinforced by weaving carbon or graphite fibers in three or more directions or by using two-dimensionally woven carbon or graphite fabrics as flat laminates or three dimensional involute patterns, or by simply molding shorter chopped fibers. These preforms are subsequently impregnated, carbonized, and graphitized during the densification process using resin matrix precursors. Some carbon–carbons have been used as freestanding components that can operate above 2000°C.

4. *Tungsten*

Tungsten melts at approximately 3760°C and is relatively insensitive to oxidative gases, thereby providing essentially a noneroding throat material. Zero

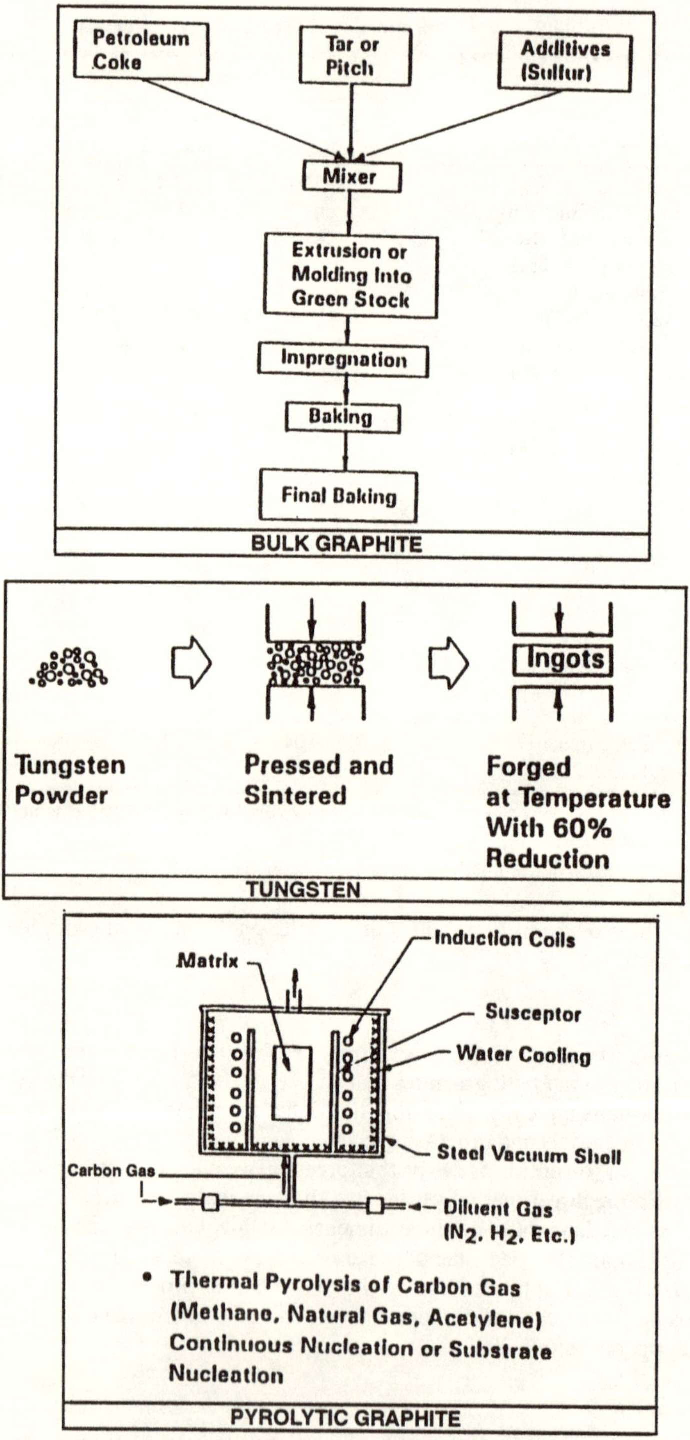

Fig. 13 Typical throat materials manufacture.

erosion maintains a more constant motor operating pressure and a higher average nozzle expansion ratio, thereby providing maximum motor performance. A typical tungsten throat may even decrease throat area with burntime, as much as 7% or up to 0.070–0.090 in. on a diameter.

Tungsten is relatively expensive due to fabrication costs and is brittle at temperatures between room temperature and 800°C approximately, and therefore is susceptible to thermal shock cracking. Component strength can be increased by starting with forgings with aligned grain flow for improved strength in the critical directions. Incorporation of an additive, such as thoria, improves thermal shock resistance by lowering the ductile-to-brittle transition temperature and increasing its strain capability at elevated temperatures. Because of its carcinogenic nature, alternatives to thoria, such as carbides, are being evaluated.

Because of tungsten's increased ductility above 1000°F, the insert must be thick enough in cross-section that sufficient material remains supported by other nozzle components to preclude the insert from being extruded out of the nozzle. The larger the cross-section, however, the more susceptible the tungsten is to thermal shock cracking. Recent insert designs have incorporated a series of radial grooves to essentially precrack the insert and reduce the internal axial stresses to allowable levels. Providing support for the hot tungsten insert and sufficient insulation to maintain required nozzle structures temperatures (approximately 200°F) complicates the design with additional costs. Tungsten is seldom justified in any large motor system. However, the loss of performance associated with even small changes to the throat diameter in a tactical sized motor sometimes justifies the added costs.

D. Adhesives

Adhesives are used quite often to assemble the various components to themselves and the structural shell. Several families of strength structural adhesives have been utilized in rocket motor nozzles, some of which are epoxies, silicones, polysulfides, and phenolic. The adhesives must have sufficient capability over the full operating temperature range, usually from −65°F to +250°F. Restrictions since the mid-1980s prohibit the use of asbestos, hence the designation NA (nonasbestos) noted on some vendor designations.

Although it is not desirable to rely on adhesives for primary structural loads, they are often used in this capacity in tactical nozzles. High-strength, high-elongation EC-3448 adhesive (from 3M Corp.) and a suitable primer have been used as the primary bond for nozzle exit cones.

Cohesive strength between the adhesive and the substrate is a function of proper surface preparation such as cleaning and wettability of the adhesive. Primers compatible with the adhesive provide surface wetting and can increase bond strengths significantly.

E. Trade Studies

As discussed earlier the nozzle designer must decide on the appropriate materials that will satisfy all design requirements and provide the most efficient and cost-effective solution to the design problem. Relative erosion resistance and insulative capabilities of various materials for both aluminized and nonaluminized propellants are presented in Table 6. The relative insulating capability is based on

Table 6 Relative erosion resistance and insulative capabilities of nozzle materials

Erosion resistance, erosion rates, mil/s	
Aluminized propellants	non-aluminized propellants
20 Silica phenolic	27 Carbon phenolic
12 Carbon phenolic	13 Carbon/carbon
5 Carbon/carbon	10 Bulk graphite
4 Bulk graphite	6 Silica phenolic
2 Pyrolytic graphite	4 Pyrolytic graphite
0 Tungsten	2 Tungsten

Relative insulative capabilities

Material	Thermal diffusivity, $k/\rho c_p$, (ft^2/s)
Low-density carbon phenolic	3.62×10^{-6}
Silica phenolic	3.72×10^{-6}
Carbon phenolic	1.23×10^{-5}
Pyrolytic graphite (across ply)	1.34×10^{-5}
Pyrolytic graphite (with ply)	1.79×10^{-3}
Tungsten	5.45×10^{-5}
Carbon/carbon	2.62×10^{-4}
Bulk graphite	1.30×10^{-3}

thermal diffusivity, which combines the effects of a material's thermal conductivity, density, and specific heat.

To compare the relative costs and nozzle cost impacts of different materials is difficult because of changing costs with time. As an example, raw material costs of carbon or graphite phenolics shown in Fig. 14 have increased from the \$50 per pound shown in 1988 to approximately \$75–\$90 per pound in 1994. In addition, the cost is highly dependent on the size, shape, desired strength, consistency, and orientation of the finished component. For example, there are many grades of bulk graphites, with extremely varied costs. These grades not only vary in density but also in material strength, consistency, and overall quality. Likewise, a tape-wrapped carbon phenolic component can cost twice as much as one that is compression molded using chopped squares. Of course, this results in a more

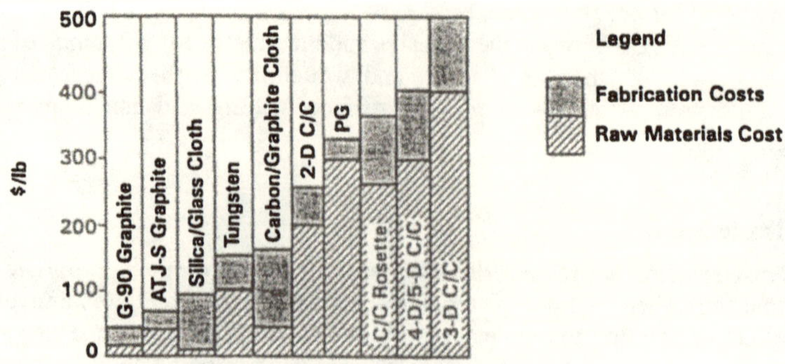

Fig. 14 Relative costs of nozzle materials.

random ply orientation, but choice of grade depends on the desired properties and/or the component performance in a given environment.

F. Quality Control Provisions

Nozzle quality is assured by 1) raw material controls, 2) process controls, 3) nondestructive testing, and 4) nondestructive verification components or of completed nozzle assemblies where possible.

Controlling key raw materials properties will ensure that quality components can be manufactured economically with demonstrated processes. Sample testing of components assures that their performance would be in the range of previous demonstrated or qualified parts that exhibited similar sampled characteristics. Total quality management and concurrent engineering have been demonstrated to produce improved reliability with lower costs on recent tactical nozzle programs. Concurrent engineering involving the customer, designers, vendors, manufacturing and quality personnel should be practiced even during prototype or demonstrated nozzle development. Examples of some nondestructive and destructive tests used on typical tactical nozzle programs for raw material, component, or metals evaluation and acceptance are summarized on Table 7 and discussed as follows:

1. Insulators or Ablative Raw Materials

Insulators or ablative raw materials, such as reinforced phenolics, are usually tested for resin content and resin flow to ensure quality components can be manufactured with established process parameters. Sometimes cured panels are tested

Table 7 Nozzle materials quality control summary

	Ablatives			
Tests	Raw Materials	Components	Throat materials	Metals[a]
Nondestructive				
X ray	NA	X	X	X (forgings)
Alcohol wipe	NA	X	X	NA
Ultrasonics	NA	X	(sometimes)	X (forgings)
Dye penetrant	NA		NA	X
Hardness	X[b]	X	NA	X
Destructive				
Density	X[b]	X	X	X
Volatiles content	X	X	NA	NA
Resin content	X[b]	X	NA	NA
Resin flow	X[b]	X	NA	NA
Acetone extraction	X[b]	X	X	X
Tensile strength	X[b]	X	X	X
Compressive strength	X[b]	X	NA	NA
Interlaminar shear	X[b]	X	NA	NA
Flexural strength	NA	NA	NA	X
Percent elongation	X[b]	X	NA	NA

[a]See Ref. for more requirements.
[b]Sometimes cured panels are tested by vendors.

for density, residual volatiles, resin content, tensile or compressive strength, interlaminar shear strength, flexural strength, or acetone extractibles. Tests required may vary depending on type of raw material, vendor, and program requirements and previously have been conducted per standard American Society for Testing and Materials (ASTM) or equivalent test methods.

2. *Components*

Components are subjected to x-ray, alcohol wipe, and tag-end testing. Tag-end tests consist of density, resin, and volatile content, acetone extractible, tensile/compressive/flexural strength, interlaminar shear, and hardness tests as noted in Table 7. Occasionally high-rate-production programs will select one component out of 10 or 20 and conduct a destruct test for lot acceptance. Defect criteria for ablative components does not usually allow cracks or delaminations. Sometimes minor delaminations and voids in qualified sizes are allowed depending on location in the component.

3. *Throat Materials*

Graphites and carbon–carbon materials are generally subjected to 100% x-ray and alcohol wipe inspection, as well as tag-end tests for density, tensile and/or compressive strength tests as noted in Table 7. Cracks in throat materials are not acceptable, whereas low-density regions and voids are usually limited in size and quantity depending on location within the component relative to critical stress regions or expected eroded surfaces.

4. *Elastomeric Insulators*

Elastomeric Insulators are usually x-ray and hardness checked for inclusions, delaminations, or voids. Some inclusions or voids can be tolerated depending on size and location, whereas separations are generally considered undesirable.

5. *Interfaces or Bondlines*

Interfaces or Bondlines are x-ray and alcohol wipe inspected where possible or meaningful to check for voids or unbonds. Continuous unbonds that might allow gas flow to critical structural components are unacceptable, whereas intermittent unbond indications or voids can be tolerated, as long as they are qualified during nozzle development and qualification phases of a program.

6. *Destructive Sample Testing*

Destructive sample testing is usually done by one of three methods: 1) random selection and destructive testing of one actual component of every so many produced, 10–20 being typical; 2) destructive testing of an excess part of the component or tag end; or 3) destructive testing of an extra test slab of the same material that has been processed with a batch of components hydroclaved with them, or heat treated with them, for example. Method 1 has the advantage that an actual part is tested but the disadvantages are the expense and the statistical possibility that a part not tested is bad. Method 2 has the advantage that a test is conducted for each part, but it is often questioned whether the tag end is truly representative of the part. Method 3 is likewise questionable as to whether the test slab is truly representative. Method 1, being the least expensive, is practical only in production after development and qualification testing has been completed and statistical limits have been established. Methods 2 and 3 are normally used in concept demonstration and during development and qualification phases of a program. If safety factors are large enough, destructive testing is often not included in test and development programs.

IV. Design Sequence

Because of the wide operating requirements and diversity of materials utilized for tactical nozzles, a specific design usually requires support from numerous disciplines requiring their own unique expertise. Typically a design requires involvement by persons with experience in 1) materials and processing (reference Sec. III); 2) aerodynamic design and analyses (both internal and external), 3) structural analysis of conventional and specialized metals, as well as ablative composite or high-temperature metals; 4) manufacturing knowledge of metallics and composites; and 5) reliability.

Armed with a working knowledge of the listed disciplines, as well as previous experience, a designer can prepare a preliminary design that can be used for experimental evaluation or submitted to the necessary disciplines for detailed analyses. There are no industry standard practices other than those suggested in Ref. 1. Approaches and methods vary with each company based on corporate experience.

This section will not attempt to cover details necessary for all materials and processes available, nor complete internal and external aerodynamic and structural analyses, nor all the various methods required to machine nozzle components and machines available, nor all methods available to assess a nozzle design's expected reliability, but will present a systematic step through a preliminary design. As each particular discipline is discussed, typical industry standard methods or programs used for detailed analyses, etc. will be referenced.

Designing a tactical nozzle involves three to five basic steps or processes as outlined on Fig. 15. Detailed thermal and structural analyses in step 5 will only introduce the reader to typical analytical methods used in the industry and their reference sources.

A. Preliminary Design

1. *Step 1: Defining Preliminary Design Requirements*

Typically a designer prepares a summary of basic or design driver requirements, such as those noted in Table 8 before starting the design sequence. From these requirements, the designer then will know what type of nozzle design, i.e., fixed with or without TVC, movable TVC, external, submerged, and so forth, is required, or whether the design requires a blast tube.

2. *Step 2: Defining the Internal Aerodynamic Profile*

This is when the designer prepares the aerodynamic profile of the nozzle inlet, throat, forward exit cone region, and finally the supersonic exit cone portion.

3. *Nozzle Inlets*

Fixed external. To provide smooth gas flow into the throat, a blend radius of one to two times the throat radius has been used. With an external nozzle design, such as that shown in Fig. 1A, this radius is joined tangentially to the conically shaped inlet section with a converging angle typically between 30 and 60 deg.

Submerged fixed or movable. Submerged, fixed or movable nozzle inlets use the same $1.0R_t$ blend radius immediately upstream of the throat combined with a longer entrance contour consisting of a 3:2 ellipse or hyperbolic spiral. The length of this inlet is usually one to 1.7 times the throat radius for smooth gas flow transition. The radius at the nose tip typically is 1.3–1.7 times the throat radius and incorporates another blend radius into the elliptical or hyperbolic portion with a small radius.

Downstream throat transition region. Smooth gas flow transition from the throat region into the supersonic exit cone is provided by a blend radius between 0.5 and 3.0 times the throat radius, depending on design envelope restrictions for

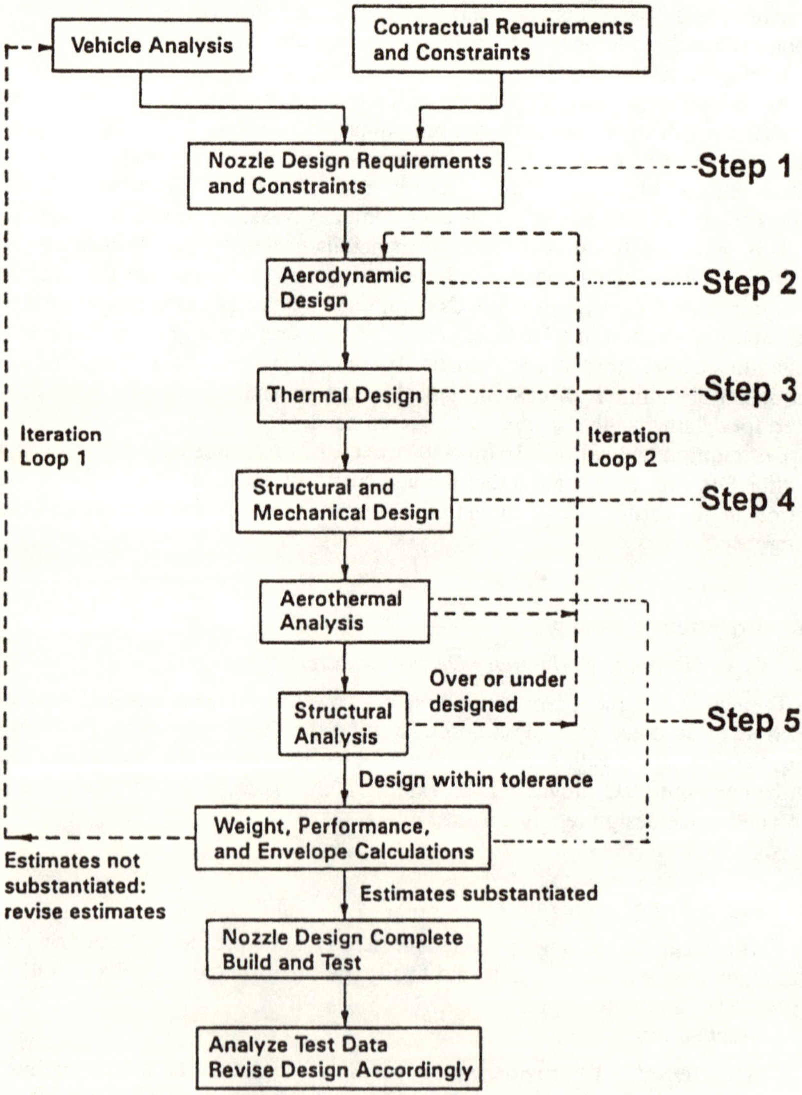

Fig. 15 Nozzle design sequence.

Table 8 Preliminary nozzle design requirements list

Requirement	Value(s)
Nozzle parameters:	
Fixed	
Without TVC	Yes or No
With TVC	
Jet tabs, ft-lbf	Need pounds-force turning moment defined
Jet vanes, ft-lbf	
Movable, deg	Need vector angle, deg and required duty cycle, deg/s
Throat diameter, in.	Obtained from motor ballistician
Throat type	
Eroding	Need minimum/maximum erosion rate, in./s (mm/s)
Noneroding	
Expansion ratio (dimensionless)	Determined from motor ballistician and/or envelope constraints
Motor parameters:	
Pressure	
Average, psi	Obtained from motor ballistician
Maximum, psi	Obtained from motor ballistician
Burntime, s	Obtained from motor ballistician
Propellant type	
Percent solids, etc.	Obtained from propellant
Safety factors:	
Thermal	Defined by program
Structural	Defined by program
Envelope	Obtained from motor design requirements
Special requirements:	
Weight, lbm, maximum	Obtained from weight allocation from motor performance optimization
Temperature during/after motor burn, (°F)	Obtained from motor design requirements
Operational temperature, (°F)	Obtained from motor design requirements
Removable/replaceable during service life	Obtained from motor requirements

insulators and structural elements immediately aft of the throat. Occasionally, a short cylindrical section is employed in the throat of approximately 0.1–0.5 in. in length.

Exit cone. The supersonic exit cone can either be a straight conical design, usually 15–24 deg, or contoured to achieve more performance in the same or shorter length. Most tactical nozzles have expansion ratios exit area divided by the throat area of 10 : 1 or less because of their low-altitude operating conditions. Contoured designs can either be defined by a method of characteristics[15] or a large radius tangent to the throat downstream blend radius and the exit plane as discussed in Ref. 4. As shown in Fig. 16, a contoured exit cone profile achieves the same diameter and expansion ratio in a shorter length.

Typical initial angles θ_i range from 18 to 24 deg with up to 15 deg of turnback (i.e., $\theta_1 - \alpha$) demonstrated. Contour turnback is determined by exit cone liner

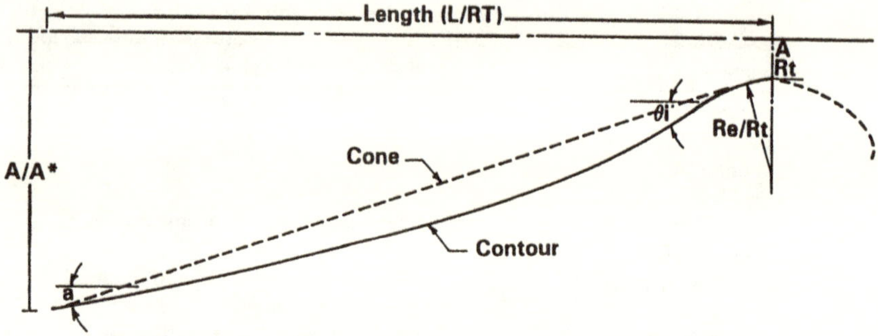

Fig. 16 Nozzle exit cone design parameters.

material selection and acceptable erosion with the particular motor propellant type and motor operating pressure and burn times. Exit angles α as low as 9 deg with an initial angle of 24 deg have been demonstrated on a performance driven tactical nozzle. The preliminary aerodynamic profile has been defined and layed out as shown in Fig. 17.

4. *Step 3: Sizing the Inlet-Throat-Exit Cone Ablatives*

With the preliminary aerodynamic profile completed, the designer may address the insulative and ablative (or liner) materials, including the throat. Materials are selected for two primary reasons: their ablative resistance and thermal insulating capacity. Secondary considerations include strength and ductility (at all temperatures), density, cost, and processibility. The material selected for the throat, more than any other, is generally selected for its erosion rate. A material's erosion rate is dependent on several factors, including propellant composition, flame temperature, operating pressure, flow condition (Mach number), geometry (diameter), and support. For composite materials, erosion rates, and to a lesser degree thermal conductivity, vary as a function of the materials' ply orientation to the flow surface. The general rule is that a material, oriented with its ply or grain direction less than or equal to 75 deg downstream to the flow surface, will offer adequate erosion resistance without creating an internal thermal-structural problem, especially with carbon phenolic composites.

With all of these factors in mind, the nozzle designer must lay out a configuration that puts materials in the proper position and orientation to ensure a relatively smooth flow surface throughout motor operation. In addition, the chosen materials must provide sufficient insulation to maintain the integrity of the nozzle structure and keep backside temperatures within requirement limits. In many cases, the material required to provide the desired erosion resistance does not provide sufficient insulation capability, requiring a second material with improved insulative properties outboard of the ablative.

As previously mentioned, there are other factors that must be considered, including material cost, processibility, and reliability. Some materials such as carbon cloth phenolic (CCP), have a long history of use in solid rocket motor nozzles. Its properties are well documented and processing parameters well understood, but material prices are very high, usually \$50–100 per pound. Silica cloth phenolics are lower in cost, approximately \$25–40 per pound, and have demonstrated adequate erosion resistance for low and nonaluminized propellants.

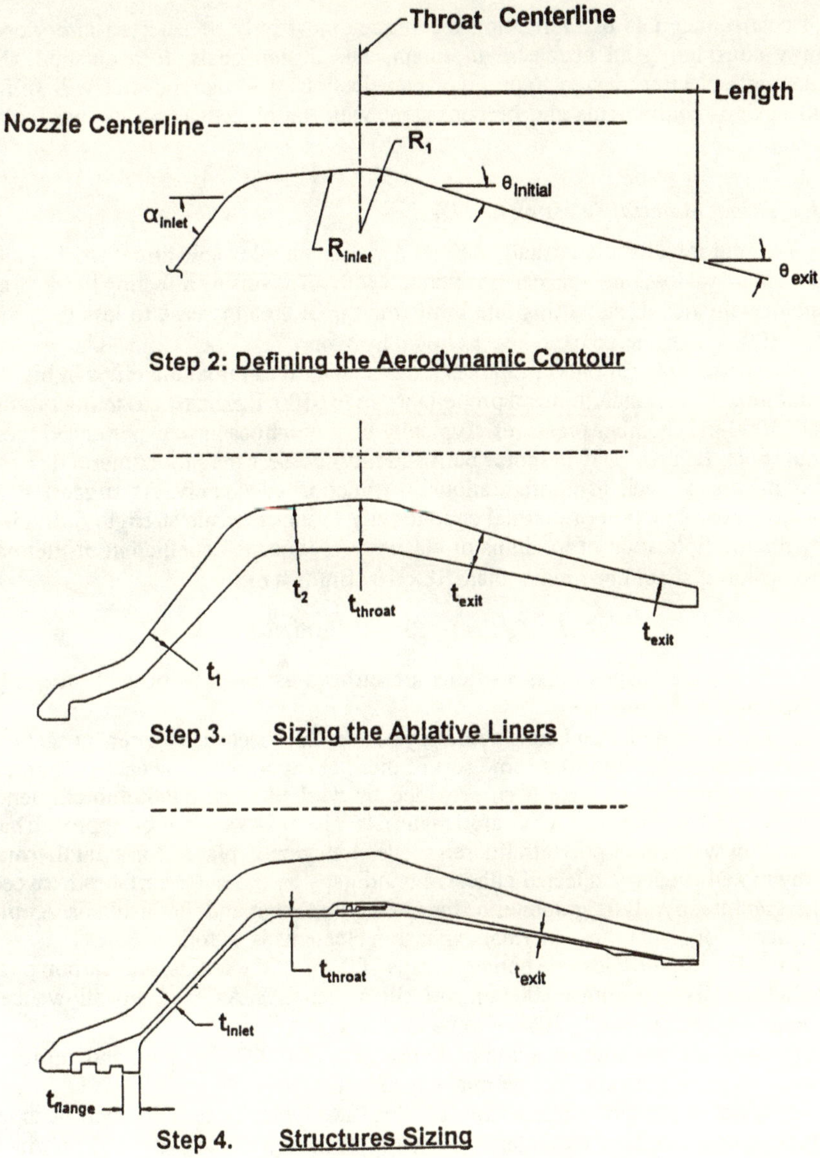

Fig. 17 Preliminary nozzle design sequence.

Phenolic components are typically fabricated using one of three methods: 1) tape wrapped on a mandrel using straight or ±45-deg bias tape, 2) compression molded using precut patterns, or 3) compression molded using chopped squares or macerated fibers. The first two methods, which are the most labor intensive and costly, yield a component with known and repeatable ply orientations optimized for material strength and erosion resistance. In contrast, the third method yields a more random, less repeatable orientation. Thus, the use of costly materials or fabrication techniques must only be selected after carefully considering all nozzle requirements and design goals. In a nutshell, the material selection design trade involves selection of a material that will fulfill all design requirements and be consistent with the objectives of any particular design.

5. *Throat Material Selection and Design*

The motor ballistician usually defines a maximum allowable throat erosion rate (i.e., material loss rate) that can be tolerated without adversely affecting the desired motor ballistics. Usually this rate limits the throat area increase to less than 5% and 10% is considered excessive as noted by Wong.[4]

Polycrystalline (or bulk) graphites have been used as throat materials in highly aluminized and nonaluminized propellants up to 6500°F exhaust gas temperatures and 1000-psi chamber pressures. Typically bulk graphites have experienced thermal shock failures early in motor burn and must possess the proper thermal stress parameters, as well as configurations, to function adequately. As suggested in Ref. 4, a combination of thermal conductivity K times tensile strength S divided by the multiplication of modulus of elasticity E times the coefficient of thermal expansion α should be greater than 20×10^3 Btu/ft-h or

$$K \cdot S/E\alpha \geq 20 \times 10^3 \text{ Btu/ft-h}$$

Usually square shaped cross sections are utilized as much as possible for bulk graphite components.

Pyrolytic graphite and tungsten alloys have also been used often in tactical nozzle throats, all requiring some sort of thermal expansion allowances. Thermal expansion provisions have been provided by washers of polytetrafluorethylene, silicone rubber, adhesives, or Grafoil materials. The most widely used approach has been thin washers of polytetrafluorethylene strategically placed for axial thermal growth and properly selected adhesive bond gaps on the outer surfaces between the graphite, pyrolytic graphite, or tungsten component and the insulator. A rule of thumb for sizing these thermal expansion elements is as follows:

a) Bulk graphite backside bond use gaps of 0.4% of the radius with carbon phenolics, and 0.4% R minus 0.005 in. with silica phenolics. Axial growth allowances should be approximately 10% of axial thicknesses.

b) Pyrolytic graphite backside bond use gaps of 0.002–0.006 in. and approximately 80% of the axial free thermal expansion.

c) Axial spacers of polytetrafluorethylene are typically 0.06 in./inch of axial thickness for pyrolytic graphite; 0.01 in./inch of bulk graphites; and 0.015 in./inch of carbon–carbons.

Recently, tactical motors utilized reduced or nonaluminized propellants that allow the use of noncarbonaceous silica or glass phenolic composites to be used in the throat region, as well as the inlet and exit cone areas.

Predicting the material loss rate or erosion rate can be an extensive procedure either using Aerotherm's ACE computer program[7] or scaled from rates measured for the same throat material from similar nozzle tests with similar propellants. Most often the designer scales a throat erosion rate using a simplified equation involving throat diameters and pressures (from Ref. 4) as

$$\dot{e}_r = \dot{e}_r \text{ known} \left(\frac{P\text{new}}{P\text{known}}\right)^{0.8} \left(\frac{D\text{throat known}}{D\text{throat new}}\right)^{0.2}$$

This equation can be used directly for silica phenolics without any corrections for chemical composition of the propellant exhaust products. Correction for carbonaceous materials needs a factor that is too extensive to cover here, but a more detailed discussion can be found in Ref. 8.

Char, or in-depth thermal profiles, can be determined either from Aerotherm's charring material and ablation (CMA) computer program (Ref. 8) or scaled from similar nozzle test results that used the same materials in similar propellant environment. A simplified scaling equation uses char depth as a function of burn time and exhaust gas products flame temperature, or

$$\text{char depth}_{\text{desired}} = \text{char depth}_{\text{known}} \left(\frac{\text{burntime} - \text{new}}{\text{burntime known}} \right)^{X} \times \left(\frac{\text{new flame temperature}}{\text{flame temperature} - \text{known}} \right)^{Y}$$

where X and Y coefficients are empirically derived from motor test history. This scaling equation is valid for the throat and inlet regions where the chamber exhaust gas temperatures are fairly constant and radiative heating is a strong factor. Results in the exit cone at comparable expansion ratios may be slightly conservative due to decreasing gas temperatures.

6. *Step 4: Structural Components Sizing*

Designing the nozzle structures requires definition of the loads and the different combinations that are possible. The primary sources are internal gas pressures and thermal expansion of the liner components. Secondary loads from TVC mechanisms, external aerodynamic fin-reaction loads, and vibration-induced loads also need to be evaluated.

Structural designs are selected for specific applications considering the magnitude of the load, weight limitations, if any, and the complexity of the nozzle configuration.

Preliminary sizing of structural components can be done using simple standards of strength of materials equations and procedures such as those defined by Roark.[5] Definition of metals, their capabilities, and recommended uses can be found in the latest edition of MIL-Handbook-5E.[9] Most structural components are sized to minimize deflections of insulator components.

7. *Step 5: Final Nozzle Design and Detailed Analyses*

With the preliminary liner and structures sized, they are combined with the preliminary aerodynamic profiles defined in step 2 and compared to interface envelope requirements defined in step 1 and weighed if required. Adjustments may be necessary to smoothly blend and interface as required.

Once the basic layout and materials have been selected, the designer must review how various components are assembled. High-strength, structural adhesives with excellent high-temperature capability are often used to bond insulating components to the structural shell and to each other. If backside temperatures become too hot during operation, a good insulating material will be incorporated outboard of the ablatives, sometimes tape wrapped directly over the ablative and cured in place. In some cases, positive mechanical retention is added to ensure that the liner materials are not expelled. This is accomplished by incorporating a mechanical stop into the nozzle structure or using metallic or plastic pins that penetrate through the structure directly into the plastic liner.

As the nozzle plastic materials are heated, they expand, and if this expansion is constrained, large thermal stresses can develop. Thus, expansion relief must be incorporated into the design between components to minimize this effect. Adhesive bond gaps or materials such as polytetrafluorethylene, polypropylene, or foam that burns out shortly after motor ignition are used to provide the desired expansion gaps.

Pyrolyzing materials pose another problem that must be considered in the design layout. As the materials char, they go through a pyrolysis process that generates a significant amount of gases or off gassing. This off gassing must be allowed to vent or significant forces may be generated that lead to local material failures in the form of spalling or excessive erosion. Venting is accomplished through the mentioned gaps and/or orienting the material with ply directions that allow these gasses to be vented between the laminates as discussed earlier in step 3 of this section.

B. Detailed Analyses

1. Aerodynamic Analysis

The objective of the aerodynamic design phase is to optimize the nozzle contour (inlet and exit) within the motor/nozzle constraints to achieve optimum nozzle performance or maximum I_{sp}. This is accomplished analytically using the industry standard program, SPP.[6]

Basically, SPP calculates the maximum or ideal theoretical performance for a rocket motor and then degrades it by subtracting individual calculated losses that are known to occur within the motor and nozzle. The assumption that all losses can be treated independently from each other has been substantiated throughout the industry. The only constraint is that for any given number of individual losses defined, each physical source of nonideal performance loss is accounted for once and only once. A description of the SPP program and its modules is presented in Fig. 18, as well as a description of the various performance losses calculated and their typical ranges of values. Although there are several variables that control the final nozzle contour, experience has identified the following parameters as the most influential on overall nozzle performance (see description shown earlier in Fig. 16: exit cone length (L) in relation to throat radius (R_T), (L/R_T); expansion ratio (A_e/A_t) or A/A^*; exit cone initial (or entrance) angle Θ_1; exit cone exit angle (a); and exit cone initial blend radius in relationship to the throat radius (R_e/R_T).

2. Thermal Analysis

Thermal analysis determines surface recession or erosion rates as well as in-depth thermal response of the nozzle ablatives and insulators. Erosion rates can be affected by mechanical as well as thermal failure mechanisms depending on component and material configurations and flow conditions. Thus, experienced analysts depend heavily on past experience to choose the correct modeling technique for any given combination of propellant, operating conditions, material, and component geometries.

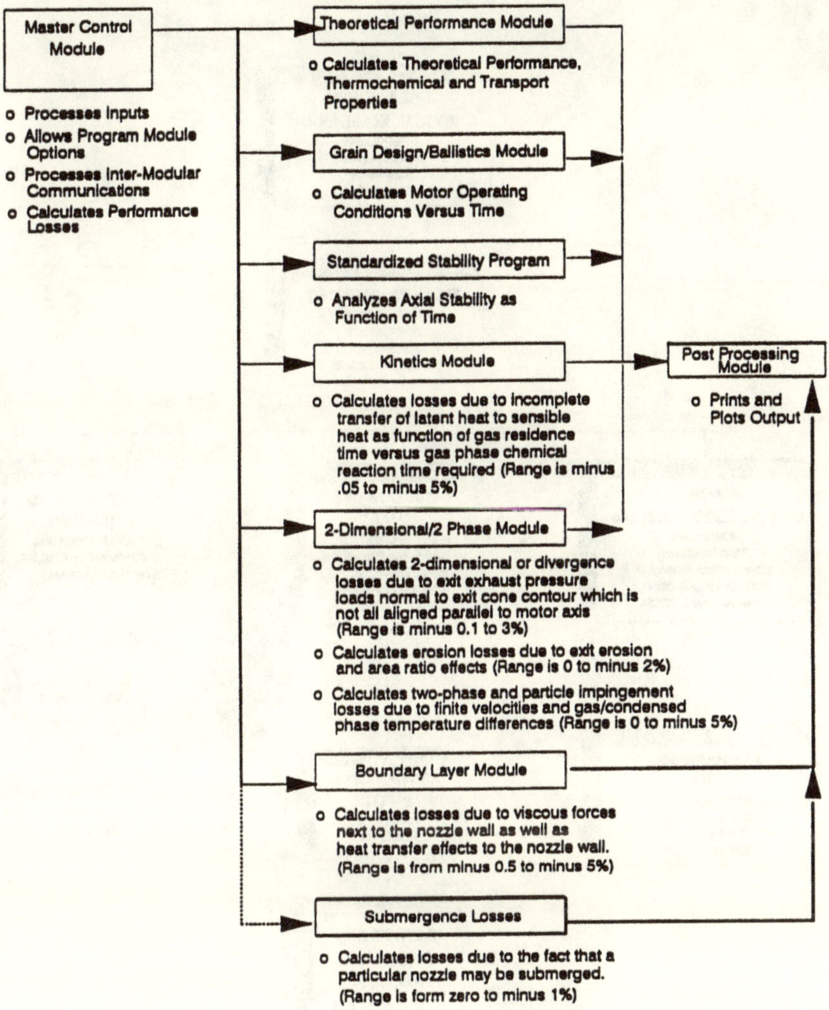

Fig. 18 Solid propellant rocket motor performance. Prediction computer program (SSP) summary.

There are several computer codes available to predict thermal responses in nozzles. A flow diagram for a typical aerothermal analysis is presented in Fig. 19. A description of the computer codes used for the required input and output of each is presented in Table 9. This section presents an overview of the basic principals involved thereby establishing a general understanding of the ablation/heat transfer solution methodology.

Thermal analyses consists of defining heat transfer boundary conditions that are input to one- or two-dimensional axisymmetric analytic codes that predict surface recession (i.e., erosion) and in-depth thermal responses.

Boundary conditions. Radiative heat flux to the internal surface is defined by

$$q_r = \varepsilon \sigma T_S^4$$

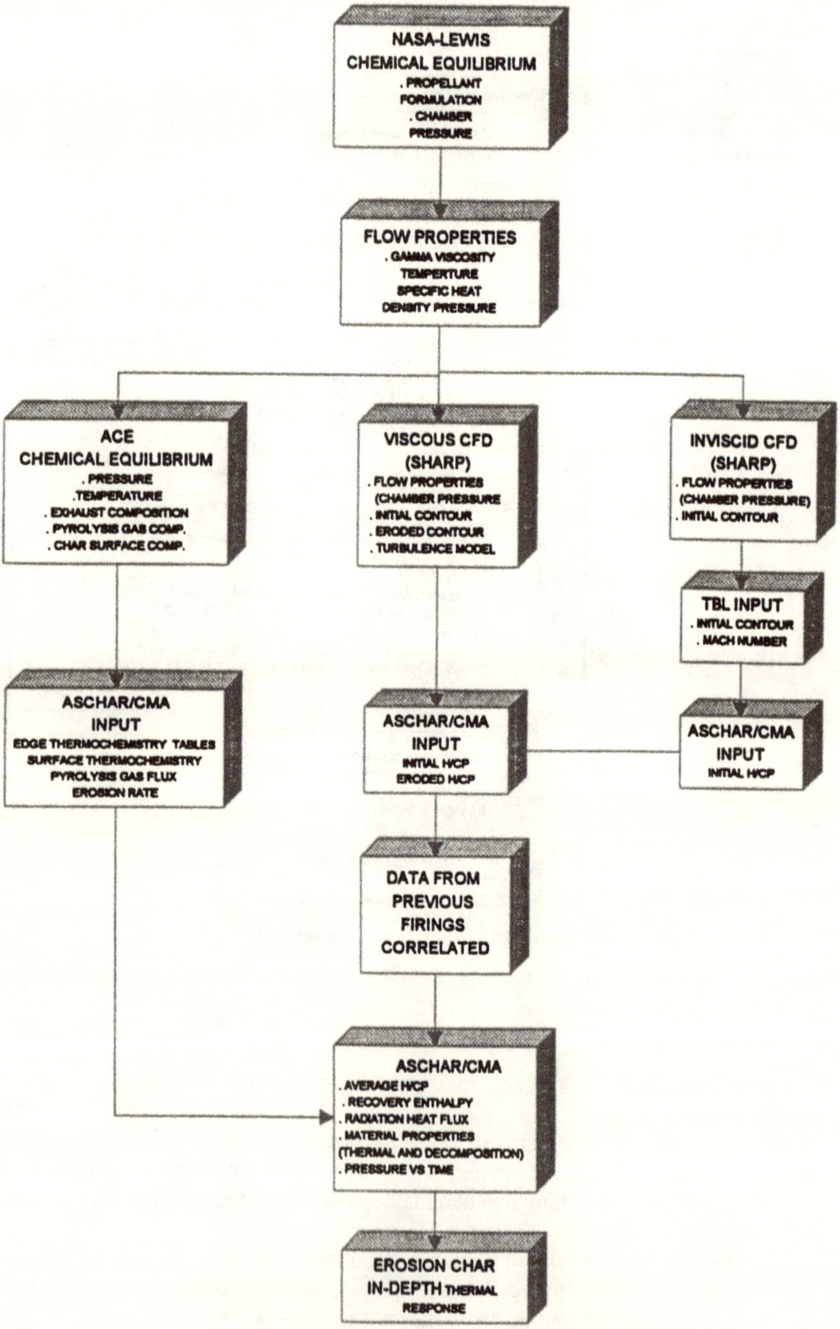

Fig. 19 Thermal analysis methodology.

Table 9 Aerothermal analysis codes

Program	Input data	Output data	Calculation methodology
1) NASA-Lewis chemical equilibrium	Propellant composition, pressure, heat of formation	Exhaust gas composition and properties	Minimizing free energy based on JANNAF thermodynamic data
2) SHARP	Flow geometry and gas properties from No. 1 Flow boundary conditions	Pressue, temperature, density, and velocity components Heat transfer coefficient	Full Navier–Stokes Flow solver and two-equation turbulence model
3) Aerotherm chemical equilibrium (ACE)	Surface pyrolysis gas composition Propellant exhaust gas composition from No. 1	Nondimensional ablation rate of surface material as a function of surface temperature	Equilibrium or kinetically controlled surface chemical reaction
4) Aerotherm charring material and ablation (CMA)	Material thicknesses and thermal properties Boundary conditions from Nos. 4 and 5	Surface recession and in-depth temperature history of a charring, one-dimensional, axisymmetric body of several materials	Finite element surface and in-depth material and energy balance Finite difference method
5) Aerotherm axisymmetric transient heating and material ablation (ASCHAR)	Same as No. 6 (CMA) two-dimensional finite element grid	Same as No. 6, except two-dimensional axisymmetric bodies	Same as No. 6 (CMA)

where

q_r = incident radiative heat flux, Btu/ft^2-s
σ = Stefan–Boltzman constant, 4.76 E-13 Btu/ft^2-s-R^4)
ε = emissivity of combustion products
T_S = combustion products static temperature, °R

Radiation source emissivity ε can be defined by the following correlation:

$$\varepsilon = 1 - e^{(-0.808\rho DN/16)}$$

where

ρ = local gas density, lb/ft^3
N = percent aluminum (mass) in propellant formulation
D = local flow diameter, in.

Recovery enthalpy is the driving potential for convective heat transfer. This parameter is calculated using local static enthalpy and Prandtl numbers obtained from Aerotherm's Chemical Equilibrium (ACE) program by

$$h_r = h_s + Pr^{\frac{1}{3}}(h_c - h_s)$$

where

h_s = local static enthalpy, Btu/lbm
h_c = chamber enthalpy, Btu/lbm
Pr = local Prandtl number

Convective heat transfer coefficients are calculated. The convective heat transfer coefficients are calculated from the SHARP flow solution using the Reynolds analogy, in which the Stanton number St is related to the skin friction coefficient C_f as follows:

$$St = C_f P_r^{2/3}$$

where Pr is the laminar Prandtl number and C_f is related to the wall shear stress τ_w, density ρ, and the parallel to wall velocity u as follows:

$$C_f = \tau_w / \rho |u|^2$$

where u is assumed to have either a parabolic (laminer sublayer) or a logarithmic law of the wall profile depending on the local Reynolds number.

These local heat transfer coefficients are then combined with the enthalpies predicted by the ACE program,[7] and the resulting values can be input either to the one-dimensional charring materials and ablation (CMA) or two-dimensional axisymmetric transient heating and material ablation (ASCHAR)[8] computer programs for erosion, char, and in-depth thermal profile definition.

Thermal analyses. The CMA and ASCHAR computer codes handle a very complex physical phenomena schematically represented in Fig. 20. Carbon phenolic and carbon–carbon materials are modeled as diffusion controlled equilibrium reactions directly affected by the rates of mass transfer to heat transfer Stanton numbers. Elastomeric materials behave in a similar mode. Silica phenolic

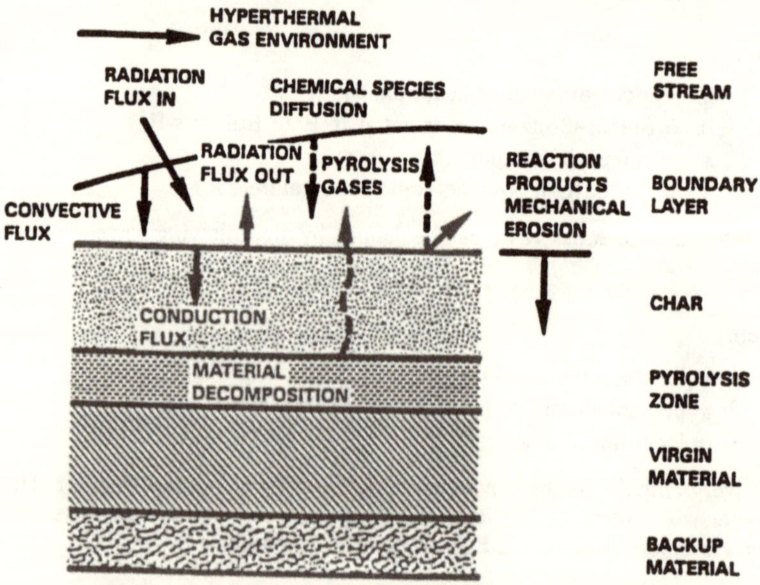

Fig. 20 Aerothermal analysis, general physical problem.

materials are modeled using a fail temperature, which melts the material away. High-temperature refractory metals, such as tungsten and its alloys, are modeled with surface failure temperature or melting erosion mode. The ratio of mass transfer to heat transfer can be approximated by the following equation.

$$Cm/Ch = (Pr/Sc)^{2/3}$$

where

Cm = mass transfer Stanton number
Ch = heat transfer Stanton number
Pr = Prandtl number
Sc = Schmidt number

Each material is modeled with a set of temperature dependent properties that include the following: density (initial and charred), specific heat, thermal conductivity (in-plane and across-ply), emissivity, heat of formation, and chemical decomposition data.

Output. The results of the aerothermal analyses provide a description of the eroded nozzle surface, progression of the char layer, and in-depth temperatures through the thickness of the nozzle for any time during motor operation. The in-depth temperatures are used in subsequent structural analyses to determine the thermal stress state in the nozzle components at any given time. This will discussed further in the following section. Heat transfer analysis is also conducted to determine the maximum temperature of the nozzle structure at the conclusion of heat soak.

The aerothermal analysis provides an indication of the erosion and thermal conditions in the nozzle. Several iterations may be required utilizing data obtained from actual nozzle testing such as backside temperatures from thermocouples and posttest measurements of the char depth and eroded contours. This database is continually updated and conducted to improve the material and thermochemical models.

3. *Structural Analyses*

Structural analysis utilizes data and information from all the preceding sections to evaluate the structural integrity of the nozzle assembly. This section provides a synopsis of the required inputs (loads and boundary conditions), material models required, and structural mechanics models available for use.

Overall structural analyses, described in Fig. 21, consists of 1) evaluating the primary metallic or composite structures that operate within well-defined temperature-dependent materials properties ranges, such as those defined in handbooks,[9,10] and 2) thermal-structural evaluation of ablative or insulator composites that operate under high temperatures and temperature gradients. These two tasks are briefly discussed in the following sections.

Primary structures analysis. Evaluation of most tactical nozzle primary structures, usually composed of high-strength stainless or alloy steel, is a relatively straightforward problem involving standard strength of materials calculations. Linear elastic material models are generally used with standard failure criteria such as ultimate strength or Von Mises equivalent stresses. The structure is often analyzed separately from the nozzle insulating components to allow for finer mesh concentration in critical structural areas, incorporation of asymmetric loads

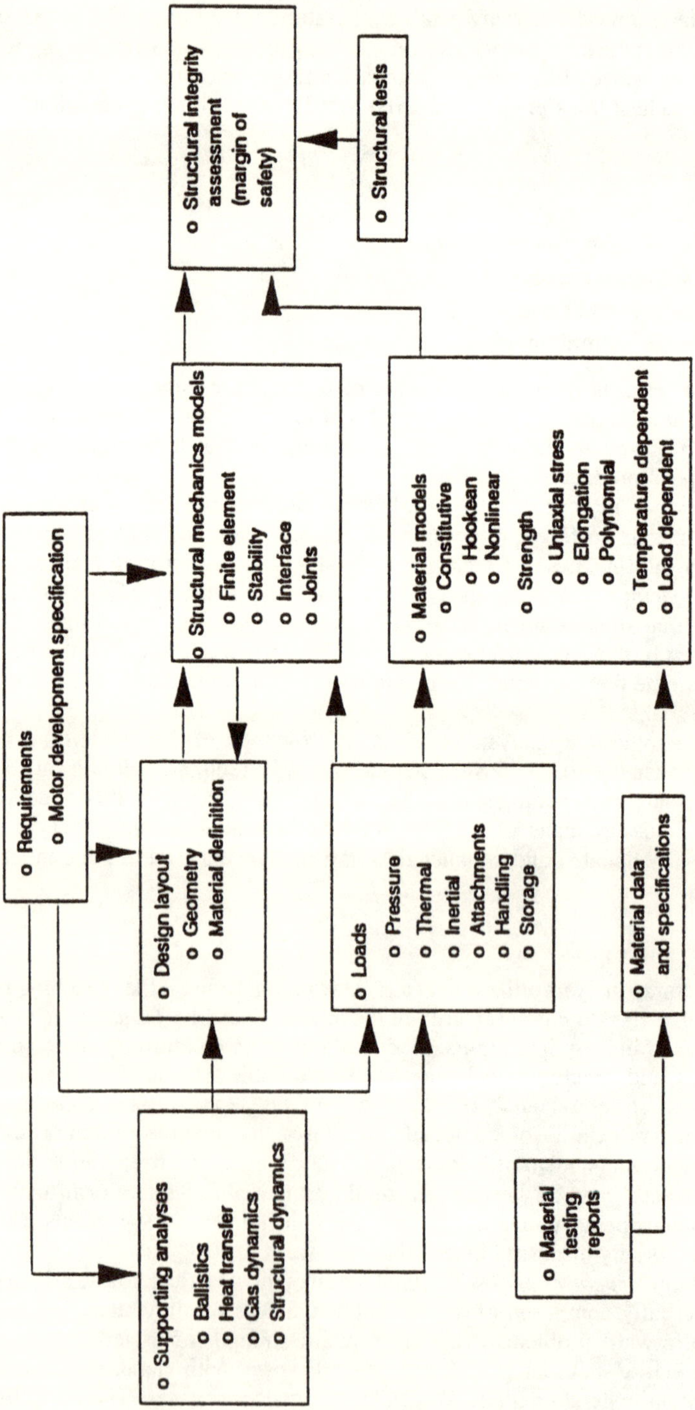

Fig. 21 Thermostructural analysis methodology.

or geometry, and possibly even nonlinear effects. The insulating components are often simulated in these models as a low modulus, isotropic "mush" material to allow the correct pressure loading distribution to be applied to the structure. This method allows for the use of more standard codes such as ABAQUS[11] in evaluating the structure, without concern for the orthotropic insulating materials and their complex, thermally induced stress states.

Ablatives/insulator thermostructural analyses. Typical ablative/insulator composite thermostructural analysis is conducted using finite element (FE) analytical techniques. At present, there are no industry standardized codes. General procedures are fairly common, however, and company proprietary finite element codes are similar in many ways. Examples of such codes are listed in Refs. 11–14. The NASA Solid Propulsion Integrity Program (SPIP) is currently attempting to develop an industry standardized code specifically for nozzle insulator and ablatives thermostructural analysis. This code will probably not be available until the late 1990s.

Analyzing ablative and insulator composites is much more difficult due to erosion, resin decomposition, and in some cases extreme temperature gradients throughout their cross-sectional configuration during motor operation. Temperatures within an ablative component may vary from over 5000°F at the inner-gas surface to −65°F at the outer structural interface. Temperature gradients may be as high as 2000°F difference per each 1/10 in. of thickness. Heating rates in excess of 1000°F/s at the gas surface are also common.

Ablative and insulator composites are anisotropic and nonhomogenous because of their combination of resin and/or fibrous or woven reinforcements. Sometimes composites exhibit nonuniform strength and modulus properties in different directions that also vary with temperature, heating rate, loads, and pressure levels.

Asymmetric loads may be introduced through structural components to the ablative insulators due to TVC loads associated with a movable nozzle. Dynamic loads due to accelerations or vibrations are handled by superimposing a dynamic load factor onto the appropriate static load or stress. These dynamic load factors are derived from separate dynamic analyses.

The general shape of rocket nozzles lend themselves nicely to two-dimensional finite element axisymmetric models. Typical input consists of a description of geometry defined by a finite element grid, material, properties, boundary conditions, thermal environment, and pressure distribution as discussed later. Analysis output is typically expressed in terms of elemental stresses, strains, and nodal displacements that are compared to material allowable strength, using suitable failure criteria, to determine adequacy.

The nozzle *geometry* is represented by a finite element grid. Care must be taken to utilize minimum structural material thicknesses and account for all material interfaces and bondlines. The FE grid used for a typical tactical nozzle exit cone is shown in Fig. 22. Throughout motor operation, the nozzle geometry changes as the materials near the flow surface erode. Most nozzle codes automatically adjust for this and apply appropriate pressure loads throughout motor operation based on input from the thermal model. Structural and thermal analysis finite element models need to be coordinated for efficient transfer of thermal gradients and eroded surface conditions.

Boundary conditions are applied to the FE model to simulate the nozzle support and mechanical loads. Interfaces and joints are simulated to transmit load into the correct direction (axial, radial, etc.). The motor pressure load is applied to

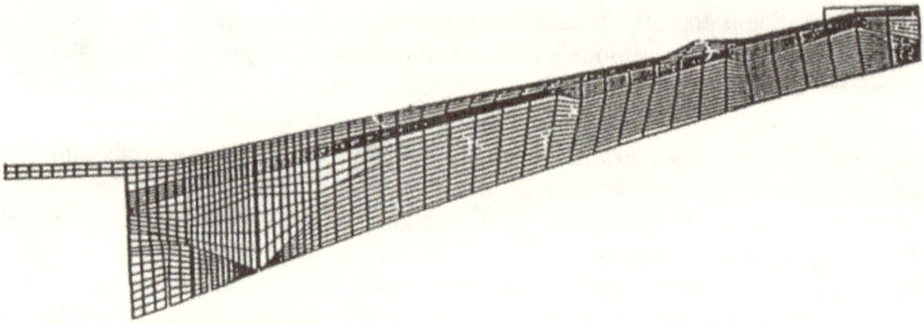

Fig. 22 Typical exit cone assembly thermostructural analysis finite element grid.

the internal contour of the nozzle and adjusted along the length of the nozzle to simulate the pressure distribution resulting from the various flow conditions along the nozzle. This distribution is typically obtained directly from the results of the aerodynamic analysis or may be estimated using standard one-dimensional isentropic flow tables. Pressure levels may need to be adjusted throughout burntime to match motor variations.

Nozzle *material properties* are generally obtained from databases generated by individual companies or agencies such as NASA, the Air Force, or the Navy. It takes a great deal of time and money to fully characterize a material to obtain statistically meaningful temperature-dependent structural properties, including those at elevated temperatures suitable for use in analytical models.

Most ablative or insulator materials are orthotropic, with properties dependent on the orientation of the material layup or grain direction. Some are more random, as is the case for molding compounds. Phenolic ablative composites exhibit linear-elastic temperature-dependent properties up to 500–700°F, but nonlinear behavior is exhibited above this temperature range. Metals exhibit linear-elastic properties below their yield strength, and nonlinear behavior above, as discussed in MIL-Handbook-5E.

Carbon–carbon materials, because of their multidirectional reinforcing construction exhibit multimodulus materials properties as a function of temperature. Sometimes a Jones–Nelson multimodulus materials properties model is used for these materials.

Typical ablative composite materials properties models consist of modulus, Poisson's ratios, and shear modulii as a function of temperature. Bondlines are modeled to transmit compressive and shear loads across them to simulate the correct degree of material interaction. As the temperature increases, the material properties are degraded accordingly.

Thermal environment inputs are obtained directly from the results of the thermal analyses. The eroded depth is simulated and in-depth temperature responses are applied as elemental temperatures. These thermal gradients, such as those mentioned earlier in Section 3, and material properties are used to calculated the thermal stress field of the finite element model. This input is available for any distinct time during motor operation using time increments down to 0.1 s. Structural analysis is generally conducted at discrete time intervals to identify the maximum stress state for each component. These generally include but are not limited to the following: 0–2.0 s ignition transient, maximum thermal gradient inducing thermal shock; time when pressure is maximum; end of burn when maximum material degradation has occurred and pressure has decayed to ambient. The term thermal shock refers to the large tensile stress field that develops due to the steep thermal gradient through the component thickness. This is especially critical for brittle materials, such as tungsten and graphites. At motor ignition, the inner diameter of materials exposed to the flame front heats very rapidly, almost instantaneously, while the outer portion of the material remains cool. As the inner material tries to expand, the growth is restrained by the outer material resulting in potentially very large tensile stresses in the middle of the part. The thicker the cross-sectional area, the more susceptible these materials are to thermal shock.

For *output* the finite element programs calculate elemental stresses and strains in the principal directions of each material and at all locations in the model. These are compared to the capabilities of the materials at the representative temperature to calculate a structural margin of safety. The standard failure criterion is ultimate material strength, at temperature, in all principal material directions. The non-linear nature of some materials, however, requires the use of an ultimate strain criterion. Other criteria have been developed for composite materials that are more representative of potential failure modes. The following three failure criteria have been established relative to the material(s) defined.

1) For metals uniaxial equivalent stresses using Von Mises yield theory are compared to allowables from handbooks or component data. Exceeding the yield strength locally or in regions surrounded by compressive stresses is not necessarily unacceptable. Generally acceptable metallic structural temperature limits are 450°F for steels, 250°F for aluminums, and 200°F for epoxy composites.

2) For phenolic ablative/insulator composites the maximum fiber stress criterion defined by the lamina stress state is the primary failure criteria and is compared to fiber strength allowables to assess possibility of fiber rupture in the noncharred temperature region less than 500°F. A secondary failure criteria is delamination between plies that could allow gas flow to structural components and burn through or excessive heating due to radiative heating and subsequent failure. Quadratic failure boundary criteria, such as that shown in Fig. 23, which combine interlaminar tensile or compressive strengths and interlaminar shear strengths, have proven successful on some tactical programs.[15]

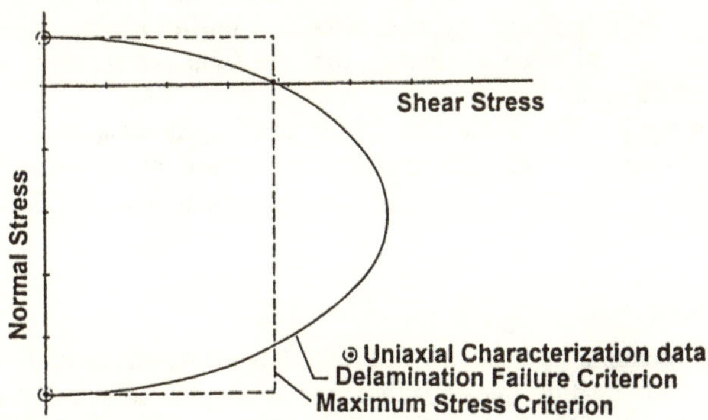

Fig. 23 Phenolic composite quadratic failure boundary.

3) Epoxy structural adhesive bondlines are usually limited to temperatures less than or equal to 200°F because of modulus degradation. Phenolic bonded interfaces are limited to 600°F. Structural bondlines in compression are evaluated using predicted shear stresses compared to allowables, since normal compression failure would still transmit the loads or retain the abaltive or insulator components. Some bondlines are considered only for thermal expansion capability and structural retention during manufacturing and handling.

As stated earlier, the stress output and subsequent margin of safety calculations are merely tools to aid in the overall assessment of the nozzle integrity. The nature of phenolic materials and the manner in which they char and ablate are in reality a series of localized material failures. It is by controlling or limiting their local failures to the char region and assessing the integrity of the remainder of the component that determines the survivability of the nozzle assembly.

C. Mass Properties

Sometimes the design sequence requires several iterations to achieve a nozzle design that provides adequate performance and has sufficient thermal protection and predicted positive structural margins. The nozzle mass and c.g. are then calculated and compared to required limits. The position of the nozzle at the extreme aft end of the motor has a major impact on the c.g. and subsequent performance characteristics of the missile system. Nozzle weight losses due to ablation and the change in density of the charred material are also calculated and submitted for motor performance evaluations. These weight losses are generally less than 5–12% of the original ablative/insulator weights.

D. Development Testing

1. Laboratory Bench Testing

As the nozzle design matures, the details are documented by engineering drawings, acceptance requirements, material specifications, and sometimes process definitions for critical process operations either by specifications or often contained directly on the engineering drawings. As prototype hardware is fabricated, the first step in the evaluation process typically involves laboratory bench tests to evaluate structural capability. Although these tests usually only duplicate pressure-induced physical loads, they are used to verify the structural margins of the key structural components and bondlines. The tests are also used to verify structural analyses by monitoring strain gauge responses during testing. The benefits of bench testing tactical-sized systems is that the tests can reasonably be run on actual-sized hardware.

Bench testing is particularly useful in evaluating movable nozzles. It is very valuable to obtain early verification of the nozzle response and maximum torque required to drive the nozzle. These early data are essential to successfully size the actuation system.

2. Static Tests

Although attempts have been made in a laboratory to simulate the extreme temperatures and heating rates nozzles are subjected to in a rocket motor, there is currently no substitute for a static test. Static tests provide the ultimate verification of the rocket motor and nozzle design. The relative inexpensiveness of tactical-sized systems makes static tests a very cost-effective method of evaluation, especially considering the extent of data that can be obtained.

Ballisticians can verify the actual performance of the motor, calculate specific impulse, and evaluate the effect of throat erosion on motor pressure–time trace.

Strain gauges and thermocouples applied to the nozzle during static test are used to record deflections and backside temperatures during and after motor burnout. These data are used to correlate the thermal and structural models.

The nozzle is removed from the motor following static test and sectioned to allow inspection of subsurface conditions of components. Actual erosion and char lines can be measured and correlated to thermal model predictions. Components are also inspected in detail to identify any anomalous cracks or delaminations that may have developed.

V. Thrust Vector Control

Most tactical motors require some steering to meet flight maneuvering requirements. In addition to, or replacing, traditional external aerodynamic control fins, steering control has been demonstrated by deflecting the nozzles hot exhaust gases, or the motor thrust, as shown in Fig. 24.

Both aerodynamic and mechanical techniques have been used to redirect the motor thrust and provide steering forces. This section limits itself mainly to mechanical thrust vector control systems, with a brief description of two major aerodynamic techniques, jet tabs and jet vanes. Aerodynamic techniques have demonstrated very rapid response rates, but also suffer motor thrust losses at large TVC vector angles as discussed later in Sec. V.B. The higher losses with the aerodynamic jet tab, jet vane, and bleed control concepts are a result of the physical creation of side force by creating a shock pattern in the exit cone thrust, i.e., the higher the TVC angle required, the larger the percentage of thrust involved. Movable nozzles exhibit much lower thrust losses while providing TVC, because they redirect the motor and nozzle thrust without converting dynamic energy to side forces.

This section will briefly describe the different types of TVC nozzles, compare them in tradeoffs, and present examples of typical tactical systems evaluated or used in production programs.

A. Classification of TVC Systems

TVC systems are classified primarily by nozzle type, either fixed or movable, and second by the method of providing actual thrust vector control, such as fixed liquid injection thrust vector control (LITVC) or movable-flex bearing thrust vector control. A complete summary of all possible types of TVC systems is presented in Fig. 25.

1. Fixed Nozzles TVC Systems

Thrust vector control with fixed nozzles is provided by injection of a liquid, or by hot chamber gases, or by mechanically deflecting tabs or vanes in the exhaust stream.

Secondary injection. LITVC is achieved by injecting a reactive liquid, such as nitrogen tetroxide, strontium perchlorate, hydrazine, Freon, or hydrogen peroxide, into the supersonic exit cone region. Side force is achieved from recovered

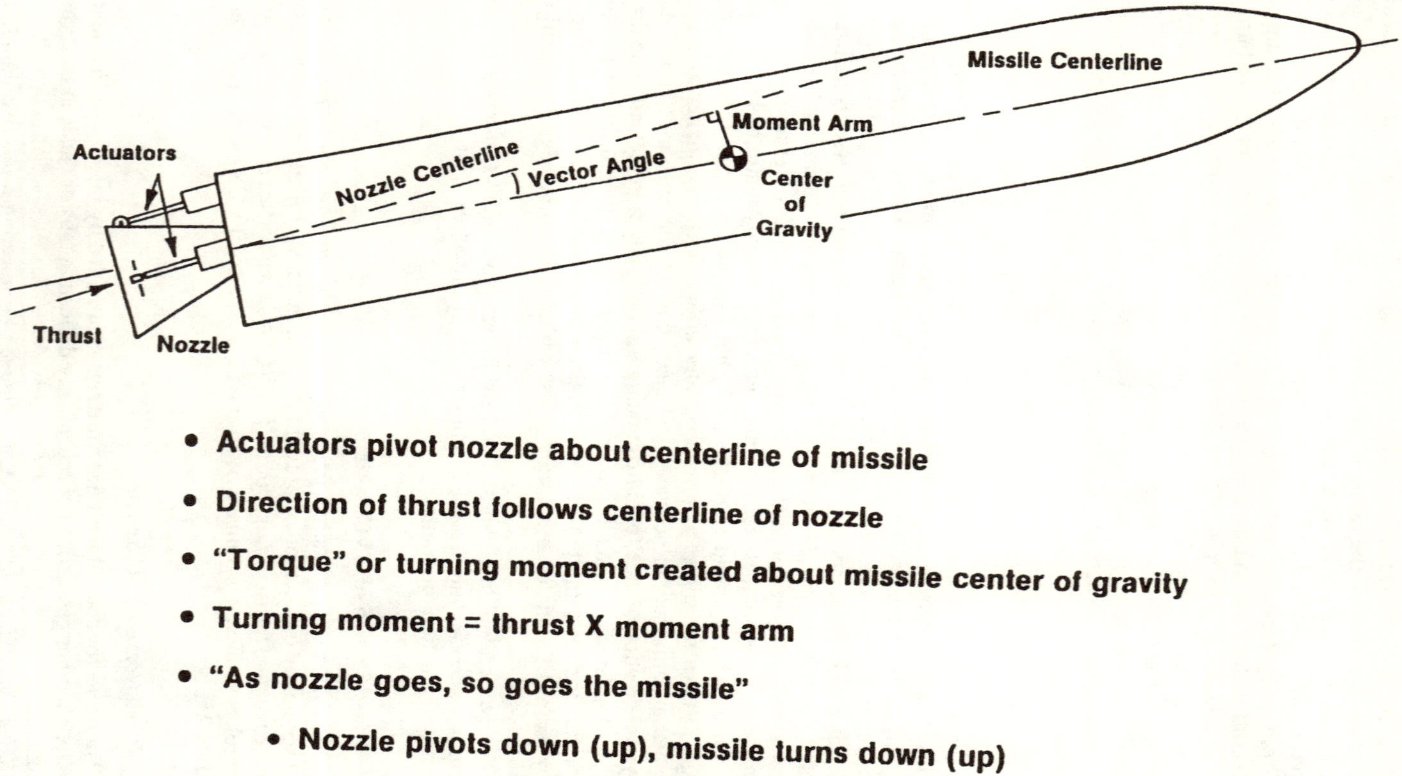

Fig. 24 Missile thrust vector control.

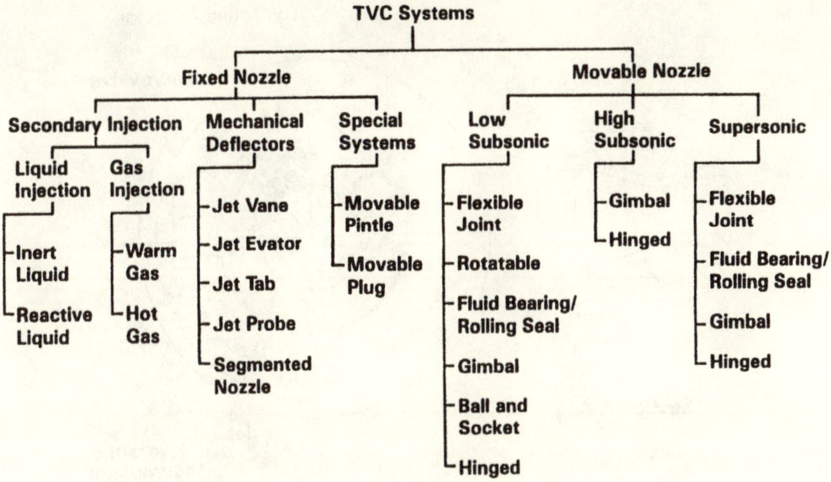

Fig. 25 Classification of TVC systems.

momentum of the injectant, pressures on the exit cone wall from the shock wave created, and exit cone wall pressure due to the added weight of the injectant mass to the exhaust flow, as shown schematically in Fig. 26. A more complete description of LITVC design can be found in Ref. 16.

Hot gas injection thrust vector control (HGITVC) creates side loads similar to LITVC systems by routing some of the chamber gases to injection locations in the exit cone. This concept has an added performance loss over and above LITVC with decreased chamber pressures and mass flow out through the nozzle throat. These systems have previously been very heavy and inefficient, but recent advances of lighter weight carbon–carbon materials have generated renewed interest.

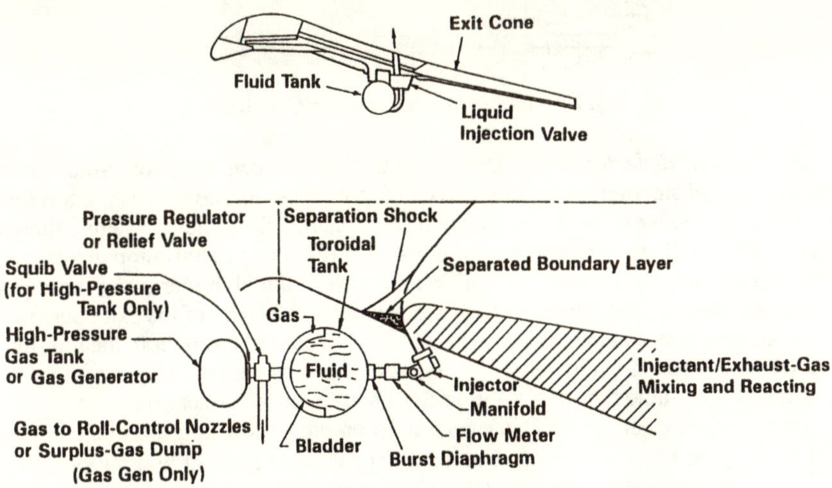

Fig. 26 Secondary injection thrust vector control.

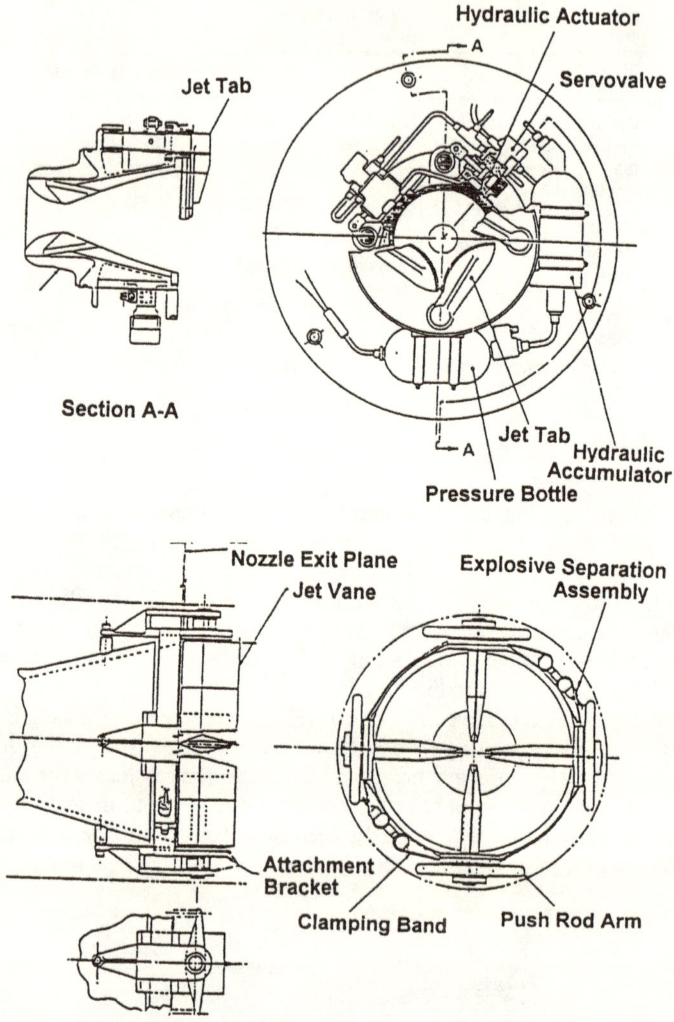

Fig. 27 Jet tab and jet vane TVC systems.

Mechanical deflectors. Jet tabs or vanes[17] have been used for rapid thrust vector control steering, especially early in flight when missile speeds are too low to achieve effective control with external aerodynamic fins. These systems, shown in Fig. 27, usually require very heavy tungsten or refractory metal components and, therefore, resulting heavy actuation systems due to the high inertias at the rapid response rates desired. These systems are usually attached aft of the exit plane as a separate package, so they can be jettisoned and not carried throughout the mission. Jet tabs are inserted in and out of the exhaust plume as needed, whereas jet vanes remain in the plume constantly and vector about their own retention shaft.

More recently, jet tab TVC has been used on one production missile, whereas jet vanes have been incorporated onto an antisubmarine rocket booster system.[18] A more detailed discussion of jet tab and jet vane TVC systems and design can be found in Ref. 19.

2. *Movable Nozzles TVC Systems*

Movable nozzle TVC systems are generally categorized by 1) type of hot-gas sealing concept, such as flexible bearing, o-ring, fluid bearing, or bellows; 2) type of pressure blowout load retention, such as flexible bearing, gimbal, or trapped ball and socket; 3) submerged or desubmerged; and 4) location of splitline between the fixed or stationary and movable portions.

Recently the most common tactical TVC system has been a trapped ball (both submerged and desubmerged) because of their low torque and small envelope packaging ability for TVC vector angles from 10–20 deg. Fluid bearing TVC systems were demonstrated with 15–20 deg capability, but require larger envelopes and result in heavier systems than trapped ball or flex bearing concepts. Recently, improved materials and design/analytical capabilities have resulted in smaller, more compact, and low torque ±8-deg TVC flexible bearing design. Because of the complexity involved in designing and analyzing either a trapped ball or flexible bearing TVC mechanism, this section will address only basic sizing considerations for these two TVC concepts.

Subsonic splitline. Subsonic splitline designs located in a submerged region with a Mach number less than 0.10 are by for the most common type of tactical movable nozzle design. They require special design considerations to ensure propellant exhaust gases do not cause excessive heating of critical structural components or other serious problems. Designs must keep adjoining insulators from swelling and interfering with nozzle movement. A design must also prevent the splitline from filling with semimolten exhaust products that could solidify causing torques to increase or interfere with nozzle movement.

Subsonic splitline gap widths are typically 0.02–0.05 in. wide and in excess of 1 in. in length (approximately), depending on materials and burntimes.

Supersonic splitline. Supersonic splitline designs, such as that described in Ref. 20, contribute their own unique considerations, such as loads and splitline flow due to turning the supersonic exhaust in the exit cone. Until full flow is developed in the exit cone, there can be an instantaneous separation load between the fixed and movable sections that must be reacted by external support. Once exit cone flow is established, a net compressive load exists across the gap between the fixed and movable sections. The external support has typically been two passive (noncontrolled) actuators or gas bearing mechanisms similar to shock absorbers.

Trapped ball. Trapped ball TVC nozzles consist of a spherically shaped movable ball section that rides on and is supported by a mating spherically shaped fixed portion, as shown in Fig. 28. This figure also illustrates submerged and desubmerged trapped ball concepts. Movable section blowout loads are reacted by a low-friction material, typically a thin layer of polytetrafluorethylene.

Figure 29 briefly describes trapped ball TVC sizing considerations, and typical design capabilities. As noted, the socket is usually designed to minimize deflections so that a true spherical surface is maintained for the low liner friction material. As the contact angle decreases, the point bearing load increases due to decreased projected axial bearing area. Most trapped ball designs have demonstrated contact angles between 18 and 41 deg. Thin 0.015-in.-thick, polytetrafluorethylene materials have demonstrated contact pressures from 12,000–25,000 psi. Friction coefficients decrease as bearing pressures increase. Because movement and friction generate heat, nozzle actuation torques also decrease with increasing temperatures up to a point where bondlines and polytetrafluorethylene materials begin to

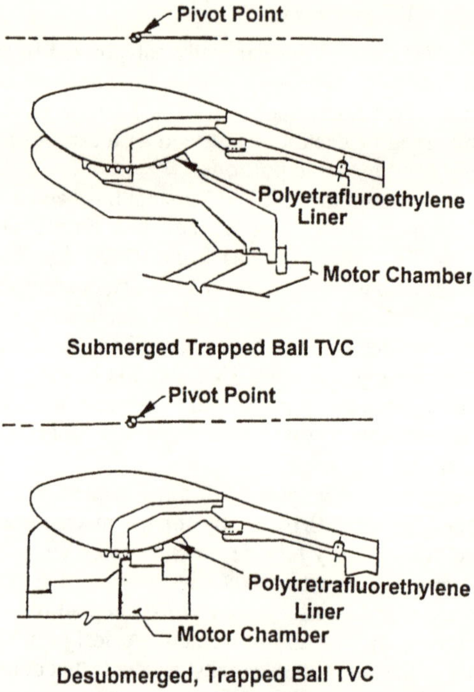

Fig. 28 Comparison of submerged and desubmerged TVC nozzel designs.

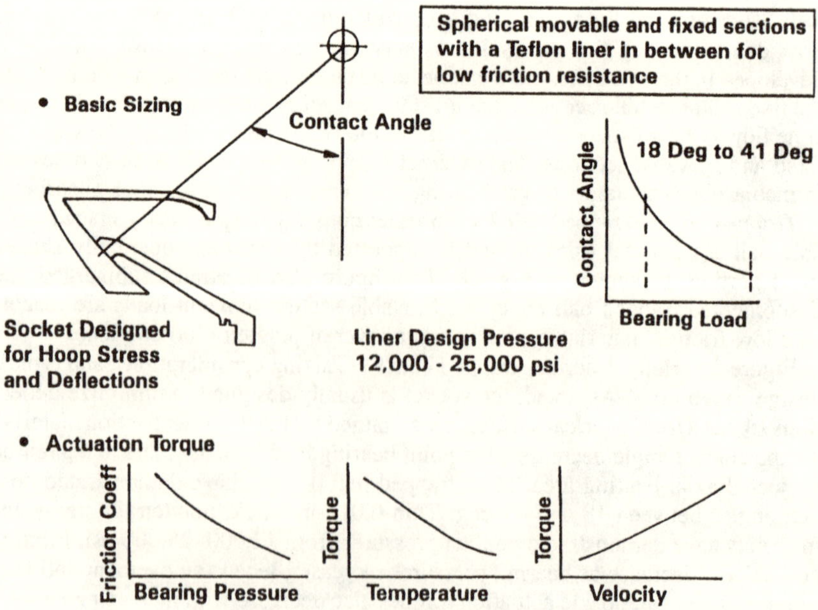

Fig. 29 Trapped ball TVC mechanism sizing.

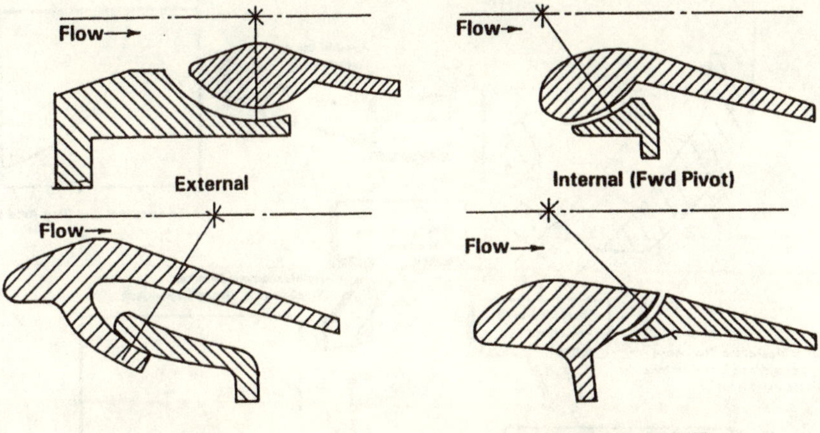

Fig. 30 Types of movable TVC nozzles.

physically degrade. Torques increase with velocity mainly due to inertial effects, not friction coefficients. Specific designs depend on size, materials, and design details too numerous to cover here. Friction coefficients range from 0.015–0.025 for breakout starting motion from rest to 0.01–0.015 during operation. Trapped ball TVC concepts can be designed with several different splitline and pivot point locations as shown in Fig. 30 for tailored TVC system depending on system and envelope requirements. Vector angles as high as 35 deg have been demonstrated on a trapped ball TVC system used for a crew ejection system.

Flex bearing. Flex bearing TVC systems that are used almost exclusively on larger strategic and very large commercial launch vehicle nozzles have recently been pursued for use on tactical systems requiring TVC angles of 5–15 deg. Flex bearings consist of alternating layers of spherically shaped reinforcements and elastomer pads vulcanized together and to the forward and aft end rings. Thrust vector is achieved by shear deformation of the elastomer pads. Reinforcements consist of metallic or composites, and demonstrated elastomers have been silicones, natural compounds, as well as synthetic polyisoprenes. Silicones exhibit very little modulus change over typical tactical motor operating temperature ranges from −65 to +165°F. Natural rubber compounds exhibit low shear modulus values at moderate temperatures from +45 to +145°F, but cannot be used at temperatures below 20°F. Polyisoprenes are good for temperatures as low as −25°F, but not above +145°F. Flex bearing torques increase with vector angle, pressure, and age, as shown in Fig. 31. Aging effects on silicone flex bearing torque are nearly zero.

B. TVC System Tradeoffs

Generalized pros and cons of each TVC system are compared in Table 10. In addition to the specific system requirements addressed in Secs. II.A and VI.A, special considerations that need to be considered are now discussed.

Axial thrust losses occur anytime a nozzle is vectored from the missile centerline. A subsonic movable nozzle has the lowest demonstrated thrust losses with those limited to the geometric cosine function of the vector angle. The other

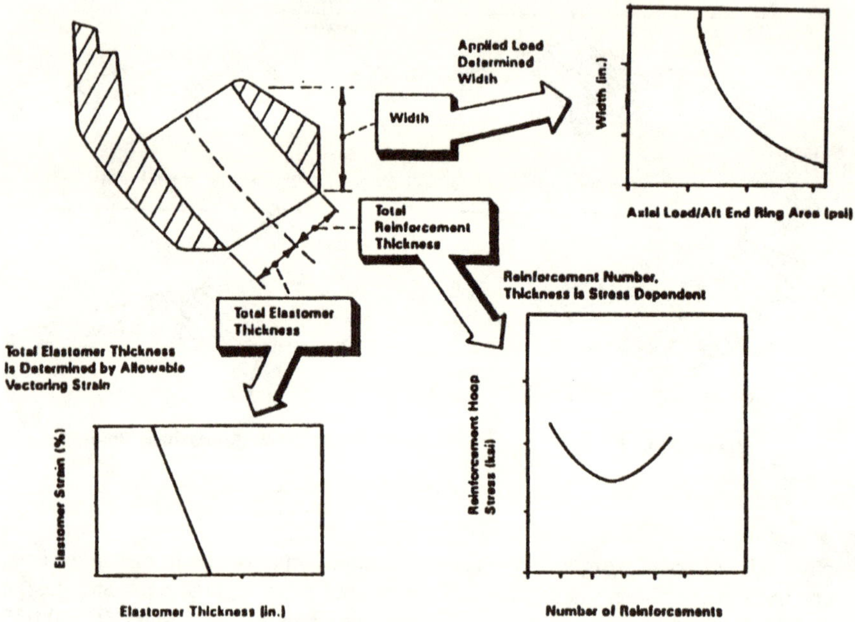

(a) Basic sizing

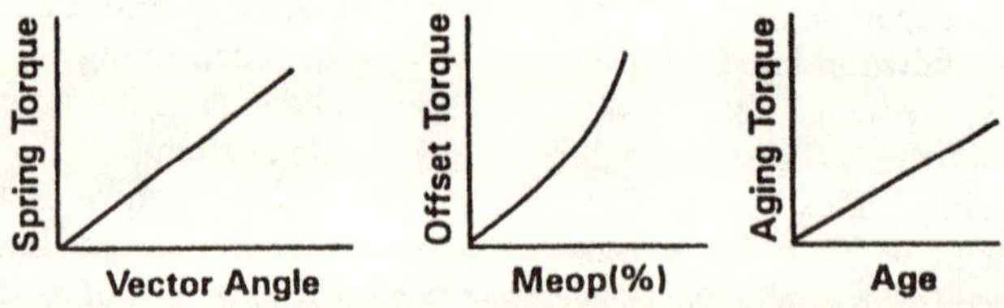

(b) Actuation torque

Fig. 31 Typical flex bearing TVC design sizing parameters.

extreme is jet tabs and vanes, with resulting large impingement shock losses. Figure 32 compares thrust losses for the various systems as a function of TVC vector angle.

Roll control cannot be achieved with a single movable nozzle but requires two to four. Omnidirectional trapped ball nozzles must incorporate some sort of antirotation feature to keep the movable portion of the nozzle from spinning when actuated. Flex bearings do not typically require TVC antirotation devices.

Another factor that must be considered, and probably the most important, is demonstrated reliability. TVC has been around since the early days of rocketry, as demonstrated by Goddard's crude jet vanes in the rocket exhaust nearly 60 years ago. In the 1990s, however, only two tactical TVC systems have high demonstrated reliabilities with low performance losses: subsonic splitline cold trapped ball and flex bearings. Thus, as new designs are conceived for future missions, it is the demonstrated success of the subsonic trapped ball nozzle that is the current design of choice.

Table 10 TVC system general comparisons

TVC system type	Pros	Cons
Flex bearing	Well demonstrated/ characterized ±8–12-deg capability Reliability/full checkout/ low development risk Wide operating temperature range	Higher torques above 15-deg TVC Larger packaging envelope above 15-deg TVC
Trapped ball	Well demonstrated/ characterized ±20-deg capability Low development risk Good packaging	Requires continual positive load to maintain seal capability Requires antirotation device for omnidirection capabilities
Supersonic splitline	Low weight Good packaging Vector capability	Not fully developed Possibly high torque Splitline flow
Jet vanes	Provide roll control 10-deg vector capability Packaging independant of nozzle	High-thrust losses Moderately heavyweight
Jet tabs	Low risk/well characterized Good packaging Rapid response	High-thrust losses High torque Moderately heavyweight
LITVC	Rapid response	±6-deg TVC capability
HGSITVC	Rapid response	Vector capability limits unknown Not fully developed

Jet vanes first used by the Germans on the V-2 rocket continued to be used exclusively after World War II on several operational and experimental tactical missile systems. In the 1960s, jet vane TVC was replaced by movable gimballed nozzles or jet tab fixed nozzle TVC that was developed in the 1970s. Jet vane and jet tab TVC systems remain strong candidates for future systems if the performance losses are tolerable. As flex bearings exhibit more demonstrated history on tactical-size nozzles, they have the associated reliability history of being demonstrated on over 3700 TVC systems without any failures.

Any of the aforementioned systems are viable for future tactical TVC missiles and like many aspects of a TVC nozzle design, the best choice is mission and systems requirements dependent.

VI. Summary

It should be apparent that the design of a tactical solid rocket motor nozzle is a highly iterative procedure involving many distinct steps. A successful design is one that uses the information derived from each step of the process, continually updating previous data, and arriving at a fully integrated design. Analytical

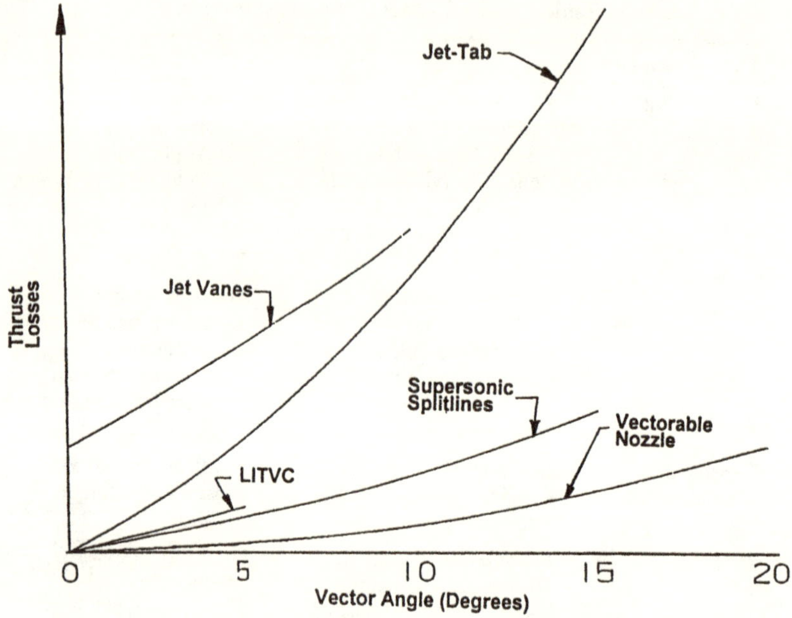

Fig. 32 TVC system thrust losses.

techniques, materials, and material properties are constantly being improved and updated, resulting in a dynamic environment encompassing the entire history. The successful nozzle design is also one that not only meets all requirements and design objectives, but is reliable, affordable, and producible.

VII. Acknowledgment

The authors wishes to express his appreciation and thanks to the following persons for their assistance and help in preparing this chapter: Larry Smart and Roy Loh of Thiokol Corporation, Gas Dynamics and Heat Transfer, for contributing input to the thermal analyses; Gordon Jensen, Chief Combustion Systems, Ramjet Engineering, United Technologies, Chemical Systems Division, for review, suggestions, and format requirements; Larry Sayer, Director of Design Engineering, Thiokol Corporation, for review and suggestions; and Vera Rawlins and Carolyn Robinson, Secretaries, Thiokol Corporation Design Engineering, for indescribable patience and work in preparing the final chapter input.

References

[1]Ellis, Russell A., "Solid Rocket Motor Nozzles," NASA Space Vehicle Design Criteria (Chemical Propulsion), NASA SP-8115, June 1975.

[2]Sutton, P., *Rocket Propulsion Elements,* Wiley, 1963, Chaps. 2.3–4.

[3]Potts, M. F., *Machine Design Magazine,* Northwestern Univ. Mechanical Engineering Dept.

[4]Wong, E. Y., "Solid Rocket Nozzle Design Summary," AIAA Paper 68–655; June, 1968.

[5]Roark, R. J., *Formulas for Stress and Strain,* McGraw–Hill, New York, 1965.

[6]Dickerson, G. R., Coats, D. E., Dang, A. L., and Gunn, S. S., "The Solid Propellant Rocket Motor Performance Prediction Computer Program (SPP), Version 6.0," Software and Engineering Associates, Inc., AFAL-TR-87-078, Dec. 1987.

[7]Anon., "User's Manual, Aerotherm Chemical Equilibrium (ACE) Computer Program," Aerotherm Corp., May 1969.

[8]Anon., "User's Manual, Aerotherm Axi-Symmetric Transient Heating and Material Ablation Computer Program (ASCHAR)," Aerotherm Corp., AFRPL-TR-72-24, Aerotherm Rept. UM-72-26, Jan. 1972.

[9]Anon., "Military Standardization Handbook-Metallic Materials and Elements for Aerospace Vehicle Structures," MIL-HDBK-5E, 1987.

[10]Anon., "Military Handbook-Polymer Matrix Composites," MIL-HDBK-17-1D, 1993.

[11]Anon., "ABAQUS Theory Manual, Version 5.4," Hibbit, Karlsson and Sorensen, Inc., Pawtucket, RI, 1995.

[12]Crose J. G., Holman, R. L., and Mack, T. E., "NIFD Computer Program (ROSAAS)," PDA Engineering, TR-9204-00-01, Oct. 1976.

[13]Anon., "I-Deas User Guide, I-Deas 4.1," Structural Dynamics Research Corp. 91688.

[14]Anon., "MSC/NASTRAN, Version 67," MacNeal–Schwendler Corp., Los Angeles, CA, 1993.

[15]Simmons, B. D., and Berndt, W. E., "Tensor Polynominal Failure Theory Application," AIAA/SAE/ASME 20th Joint Propulsion Conf., AIAA Paper 84-1309, June 11–13, 1984.

[16]Anon., "Solid Rocket Thrust Vector Control," NASA Space Vehicle Design Criteria (Chemical Propulsion), NASA SP-8114, Dec. 1974.

[17]Anon., "Solid Rocket Motor Thrust Vector Control (TVC)," AIAA Lecture Series, Reno, NV, Jan. 9–12, 1995.

[18]Ripley-Lotee, M.S., and O'Neil, S. M., "Jet Vane Thrust Vector Control—A Technology with New Horizons," JANNAF Joint Propulsion Conf. NWC/CL, 1980.

[19]Anon., "Jet Vane and Jet Tab TVC Technology, a Bibliography," Chemical Propulsion Information Agency, CPIA-LS-88-04, May 1988.

[20]Kearney, W. J., and Prins, W. S., "Supersonic Splitline Thrust Vector Control," Atlantic Research Corp., AFAL TR-88-111, Oct. 1988.

Chapter 7

Solid Rocket Case Design

M. Chase* and G. P. Thorp†

Royal Ordnance plc, Summerfield, Kidderminster, England, United Kingdom

Nomenclature

B = bending moment, Nm
Bc = critical bending moment, Nm
d = internal diameter of case, m
d^* = external diameter of charge, m
E = Young's modulus, Nm^2
I = second moment of area of tube, m^4
L = length of end ring bond, m
l = length of case, m
M = freestream Mach number
p = minimum chamber pressure at burst, Nm^{-2}
p' = chamber proof pressure, Nm^{-2}
p^* = maximum expected operating pressure, Nm^{-2}
S_H = hoop stress, Nm^{-2}
S_L = maximum longitudinal stress, Nm^{-2}
S_s = maximum shear stress, Nm^{-2}
T_a = freestream ambient temperature, K
T_r = recovery temperature, K
t' = thickness of case, m
W = strip width, m

I. Introduction

TACTICAL missiles vary considerably in mass, size, shape, and purpose. For example, some are light enough to be readily carried by one person whereas others are more massive requiring large custom-built transporting, storing, and launching facilities; some are high aspect ratio darts whereas others are only slightly greater in length than diameter; some are used at close range for

*Manager of Materials, Rocket Motors Division.
†Chief Engineer, Rocket Motors Division.

line-of-sight targets whereas others are guided well beyond the horizon. It follows that in nearly all instances each tactical missile has to be individually designed.

A tactical missile consists of three basic units: the warhead, the guidance system, and the rocket motor propulsion unit. This chapter deals with a subassembly of the propulsion unit known as the rocket motor case and is intended to draw attention to the various aspects of design, materials, and fabrication techniques which need to be considered in each individual instance.

Basically, the rocket motor case has to provide a leak-free chamber that will adequately contain the stresses imposed by the high-pressure gases evolved during the burning of the propellant charge. Such chambers usually have, or at least closely approach, those structural properties that are required for the missile case; hence, it is not unusual for the motor case to be integral with the missile. However, the motor case can also be discrete, not forming part of the missile's load bearing structure, either being housed within the missile or, alternatively, being attached to the rear (in tandem) or the side (wrap round) of the missile and having the capability of being jettisoned during flight when burning is complete.

A rocket motor case can, therefore, be defined as a leak-free structural chamber of a size and shape determined by the missile requirements including a nozzle and igniter and having provision for attaching the fore and aft missile subassemblies, the latter possibly requiring a transfer or blast tube to duct hot gases past the guidance system. It must withstand the pressures and high temperatures generated by the burning propellant charge, the loads imposed by flight maneuvering, the loads imposed by propellant change of shape when stored at various temperatures, the effect of external aerodynamic heating during air carriage and/or missile flight and the effects of mechanical shock, vibration, and corrosion during storage and handling. Provision must be made for efficient and practical internal assembly and sometimes for possible maintenance dismantling.

Above all, the designer recognizes that the mass of the case is a penalty which reduces the performance of the missile and, therefore, has to be kept to a minimum; this aspect greatly influences choice of case materials and manufacturing techniques. During the design phase a number of possible solutions may be apparent and many factors need consideration before the design of an individual motor case can be finalized. Consequently, it is only possible to give concepts and guidelines and not precise design details. The process of optimization involves consideration of material properties, methods of manufacture, inspection, and proof, as well as interactions with other rocket motor and missile components.

In the three sections that follow, design, materials, and fabrication are considered separately, each section is subdivided into structural and thermal aspects. Mention is made of end closures and blast tubes; nozzles are considered in Chapter 6. A relatively recent requirement is that of insensitive munitions, and the importance of low vulnerability to fuel fire and fragment attack is discussed in this context. The need for development and qualification phases to follow the design phase and for quality assurance procedures to be incorporated throughout are highlighted. Materials and techniques mentioned are all considered to be within the limits of good engineering practice and, with proper precautions, safe to handle. Other techniques of doubtful safety, such as welding to or winding over cases already filled with live propellant, have been omitted.

In the fourth and final section a specific rocket motor case design is given as an illustration of the steps to be undertaken during design.

II. Design Considerations

A. Performance

A performance requirement for the rocket motor in the form of a minimum acceptable total impulse level will normally be specified by the customer, accompanied by constraints of mass, of length and diameter, of performance variation over a given temperature range, and of environmental conditions. The rocket motor case designer's first consideration is mass, since mass of any inert component reduces the available impulse. Choice of case material will be greatly influenced by this aspect. Closely following mass will be consideration of size; larger diameters and, to a lesser extent, greater length, will reduce available impulse by aerodynamic drag and, in the instance of missiles transported underslung from aircraft, also reduce aircraft performance. In addition to general choice of case materials, overall diameter may be affected by a necessity to apply external insulation to the case, and overall length will be affected by the type of guidance system chosen. Should the guidance system require a transfer (or blast) tube to duct the hot gases past the guidance system, a further loss in performance occurs due to the need to increase motor pressure (for subsonic blast tubes) or through resulting thrust losses (for supersonic blast tubes).

Consideration of environmental conditions will be influenced by the application; land use often will necessitate wide temperature extremes, particularly for solar radiation, and air use will require more stringent vibration environments. Interfaces with the missile must be considered, such as attachment points, fittings for wings, launch attachments, and cable conduits; all affect performance and their effects need minimizing during the design and development stages of the motor. Performance as an insensitive munition, if specified, must be considered as an integral part of the motor design and may also affect the available impulse.

B. Structural

1. General

For design purposes the motor case may be divided into a number of components, as shown in Fig. 1, each of which reacts with adjacent components. The body tube carries the main loading comprising internal pressure stresses, missile bending stresses, and localized stresses from wing and launcher attachments. The end rings provide attachment points for the end closures and interface points with the missile components. The end closures complete the pressure vessel and may provide an igniter attachment boss at the forward end and a nozzle attachment boss, with or without an intervening blast tube, at the aft end.

Not surprisingly many of these components are made from readily available engineering materials, usually good quality steels and aluminum alloys and occasionally titanium alloys. These are homogeneous materials exhibiting isotropic properties and they are the subject of Secs. II.B.2–6. Sealing of joints between components is discussed in Sec. II.B.7. In addition, other types of construction are available which make use of higher specific strength materials in which, in some

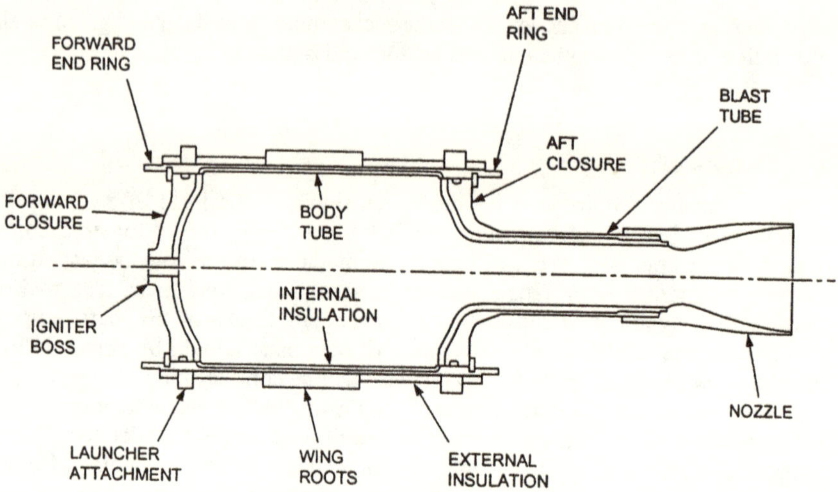

Fig. 1 Rocket motor case.

instances, the structural properties of the components are anisotropic. Design with these materials is discussed in Sec. II.B.8.

2. *Body Tube: Pressure Considerations*

Factors that have to be taken into account when designing the body tube component are internal pressure, motor and missile stiffness requirements, attachments and their loading, and, for certain missiles, external pressure. Of these, the overriding consideration is usually the internal pressure which is associated with the gas temperatures and velocities and the burn time of the propellant charge. All of these may vary considerably from design to design and may, for example, cover the ranges 20–250 × 10^5 Nm^{-2} pressure, 2500–3600 K temperature, 0–2500 ms^{-1} velocity, and 1–60 s time. In general, however, the structural material is insulated from the hot gases in order to limit its temperature rise, and this ensures its room temperature properties are virtually unaffected, thus allowing the initial design to be carried out on the structure based solely on the internal pressure and ignoring gas temperature, gas velocity, and burn time, and utilizing the well-defined room temperature properties of the chosen materials. Occasionally, however, uninsulated structural materials are used and, in these instances, heat transfer calculations have to be carried out and the structural properties evaluated at high temperatures.

The maximum pressure expected during the motor firing, the limit load, will be supplied by the charge designer. It is typically defined in the U.K. as that pressure that will not be exceeded on more than 1 firing in 1000 with 95% confidence at the upper firing temperature. At the initial design stage an allowance for statistical variation will have been included in the limit load. In established designs this load will be accurately known from firing records. The limit load is multiplied by a proof factor of 1.125, and this results in a pressure at which all cases will subsequently be proof-pressure tested. Where case failure would endanger personnel or launch apparatus, as is usual, a safety factor of 1.33 is imposed on the limit load to provide the designer with a minimum burst level for the case design. To minimize mass it is necessary to work as closely as possible to the safety factored value. Definitions and safety factors vary slightly from country to country; the preceding comments refer to those factors currently applied in the U.K.

For an initial estimate of tube thickness, based on internal pressure, the simple formula for hoop stress in a thin-walled pressure vessel is used, Fig. 2a, hoop stress is equal to load/area, i.e.,

$$S_H = \frac{pdl}{2t'l}$$

$$= \frac{pd}{2t'}$$

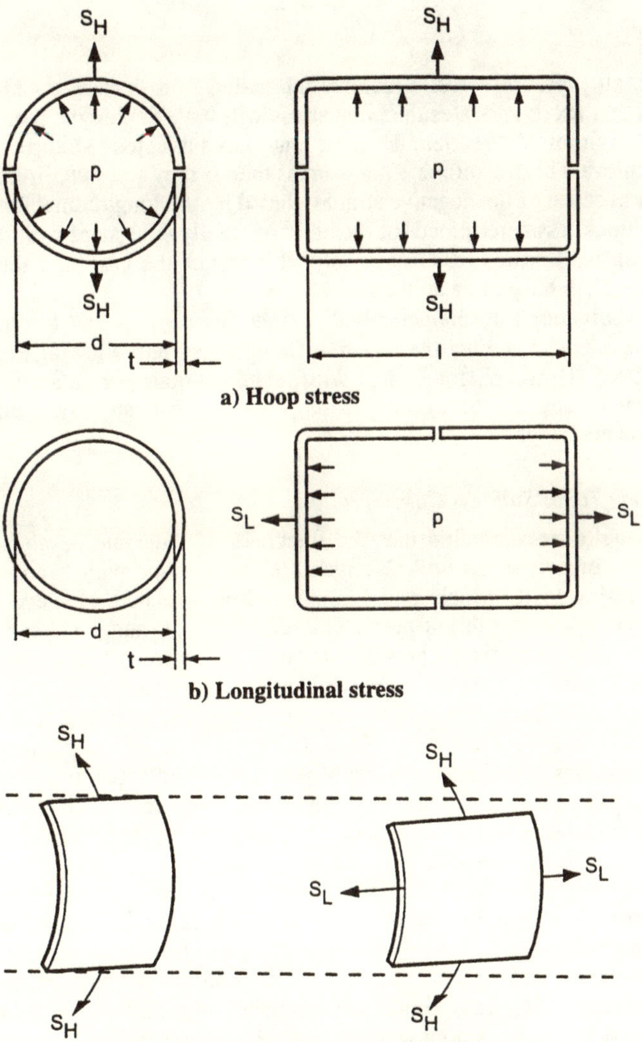

c) Hoop stress induced by longitudinal stress

Fig. 2 Stress considerations.

In all instances the weakest structure is considered, that is, the thickness is that at the minimum end of the thickness tolerance band and the material strength is the minimum specified. The resulting thickness may give a first indication of the material that should be considered. For example, a small diameter or a low-pressure motor may require steel that is too thin to be handleable thus leading to consideration of aluminum alloy.

It should be noted that the longitudinal stress in the pressure vessel is only half the hoop stress, Fig. 2b, longitudinal stress is equal to load/area, i.e.,

$$S_L = \frac{p\pi d^2}{4\pi d t'}$$

$$= \frac{pd}{4t'}$$

Hence, all pressure vessel structures fabricated from isotropic materials from a design that accommodates the hoop stress will have double the required capacity in the longitudinal direction. Isotropic materials subjected to bidirectional loading will achieve a higher ultimate hoop stress than if they were unidirectionally hoop loaded because of the negative strain induced by the longitudinal load, Fig. 2c.

The thickness determined for the body is usually consistent with that described as a thin wall structure. Occasionally this is not the case and the appropriate formulas then have to be applied.

For short length-to-diameter bodies, theories suggest that the thickened end sections interact to reduce the maximum hoop stress, thus allowing thinner sections to be used. However, for bodies with length-to-diameter ratios as low as 1:1, practical testing, including burst tests, does not indicate any significant effect from the end rings.

3. *Body Tube: Stiffness Considerations*

It is quite usual to find that the thickness of the material chosen based on pressure considerations will also be sufficient to cope with handling and flight maneuvering loads which are generally stiffness related. However, care has to be taken to analyze this aspect at the design stage and, in this respect, finite element stress analysis is a powerful aid used extensively in motor case design for evaluating stresses and strains allowing the design of more optimized components which could require less testing. The structure or continuum to be analyzed is modeled as a set of elements of finite size connected only at points called nodes. Each node has specified freedoms of translation and rotation in response to the neighboring elements or boundary conditions. The stresses are obtained from the field of nodal displacements together with the appropriate constitutive equations for the material properties.

Significant advances are still being made in the basic computer stress analysis programs, some programs being very general whereas others have been developed for a specific industry, but however sophisticated these programs, in general they can only analyze a structure; they are design aids and cannot replace designers and stress analysts. To optimize a component, an initial design is evolved based on a classical text book approach coupled with general design expertise. The stress analyst will then define the elements that, from experience, he considers best fit the component. The resultant analysis will show areas of high or low stress or strain.

If it is considered that changes are necessary to the design, then the designer and stress analyst will decide the modifications to be made prior to carrying out a further analysis.

As programs become simpler to use, there is the temptation to design the ultimate component. When deciding whether a design is satisfactory, the limitations of the program, imposed by machining accuracies and material properties, and possible advantages, particularly in terms of mass, must all be taken into account. Programs for rocket motor design should be able to deal with almost any stress problem featuring almost any geometry, complete anisotropy, various material constitutive formulations, regions of differing materials in the same body, and thermal, pressure, acceleration, or point loadings in combination. Both two- and three-dimensional problems should be handled, and the analysis should not be limited to small strains or small deflections.

A major finite element program will offer about a hundred different element types ranging from those with very general applications, such as brick elements, to special elements exemplified by those that simulate a crack tip. The analysis tends to yield relatively inaccurate results if elements exhibit long, thin aspect ratios or corner angles near to zero or 180 deg. Usually these situations can be avoided, but difficulties can be experienced with long, thin adhesive layers in bonded structures.

The two elements most frequently used for the analysis of components are the quadrilateral, with its degenerate form the triangle, and the hexahedron, with its degenerate form the triangular prism or tetrahedron. The quadrilateral is used as a plane element in either plane strain or plane stress or as an axisymmetric element under axisymmetric loading. For many applications an adequate initial solution may be obtained by making an analysis using slight modifications to the geometry to bring it within the scope of this element. Applications include motor bodies, end rings, and reasonably axisymmetric end closures in the axisymmetric mode. The hexahedron is used as a solid brick element in a three-dimensional analysis. A particular application would be a closure for a multinozzle motor or with nonsymmetric features and launch hooks.

Idealization is the process of representing a real body or structure in terms of its geometry, material properties, loading, and constraint by a mathematical model that is amenable to analysis by the finite element method. Very often, an insignificant modification to geometry, loading, or support will result in a very great reduction in the computing and data preparation times. The term insignificant depends on behavior under load and also on which aspect of the result is under investigation. On the other hand, the model must possess adequate freedom or flexibility for the results to reflect the essential features of the real situation. The idealization process requires judgment and experience, and application of the finite element method is, therefore, not an exact science.

Finite element analysis should never be used uncritically or without insight. It is vital to ensure that the elements chosen do not lack a degree of freedom that must be taken into account and that they are sufficiently numerous for the model not to be significantly too stiff. Finite element models, in general, err on the side of being stiffer than the real item. Most finite element analyses used in rocket motor design are applied to a geometry of a type that has been analyzed previously and the expected behavior is known. In the case of a problem that is new to the user, it is good policy to do an initial run with a coarse division into elements to determine the general behavior. The solution should be examined with a view to understanding the deformation, shown by plotting the deformed shape, and the stress output. It is possible for a solution to appear feasible and yet be quite wrong, and so further checks should be made. The solution should reflect the loads and constraints that were intended in the input and, in the case of a compound body, the material properties should be seen to be distributed as intended. It is usually a simple matter to check the load balance in at least one direction by ensuring that the reactions given in the solution are equal and opposite to the total load that was intended to be applied in that direction. This check should never be omitted. If further verification is desired, it may be possible to make a very rough approximation to the actual problem and apply simple engineering formulas. Having determined that the coarse solution is behaving correctly, a finer division into elements may be made, using the results of the coarse solution to concentrate the elements in regions in which any stress is changing relatively rapidly with position. The fineness of division into elements depends on the accuracy required and is a matter of experience backed, if necessary, by trial.

4. *End Rings and Attachments*

The main function of the end rings that may or may not form an integral part of the body tube is to provide attachments for the motor end closures and for the missile. The attachment features inevitably require these end rings to be thicker than the main body tube which, when a flush outside diameter is required, necessitates the inside diameter being smaller than that of the main tube.

The minimum inside diameter at one end of the tube, at least, will be governed by the charge manufacturing method. The two basic types of charges are cartridge loaded and case bonded. With a cartridge-loaded charge, its outside diameter must be slightly smaller than the inside diameter of the end ring, hence, the inside diameter should be similar to the inside diameter of the body lining to ensure minimum motor length. With a case-bonded charge, no such restriction exists, and the diameter would be governed by manufacturing considerations, such as core extraction and ease of propellant filling. In this instance ease of body lining is also an important criterion.

In addition to the pressure and missile loads, the end rings must also withstand the bending moments imparted by the end closures so there is an interaction between the design of these components. The method of missile attachment required is also an important factor in the design of the end ring. Other factors, such as access holes for igniter cables and safety and arming mechanisms, also have to be taken into account.

Attachment features such as launcher hooks and wing fixings require detailed consideration and could influence the choice of material. For example, aluminum alloy could be acceptable for wing fixings but steel would be necessary for launch hooks to withstand the mechanical environments. These features may be incorporated into the case by machining integrally with the case, especially suitable where attachments extend most of the length of the case and where attachment loading is high, by welding or by adhesive bonding. Sometimes local thickening of the tube is necessary to withstand the additional loads imposed. The optimum structure, that is one in which the loads result in a constant maximum stress in any radial location, will vary in thickness. The thickness will be a maximum in the neighborhood of the load and a minimum at a point farthest from the applied load taking into account the degree of machining and mass. In deriving the optimum thicknesses throughout the ring, initial estimates are carried out based on the theory of curved beams followed by detailed stressing and modifications to the design until an acceptable solution is obtained.

5. End Closures and Methods of Attachment

Three basic forms of end closure are possible. The first is integral where the closure and case are a continuous structure. In this instance there may be a small central hole to house an igniter or to ease some aspect of motor manufacture, Fig. 3a. The second is part integral where the outer part of the closure is integral with the case but the closure has a relatively large central hole, Fig. 3b. Third, the closure is separated and mechanically attached to the end ring, Fig. 3c.

An integral closure will generally be the lighter for any given material and, hence, with back extrusions, forgings, or for bodies with welded ends, this type of closure can be used. However, important criteria for the choice of closure will

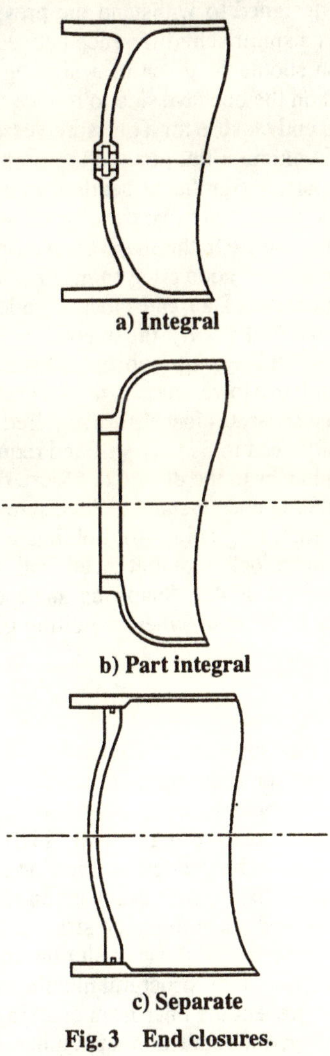

a) Integral

b) Part integral

c) Separate

Fig. 3 End closures.

be body lining and charge manufacture; the lining of relatively long motors with sheet material is usually simplified with access from both ends; a cartridge-loaded charge requires a separate closure at one end for loading, and manufacture of a case bonded-charge will necessitate a relatively large opening at one end for ease of propellant filling and core removal. For small diameter motors, that is, up to about 150 mm, there will be little difference in mass between separate and part integral closures and body manufacturing considerations become the criteria.

When using a cartridge-loaded charge with an integral closure, this closure could be at either end of the motor. However, lining of the aft closure is a very critical operation and is more difficult if this is the integral closure at the end of a long tube. For this reason the forward closure should be the integral closure. In general, closures are designed so that motor pressure is on the concave side, this generally results in a lighter closure and lower bending moments on the end ring.

The operation of a dual chamber motor with a single internal closure may require the closure to be designed to withstand the pressure in the two chambers independently. If there is a significant difference between the two operating pressures, then consideration should be given to designing the closure such that the higher pressure operates on the concave side to reduce the effect of buckling.

The ideal shape for an end closure for a pressure vessel is a hemisphere, but for maximum utilization of volume a flat plate is the optimum. However, the latter is much heavier and imparts a significant bending moment into the end ring. In practice a compromise between these two extremes is sought. From experience it has been found that an end closure in the shape of an ellipsoid with a ratio of minor to major semiaxes of 0.4 gives a good compromise between mass and volume.

Because of the deformation of an end closure under pressure loads and the bending moments imparted to the body, the overriding design criterion is usually strain, rather than stress. As stiffness is proportional to thickness cubed, a light alloy closure will generally result in a lower mass structure and, consequently, aluminum alloys are often used when separate closures are required. Closures for multinozzle side vented motors usually need to be very stiff and require special consideration. It may be necessary to add ribs to the design to ensure the required stiffness.

Methods used for attaching end closures include screwing to the case (Fig. 4a), holding with a screwed lock ring (Fig. 4b), holding with a snap ring (Fig. 4c), bolting (Fig. 4d), and using a lock wire that mates both with a slot in the end ring and a groove in the closure (Fig. 4e). Snap ring and lock wire methods give the maximum entry diameter to the case, whereas bolting gives the minimum.

6. *Blast Tubes*

Blast tubes may be either supersonic, where the choke is at the forward end so that the gas flow is supersonic throughout the tube but the pressure is relative low, or subsonic, where the choke is at the aft end so that the gas flow is subsonic throughout the tube but the pressure is relatively high. For the supersonic tube the maximum stress is likely to be the design criterion for the structural member. Subsonic tubes, however, are likely to be designed on strain. This is because blast tubes have to be insulated, and since the hoop strain capabilities of molded rigid insulants are near 0.2%, compared to those of aluminum alloys at 0.35% and steel at 0.5%, it is necessary to design the structural member of subsonic blast tubes on dilation considerations to prevent the liner from cracking under the high pressure. In either instance, however, on small diameter tubes a thickness less than the

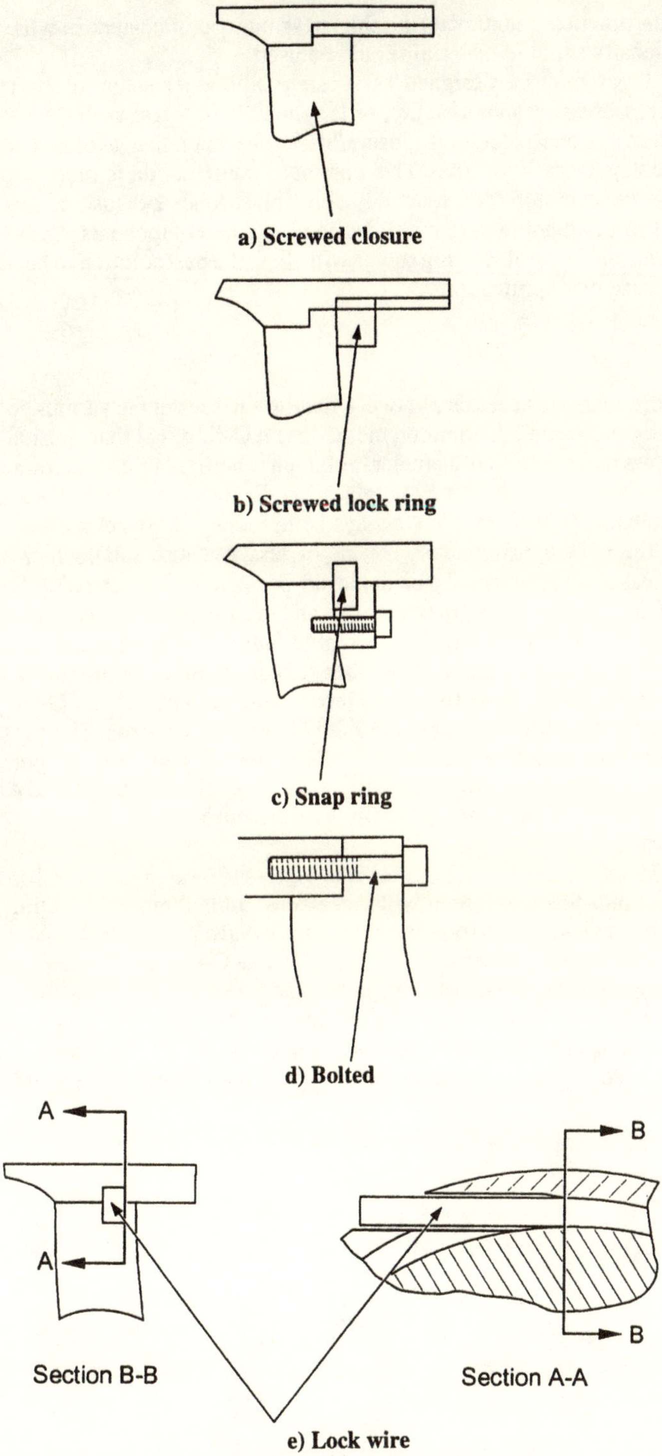

Fig. 4 Closure attachments.

minimum practical manufacturing thickness may be indicated, in which case the lowest density suitable material should be used.

Blast tubes may be designed as separate components that are assembled to the end closure after manufacture, or they may be integral with the closure. The latter allows a one-piece end closure/blast tube liner that avoids a joint at the critical entry to the blast tube. This one-piece construction is more applicable to small diameter motors and relatively short blast tubes because of the problems involved in the manufacture and handling of large components. Only a detailed consideration of manufacturing routes will allow the best solution to be determined for any particular application.

7. *Sealing of Joints*

Pressure seals are necessary between mating surfaces of the various components of the pressure vessel. A common method is an O-ring seal that is usually located in a groove on an external diameter, although where other considerations dictate a thick body attachment, it may be placed in the inside diameter. Face seals have also been used. O seals may not always be necessary in rocket motors with short burning times of approximately 25 ms or less, because satisfactory sealing of motor pressures can normally be achieved over such time scales by face to face seals, the maintenance of tight tolerances, and the use of proprietary thread sealants on threaded joints. However, if catastrophic failure of a system could result from even a small amount of gas leakage, then it is only prudent to use an O seal. With screwed joints it is unwise to rely on thread sealant, whereas the O-ring seal is a more positive method with a proven life in excess of 15 years. The most suitable material for all round properties, including compatibility, low temperature, and high temperature use, is silicone rubber. Other materials such as EPDM and butyl have been used. Viton is not generally used because of its poor low-temperature properties.

All joints should be leak tested at a low pressure, say about 5×10^5 Nm^{-2}. If an O seal holds this pressure, it will seal against normal motor operating pressure. One method of testing is to pressurize the motor and check for leaks. An alternative method which is used extensively is to use double O-ring seals, Fig. 5, and to leak test by applying pressure between joints and checking for pressure decay over a period of about 2 min. This is a severe test which detects very minute leaks because of the small volume of gas between the seals. After a successful test the leak test hole is plugged and sealed to give added protection should the inner

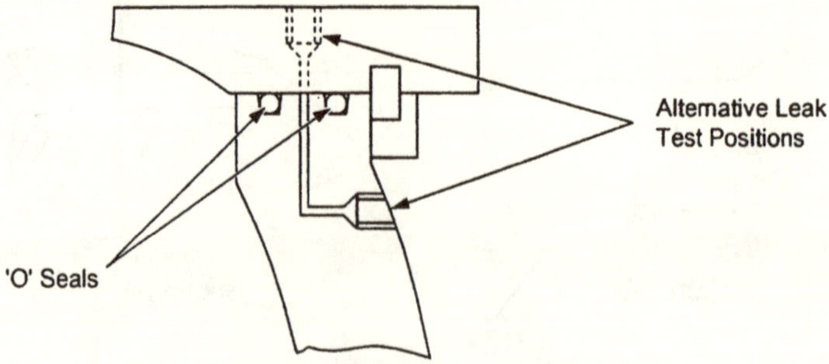

Fig. 5 Double O-ring seals.

seal fail, although this plug cannot itself be leak tested. This technique is more expensive than a single O-ring seal because two grooves require machining and it is necessary to machine holes for the leak test itself. However, no account has to be taken of the effect of pressure on the internal components of the motor, and this may be considered advantageous.

The use of O-ring seals with a threaded component has the disadvantage that when assembling there is movement between the component and the O-ring seal, and there is more likelihood of damage to the seal. When designing the mating surfaces it is essential to ensure that there are no sharp corners that can damage the O rings. In addition, the minimum length of mating surface is desirable to assist assembly and to ensure minimum movement of the O-ring seal in the compressed state.

8. *Body Tube: Composite Materials*

Strip laminate. During the 1950s the structural material with the highest specific strength known in the engineering field was a carbon steel used for razor blades and saw teeth. This material proved difficult to machine and to weld and was only available in thin sheet or strip form. Consequently, a method of using it to form tubes was devised which used only adhesive bonding to hold it together structurally. The process was called strip laminate winding, a Royal Ordnance plc, Rocket Motors Division process. The process consists of winding multiple layers of steel strip in a helical fashion on to a mandrel and bonding them together and to end rings with adhesives. The process is particularly suitable for cases with length to diameter ratios of 4:1 and greater and for thin-walled straight components. Minimum thickness achievable with three layers is approximately 0.5 mm (0.020 in.). In most respects the mechanical properties of tubes made by the strip laminate process are identical to those of homogeneous tubes, the adhesive layers being very thin.

The principles of rocket motor case construction by this method are that the hoop loads are taken by the layers of the steel strip, the longitudinal loads in the tube are transmitted by shear in the strip adhesive, and then to the end ring by shear in the end bond adhesive. The cross section of a three-layer strip laminate tube and end rings, including shear paths, is shown in Fig. 6.

For the strength and rigidity of the tube, the usual principles of design are applied based on the total metal thickness of the tube wall, provided that the strip

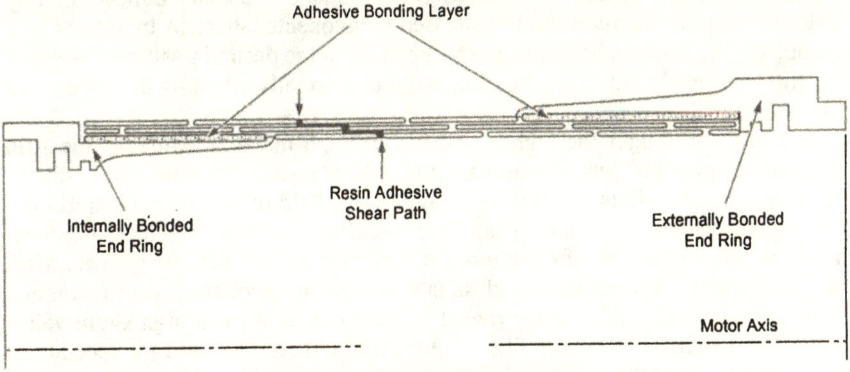

Fig. 6 Strip laminate case.

overlap pattern has been chosen so as to give adequate longitudinal shear strength in the adhesive between the layers. This would only be a problem if a narrow strip was being considered. In general, the length of shear path in the tube adhesive is about one-half times the length of shear path in the end ring to tube adhesive bond, this being a consequence of the construction rather than a design criterion.

The number of layers used should not be fewer than three to avoid distortion under pressure at adjacent strip edges and to obtain the maximum material strength, but may exceed this by any desired amount. However, it is advantageous to make the number of layers as small as possible, consistent with reasonable strip thickness, as the winding time is thus reduced. The overlap pattern chosen depends on the number of layers of strip. For a three-layer tube, the pitch is simply one-third of the helix pitch. For a four-layer tube, a pitch of one-quarter of the helix pitch has been used successfully, as have alternative pitches of one-half and five-eighths. An overlap pattern with alternate pitches of one-half and three-quarter helix pitches has been successfully used on a 13-layer tube. Experience indicates that where the thickness demands a large number of layers of strip, there should be at least four layers before the pattern repeats itself.

The end fittings, which may be in the form of attachment rings or complete closures, are provided with cylindrical extensions to provide the necessary bonding surface. These extensions may be designed to fit the strip laminate tube internally or externally. Internal bonding has the advantage that a flush external diameter can be provided. Nevertheless, if a flush internal diameter is required, external bonding must be used with a consequent local increase in external diameter. Center rings or closures can also be provided to join two tubes of the same or differing diameters or thicknesses together. Figure 6 shows examples of internally and externally bonded end fittings.

Care must be taken when designing an end joint to ensure that stress concentrations at the end of the bond are avoided so that no peeling effect is produced; this is particularly important with internally bonded fittings. Attention must also be paid to the effect of any bending moments that may be present in the end region; this effect normally takes the form of an additional shear stress, positive or negative according to the position in the end bond.

In general, the end ring bond is designed to be stronger than the tube so that the body will fail in hoop. The end rings normally provide attachment locations forward and aft of the motor for other missile items. Launcher feet or fin attachments can be incorporated into the end ring design or, alternatively, bonded directly onto the motor body. They may be attached as individual components bonded directly onto the tube or, for more highly stressed components, attached by means of a complete ring that can be bonded onto the body in the desired position. (Welding to strip laminate bodies is not possible without thermally affecting the strength of the strip steel.)

Although no longer the highest specific strength material available, the strip laminate process has certain manufacturing advantages that maintain its competitiveness, and it exhibits specific advantages in the field of insensitive munitions.

Filament winding. In the 30 years between 1950 and 1980 much progress was made in the production of continuous filaments with excellent properties. Most of the initial work centered on glass but, though of excellent specific strength, it is not universally suitable for rocket motor cases as it has a high strain value. Effort then switched to boron fibers but these could only be used for specialized components due to their extremely high cost. More recently a wide range of carbon fibers and a smaller, but equally competitive, range of aramid fibers have become commercially available and are excellent structural materials for rocket motors. Being filamentary in nature they have to be locked in position using a resin matrix. This means that the mechanical properties are high along the filament but only very limited at 90 deg to the filament; these properties are related to the adhesive properties of the matrix resin.

This extreme anisotropy is very suitable for rocket motor tubes, giving the designer the opportunity of avoiding the unnecessary excess longitudinal strength which occurs in a tube fabricated from isotropic material, Sec. II.B.2. For instance, in a simple pressure vessel one-third of the fibers could be arranged to lie in the longitudinal direction and two-thirds in the hoop direction, Fig. 7a. However, this is not a very practical arrangement as it is difficult to lay the fibers in the longitudinal direction at 0 deg to the axis. A better theoretical solution would be to lay the fibers in a single helix with the winding angle (54.75 deg) calculated to give the 2:1 stressing required in the hoop:longitudinal directions, Fig. 7b. However, with no fibers at a low angle to the axis, the longitudinal stiffness of such a tube is low. The practical solution to the problem is to wind some of the filaments at as low an angle as is convenient, typically 30 deg, and the remainder at 90 deg; the proportions are adjusted to give the 2:1 stressing required, in this instance in

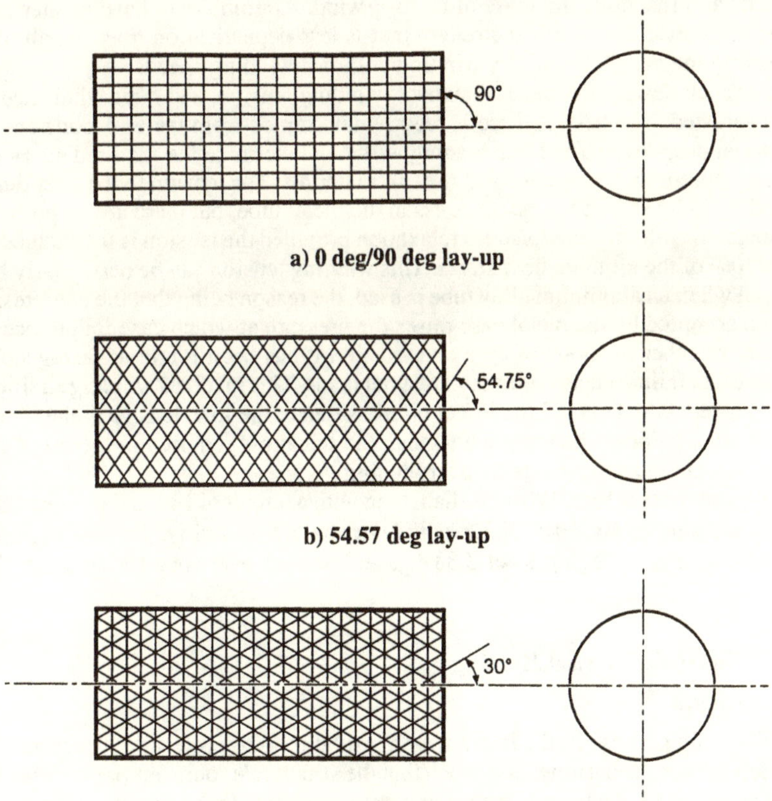

a) 0 deg/90 deg lay-up

b) 54.57 deg lay-up

c) 30 deg/90 deg lay-up

Fig. 7 Filament lay-up patterns for helical winding.

the ratio 4:5, Fig. 7c, nevertheless, the wall thickness required to hold the pressure remains the same for any angle of wind so long as the 2:1 stressing requirement is satisfied.

The high specific strengths of these fibers, together with the opportunity offered by filament winding to build a balanced hoop/longitudinally stressed case, gives this process a unique position in the field of motor case fabrication. In addition, the available option to design in localized stiffness and good low vulnerability properties are added advantages. Disadvantages of the process are the increased wall thickness required and, in some instances, increased costs. Comprehensive methods of design using filament winding techniques are given in Refs. 2 and 3.

Fiber overwind. Another elegant way to avoid the unnecessary excess longitudinal strength that occurs in tubes fabricated from isotropic materials is to design the base metal tube to meet the longitudinal loads and overwind it in the hoop direction with a higher specific strength material. In this way significant mass saving over isotropic materials can be achieved. Wires and fibers can easily be hoop wound onto a cylindrical tube, but metal wires are unlikely to provide the additional specific strength and have been discounted. Fibers with high specific strength are glass, carbon, and aramid. With the glass and carbon fibers, a resin matrix is required because of the relatively poor abrasion resistance and the need to transfer some stress between fibers. The use of a resin system reduces winding speeds and increases the mass of the overwind. Aramid fibers have greater abrasion resistance and nominal strength that is less dependent on fiber length, thus, allowing the possibility of dry winding with fast winding speeds.

For each design, the metal thickness, winding tension, and fiber winds need to be evaluated. The first of these is readily determined from the end loads and the material properties. The hoop loading will then determine the required overwinding by summing the capability of the two materials. The tension in the fiber during winding induces a compressive stress in the metal tube, but there are no problems associated with fiber creep/stress relaxation provided the tension is maintained below 30% of the ultimate fiber stress. This winding tension can be particularly beneficial when an aluminum alloy tube is used, the reason being that the compressive strain accepted by the metal case raises the pressure at which case failure occurs.

Unlike other methods of case construction it has been found advantageous to ensure that failure occurs as a result of axial loads. This involves winding additional hoop layers which have little effect on mass due to the low density of the fiber. If there are significant bending loads, then the thickness required in the metal case may obviate the gains made by overwinding.

Typical masses for a 120-mm-diam tube with a length of 1 m and designed for a burst pressure of 40×10^6 Nm^{-2} with hoop: axial load ratio of 2:1 are aluminum alloy 4.33 kg, maraging steel 3.88 kg, and aramid overwound aluminum alloy 2.74 kg.

C. Thermal: Internal Heating

1. General

The main purpose of the internal case insulation is to protect the structural case material from the burning gases such that the structure's room temperature properties are retained. Additionally, the temperature external to the case must not exceed that which can be tolerated by adjacent equipment. These temperature limitations apply throughout the functioning period of the missile, and this may include a period of flight after the motor has finished burning when heat continues to dissipate through the case. Internal insulants are usually elastomeric and are discussed in Secs. II.C.2–4, which follow. Nevertheless, reinforced plastic insulants are required where erosion resistance is important, and these are mentioned in Sec. II.C.5.

2. *Motor Burning Considerations*

A simple model of the insulating liner assumes it to be divided into three zones. The first is the unattacked zone of virgin material; this is the zone adjacent to the case in which the rise in temperature is not sufficient to cause any significant change, chemical or physical, in the insulant material. The second is the char; this is the hot zone from the ablating surface to the interface with the unattacked zone where pyrolysis may be assumed to begin. The third is the erosion zone; this is the layer of material that will be lost due to the chemical and physical effects of the hot gases.

Designers first need to know the total material attacked, which is defined as the sum of the char and the erosion zone. The chemical and physical degradation reactions that cause the liner to be attacked are complex and are covered, generally, by the term ablation, Fig. 8. During ablation, material is lost both internally and from the surface, thus carrying heat away. Additionally, and most important, energy may be absorbed by various degradation processes using energies of activation, reaction, and phase change. From the designer's knowledge of how insulant materials degrade under the relevant gas conditions and time of burning, the thickness that will be attacked is calculated. In addition to the thickness allowed for attack, the designer adds the required unattacked thickness, calculated from thermal conductivity and diffusivity formulas.

3. *Considerations External to the Motor*

The customer may impose requirements external to the motor. For example, the efflux may be required to exhibit low smoke, in which instance a material is chosen for the insulant that either resists erosion sufficiently or ensures fairly complete combustion of the degraded volatile matter. Additionally, ancillary equipment may not be able to function above a stated temperature, possibly necessitating a greater thickness of insulant than previously calculated for the unattacked layer.

4. *Considerations Associated with the Charge*

Although it is simpler to specify a single thickness of insulation throughout a given motor case, weight penalties are incurred as the thickness chosen must be that of the thickest layer required. Therefore, in practice, the insulation may be tailored wherever reasonably possible, for example, by plying-up sheets of suitable thicknesses.

In a cigarette-burning (end burner) type motor the charge burns from one end and the burning progresses along the length of the motor. One end of the insulation is, therefore, subjected to hot gases throughout the motor firing and the other end remains protected by the propellant, with progressive attack along the length between the two extremes. Hence, tapered or stepped insulation is sometimes preferred if a minimum weight design is to be obtained, but it leads to fabrication difficulties with both the insulation and the charge. When a tapered or stepped

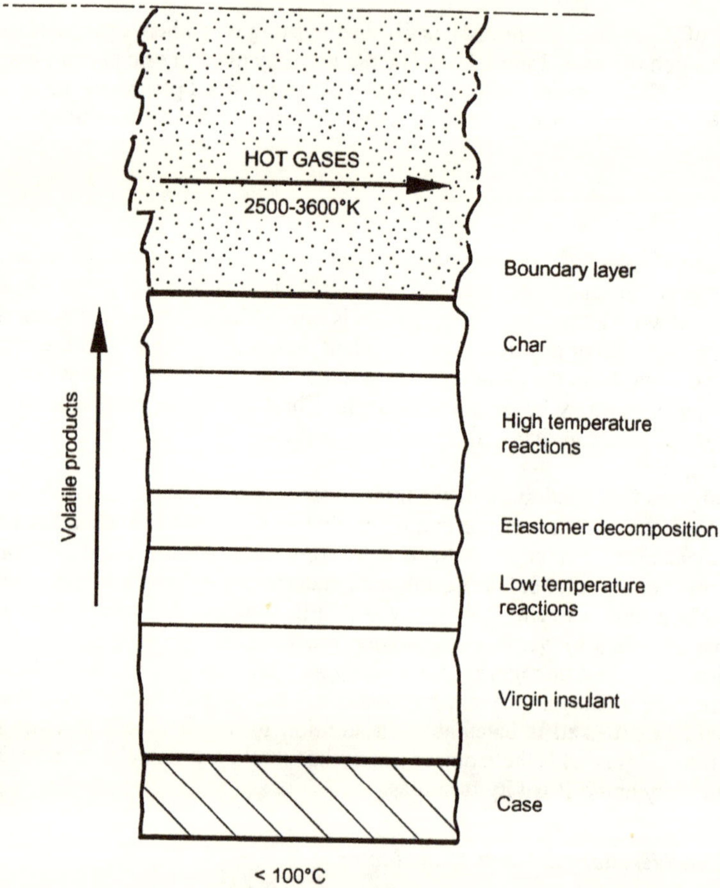

Fig. 8 Rubber ablation.

insulation is used, this is usually fabricated outside the motor case and bonded into position as a fully or partially cured component.

In a radially burning motor the charge burns along its entire length from the center to the outside, and for most of the burning time the propellant itself normally acts as an insulation for the motor case. Only a minimum of insulation material for the structural member is, therefore, necessary. A radially burning charge may, however, include "burn-back" regions (where an uninhibited end burns longitudinally for a length equivalent to the maximum burning thickness) and regions adjacent to slots, apexes of star-centered charges or other areas of reduced charge thickness. In these regions the structural member could be exposed to burning gases at a much earlier stage than in the wholly radially burning region, the burning gases may be turbulent, and the gas velocity will be high where the cross section of the port area is small. Hence, it is usual to locally thicken/tailor the insulation in these regions.

Case-bonded charges, where the insulant is in intimate contact with the propellant grain, impose physical and chemical requirements on the lining materials and necessitate certain unique procedures in the fabrication process. Physically, the material in contact with the propellant should preferably have a greater strain capability than the charge itself and still have strength to support the charge. This necessitates the use of flexible materials, usually rubbers. There is little difficulty in meeting these requirements in the temperature range 10–60°C. At lower temperatures suitable materials are available, but the method of fabrication has to be designed to suit, and considerable thought has to be given to the choice of material and the method of utilization. Chemically, the requirements are for an adhesive bonding capability that gives satisfactory bonds between insulant and structural member and between insulant and propellant (by way of an intermediate layer if required). Both bonds should be stronger than the cohesive strength of the propellant.

5. *Reinforced Plastic Insulants*

Reinforced plastics replace elastomer-based insulant for areas of high erosion. These areas certainly include blast tubes and may include aft closures and miscellaneous areas of motors where gas velocity is high. With reinforced plastics, less advantage is gained from energy absorption reactions than with elastomer-based insulants and more use is made of erosion and heat resistant chars.

D. Thermal: External Heating

1. *General Requirements*

Two fundamental chemical parameters impose temperature limitations on motors. Firstly, the burning rates of propellants used in solid propellant rocket motors are related to the propellant temperature, and in most instances the control of the burning rate, and consequently control of the missile, becomes difficult above 60–80°C, the exact temperature depending on the propellant in use. Secondly, conventional organic polymers, on which many of the adhesives and rubbers used in rocket motor hardware are based, typically start to lose strength rapidly above 100°C and frequently commence to decompose near 200°C. (Advanced polymers are slowly becoming available that, although of lower strength, may be expected to raise these temperatures but only by a limited amount, say to 200°C and 400°C, respectively.) With a few exceptions the two fundamental parameters have not inhibited the design of rocket motors to date. Nevertheless, an era is approaching when aerodynamic heating (aeroheat) is going to become increasingly severe. This is inevitable as the speeds of aircraft increase resulting in higher aeroheat temperatures due both to the carriage by those aircraft in captive flight of underslung missiles and to the increasing speed required in missile free flight in order to both attack other high-speed aircraft and to avoid interception.

It is very unlikely that burning rate/propellant temperature technology will change significantly, hence, it will be necessary to incorporate a thermal barrier to protect the propellant from aeroheat. This could be on the outer or the inner surface of the motor case, but in either instance will incur mass and volume penalties. It will also be necessary to modify or protect any organic polymer-based hardware materials. If modified, a mass penalty due to their lower strength may be incurred and, if protected, a thermal barrier, in this instance specifically on the outer surface of the case, will be required. Protection against aeroheat will clearly require the addition of a new material to rocket motors. If external insulation is used, the penalty will be either an increase in motor diameter or an equivalent decreases in charge diameter resulting in an increase in length together with increased inert mass. Increased outer diameter of air-carried underslung missiles would also increase drag affecting aircraft performance. Thus, a reduction in efficiency of rocket motors is expected when they are designed to cope with aeroheat temperatures.

2. *Aeroheat Temperatures Expected*

Skin temperatures expected on rocket motors due to aeroheat depend on the velocity of the missile, its height above sea level, and the ambient temperature. As a rough guide, skin temperatures of 80°C can be expected at Mach 1, 220°C at Mach 2, and 480°C at Mach 3 when at sea level on an average day in a temperate climate. Temperatures within the rocket motor will additionally depend on the initial temperatures and the duration of the flight. The accurate measurement of such temperatures when rocket motors are subjected to aeroheat conditions is not simple.

For captive flight conditions, arrangements have to be made to carry a missile structure underslung from an aircraft and to record real time temperature data during the flight. Recording such data is a specialized task; careful consideration has to be given to carrying experimental equipment on aircraft, and arranging for a special flight schedule covering useful speeds, altitudes, and times may be expensive. For free flight conditions a sophisticated telemetry recording system would be required.

As an alternative to real aeroheat conditions it would be possible to consider simulated conditions. However, no facility is known that covers the full range of temperatures, duration times, and air velocities required. Consequently most work has been, and will continue to be, based on calculated temperatures.

The theoretical recovery temperature for turbulent flow is related to ambient temperature and velocity

$$T_r = T_a(1 + 0.18M^2)$$

The recovery temperature is clearly very dependent on the ambient temperature, and the level at which the latter may be set is debatable. For instance, international tables[4] are available that give the average of day maximum and night minimum temperatures at various altitudes. They are useful to set the general level of temperature, but it is necessary to add a value sufficient to cover the day maxima, which are the ones of interest to the rocket motor designer. Alternatively, designers of operational aircraft use specification tables,[5] which give practical temperatures at various altitudes used by designers for operational aircraft, those relating to the day time maxima ("hot atmosphere") being of most interest to missile designers. Also, the U.K. Ordnance Board issues tables,[6] which give temperature limits at various altitudes that are exceeded for 1% of the most extreme month of the year. Again, those relating to daytime maxima ("1% hot") being of most interest to missile designers. However, these circumstances are met only very rarely and may be considered as survival rather than operational limits.

Opting for the practical temperatures used by operational aircraft designers,[5] ambient air temperatures have been plotted against altitude in Fig. 9. These ambient air temperatures have been used to calculate recovery temperatures for various speeds and altitudes, Fig. 10. Practical aircraft flight trials have confirmed that the skin temperatures does stabilize very close to these calculated recovery temperatures.

In addition, it is generally necessary to know the starting conditions since flights are frequently of short duration and the initial temperature influences the temperature at the time of engagement. A typical hot soak desert temperature is

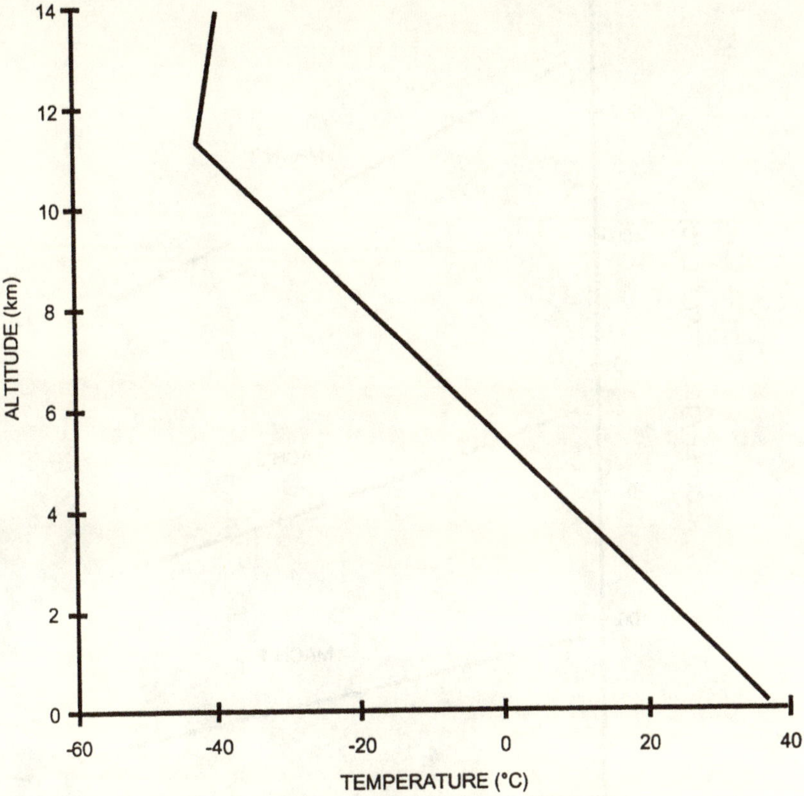

Fig. 9 Ambient air temperature vs altitude.

44°C, this being applicable to the missile when mounted in a shaded position under the aircraft. It should be noted that during unshaded storage in these conditions a skin temperature of 63°C can be reached.

Three types of flight conditions that could cause aeroheating are possible. Those missiles that are air launched are generally required to be carried by aircraft in such a manner that they may be deployed rapidly. This implies that they are underslung from the aircraft rather than housed within the fuselage. For the same reason, and to reduce drag to a minimum, the use of protective pods tends to be avoided. The result is that missiles are exposed to captive carriage conditions at cruising speed for relatively long periods, depending on the range of the aircraft, and at attack speed for shorter periods. In addition, all missiles, whether air launched or not, may be subject to high free flight speeds both during and after the burning time of the propellant. A diagrammatic representation of the approximate historical situation is shown in Fig. 11. This shows past and predicted aircraft and missile maximum speeds and the corresponding recovery temperatures calculated from Fig. 10.

3. *Thermal Requirements*

Calculations on flight time/temperature profiles should be confirmed by practical data to enable precise thermal requirements for external insulation to be

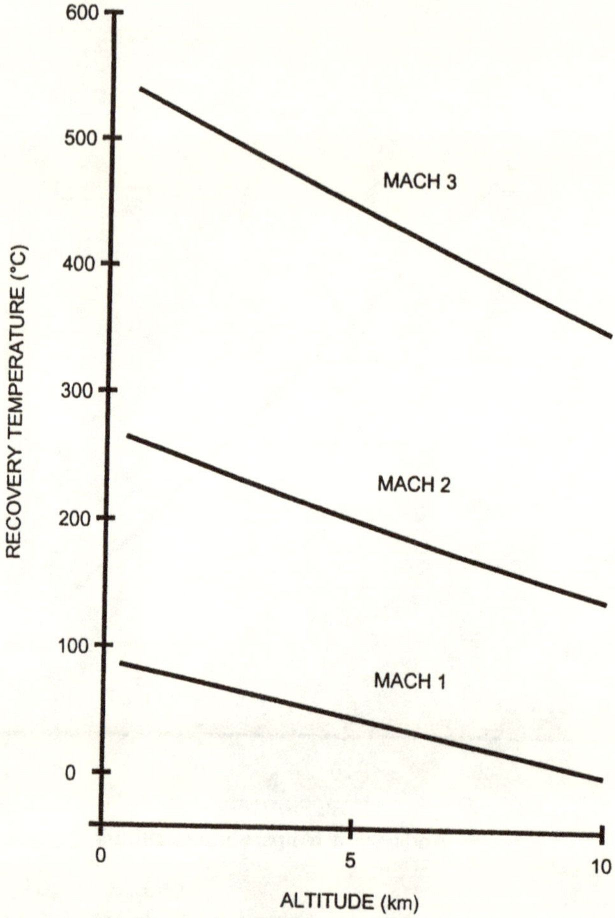

Fig. 10 Calculated recovery temperatures.

evaluated. If, however, the problem is considered from each of the three different flight conditions, it can be estimated from Fig. 11 that, in the early 1990s, the rocket motor requires protection from skin temperatures of 100°C during captive cruising (for, say, 2-h periods), from 250°C during attack missions (for, say, 20-min periods) and from 800°C in free flight (for, say, 20-s periods). During captive flight the insulant must not degrade since missiles may well be carried on numerous practice or aborted missions prior to actual use.

4. *Nonthermal Requirements*

Attaching relatively thin coatings of controlled thicknesses to the exterior surfaces of rocket motors would not be particularly difficult if the motors were simple right cylinders. Application by paint spray techniques, hand lay-up of laminar materials, molding, shrink-fit polymers, or string, tape, or filament winding would utilize existing techniques. However, almost without exception the motors in question are used on missiles that require projections from the motor body to carry

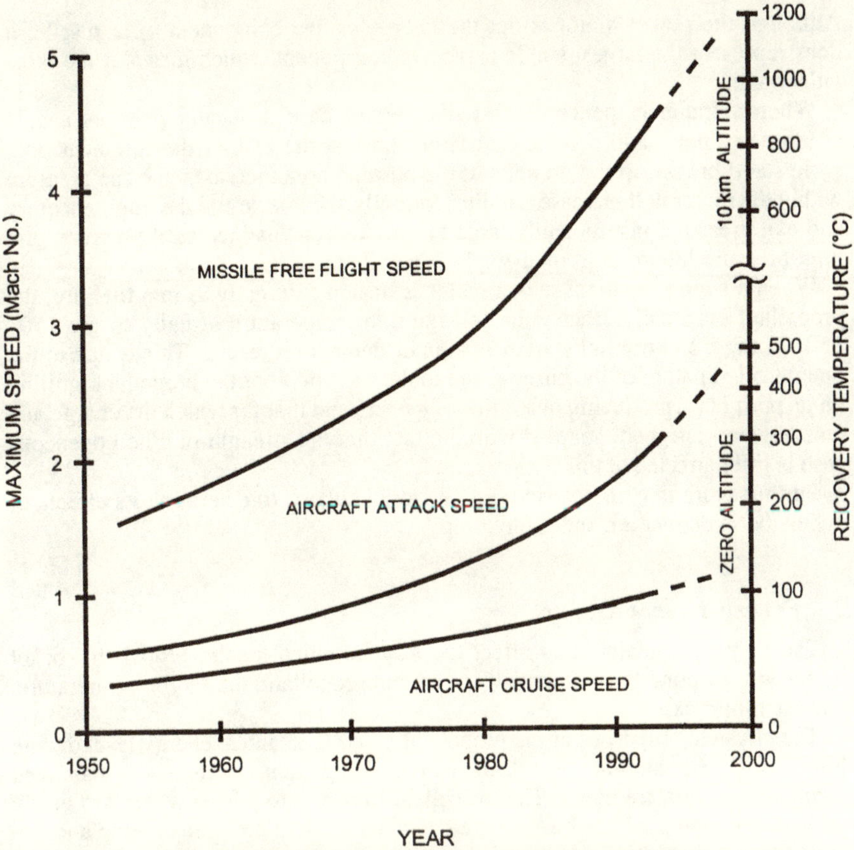

Fig. 11 Historical advance of flight speeds.

wings or launcher lugs. This requirement imposes difficulties during application to all but the paint spray and hand layup techniques.

All rocket motors, and hence their external insulants, are required to withstand bump, shock, vibration, and temperature and humidity cycling. They are also expected to withstand handling damage, erosion by rain, hail, or sand, and contact with fuels, lubricants, and de-icing agents. Whereas metal motor bodies may be expected to satisfy these environments with some confidence, low-conductivity insulants are not so resistant and require testing to ascertain confidence levels. Where insensitive munitions are specified it will be necessary to demonstrate that external insulants do not adversely affect the sensitivity of the structural case to fuel fires or fragment attack.

E. Vulnerability

1. Mechanisms

A relatively recent requirement for solid propellant rocket motors is that for insensitive munitions, that is, ideally, there is no reaction when subjected to external stimuli as defined in DOD-STD 2105.[1] Of these, considerable work has been carried out to evaluate the response to fragment or bullet attack and/or to fuel fire.

Although the rocket motor is not the only sensitive component of a missile, it does represent the largest single explosive component, which makes it the most vulnerable.

When a fragment penetrates a rocket motor case it usually punches a hole at the entry point and a somewhat larger hole at the exit. If the fragment both ignites and breaks up the propellant, the burning area increases and the pressure within the motor will increase rapidly. Normally, the nozzle and the fragment entry and exit areas are insufficiently large to provide for this increased pressure, and high-pressure failure or even an explosion will result.

When a solid propellant rocket motor is heated externally as in a fuel fire, the propellant eventually reaches its self-ignition temperature, usually in the 150–250°C range, and normally an explosion or detonation results. This is due to the uncontrolled nature of the burning and to the likelihood of the propellant igniting on its periphery, producing quantities of gas beyond that for which the charge and nozzle were designed, coupled with the fact that the strength of a homogeneous case is little affected at this temperature.

Further work is being carried out on slow cookoff, to determine its effects of sympathetic detonation and safety drop.

2. *Fragment Attack Aspects*

The major parameters that affect the way in which a solid propellant rocket motor will respond to fragment attack are the propellant, the motor temperature, and the motor case.

The physical properties of the propellant allied to its latent energy level dictate how it will respond when attacked. A brittle propellant is extensively damaged when struck by a fragment. The propellant breaks into pieces which are easily ignited, producing a large burning area which leads to a pressure buildup and, inevitably, to an explosion when retained in a strong case. A more flexible propellant generally remains intact on fragment attack. A hole is made in the propellant which ignites and, having a relatively small burning area, results in a slower buildup of pressure.

The properties of some solid propellants vary with temperature, becoming more brittle on thermal cooldown and conversely more flexible as the temperature rises. Consequently, some propellants that are flexible at room temperature can react like brittle propellants at low temperature and will then be more vulnerable to fragment attack.

Rocket motors using homogeneous cases, whether of aluminum alloy or steel, tend to explode or pressure burst when subjected to fragment attack, although with lower velocity fragments the reaction of aluminum alloy motor cases may be reduced to burning only. However, the use of composite cases reduces the effect significantly, in most instances only burning or, at low temperatures, deflagration occurring. With a strip laminate case, the area of damage at the fragment exit point is more extensive than it would be in a conventional motor case due to the strip delamination effect. When the propellant ignites and the pressure increases, the delamination of the strip continues, the pressure rise is restricted, and the motor burns out. With a filament wound composite case extensive damage results to the fiber construction, significantly weakening the body and allowing further delamination of the fibers as the pressure builds up, again restricting the motor to burning.

3. *Fuel Fire Aspects*

When subjected to fuel fire or external heating, the main parameter that affects the response of a rocket motor is the case. The autoignition temperature may affect the time at which the propellant ignites, but the resultant effect once ignition occurs is similar. Also, although the autoignition temperature for propellants varies, the difference is not significant. Homogeneous motor cases that retain their strength above the self-ignition temperature of the propellant invariably explode when the propellant ignites.

With composite cases of the strip laminate type, however, the failure mechanism is markedly different. When subjected to heat, the adhesives used for bonding in begin to lose strength at temperatures above 130°C and at 250°C have virtually no strength remaining. For a propellant that ignites at, say, 180°C, the adhesive strength will be significantly reduced. At the instant of propellant ignition only a very low pressure is required to disrupt the motor case, thus preventing a rapid pressure rise and limiting the effect to one of a mild pressure burst. Filament wound cases, in general, show similar results to those strip laminate bodies.

4. *Summary*

Experimental information indicates that the rocket motor case aspect of insensitive munitions may be summarized as follows. For vulnerability to a fragment attack, homogeneous steel is most sensitive, homogeneous aluminum exhibits medium sensitivity, and composite strip laminate or filament wound cases are least sensitive. For vulnerability to a fuel fire, homogeneous steel or aluminum is most sensitive and composite strip laminate or filament wound cases are least sensitive.

F. Development

The various phases in a rocket motor's life are feasibility, development, qualification, full-scale engineering development (FSED), and production. If there is already sufficient background knowledge it is possible that the feasibility phase may be omitted. Otherwise it consists of ad-hoc tests on components and/or complete motors that are intended to give the designer information on areas of technology which are known to carry high-risk factors or with which he or she is unfamiliar.

Full-scale motor work commences with the development phase. At the start of development the best available design together with prospective methods and specifications are committed to paper, and the first few motors are fabricated. However, it is understood that after a few development phase firings it may be necessary to revise the design either radically, should the motor fail to reach its requirements, or marginally, to improve performance or overcome minor problems.

Once development phase firings have proved successful over the temperature range specified, typically from −40 to +60°C, further firings are conducted after subjecting motors to individual environmental trials such as hot storage, temperature and humidity cycling, bump and shock, and vibration, all of the latter trials being at the upper and lower temperature limits specified. Once the design of motor has been shown to satisfactorily pass all of these trials, the final part of the development phase is to "chill" all design documents (i.e., bring the documents under formal control), fabricate more motors, subject them to all of the environmental trials (this time in a sequential mode), and fire them. Provided the latter motors (prequalification) perform satisfactorily, the development phase is considered complete.

During development all motor cases are subjected to pressure tests up to proof-pressure level. Initially, a high percentage will be burst to demonstrate the design; thereafter, a reduced percentage are burst to demonstrate the consistency of the manufacturing process. All must pass the proof test, and those taken to burst are expected to fail at a level above the minimum design figure. During pressure tests the component should be supported in a similar manner to that used in the final assembly so that loads are as representative as possible. For example, the body should have end closures reproducing as closely as possible those of the finished assembly, and they should be held by similar retention devices to those used in the motor assembly. In this way the designer accumulates evidence that the structural case design is satisfactory.

Miscellaneous tests are also conducted during the development phase to ascertain the degree of confidence in the design. The thickness of internal linings should be measured before and after firing to ensure that sufficient lining material will be present at the end of all future firings, given that batch variation in the quality of the lining material is unavoidable. For the same reason, the temperature of the structural case can be monitored during and immediately after firing. The effects of external influences will require checking, for example, mechanical tests to examine the effect of acceleration (longitudinal and lateral) and handling, of safety drop tests, of environments such as saline atmospheres and, if relevant, of aeroheat.

A typical development program is outlined in Fig. 12, at the completion of which the designer should have recorded information on all aspects of the motor case design. Tests required for specific ballistic/charge assessment, for example, shock cycling, may be added as necessary.

G. Qualification

The qualification phase follows the development phase utilizing the environmental tests in a specified sequential pattern. At the start of this phase all design documents incorporate the best design knowledge obtained from the development phase. A typical qualification program is outlined in Fig. 13. During qualification, design, specification, or method changes are to be minimized and every trial should be completed satisfactorily. Any standard less than this cannot generally be accepted and certainly not without the written recommendation of the designer and the written acceptance of the customer. As in the development phase, specific tests for external insulation and for vulnerability may be specified by the customer. At the completion of the qualification phase the designer should have confirmatory evidence that the design is satisfactory for the motor in question and the design can be "frozen" (i.e., changes only with the agreement of the customer).

H. Quality Assurance

1. Structural

All engineering production activities require quality assurance systems to ensure efficient work procedures and satisfactory material quality. Solid propellant rocket motor systems, however, are different from many engineering components in

TRIAL		PRELIMINARY				GENERAL												ENVIROMENTAL								PRE-QUALIFICATION	
NUMBER OF MOTORS		FOUR				TWELVE												EIGHT								TWO	
PROOF PRESSURE TEST		/	/	/	/	/	/	/	/	/	/	/	/	/	/	/	/	/	/	/	/	/	/	/	/	/	/
HOT STORAGE																								/	/	/	/
HUMIDITY & TEMPERATURE CYCLING																						/	/			/	/
BUMP/SHOCK	HOT																			/						/	
	COLD																				/						/
VIBRATION	HOT																	/									/
	COLD																		/							/	
BURST PRESSURE TEST		/	/	/		/	/	/																			
FIRING	HOT								/			/			/				/		/	/		/		/	
	AMBIENT				/					/			/			/											
	COLD										/			/			/	/		/			/		/		/

Fig. 12 Typical development phase program.

TRIAL		GROUP I				GROUP II				GROUP III			
NUMBER OF MOTORS		FOUR				FOUR				FOUR			
PROOF PRESSURE TEST		/	/	/	/	/	/	/	/	/	/	/	/
HOT STORAGE						/	/	/	/				
HUMIDITY AND TEMPERATURE CYCLING		/	/	/	/	/	/	/	/				
BUMP/SHOCK	HOT			/	/			/	/				
	COLD	/	/			/	/						
VIBRATION	HOT	/		/		/		/					
	COLD		/		/		/		/				
FIRING	HOT			/	/	/	/			/	/		
	COLD	/	/					/	/			/	/
BURST AFTER FIRING						/	/	/	/	/		/	

Fig. 13 Typical qualification phase program.

Fig. 13 Typical qualification phase program.

that the final product cannot be tested before delivery to the customer and they, therefore, require specific quality assurance procedures.

For rocket motor cases, the procedure is to subject all incoming materials to chemical and physical laboratory tests, to measure all possible component dimensions, to detail in writing all fabrication and test procedures, and to carry out any possible tests on subassemblies. Should homogeneous metal cases suffer light damage during handling, there is a high probability that their ductility will ensure no significant loss of strength. Light damage to filament wound composite cases, on the other hand, may result in fracture of some of the filaments not visible on the surface of the component resulting in an unquantified loss of strength of which the user will be unaware. Quality assurance test methods in this field are being actively sought.

At the end of the development phase all quality data of this type can be assessed, limits allocated to chemical and physical material tests and to subassembly tests, and fabrication and test procedure details can be finalized. If, during the qualification phase, all of the quality assurance aspects are adhered to and if all of the motors fire satisfactorily, the quality standards will have been set for future production. It is part of the designer's job to assure himself that those quality standards have been set to his satisfaction to cover all possible future motors and that the production phase commences with those standards specified and with precise instructions on how to deal with materials, components, or subassemblies which fail those standards.

2. *Internal Insulation*

Ingredients for all internal insulant materials are purchased to specifications that maintain their initial quality, and the insulant materials themselves are checked to ensure they conform to specified mechanical properties. User tests are carried out on elastomeric materials to ensure compatability with propellant and to monitor those properties, such as adhesive bondability, which may be expected to deteriorate with prolonged storage. Molded reinforced plastic insulants are required to be of at least 95% theoretical density to ensure satisfactory erosion resistance. Completed insulating liners are checked both dimensionally and for flaws.

III. Materials

A. Structural

1. *Homogeneous*

Conventional structural materials used for rocket motor cases are good quality alloys of steel, aluminum, and, occasionally, titanium. These are homogenous materials with isotropic properties; their strengths and moduli are given in Table 1. Also in Table 1 are their mass factors calculated on both stress and on strain criteria in comparison to high-tensile steel.

On a stress basis it will be seen that standard aluminum alloy would be heavier than high-tensile steel whereas the latest conventional aluminum alloy, titanium and maraging steel, offer reducing case masses; maraging steel is 65% the mass of high-tensile steel. On a strain basis the mass variation is less, the lowest being the latest conventional aluminum alloy, which would be 92% the mass of high-tensile steel. Hence, for rocket motor cases, which are usually designed on stress, maraging steel would be a first choice; whereas for end closures and blast tubes,

Table 1 Properties of metals and composites

Material	Density, Mg m^{-3}	Tensile strength, 10^{-9} Nm^{-2}	Tensile modulus, 10^{-9} Nm^{-2}	Mass factor for pressure loading relative to high-tensile steel: Stress	Strain
Standard aluminum alloy	2.70	0.45	73	1.07	0.94
High-tensile steel	7.85	1.40	200	1.00	1.00
Latest conventional aluminum alloy	2.70	0.60	75	0.80	0.92
Titanium forging	4.65	1.24	110	0.67	1.08
Maraging steel	8.00	2.20	215	0.65	0.95
Heat treated aluminum strip	2.70	0.54	73	0.89	0.94
Steel strip	7.85	2.00	193	0.70	1.04
Titanium strip	4.60	1.35	106	0.61	0.90
Glass fiber/ epoxy	2.10	0.63	24	0.59	2.23
Carbon fiber/ epoxy	1.60	1.00	73	0.29	0.56
Aramid fiber/ epoxy	1.40	0.87	36	0.29	0.99

which are usually designed on strain considerations, aluminum would give the lowest mass.

However, other criteria that have to be taken into account are insensitive munitions; economic manufacturing thickness; internal volume; the thickness to diameter ratio, where a value less than 0.005 is considered undesirable on a damage criterion; aerodynamic heating or subsequent high-temperature processing, which could affect the strength of aluminum alloys; possible stress corrosion with aluminum alloys, which can be virtually eliminated by careful design in areas of changing section (for example, by the provision of large fillet radii); and external attachments.

New materials which are currently showing improved properties over conventional aluminum alloys are aluminum-lithium alloys, which have a 15% higher modulus and a 8% lower density, making them particularly useful for components where stiffness is required and fiber reinforced aluminum which can have much increased strength and stiffness. These materials require evaluation; the second is currently expensive and is only likely to be useful for very special applications.

The majority of metallic materials need to be protected from the environmental conditions that could be encountered during service life. Although most external surfaces are painted, this is not usually enough to combat the effects of the elements such as salt spray, acid rain, and prolonged exposure to hot-wet conditions as well as mechanical abrasions resulting from handling. Hence, a number of plating and deposition procedures have been used on steel and aluminum components. Cadmium plating has been the most commonly used protection for steel surfaces and, indeed, this electroplating process is relatively simple and, hence, low cost. It is also very effective in terms of corrosion resistance. The process with high strength steel, however, can promote hydrogen embrittlement. It has a limited salt spray life and recent medical evidence has shown that cadmium is as toxic as mercury; thus, its use is likely to be severely restricted in the future. An alternative process is to employ an insoluble phosphate coating which protects by exclusion, whereas cadmium protects sacrificially. Phosphating needs to be supplemented by paint or an organic finish, but can be applied in very thin coating layers and, hence, is advantageous in regions of high-dimensional accuracy. Other options include ion vapor deposition of aluminum, mechanical plating by dipping using combinations of tin and zinc, or cladding using aluminum, electroplating using tin and zinc, and electroless nickle plating. Short-term protection of aluminum alloys is provided by anodizing and sealing.

Some materials, in particular stainless steel and titanium, can be left unprotected, although as part of the missile it is likely that they will be painted. Testing of unpainted bodies in stainless steel, which have been subjected to temperate atmospheric conditions for 3 years, has shown no detrimental effects.

2. *Composite*

Strip laminate. The properties of materials available in strip form are also shown in Table 1. A case made from steel strip would be of less mass than one for aluminum designed on stress considerations. Titanium strip shows overall mass advantages in both stress and strain designs but this material is considerably more costly than steel and not readily available in strip form. Most of the experience gained to date with the strip laminate technique has been based on the use of carbon-steel strip, rolled and polish ground, having a design ultimate tensile strength of 2 GNm^{-2}. More recently stainless-steel strip of equivalent strength has also been qualified which offers the additional advantage of increased resistance to corrosion. Aluminum-alloy strip, heat treated to give an ultimate tensile strength of 0.54 GNm^{-2}, has also been successfully used and there is no theoretical limitation on the strip material provided that it possesses sufficient ductility to enable it to be wound successfully.

For cheapness it is desirable to use strip as wide and thick as possible, as this reduces winding time. However, if excessively wide strip is used, difficulties arise due to lateral curvature and gap variations between successive turns of the helix. From practical experience the strip width should be less than the tube diameter. The ease of winding a tube of a given diameter will also influence the maximum thickness of the strip, although in practice the design thickness is the overriding criterion on smaller diameter tubes.

The adhesives used in strip laminate are based on epoxide chemistry, this being the strongest practical system available for production use. Different epoxide systems are used for strip and end rings because of the different methods of application but in both instances adhesive strengths in excess of 30 MNm^{-2} are achieved.

Filament winding. Properties of three fiber/epoxy composites are also shown in Table 1. The figures are not for fibers alone but for fiber/epoxy composites of 67:33 fiber to epoxy ratio by volume. The figures for carbon/epoxy are based on a fiber with midrange properties for strength, stiffness, and strain. All the properties given include a practical allowance for efficiency losses which are always found in the winding process.

Very considerable mass advantages are evident with these materials when designed on stress considerations, carbon and aramid fibers giving cases with masses only 29% of that for high-tensile steel. On a strain basis there are problems with glass fiber, aramid fiber is of comparable mass to high-tensile steel whereas carbon fiber is 56% of that of high-tensile steel. However, other factors need to be considered prior to choosing a fiber composite, in particular, the much increased wall thickness resulting in either increased outer diameter or loss in internal volume.

Glass fibers have been available for many years and are comparatively cheap. For rocket motor cases they suffer a major disadvantage in that they have a high strain-to-break value. This results in appreciable dilation of cases when pressurized, with a real likelihood of dimensional distortion and, for case-bonded charges, the danger of the case with its insulation liner breaking away from the propellant to allow uncontrolled burning. Carbon fibers are now available in engineering-approved grades with many choices of strength, stiffness, and strain-at-break, the latter being at a suitably low level for rocket motor case designs. Components made from carbon fiber/resin matrix components are easy both to process and to machine. Aramid fibers are similarly available, though in a much restricted number of grades. They are very suitable for fiber/resin composites though, being of very tough fiber, they are difficult to machine accurately.

There are numerous resins available for the matrix in which to embed fibers during filament winding. They must have a practical pot life and good wetting and flow properties to allow economic manufacture and, in particular, must bond to the fiber surface to give good interlaminar shear strength. Many such resins are of the epoxide type, usually modified to give enhanced properties.

Overwinding. For the overwinding process aramid fibers are used since their toughness allows them to be used without a matrix resin in simple hoop winding.

B. Internal Insulation

1. General Requirements

Rocket motor case liners have to withstand flame temperatures of up to about 3800 K as well as chemical erosion from reactive species and physical erosion from particulate matter in the high-velocity gases. There are materials that combine strength at high temperatures with good erosion resistance but, unfortunately, these are either good conductors of heat, such as refractory metals and graphite, or suffer from thermal shock, such as ceramic oxides. They are, therefore, unsuitable except in very specialized instances. In practice, use is made of polymeric materials that are normally used as insulation at much lower temperatures. They insulate well, accommodate intense heat, and resist shock. Although having no maximum service temperature in the context of the short burning times of rocket motors, they do ablate and their efficiency decreases as the severity of the environment increases.

They are generally low-density materials, simple to design, and relatively easy to fabricate.

In addition to their insulating capabilities, liner materials are often required to meet other demands of the designer, for example, the formation of smoke and condensable debris often has to be avoided. Both volume and mass are at a premium in rocket motors so, although low-density materials are currently preferred, a dense liner of minimum volume may have advantages in some designs. There must be no deterioration during storage in tropical, temperate, and arctic conditions and, as the liner is often in intimate contact with the propellant or in very close proximity to it, it is of paramount importance that all liner materials are chemically compatible with the propellant in use. Two favorable aspects to bear in mind when choosing insulant materials are the relatively nonoxidizing atmosphere of the gases and the short operating time of a rocket motor.

In practice the insulating materials used for lining cases are usually either rubbers or reinforced plastics. Rubbers have three advantages over reinforced plastics, lower thermal conductivity, higher heat energy dissipation on degradation, and mechanical properties which are a better match for adjacent propellant and load-bearing shell movements. However, reinforced plastics are more resistant to physical erosion and have to be used where high-velocity gases are likely to be encountered, such as in end closures or blast pipes or even in turbulent gas areas of cases where low-exhaust smoke is stipulated. Thus, as a general rule, elastomeric materials are used for motor cases and reinforced plastics are used for end closures and blast pipes.

2. *Elastomeric Insulants*

Rubber insulants consist of a base elastomer to which has been added curing agents to bring about vulcanization; accelerators and antioxidants, which control the rate of reactions both during vulcanization and during subsequent storage; fillers, which improve mechanical properties; and additives, either those unique to rocket motors or "fire-avoidance" products, which absorb heat energy and/or extinguish flames.

The base elastomers, which are of no technical value on their own, are chosen after research indicates they are capable of absorbing more heat energy than other elastomers. Acrylonitrile, butyl, hydroxy-terminated polybutadiene, chlorosulphonated polyethylene, and ethylene propylene diene monomers are examples of elastomers used by various manufacturers. Curing agents, accelerators, and antioxidants are added according to normal commercial rubber technology requirements. Reinforcing erosion resistant fillers such as asbestos (now banned) and silica are added to the base rubbers to improve erosion resistance, and fibers such as aramid and low-conductivity carbon may be added to give reinforcement to the char formed on ablation. Other additives which make mixes uniquely suitable for rocket motor insulants are added after research work has been carried out to ensure suitable heat absorption, flame suppression, and low-smoke properties. The names of these additives and the quantities used are of commercial value. Typical elastomeric insulant properties are given in Table 2.

3. *Reinforced Plastic Insulants*

Most plastics used in engineering applications need reinforcing to improve their mechanical strength, and this is true for rocket motor insulants. Usually,

Table 2 Typical physical property ranges for internal insulants

Property	Material: Rubber	Material: Random fiber reinforced plastic
Density, $Mg\ m^{-3}$	110–180	2000–3000
Thermal capacity per unit mass, $J\ kg^{-1}K^{-1}$	2000–3000	1000–1400
Thermal diffusivity, $10^8 \times m^2 s^{-1}$	7.0–9.5	2.5–2.7
Thermal conductivity, $Jm^{-1}s^{-1}K^{-1}$	0.25–0.40	0.30–0.70
Tensile strength, $10^{-6} \times Nm^{-2}$	6.0–12.0	30–60
Young's modulus, $10^{-9} \times Nm^{-2}$	—	8–11
Extension to break, %	50–500	0.3–0.9
Hardness, (IRHD)	75–85	—
Coefficient of linear expansion, $10^6 \times K^{-1}$	150–250	4–15

the plastic constituent is a thermosetting phenolic resin reputed to retain its basic structure even after appreciable heat degradation, thereby tending to retain the charred residue in position. For this reason the aromatic structure of phenolic resins has long made them the first choice. The degree of endothermic reaction during pyrolysis of the resin is a bonus but not a major part of the insulant action.

It is generally considered that materials having a high loading of fibrous reinforcement are needed for erosive areas. However, the higher fiber-to-resin ratios give rise to mold tool flow problems and a compromise is necessary. The fibers may be in random orientation or specifically oriented for either maximum erosion resistance or maximum strength. Where cheapness of products and simplicity of tooling are overriding factors, random fiber orientation is often acceptable. Orientation for maximum erosion resistance, that is, normal to the gas flow, may result in increased thermal conductivity through the fibers to the wall, and some compromise in the angle of orientation may be desirable.

The most common liner material, from a historical viewpoint, is compounded from a phenolic resin filled with chrysotile asbestos in either string, felt, or flock form. Chrysotile asbestos fiber is cheap and is readily incorporated into molding material, it has good erosion resistance and a low conductivity, it also absorbs some energy during pyrolysis (water molecules are driven off at about 700°C). In recent years, however, the health hazards associated with asbestos have necessitated a move to the use of other fibers. Where erosion is not severe, the cheapest alternative is glass fiber. Where greater erosion resistance is required, silica or carbon fibers may be used at the expense of cost, strength, and conductivity. It is not uncommon for powdered materials, such as carbon or zirconium oxide, to be added to the basic fiber and resin compounds in an attempt to improve the thermal properties of liner materials.

High-density molded materials are the most resistant to erosion. However, in practice values lower than theoretical density are always obtained in molded components due to the presence of voids. Experience has shown that densities in the range 95–99% of the theoretical value are essential for the best erosion resistance. Typical reinforced plastic insulant properties are given in Table 2.

C. External Insulation

The main groups of materials of interest as external insulants are listed, together with steel and aluminum for comparison, in Table 3. The lowest thermal conductivity external barrier that could be imagined would be a vacuum. As this is clearly impractical, the best starting point is to consider a layer of static air. This has a very low thermal conductivity, 0.026 $Jm^{-1}s^{-1}K^{-1}$ at sea level reducing to 0.017 $Jm^{-1}s^{-1}K^{-1}$ at an altitude of 15 km. Fortunately, certain materials entrap air enabling this property to be utilized. Those with microporosity have pore sizes approaching the mean free path length of nitrogen molecules. In such materials the majority of molecular collisions are with the pore walls, not with other molecules and, hence, thermal conduction is low. These materials have low densities but are fragile.

The pores in fibrous paper also entrap air, giving a material with a coefficient of thermal conductivity only about 1.5 times that of air. Paper can also be made

Table 3 Materials for external insulation

Material		Coefficient of thermal conductivity		
Group	Example	Governed mainly by	Typical range relative to air	Typical specific gravity
Gases	Air	—	1	0.001
Microporous materials	Min-K, microtherm	Pore size	1	0.2
Macroporous materials	Aluminum fiber paper	Amount of air entrapped	1.5	0.2
	Cork		2.0	0.3
Rubbers	Sheet rubbers	Filler/rubber ratio	10–15	1.5
	Shrink-fit rubbers		10–15	1.5
Filled paints	Rubber-based paint	Filler/rubber ratio	10–15	1.5
Fiber/resin composites	Glass/epoxy	Resin content	10–25	1.7
Fibers	Aramid	Orientation and packing	20–30	1.5
	Glass		20–30	3.5
Metals for comparison	Steel	Inherent material properties	2,250	7.8
	Aluminum		10,500	2.7

from man-made fibers of, say, alumina. Hollow microspheres are also commercially available and these provide entrapped air. However, they have to be used as a filler in a rubber or resin matrix and, as such, cannot be a very large volume constituent. Thus, microspheres have relatively poor performance and density approaching unity.

A more readily available material with a reasonably low thermal conductivity, about twice that of air, is cork. Although this is a natural product, it is commercially available as ground material reconstituted to sheet form by resin bonding. As such it has both inter- and intragranular porosity. It can readily be applied by hand lay-up techniques, bonded with adhesives, and has a very low density. Although reasonably resistant, it requires a further material to protect it from handling damage, but it provides an adequately robust surface to which that protective layer can be attached. Certain plastic foams in thin sheet form are also available in limited supply and at high cost. These offer higher temperature performance than cork, but have not as yet been fully evaluated.

Some rubber-based materials are worthy of consideration although they all have thermal conductivities in the range 10–15 times that of air. They may be applied in the form of sheet by hand lay-up techniques using adhesive bonding or as solvented sprays. Others have shrink-fit properties. Some rubbers may need a protective coating if they are of a soft variety whereas others are sufficiently robust on their own. Fiber resin composites avoiding high-conductivity fibers, the commonest of which is glass/epoxy, have composite thermal conductivities 10–25 times that of air. Their attraction for use as the external insulation of rocket motors lies more in their ability to provide a rigid protective layer for more fragile insulants than as insulants themselves.

Dry fibers, such as those used in the aramid overwinding process for case structures, have insulating properties in the range 20–30 times that of air but, of course, can only easily be applied to simple tubular structures. With the exception of rubber-based sprayable paints little experience has yet been gained from external insulants. In the immediate future it is likely that the choice of material will be determined by the application methods available which, in turn, are limited if the rocket motor case has projections such as wing roots and launch attachments.

IV. Methods of Fabrication

A. Structural: Homogeneous Materials

1. Welding

Ready availability of material and equipment in the engineering industry ensured that wrapping (rolling) of steel sheet followed by one longitudinal weld became the original first choice for medium diameter and larger tubes for rocket motor cases. End rings and attachments, such as wing roots and launcher lugs, could equally readily be fitted to the tubes by the same technique to give the finished cases. This route must, today, still be considered as a first choice for many large motor designs but has been superseded for medium and small motor designs. One variation of this technique is to form the tube from strip rather than sheet, winding it helically onto a mandrel and continuously welding the helical joint. Winding a relatively thick steel strip may induce bending across its width and this, together with the weld bead, may give rise to nonsmooth internal and external surfaces. Welding techniques are primarily, though not necessarily, used for low and medium strength steels where welding problems are negligible.

2. *Forging*

Forging is used for steel components where the length to diameter ratio is low and for aluminum alloys when the ratio is low or medium. Thus, end rings, end domes, and blast tubes can readily be made this way, but the fabrication of complete rocket motor cases, usually with one closed end, is limited by their length: diameter ratio, although end fittings and attachments can be integrally machined. The method gives good grain flow and, hence, optimum properties but tends to be wasteful of material.

3. *Casting*

Precision casting, although not as good on grain flow as forging, offers advantages in the reduced amount of machining and, hence, cost but additional inspection is necessary to ensure the high reliability necessary.

4. *Deep Drawing*

This technique is used for medium to high strength steels where the length to diameter ratio is low. Ends can be machined.

5. *Extrusion*

Extrusion, particularly for aluminum cases, is a readily available technique. It tends to be limited to the smaller sizes of cases due to limited availability of heavy extrusion presses. Again, end rings and attachments can be integrally machined. Since large amounts of metal are removed during finish machining, distortion may occur.

6. *Flow Forming*

During the 1980s many engineering factories equipped themselves with production flow-forming equipment and this must now be one of the favored production methods, particularly for high strength steels for small and medium diameter cases. One end ring may be integral; the other end ring usually has to be welded to avoid difficulty in removing the tube from its mandrel. Other fittings can also be welded. This technique can equally readily be used for aluminum. As the strength of aluminum alloys is greater in the direction of grain flow, flow forming gives the higher strength in the hoop direction, which is advantageous for a pressure vessel, whereas other methods of manufacture give the higher strength in the longitudinal direction.

B. Structural: Composite Materials

1. *Strip Laminate Winding*

The basic manufacture of a rocket motor case by the strip laminate process is fairly simple, the operations are outlined next and are shown diagramatically in Fig. 14. This is a Royal Ordnance plc Rocket Motors Division process. Steel strip is degreased, shot blasted, and degreased again. It is then coated with a suitable adhesive which is gelled and the strip is coiled. The coated strip has a life of at least 6 months. The steel strip is helically wound onto a heated mandrel so that there is a small gap between successive turns of the helix. Successive layers of strip are added until the desired thickness is obtained. Each layer is wound in the same

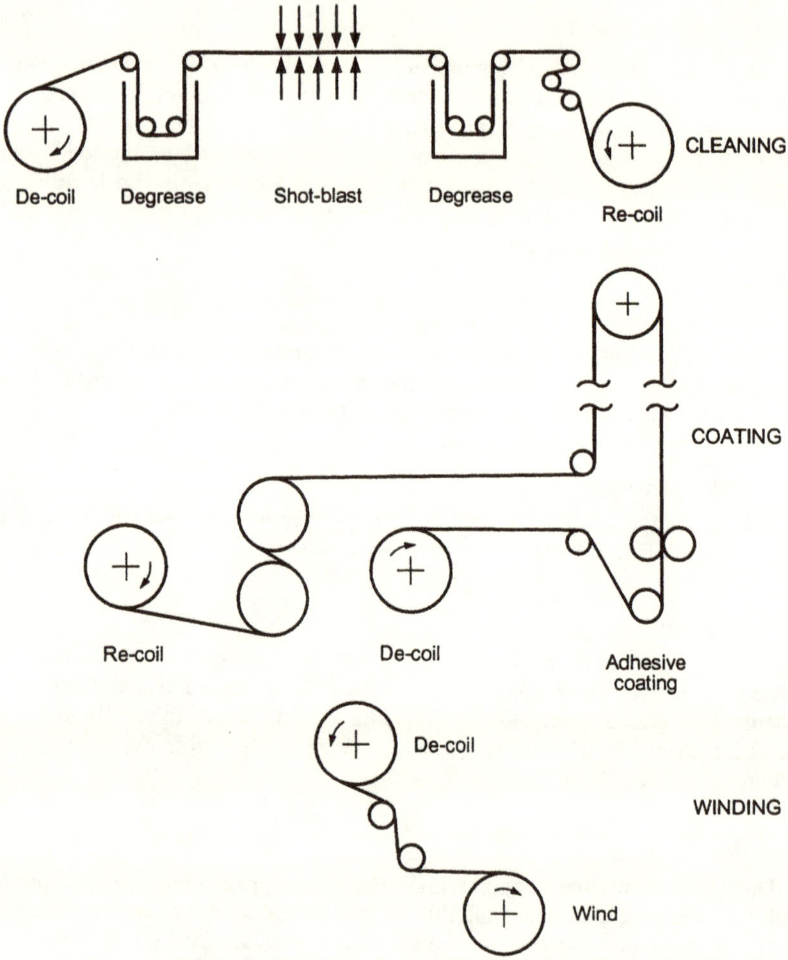

Fig. 14 Strip laminate fabrication operations.

direction but the helixes are staggered. After winding is complete the adhesive is partially cured and the tube is removed from the mandrel and cut to length. End fittings, and any other fittings that may be required, are bonded in place using a jig to ensure accuracy and the assembly is then fully cured. The finished motor case may be lined, painted, and filled in exactly the same way as a case made by any other process.

Basically, the equipment required consists of strip preparation and coating plant, a winding machine which may be a converted lathe, a curing oven, and an assembly jig. The end fittings are assembled to the motor tube in the finish machined state, and this applies also to any other fittings. This gives two of the advantages of the process, namely, a wide range of supply sources for the body fittings and a minimum of manufacturing scrap. The process is capable of producing round, straight tubes with combined bow and ovality of less than 1 in 10^4 on length and end squareness of less than 1 in 10^3 on diameter. Provided strip is available, then development bodies can be produced within 4 months.

2. *Filament Winding*

There are two basic types of filament winding machines, polar and helical. With polar machines the filaments are applied to the mandrel in a planar path across the poles where either the guide arm rotates about the mandrel, Fig. 15a, or the mandrel tumbles end over end with the filament guide remaining static, Fig. 15b. Polar winding appears to be attractive for low length to diameter ratio motor cases with integrally wound, domed end closures, but there are two disadvantages. Firstly, only a few filaments can be applied at one time, and, secondly, the filaments are unstable on the domes because they are laid down on a nongeodesic path.

With helical machines the filament and its guide eye are mounted on an oscillating carriage. As the mandrel rotates, the carriage traverses the length of the mandrel laying the filaments down in a helical pattern, Fig. 16. The carriage and mandrel drives are connected, originally by means of chains and sprockets but nowadays using computer programmed motors, so that the winding angle can be controlled and, because the filaments are laid down on a geodesic path, they are stable. Helical winding is also more suited to high length to diameter ratio cases.

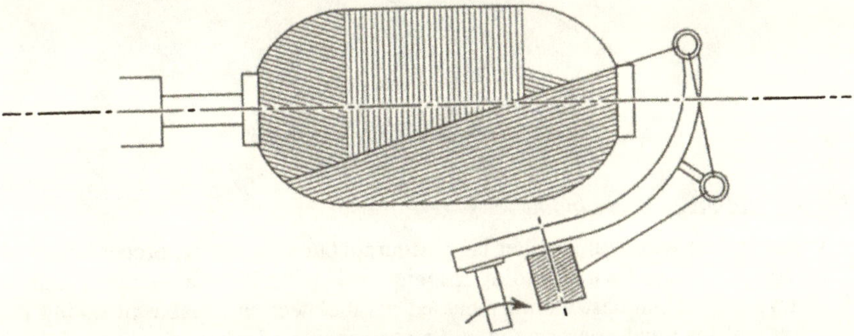

a) Rotating guide arm

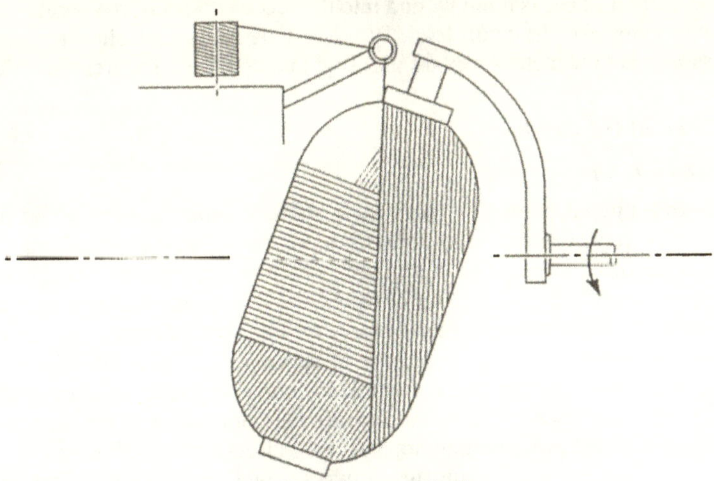

b) Tumbling mandrel

Fig. 15 Polar winding machines.

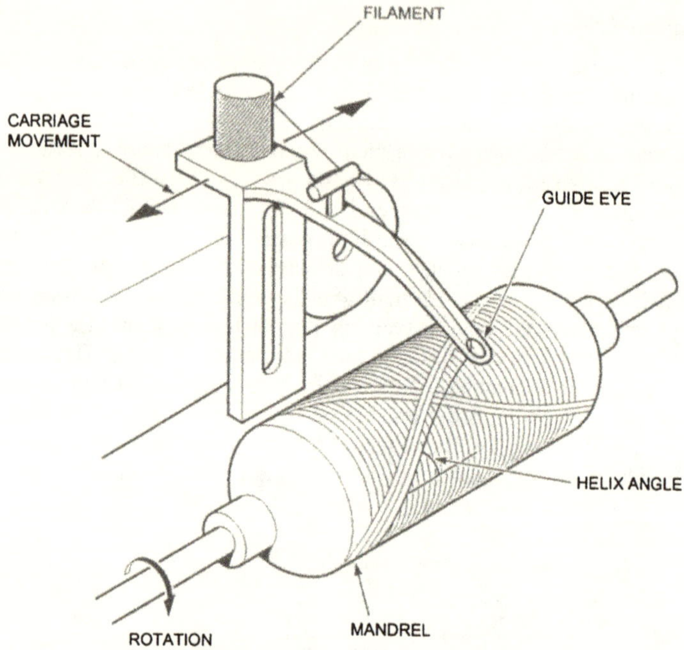

Fig. 16 Helical winding machine.

3. Aramid Fiber Overwinding

Overwinding with aramid fiber is an accurate but very simple process adapted from commonly used wire-winding techniques. Having been fabricated, often by flow-forming techniques, with a recessed parallel section capable of taking the longitudinal load and nonrecessed end rings capable of taking the hoop as well as the longitudinal loads, the metal case is mounted on a rotating mandrel, and sufficient aramid fiber is hoop wound into the recessed area of the parallel section to enable it to take the hoop load. Winding tension is controlled because it is necessary to put the metal structure into slight compression for greatest efficiency.

C. Internal Insulation

1. Hand Lay Up

The conventional method of applying a rubber insulating liner to the internal surface of a rocket motor case consists of obtaining unvulcanized material in sheets of various thicknesses, coating with adhesive, laying planar sheets into the parallel section of the case, and tailoring the flat sheets of material to fit the curve of the domed ends as closely as possible. After fitting an internal pressure bag and external constraints, Fig. 17, a pressure of approximately 5×10^5 Nm^{-2} is then applied to the pressure bag and the whole assembly is placed in an oven at about 150°C until the rubber becomes vulcanized. On completion of the vulcanization and removal of the pressure bag apparatus, the lining, although originally made up of several pieces, will be vulcanized into a single, smooth layer. Case bonding of certain propellants often requires a special adhesive film to be bonded on to the inner surface of the liner at the same time as the final vulcanization takes place.

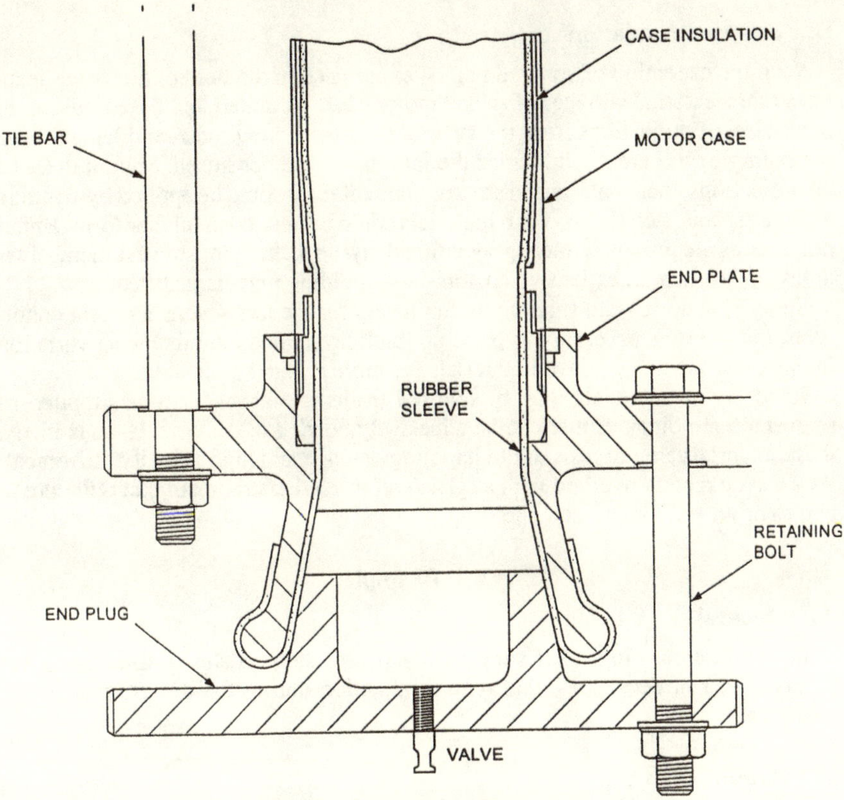

Fig. 17 Pressure bag apparatus.

2. *Molding*

Where shaped rubber liner components are required, such as domes, it may be convenient to mold such shapes. As the molding may be vulcanized into position later, the least possible cure should be introduced at the molding stage. The finished molding is, therefore, susceptible to deformation on extraction from the mold tool and during storage. Where reinforced plastic insulants are used, molding by conventional methods will be essential.

3. *Casting*

As an alternative to the methods just described, it is possible to apply some linings to rocket motor cases by centrifugally casting a filled liquid polymer. This technique is particularly suitable for cases with large length to diameter ratios or where access is limited. The premixed compound is added to the motor by a dispenser, the curing agents having been added in measured quantities immediately before application. The compound may be metered accurately and placed in relevant portions of the motor case, depending on the design. The motor is rotated at a given speed according to its diameter and the thickness of insulation required. Heat may be required for polymerization to occur. Local thickening may be built up by the use of dams prefabricated from the same material. Dome shapes are preformed by casting the material into mold tools or by spinning the case on a vertical axis.

D. External Insulation

With the exception of sprayed paints, experience in the field of attaching insulants to the external surfaces of rocket motor cases is limited, but the existence, or otherwise, of projections from the structural surface, wing roots, and launch lugs, for example, will greatly influence the fabrication methods used. Should there be no projections, then materials in strip or filament form could be applied by winding whereas specialized shrink-wrap materials could be applied in tubular form. But, if projections are present, either spray or hand lay-up techniques suggest themselves unless investment in expensive custom-built molding plant is justified.

Sprayed materials are suitable in thin layers for the less severe aeroheat conditions, but the time necessary to build up thick layers and the increasing variation in thickness preclude sprayed materials for more severe applications.

Hand lay-up materials may be tailored in flexible sheet form or supplied as preformed moldings. Either can be adhesively bonded to the case. If, as is likely, these materials are not resistant to handling damage and high velocity movement, they can be readily overlaid with a (glass) reinforced matrix resin that will cure in situ to provide a tough outer shell.

V. Example

A. General

In this section a logical if simplified series of steps is shown illustrating the points raised in Secs. I–IV that require attention during the design of a rocket motor case.

B. Requirement Data

The requirement data for the example rocket motor case design may be summarized as follows. The charge type is case bonded and radial burning. The charge external diameter is 0.120 m and the motor external diameter is the minimum possible. It has a maximum expected operating pressure of 169×10^5 Nm^{-2} with maximum burning time and total flight maneuvering time of 5 and 10 s, respectively. The operating altitude is 0–5 km and a maximum flight speed of Mach 3. There is low reaction required to both fuel fire and fragment attack.

C. Body Tube

The steps in the design of the body tube are as follows.

1) Since the charge burns radially for 5 s, the propellant will itself insulate the majority of the case for most of the burning time. The limited thickness of general internal insulation required will be assessed from the manufacturer's data bank, say, 0.001 m. The internal diameter of the body tube can then be calculated:

$$\begin{aligned} d &= d^* + (2 \times 0.001) \\ &= 0.120 + (2 \times 0.001) \\ &= 0.122 \text{ m} \end{aligned}$$

2) Since low reaction to fuel-fire and fragment attack is required, a nonhomogenous body material should be chosen and since minimum possible outer diameter is required (rather than lowest possible mass) the best available material is strip laminate steel with a minimum tensile strength of 2×10^9 Nm^{-2}.

3) Since the maximum expected operating pressure, or limit load, is 169×10^5 Nm^{-2} the proof and minimum burst pressures for the case can be calculated:

$$p' = p^* \times 1.125$$
$$= 169 \times 10^5 \times 1.125$$
$$= 190 \times 10^5\ Nm^{-2}$$
$$p = p^* \times 1.33$$
$$= 169 \times 10^5 \times 1.33$$
$$= 225 \times 10^5\ Nm^{-2}$$

4) From the information in Steps 1–3 the case wall thickness can be calculated:

$$S_H = pd/2t'$$
$$t' = \frac{pd}{2S_H} = \frac{225 \times 10^5 \times 0.122}{2 \times 2 \times 10^9}$$
$$= 0.69 \times 10^{-3}\ m$$

5) Because strip laminate motor cases require a minimum of three layers of strip, each layer must be at least 0.23×10^{-3} m thick.

6) Also, the outer diameter of the case is equal to $0.122 + (2 \times 0.69 \times 10^{-3})$ which is equal to 0.1234 m, or, more correctly, 0.1235 m, taking a small allowance for the adhesive between the layers of strip.

7) For the tensile strength of the steel to be fully utilized, it is necessary to ensure that the shear strength between layers of strip is not the weak point in the design. A suitable strip overlap pattern is chosen to ensure this, which in practice consists of the longest adhesive shear path with the minimum of repeat patterns. For a three-layer tube the overlap is one-third of the strip width, giving a shear path length of two-thirds of the strip width, Fig. 6. Hence, the width of strip required can be calculated from

end load on closure = shear load in strip adhesive

area × pressure = circumference × length shear path

× adhesive shear strength

Hence, if the strip width is W and the adhesive shear strength obtained from the manufacturer's data bank is $30 \times 10^6\ Nm^{-2}$, then

$$\pi\left(\tfrac{1}{2}d\right)^2 p = \pi d \times \tfrac{2}{3}W \times 30 \times 10^6$$
$$W = \tfrac{3}{2}dp \times 10^{-6}/120$$
$$= \frac{3 \times 0.122 \times 225 \times 10^5 \times 10^{-6}}{2 \times 120}$$
$$= 0.034\ m$$

Thus, any width of strip greater than 0.034 m will ensure that the shear strength of the adhesive between the layers of strip is not the weak point. From a production viewpoint it is economical to use the widest possible strip and, for ease of winding, it is wise to use a width that is less than the tube diameter. Hence strip of 100 mm width would be ideal in this instance.

D. End Ring Bond Length

The body tube was designed on minimum steel strip strength and thickness. Both these parameters have manufacturing tolerances, and together these tolerances add up to a possible 15% extra strength in the tube material. This factor is built into the end ring bond to ensure that failure will be in the tube and not the end ring bond if the case is subjected to burst test:

$$\text{load on end rings} = \text{strength of end ring bond}$$

Hence, if the bond length is L and the adhesive shear strength is again 30×10^6 Nm^{-2}, then

$$\text{area} \times \text{pressure} \times 1.15 = \text{circumference} \times \text{bond length} \times \text{adhesive bond strength}$$

$$\pi\left(\tfrac{1}{2}d\right)^2 p \times 1.15 = \pi d \times L \times 30 \times 10^6$$

$$L = \frac{1.15 \times d \times p \times 10^{-6}}{120}$$

$$= \frac{1.15 \times 0.122 \times 225 \times 10^5 \times 10^{-6}}{120}$$

and the minimum end ring bond strength is 0.026 m.

E. Stiffness

At this stage the stiffness of the case is calculated, using the value of EI for the material and tube dimensions. Providing it reaches the requirement, as strip laminate steel cases usually do, no modifications are necessary. In the event that it is a highly maneuverable missile and requires exceptionally high stiffness, then the addition of an extra layer of strip could be considered.

F. Bending

In most cases a strip laminate steel case designed on pressure loads will be sufficiently strong to withstand the bending load requirement. Let B be the maximum bending moment applied to the motor case. The maximum longitudinal stress S_L in the case due to B is given by

$$S_L = \frac{Bd}{I2}$$

where d is the diameter of the tube and I is the second moment of the area of the tube cross section about an axis in the plane of the section through its centroid.

The maximum shear stresses in the end ring and strip adhesive bonds due to the bending moment B, are given by

$$S_s = B/\pi\left(\tfrac{1}{2}d\right)^2 L$$

where L is the end ring bond length or the strip adhesive bond path length, respectively.

For bending loads with no internal motor pressure, the stresses are compared directly to the material strength with allowance for an appropriate safety factor, normally 1.5. For bending loads with internal motor pressure, the stresses must be added to the appropriate pressurization stresses before comparison with the material strength.

Finally, it must be verified that the motor tube will not buckle under the applied bending moment. A conservative estimate of the critical bending moment for buckling can be obtained by equating the maximum compressive stress under bending to the critical compressive stress under pure axial compression. This yields a critical bending moment B_c for buckling, given by

$$B_c = 0.3\pi E t'^2 (d/2)$$

where E is Young's modulus for the steel strip laminate case and t' is the thickness of steel in the case.

G. Internal Insulation

Although a thickness of internal insulation of 0.001 m was assumed in Sec. V.C, Step 1, for the greater part of the case it will also be necessary to locally thicken the insulation over erosive areas, for instance, adjacent to charge slots or apexes and at one or both ends where uninhibited charge faces may burn back. This extra local insulation will reduce propellant volume.

H. External Insulation

Since the missile is launched from the ground and may reach 5 km at a maximum speed of Mach 3, skin recovery temperatures of up to 550°C, Fig. 11, may be attained. The 5-s time of burning is too short for the 550°C skin temperature to have penetrated both the case and the internal lining to raise the temperature of the propellant. Hence, external insulation would not, in this instance, be required to protect the propellant itself from aerodynamic heating.

The steel strip laminate case, however, would be appreciably heated (aerodynamically) during the total 10-s time of maneuvering. This will have the dual effects of reducing the stiffness of the steel and, as it is a strip laminate case, reducing the strength of the adhesives and, hence, both the strength and stiffness of the case. Protection against aerodynamic heating would, therefore, be required. Suitable material would be chosen by the manufacturer using a thickness derived from data bank information. Since it is always likely that a good insulant plus a protective material will be required, and for practical reasons it is likely that 0.001 m thickness of each are the minimum that can be handled, the outer diameter of the finished case is likely to be at least $0.1235 + (4 \times 0.001) = 0.1275$ m.

References

[1]Badham, H., and Thorp, G. P., "Design Methods in Solid Rocket Motors," Paper 6, AGARD Lecture Series, No. 150.

[2]Evans, P. R., "Composite Motor Case Design," Paper 4, AGARD Lecture Series, No. 150.

[3]Denost, J. P., "Design of Filament Wound Rocket Cases," Paper 5, AGARD Lecture Series, No. 150.

[4]International Standard Atmosphere (ISA) Tables.

[5]Specification MIL-STD-210A.

[6]Ordnance Board Proceedings 41849.

Chapter 8

Solid Rocket Plumes

Andrew C. Victor*
Victor Technology, San Rafael, California 94901

I. Introduction

EXHAUST plumes have been a concern in rocket propulsion technology for over four decades. The early concerns involved rocket plumes that caused jet engine flameout and plume interference with microwave guidance signals for beam-rider and semiactive systems. Subsequent concerns involved exhaust detectability due to primary and secondary smoke and effects of smoke on some kinds of missile optical guidance systems. The recent emergence of autonomous optical detection and tracking systems has brought a strong concern with increased missile and launch-platform vulnerability due to the entire spectrum of plume emissions.[7] This becomes even more important as other missile signatures, such as body emissions, body scattering, and body radar cross section (RCS), are reduced by application of new materials technology and nonaxisymmetric geometry.

Over-riding all exhaust plume measurements and predictions is a requirement to quantitatively specify required signature levels for current and future missile missions. Straightforward as this sounds, it has never been done adequately for any tactical missile plume. Yet it obviously must be done to set realistic design goals for propulsion systems and to prioritize research and development (R&D) efforts for propellant development and system design as well as plume detection, identification, tracking, targeting, and control. In addition, required signature levels must be identified with specific operational time frames (outyears) to correlate with projected enemy detection capabilities. It is also critical that signature be scored or ranked as to its value for each mission and the relative value of each type of signature [for example, infrared (IR), ultraviolet (uv), smoke, flash, RCS]. Other plume effects (for example, radio frequency- (rf-) guidance interference and plume-impingement effects) are over riding performance issues that must be solved for successful deployment. Once a set of prioritized signature goals has been established, it may be possible to prioritize the R&D required to reach the goals.

The most important factors responsible for intolerable levels of tactical missile plume signature are 1) combustion (afterburning) of plume fuel species, such as

*President.

carbon monoxide (CO) and molecular hydrogen (H_2) with atmospheric oxygen, which is a major contributor to high temperature and resulting IR, uv, and rf signature effects; 2) base recirculation and burning, which can greatly increase the afterburning effect in plumes of some propellants in high-velocity missiles; and 3) solid and liquid particles in the flowfield.

Those missiles for which reduced and minimum smoke propellants have been developed over the past decades have over riding identified needs for signature control. In general, controlled smoke propellants are intended to prevent timely countermeasures by target defenses and/or to prevent interference with some optical guidance, detection, or tracking systems. For missiles launched from covert sites, there are compelling reasons to reduce launch plume signature to undetectable levels.

In general, pertinent ranges for detection will be of the same order of magnitude as the related engagement ranges. When the ranges are short, time becomes an important parameter. Time is also involved in the short burn durations of many rocket motors, limiting the time available for detection of radiative emissions.

Body and exhaust plume IR radiation have been used for decades for detection and targeting by some missile guidance systems. For complex engagements involving multiple launch platforms and multiple targets, IR signatures may be used in the near future to detect and target launch platforms on the basis of rocket exhaust signatures. In such scenarios one is concerned about highly capable detection systems and potential detection ranges out to hundreds of kilometers.

Under clear sky, high-visibility daylight conditions, plume smoke is visible to the unaided eye, by reflected sunlight, from ranges comparable to those for emitted signatures using advanced IR detection capabilities. In addition, plume smoke persists and creates an "arrow" from the missile back to its launch point. Primary smoke is the result of particulates in the rocket motor effluent. Secondary smoke is the condensed vapor contrail of a missile plume; and secondary smoke formation depends on both the nature of the gaseous rocket motor effluent and the prevailing ambient atmospheric temperature and humidity conditions. Missile contrails form at all conditions that support jet aircraft contrail formation. In addition, the HCl product of ammonium perchlorate combustion causes an extremely persistent (and often continuously growing) contrail to form at conditions much warmer and much less humid than required for formation of aircraft contrails. It takes some time for full development of a secondary smoke contrail, depending on the ambient conditions and the exhaust specifics. This time delay can affect the importance of secondary smoke to specific scenarios. In addition, the optical density (and, hence, visibility and transmissivity) of the secondary smoke contrail of a given motor depends strongly on ambient conditions. Research to understand and control physical processes (propellant chemistry, fluid dynamics/combustion interactions, signature) is critical for plume signature reduction.

Much of the effort of the past two decades in plume technology has been directed toward developing and refining analytical prediction methods. Most of these methods are now exercised as digital-computer programs. When the computer programs accurately predict plume effects, the considerable time and expense associated with flight testing to determine plume properties of specific missile exhausts can be avoided. However, it is necessary to prove the validity of any computer program by comparing its calculated results with measured data. The advantage of the computer program is that once its validity is demonstrated, it is possible to apply it to many situations for which measurements have not been made. A strong caution must be made here; validation of a computer program within a limited range of rocket motor propellants and flight environments does not guarantee validity outside that range. In fact, there are many instances of "verified" computer programs failing in subsequent comparisons with data.

Plume predictions are routinely made for IR signature, smoke visibility and obscuration, and rf-guidance signal attenuation. In addition, the analytical techniques available can be used to predict plume-impingement effects, and, with various modifications, rf-guidance signal noise-modulation and RCS, and emitted visible and uv signature. Despite this apparent state of the art there are clear areas of incompleteness in the modeling which show up as limitations in accuracy of some predictions and obvious failures of some others.

One of the major benefits of predictive modeling has been the insights it provides for understanding the effects of different rocket motor and mission variables on exhaust plume properties and effects. Even when computer programs fail to accurately predict missile plume effects, they may provide important information about the relative importance of many of the different variables that affect exhaust plume properties.

The factors that must be controlled in solid rocket motor design, development, and production include performance (range, velocity, miss distance), plume effects, service life (aging, material compatibility), safety (transportation, storage, insensitive munitions), producibility, and cost. Some of these factors are synergistic in their influence. This chapter is concerned primarily with only one of those factors, plume effects, and how it is influenced by all of the other factors.

The purpose of this chapter is to review the historical background of solid rocket plume problems, the operational implications of plume signature and other plume effects, the science and technology of plume flowfields and plume effects, and to present some design guidelines for treating plume effects in missile system design. Prediction of plume flowfields and plume effects is very complex and state-of-the-art computer models are required to obtain the most reliable results. Promulgation of the latest computational techniques in plume technology is carefully controlled by their originating nations. Measurements of exhaust plume flowfield properties and plume effects are expensive, particularly for flight conditions, and the data are controlled even more carefully than the computer codes. Therefore, the usual practice has been to refine the computational techniques by comparison with available measured data (unfortunately, usually a fairly limited data set) and then use the computations to predict plume effects at other conditions. Occasionally, a target of opportunity will provide confirmation of the computational techniques or data that inspires improvement of the models.

II. Background

Plumes of early versions of the 2.75-in. folding-fin aircraft rocket (FFAR) rocket motor caused engine flameout of launching jet aircraft. This problem was alleviated by coating the combustion stabilizer rod in the motor with potassium sulfate. Potassium compounds are very effective afterburning suppressants for double-base propellants. These and other compounds have also been demonstrated to be effective when incorporated into the propellant mix.[2]

Introduction of radar-guided missiles in the 1950s soon led to problems of plume-guidance interference.[3] These problems were linked to the presence of ionized species in plumes.[4] There were no easy solutions to this problem. The increased energy available with aluminum-filled, ammonium perchlorate- (AP-) oxidized composite propellants, as well as their improved mechanical characteristics and service life over double-base propellants made them appear to be critical components of the longer range radar-guided missiles. The chemistry of these propellants was such that afterburning could not be suppressed by simple chemical means. Other approaches tried were reduction of the aluminum content, increased transmitter power, and semiactive homing as an improvement over the earlier beam-rider homing. Finally, in the 1960s, fairly significant improvements were obtained by mounting multiple (two or more) receiving antennas on the missile and mounting them forward, on the sides of the missile instead of on the base, as had been the previous practice. This opened alternate paths to the missile antennas that in most situations reduced the energy-robbing effects of plume-signal attenuation to tolerable levels. However, with the more sophisticated semiactive guidance systems, which used frequency-coded information on the radar beam to both illuminate the target and provide baseline velocity information to the missile, it was soon discovered that modulation, or noise, generated by the plume-signal interaction was still present and could disrupt guidance. Missile onboard rf-guidance transmitters are intended to eliminate this problem in the autonomous terminal-guidance phase as well as free the launch platform from further hazardous involvement in the encounter.

The problems of radar-guidance interference must still be considered with every new rf-guided missile and with any related changes to current missiles. The ionized species in the exhaust plume can also increase the apparent missile signature from search, detection, and tracking radars. This RCS enhancement is of particular concern if the missile-body RCS is reduced by the use of composite materials or specific low-signature configurations.

Improved propellant binder systems (polymers), available in the 1960s, made it possible to eliminate aluminum from composite propellants and still achieve required rocket motor performance. This led to the wide use of "reduced-smoke propellants" that contain AP in a fuel-rich binder system with small percentages of additives. With these propellants, the persistent primary-smoke trail that always follows missiles with aluminized propellants was eliminated. However, it was soon discovered that these missiles left water-vapor (secondary smoke) condensation contrails under some atmospheric conditions that were much more persistent than those emanating from aircraft jet engines or from double-base rocket motors.

With new missile operational requirements demanding the energy available from composite propellants as well as the low observability that was possible with unfilled double-base propellants, there was a clear need for "no-signature propellants." However, since a rocket motor must have effluent, "minimum-signature propellants" constitute a more realistic goal. Smokeless propellants are an aspect of no-signature propellant development, and the best that could be done was the achievement of minimum-smoke propellants that minimize both exhaust particulates and enhancement of water-vapor condensation. In the area of plume smoke, new guidelines for terminology have been recommended by an AGARD advisory group to replace the rather ambiguous terms *reduced smoke* and *minimum smoke* with more quantitative classifications.[1]

Since it is also possible to reduce exhaust plume signatures by changing certain aspects of the missile body, the real goal should be minimum-signature missiles for which the propellant and even the rocket motor are only one, albeit important, part. The exact goals for minimum-signature missiles will vary depending upon realistic mission requirements and the operational threat environment. Reducing any one of the signature effects discussed in this chapter, or any combination of them may be the goal of a particular minimum-signature rocket-powered missile. In practice, other factors including cost, service life, and insensitive munitions requirements will affect the degree to which the goal can be achieved. All of these factors must be considered in optimizing tradeoffs.

III. Operational Considerations

This section examines some of the factors that link operational requirements to solid rocket exhaust plume considerations. These factors fall into two major areas.

1) Physical interactions: This covers such physical interactions of the plume as gas and particle impingement on surfaces and gas dynamic interactions (such as those between two rockets on closely parallel or serial trajectories or the effect of rocket exhaust on aircraft jet engine performance, including flameout).

2) Radiation interactions: This covers the areas of plume signature and guidance interference effects that are caused by emitted plume radiation or interactions of plumes with radiation from other sources. Our interest in these interactions can be grouped into the following two categories: a) detection requirements which are related to threat (enemy) plume signatures and determine our requirements for detection capability and countermeasures and b) signature requirements which include requirements for "stealth," for missile guidance, and any other operational parameters that are affected by our missile plumes and thus affect the design of our own rocket motors.

Operational requirements regarding exhaust plumes are usually a critical part of a weapon system's design requirements. The rest of this section gives specific examples of such requirements in relation to the foregoing factors.

A. Physical Interactions

In this area we are concerned with the design of a rocket motor to assure that the physical interactions of its plume do not cause operational problems. Three problem areas are considered as examples.

First, consider plume-impingement situations such as are often encountered when launching a missile from an aircraft, ship, or land-based platform; our concern might even be a small shoulder-launched battlefield missile. The geometry of the situation is well known. The first step in solving the design problem is to determine how the missile plume fits into the known launcher geometry. Existing rocket exhaust plume flowfield computer codes are ideal for determining this. The next step is to define the temperatures, pressures, and particle impingement induced by the plume onto launcher surfaces and to determine whether or not there will be a threat to the launch vehicle or person. To accomplish this, one must use the plume flowfield computer output parameters of gas density, species, temperatures, particle sizes and number densities, and gas and particle velocities and then determine impingement forces and heating rates on involved surfaces. The degradation of launcher surfaces can then be determined analytically. Experimental verification of the predictions can be obtained by submerging well-instrumented materials into the exhausts of statically fired rocket motors and comparing the results with code predictions. If the predictions are good, they can be extrapolated to other, nonstatic, rocket motor firing conditions with reasonable assurance of accuracy. Even fairly simple prediction methods have been successful for surface-launched missile launcher impingement problems.[5,6]

Second, consider a barrage rocket system; typically such a system uses unguided rocket powered missiles launched in parallel or series from launchers that contain substantial numbers of weapons. In either parallel or series firing modes, the gas dynamic effects of the exhaust jet from one rocket can affect the trajectories of neighboring rounds. The problem is to determine the magnitude of that interference and whether it causes trajectory errors greater than allowable by the aim-point-accuracy/miss-distance/kill-probability requirements for the weapon system. The pertinent analysis can be done with one of the existing rocket exhaust plume flowfield computer codes. If it is determined that an operational problem exists, it may be possible to redesign the rocket nozzle or the missile exterior to reduce the problem. Another approach might be to devise a tactical firing strategy that minimizes the interference problem.

Third, another very serious problem involving physical interactions is jet engine flameout caused by aircraft jet engine ingestion of a rocket exhaust. There is a history of this problem with afterburning rocket exhausts. In fact, the use of a potassium sulfate "salt-rod" in some early air-launched rockets was implemented just to eliminate afterburning that caused this problem. This problem has also been encountered more recently with air launch of some air-to-air tactical guided missiles, particularly if gas dynamic or autonomous missile guidance effects cause the rocket to cross the aircraft path with resulting engine ingestion of hot exhaust gases. At least three solutions are obvious: 1) eliminate afterburning in the rocket exhaust (not easy with modern high-performance composite propellants),[7] 2) relocate the missile launcher on the aircraft, or 3) delay the start of autonomous guidance of the missile until it is farther from the launch aircraft. Even some of the older rocket exhaust plume computer codes [such as the low altitude plume program (LAPP)[14] developed in the early 1970s] have been applied very successfully to modeling this problem.

B. Radiation Interactions

1. Detection Requirements

This area concerns information on the plume signatures of threat (enemy) weapons, both aircraft and missiles, that may be difficult to obtain because of the manner in which such information is controlled. Information needed to predict the signatures of threat aircraft and missiles is also difficult to obtain. Information in this area is usually highly classified in every nation and handled on a need-to-know basis because it has direct implications on weapon design, tactical procedures, and critical technology. However, certain principles that are obvious will be discussed here.

The objective is to perform some military operation at minimum risk and with maximum chance of success. Quite obviously, sensing radiation from the exhaust plume of a target aircraft or missile, because its plume emits radiation, is an ideal means of locating, identifying, and tracking, and possibly targeting (or terminal homing). Therefore, we can start the design of a passive "detection system" based on this method. A number of unknown parameters are immediately evident. 1) At what distance must we detect the target? 2) At what aspects (relative flight paths, angles, etc.) must we detect the target? 3) Over what range of atmospheric conditions must we detect the target? 4) Over what range of absolute and relative velocities must we detect the target? 5) What sensor capabilities do we have and how can they be used? 6) What are the target signature characteristics? 7) Can the missile onboard sensor(s) distinguish between plume and target?

It is possible, using modern exhaust plume computer codes and other analytical techniques, to get answers to all of these questions. However, one must be cautious because all of the analytical techniques are based on assumptions that may hide parts of reality. For example, the strongest feature of current plume computer codes is their prediction of steady-state levels of total and spatially resolved radiation. However, the distinction between the target and its plume may be more apparent in the time-dependent regime due to such features as turbulence structure and Doppler shifts in radiation, than in the steady-state regime.

For target detection and tracking, the total steady-state radiation levels are of primary importance. The sensor/detector portion of the guidance system must be capable of distinguishing, at the required maximum operational launch range, the target plume from background radiation and from spurious local radiation sources. Intervening atmospheric attenuation of the target signature must be accounted for when the maximum operational launch range is determined—this includes accounting for possible variations in atmospheric attenuation over the missile launch envelope. All signatures encounter some degree of atmospheric interference as shown in Fig. 1. The guidance system must be capable of locking on to the target plume and keeping lock as the missile approaches its target, this must include algorithms to overcome any momentary loss of lock.

As the missile approaches its target, the field of view of the sensor/detector subsystem will become filled with the target plume and an autonomous decision must be made to transfer lock to a particular part of the signature. Ideally this will involve transferring lock from the plume to the target body itself. This may require inclusion of a bias or shift of lock from the location of maximum plume

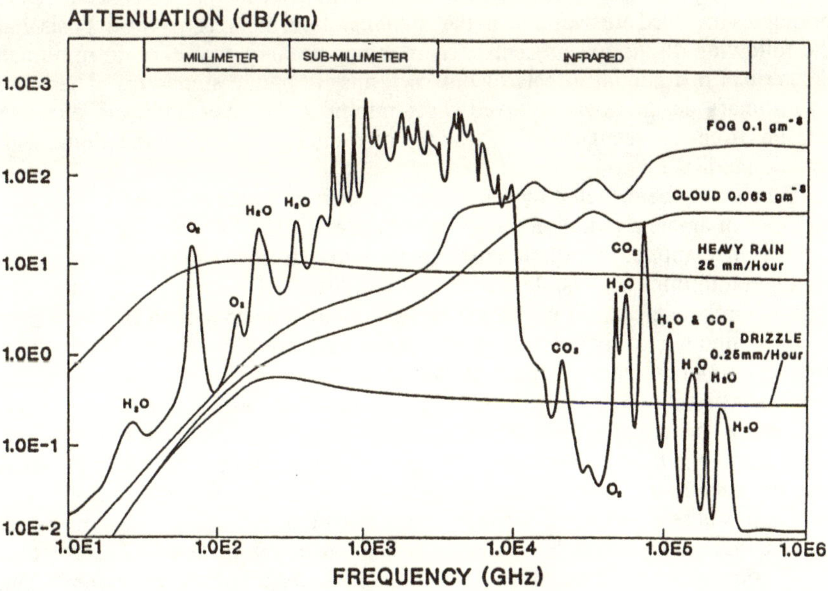

Fig. 1 Atmospheric attenuation.

signature to the expected location of the target, which is not too difficult to do if the missile is using a proportional guidance scheme. In fact, this bias could be built into the missile guidance system to operate from launch onward. The bias could be based on plume signature computer calculations; however, calculated values would be strongly dependent on target type, target velocity, and target altitude. The danger of this approach is clear; the range of variation can be large if the bias is preprogrammed, and the resulting average miss distance could be large. If distinguishing characteristics of the target–plume interface can be identified, transfer of lock would ideally involve switching from the total plume or location of maximum emission in the plume to the characteristics of the target–plume interface. Distinguishing features of this interface may be time dependent rather than steady state. The plume gases are at their maximum velocity very close to the target; therefore, some guidance solution based on Doppler shift in the signature might be used. Also, the plume will have a turbulent structure not exhibited by the target surface. This turbulence might be used to distinguish between the target and its plume, however, the turbulence will probably be a minimum near the target and a maximum at the location of maximum plume temperature and afterburning; this would tend to mitigate against using turbulence to distinguish between the target and plume. The target signature near the nozzle may be expected to be a combination of hot body radiation and scattered ambient radiation that could distinguish it spectrally from the plume.

2. *Signature Requirements*

This area is related to the design of solid rocket motors. The design objective is to minimize the signature of a missile exhaust plume for one or more reasons. Detailed information in this area is usually highly classified to prevent its acquisition by potential threat nations. However, as with the previous section, the principles involved are available in the open literature, and it is from that basis that the following discussion proceeds. There is one primary goal here: to minimize deleterious operational effects of our own missile plume signatures. There are two primary design goals involved: 1) to minimize the detectability of our own missiles in certain scenarios and 2) to minimize any interference of the plume with missile guidance.

Plumes may be detected by sensing either their emitted radiation or their disturbance of ambient radiation (either of natural or artificial sources). Plumes emit significant quantities of radiation over a wide range (from the *uv* at the short wavelength end through the visible and IR and into the short rf spectrum, particularly that region known as millimeter waves, at the long wavelength end). There may be trace amounts of radiation at shorter and longer wavelengths, however, these are usually not considered viable for tactical purposes.

Particles in plumes, often referred to as plume smoke are a major cause of concentrated or disturbed ambient radiation, which is usually apparent as it is scattered from the plume to some detector. The smoke trails of plumes are easily detected visually because they scatter sunlight and the less intense skylight. Scattering of the Earth's albedo (reflected radiation) may also be detectable, particularly when the ground is very bright (for example, when covered with snow). There is also an IR component of scattered sunlight, usually less than 5 μm in wavelength, that may cause detection or interference problems. Plume smoke can interfere with guidance systems that operate in the visible or near-IR regimes if propagation through the plume is required in tactical scenarios.

The gas-density variations in plume wakes may cause disturbances in ambient rf fields. Although such disturbances can be detected (as can clear-air turbulence, which is the same phenomenon), it is not obvious that such detections can be useful in tactical battlefield conditions.

Electronically charged species in plumes (particularly free electrons, although ions may also contribute significantly at some wavelengths) attenuate, scatter, and "focus" impinging rf radiation in the megahertz and gigahertz regimes. For very highly ionized plumes, this can cause a detectable increase in the missile RCS. Even moderate levels of ionization can cause rf interference for beam-rider and semiactive guidance systems and may result in serious guidance problems unless appropriate design compensations can be made.

The temperature of the plume is the single most important factor contributing to signature levels involving emitted radiation and rf interactions. Fine solid and liquid articles exhausted from the rocket nozzle form particle laden contrails (primary smoke) detectable for many kilometers behind a flying missile. Water-soluble gaseous species and water exhausted from the nozzle can result in the formation of water-droplet contrails (secondary smoke) that persist for many kilometers over a wide range ambient conditions, although there are conditions when the water droplets do not form. The criteria for formation of secondary smoke involve the nature and concentrations of the exhaust species, the temperature and behavior of the plume (does it afterburn, for example), as well as the ambient temperature and atmospheric moisture content (i.e., relative humidity).

In the following discussion, methods are given for arriving at solid rocket motor or propellant design requirements to achieve operational goals.

The first step is to determine the range from which nondetectability is needed and the radiation wavelengths involved. No guidelines have been established for this. It would be desirable to have a plume and missile that is undetectable at all ranges, however, this is clearly not possible. The following guidelines are believed to be appropriate.

1) Initial detection of supersonic missile plumes at ranges of less than 6 km is not important because the short engagement times available do not permit effective countermeasures.

2) Detection ranges greater than 75 km are ambiguous in one-on-one engagements.

3) Plume RCS should be controlled to keep detection probability of the plume less than that of the missile alone.

4) The need for guidance interference control is determined by the "marginal excess capability" of the guidance system without interference (that is, the ability of the guidance system to operate beyond the maximum missile envelope). For example, if the guidance signals have a power margin of 10 dB at the maximum operational range, then plume signal interference up to 10 dB might be tolerable. This is most applicable to rf-guidance interference. Plume induced rf noise must be kept to levels that will not be interpreted by the guidance system as false targets or that will not obscure real target signatures.

5) Plume smoke must be controlled for those missions in which it can compromise the launch platform or interfere with electro-optical guidance. It is also important, but less critical, to avoid detection of the missile itself. There will

always be possible scenarios in which secondary smoke can form, and the probabilities of such occurrence should be known. For this reason, AGARD working group PEP/WG-21[1] has provided a climate database which permits the prediction of the probability of occurrence of secondary smoke on a common climatic basis.

6) Missile plume signatures that can contribute to the detectability of an otherwise covert and vulnerable launch platform may need to be controlled at all costs. This is particularly true of plume smoke trails since they persist for long times.

The following fictitious example of a plume-guidance interference requirement shows some of the many factors that may be involved in even a fairly simple scenario.[1]

The missile is guided along a line of sight. The guidance system minimizes the angle formed by the target, the launch platform, and the guided missile. Designation of the target is done by an operator who must continuously designate the target until the missile hits it. The missile is located during flight by a beacon that radiates in the near-IR range.

The exhaust plume can disturb the guidance system several ways. First, the plume can obscure the target to the operator due to smoke. The target is seen in a natural environment by a contrast defined as

$$C = (B_t - B_b)/B_t$$

where C is contrast, B_t target brightness, and B_b background brightness. The attenuation (which must be less than 10) is defined as

$$A = C1/C2$$

where A is attenuation, $C1$ contrast without motor exhaust, and $C2$ contrast with motor exhaust.

Second, the plume can obscure the missile to the operator due to smoke. With the same definitions as the previous paragraph, the attenuation must be less than 20.

Third, the plume can disturb the guidance system by saturation of the light amplifier used for night firing, due to plume radiation. To avoid saturation of the light amplifier the spectral radiance of the plume must be less than 1×10^3 $W/m^2/sr/\mu m$.

Fourth, the disturbance can occur by a higher level of radiation from the plume than from the beacon. This point is satisfied if the previous one is satisfied.

Plume smoke can also reveal the missile trajectory and the launch position. This point has not been quantified for this example at the present time.

Detailed design guidelines must be worked out for each individual missile system on the basis of how and where it is to be used, and the relative value of the different contributing factors: range, velocity, guidance type, mission value, launch platform value, and technology availability.

IV. Technical Considerations

This section describes major plume properties and effects of concern, starting with the flow of rocket motor combustion products through various regions (from the combustion chamber through the nozzle and into the atmosphere as far as necessary for the properties and effects of interest to manifest). Critical design features of solid rocket motors, including design principles and contributions of propellants, liners, insulation, nozzles, and igniters to plumes are included. Plume effects are described from a theoretical basis. This basis is necessary since it is the only way to approach understanding of the phenomena involved. Existing data on plume effects tend to confirm the theoretical principles, if not the fine details of calculations in all cases.

A. Solid Propellant Rocket Motors

Solid rocket motors appear to be rather simple devices. Generally, they have no moving mechanical parts (although some modern systems are fitted with moving nozzle thrust vector control systems and adjustable pintel nozzles have been designed and tested). However, this apparent simplicity is deceptive. It is more appropriate to visualize a solid rocket motor as an engine which has all of the design requirements "frozen" into the necessary chemical and physical forms. These points are covered in Chapters 4–6. The design is "activated" upon ignition of the motor.

Specific impulse is usually considered to be the single most important propellant property related to performance, and this certainly is true for weight-limited rocket motors. However, for volume-limited systems, propellant density is almost equally important, and the density may be traded against specific impulse in propellant selection to improve performance, although not on a one-to-one basis. When propellants with reduced exhaust plume signature effects are selected for operational use, the specific impulse invariably suffers.

1. Solid Propellant Ingredients

To provide the necessary energy for missile propulsion, a rocket motor must contain fuel and oxidizer ingredients that burn to produce the high-velocity nozzle gas and other effluents that propel the missile. In a solid rocket motor the fuel and oxidizer are bound closely together in an elastomeric binder matrix that provides the necessary structural properties over a wide range of required performance and storage temperatures (as broad as from $-54°C$ to $+71°C$). The binder is typically a fuel, although, in some propellant types, the binder may also have some oxidizer molecules. There is increasing interest in polymeric energetic binders which have oxidizing capabilities.

Two basic types of propellants are in wide use today: double-base propellants and composite propellants. Although the molecules in these propellants differ widely, both propellant types are based on carbon, oxygen, hydrogen, and nitrogen atoms; and exhaust products that result from chamber combustion are close to equilibrium concentrations of the involved chemistry at nozzle exit conditions. The chamber products that enter the nozzle throat are probably very close to equilibrium. Some nonequilibration occurs during the pressure drop of the nozzle expansion, however, the effect is small on major species like CO, CO_2, HCl, and H_2O that strongly influence plume infrared signatures. Minor species, like OH, H, H_2, e^-, and other ionized species that influence ultraviolet and radiofrequency signatures, are more strongly affected and may diverge significantly from equilibrium.

Double-base (DB) propellants are the oldest type currently in use. The binder system is not dependent on curing of a polymer system but on the capability of the nitropolymer [nitrocellulose (NC) is often used] to absorb and desensitize nitroglycerine (NG). Other nitroplasticizers than nitroglycerine may be used. Nitroplasticizers may also be used in combination with polymeric isocyanate curing systems with polyurethane or polyester polymers. Double-base propellant grains are manufactured both by extrusion methods and casting methods. Addition of stabilizers to the propellant is required to prevent decomposition of the nitroglycerine. Ballistic modifiers, i.e., burning rate catalysts, are usually added to tailor double-base propellants and achieve desired motor performance. These catalysts are usually lead compounds, such as lead resorcilate, citrate, oxalate, corborate, or others. A refractory, such as zirconium carbide, may be added to control combustion instability. All metal-based additives will contribute to the exhaust primary smoke signature, although the small amounts used may not cause serious effects. Afterburning in the exhausts of double-base propellant rocket motors can be reduced or eliminated by the addition of certain solid additives (for example, potassium sulfate), which slightly increase the particle density in the exhaust and may enhance condensation of moisture for contrail formation.

Composite modified double-base (CMDB) propellants have been developed. These may contain, in addition to NC and NG, solids such as AP, an oxidizer; aluminum (Al), a fuel; or nitramines, such as HMX or RDX. The last two tend to be fuel rich in rocket motor combustion but are themselves capable of sustaining combustion and are, in fact, both detonable high-explosive molecules.

Elastomer modified cast double base (EMCDB) propellants, such as those developed at Royal Ordnance, Summerfield, improve on the mechanical and low-temperature properties of conventional DB propellants by including a urethane-linked elastomeric binder, introduced in the casting fluid as caprolactone polyols and appropriate isocyanates.[2] Solid fillers, such as used in CMDB propellants, have also been added to EMCDB propellants.

Composite propellants, the other general class of propellants, support the oxidizer (usually AP) and the fuel (usually Al, although in reduced smoke propellants the binder serves as the fuel) as fine powders in a rubbery, cross-linked, isocyanate-cured matrix, called the binder. The matrix may comprise as little as 9%, by weight, of the propellant. Burning rate tailoring can be accomplished, within limits, by adjusting the particle size distribution of the AP. To achieve highest burning rates, iron, copper, or chromium containing additives are used, however, these additives increase the hazard sensitivity of the propellants (see Chapter 9). AP combustion releases HCl into the exhausts of composite and CMDB motors. This increases the propensity to form secondary smoke. The chlorine present in composite propellant combustion gases spoils the action of afterburning-suppression additives. Thus, exhaust afterburning for AP-containing propellants cannot be suppressed by additives (such as potassium sulfate) in the propellant or motor combustion chamber. The combustion of Al or other metals results in fine oxide particles in the exhaust, creating primary smoke in approximate proportion to the amount of metal in the propellant.

Other particulate oxidizers that might be used in composite- or DB-propellant solid motors include potassium perchlorate (KP) and ammonium nitrate (AN). Requirements for insensitive munitions may lead to future use of these oxidizers for special applications. Iodine pentoxide and lead nitrate have also shown potential as oxidizers in high density propellants[8]; however, the former is a scarce and expensive material and is degraded to a very corrosive acid by small amounts of water, and the latter is increasingly prohibited by environmental protection laws in the U.S. Fluorine-based oxidizers have also been used in developmental systems, and some fluorocarbon binders were successfully employed years ago. The use of KP in propellants will greatly increase ionization in the plume and may prohibit its use in certain applications. KP is more dense but less energetic than AP, and there are problems obtaining useful burning rates with KP oxidized propellants unless some AP is added. From a plume signature standpoint, KP has more disadvantages than AP. AN has no disadvantages from a plume signature standpoint; however, nickel or potassium stabilizers (which are often added to prevent undesirable phase transitions) can result in plume signature difficulties or chemical compatibility problems with other propellant ingredients. Finally, the nitramines, such as HMX or RDX have been used in composite propellants as well as in DB propellants as mentioned earlier. These molecules are fuel rich in rocket motor combustion although they are themselves capable of sustaining combustion and are, in fact, both detonable high-explosive molecules. The nitramines are used in some strategic rocket motors and in several composite-propellant and minimum-smoke propellant tactical rocket motors. The explosion (detonation) hazard that these ingredients cause when used in propellants is a concern discussed in Chapter 9.

A wide variety of particulate metallic fuels may be used in solid propellants, depending on the application; zirconium hydride has application in high-density propellants, and boron and boron hydrides are also potential fuels. Insensitive munitions requirements may lead to use of these fuels to maintain high-motor energy while reducing sensitivity by increasing propellant density. Boron has been used as a ramjet fuel. Other fuels based on lithium or beryllium were considered in the past, but are unlikely to be used; lithium because it is very hydroscopic, and beryllium because it is extremely toxic.

Composite propellants, which produce only small amounts of exhaust particles or droplets (smoke), can be made by eliminating AP and Al and by using particles of AN and/or nitramines (HMX, RDX, and, perhaps in the future, other high-energy explosive molecules). Ingredients added for phase stabilization or combustion stability or derived from the combustion of metal-based ballistic modifiers will contribute metal oxides to exhaust smoke or increase ionization of the plume.

Future use of increasing amounts of energetic binders and plasticizers in solid rocket propellants is anticipated. This should not quantitatively change exhaust-plume considerations. The effects of additional nitrogen that such binders contain may result in somewhat cooler plumes. The reduced solid content possible in propellants with energetic binders may result in qualitative differences in plume signatures and other plume effects.

2. *Other Factors*

Nozzle, liner, insulator, and inhibitor materials contribute to the motor exhaust when these materials are exposed to combustion in the motor chamber. Liners for internally perforated motor grains are often the rubbery binder matrix material of the propellant. When they burn (being fuel rich) during motor tailoff or during the combustion of end-burning grains, significant quantities of soot may be exhausted. Liners, inhibitors, and insulators may have fibrous inorganic materials added to improve insulating behavior and leave a charred insulating layer even after organic constituents burn away. Asbestos (hydrated magnesium silicate) has been a common constituent, however, laws preventing its use because of carcinogenic hazards are forcing a search for replacement materials. Silicones and bulk aramids are being considered for use in insulators. Inhibitors and liners (often the same material) may contain a variety of metal-based oxides (for example, calcium silicate [Wollastonite], antimony oxide, aluminum oxide hydrate, titanium dioxide, etc.)

or carbon black. All of these ingredients can contribute to the exhaust signature, although the contribution will depend upon how well the components do their job; if char layers form, as is desired, very little contribution to the motor effluent will come from these ingredients. Nozzles are increasingly fitted with graphite inserts which contribute little to the exhaust signature.

Igniter combustion products will contribute momentarily, but often significantly, to plume effects at the launch location. Common igniter ingredients (including potassium nitrate, sulfur, charcoal, boron, KP, and Al) will contribute to exhaust smoke, flash, and ionization. Recently developed "smokeless" igniters have been designed specifically to reduce or eliminate signature effects.

B. Plume Properties in Different Flowfield Regions

In the following discussion, the flowfield is considered in four regions. These are combustion chamber, nozzle expansion, base region, and exhaust plume. These four regions have been selected because of the unique contributions each makes to exhaust plume phenomena. Figure 2 shows the latter three of these regions, although only the last, the exhaust plume region, is shown in any detail and that detail is fairly complete only in regard to shock structure, not afterburning, which results in a broadening of the mixing region surrounding the outer region of the plume.

1. Combustion Chamber

Propellant combustion in solid rocket motors creates a large number of different molecular species. In general, solid rocket propellants are fuel rich (to minimize the molecular weight of product gases and thus maximize the impulse) creating a surplus of oxidizable molecules following chamber combustion. These molecules may subsequently react with atmospheric oxygen in the exhaust plume to cause afterburning.

Hydrogen and oxygen are atomic constituents of all current solid propellants. Complete combustion of these two constituents forms water, which, although a gas at the high chamber temperature, may condense to form droplets far downstream in the exhaust plume if ambient atmospheric conditions are propitious.

The chlorine present in the AP of composite propellants reacts with hydrogen in the combustion chamber to form hydrogen chloride (HCl). This acidic molecule readily combines with water at lower temperatures, causing saturation vapor pressures well below those for water alone and the consequent formation of droplets in the exhaust plume at higher ambient temperatures and lower ambient humidities, i.e., HCl effectively raises the dew point.

Aluminum or other metals, added to some propellants to increase impulse, burns with oxygen and results in increased combustion temperature. Some of these metals form compounds which condense upon cooling and form the particles observed in the exhaust plume as primary smoke.

As might be expected at the high temperatures of chamber combustion, ions (charged molecules) and free radicals (uncharged molecular fragments that are unstable at normal conditions because of the presence of an unpaired electron in the outer orbital) are created in the motor combustion process. Those free radicals that survive subsequent nozzle expansion increase the reactivity of the exhaust plume since their reaction rates are higher than those of stable molecules. Free

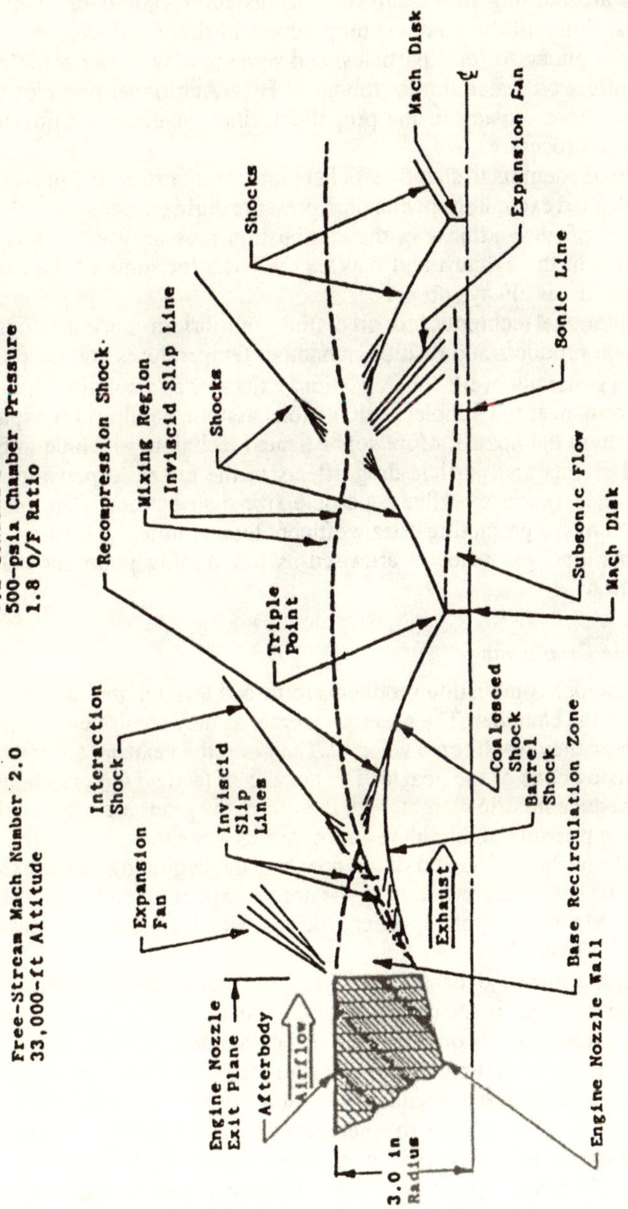

Fig. 2 Sketch of rocket exhaust plume flowfield.

(unpaired) electrons are also created in the combustion chamber, and some of these survive nozzle expansion.

Solid rocket propellants contain a number of additives at low concentrations. Included are burning rate catalysts, anti-instability additives, and afterburning inhibitors. Some of the combustion products of these additives may condense in the exhaust plume to form particles, and some may be water soluble salts which have an effect on water similar to that of HCl. Additional particles in the plume may come from erosion of the propellant, liner, insulator, or nozzle during the combustion process.

In spite of attempts to stabilize rocket motor combustion and prevent oscillatory or uncontrolled excursions of chamber pressure during combustion, there is always some level of unsteadiness in the combustion pressure that results in flowfield fluctuations in the exhaust and may be the basis for some of the exhaust plume turbulence that is always observed.

Computational techniques for predicting equilibrium concentrations of chamber combustion products and resulting chamber temperatures and propulsive performance are generally available.[9] Variations exist in the calculations used for propulsive performance; the simplest calculations assume equilibrium expansion of effluent through the nozzle. More sophisticated techniques include nonequilibrium chemical effects and particle drag effects in the nozzle expansion computation and even generation of effluent particle size distribution.[10] The results of such computations for predicting nozzle effluent temperatures, pressures, species concentrations, and gas velocity are used as the starting point for exhaust plume computations.

2. *Nozzle Expansion*

The chamber combustion products are forced through the motor nozzle by the pressure in the chamber. The gases accelerate as they are driven toward the nozzle throat where they reach sonic velocity. The gases then expand as they flow through the expansion cone of the nozzle. The nozzles of tactical motors are usually fairly short, to reduce missile weight, and, therefore, the gases are often underexpanded (that is, at a pressure greater than atmospheric) when they emerge from the nozzle into the atmosphere. The end of the nozzle is the beginning of the exhaust plume. The exhaust gas accelerates during the nozzle expansion and may reach velocities as high as Mach 3 in typical tactical rockets, which for some exhausts may be of the order of 3 km/s.

During the nozzle expansion the pressure rapidly drops about two orders of magnitude from that in the chamber [of the order of 100 atm ($\sim$10 MPa), although the exact value depends on the motor design and the propellant soak temperature at ignition] to approximately ambient atmospheric pressure. During the rapid nozzle expansion there is a tendency for some chemical species concentrations to "freeze" rather than continue to react toward their equilibrium concentrations. The departure from equilibrium is usually not very great for the major species (such as H_2O, CO_2, CO, HCl, and H_2), however, for minor species, including free radicals (such as O, H, and OH) and ionic species the departure may be substantial, since equilibrium concentrations of these species drop very rapidly with decreasing temperature.

The nozzle expansion process may be computed, as indicated at the end of the previous subsection, as part of the chamber and performance calculation. It is also possible to model the nozzle chemical kinetics with a confined flow model of the same general type used subsequently to model the plume free flow, starting either in the sonic throat or in the chamber, upstream of the throat.

3. *Base Region*

In general, the exhaust plume of a rocket motor fired statically can be calculated without regard to the shape of the rocket motor itself. However, for missiles in flight, the shape of the missile, the angle of attack, the velocity, and altitude all interact and create a flow separation phenomenon that occurs in the base region of the missile and is referred to as base flow. Base flow can have a profound effect on the downstream exhaust plume.

In the simplest cases of base flow effects on exhaust plumes, the effects are purely gas dynamic and result in modifications to the pressure field at the base of the missile. In more complex situations, recirculation and mixing of exhaust and atmospheric gases in the base region result in ignition and combustion. This can change the distribution of heat and all related properties in the exhaust plume very drastically from the their values at static firing condition or, in fact, even from very similar conditions without base combustion. The effect of base combustion on the exhaust plume is most significant for fairly cool exhausts that do not ignite unless there is some additional heat source or flame-holding action, such as that caused by base flow mixing. The effect is dramatically apparent in optical emissions and in measured exhaust plume-microwave attenuation.

Base combustion will distort the downstream exhaust plume flowfield from the geometry it would otherwise have. It also changes all local downstream plume properties.[11–13] The base ignition and combustion phenomena may be unsteady under some conditions and result in very dramatic fluctuations in apparent plume size and all plume properties dependent on temperature. The base flow will be strongly affected by the angle between the missile velocity and the nozzle centerline (angle of attack).

These remarks on base flow are based on flight and wind-tunnel observations. Computations of base flow with chemical reactions are difficult and very time consuming; however, functional computer programs have been developed and are known to be operational in the U.K., U.S., and France. These programs seem to be capable of accurately predicting measurable features of the base flow region, such as static pressure, as well as the effect on downstream plume properties. For practical purposes, the base flow region can be neglected for situations where 1) the nozzle lip is very thin compared to the nozzle exit radius, 2) the missile velocity is very low compared to the nozzle flow velocity ($V_m/V_n < 0.1$), and 3) no important chemistry occurs in the base region to modify the downstream flowfield or chemistry.

The considerations here apply generally, although they are most easily visualized and computed for axisymmetric missiles and exhaust plumes.

4. *Exhaust Plume*

Previously in this section we introduced all the upstream contributions that influence the exhaust plume except one, the atmospheric flowfield. Freestream atmospheric flowfield interactions with the missile body influence the flowfield farther downstream, where atmospheric flow interacts in the missile base region and, subsequently, with the nozzle effluent. Thus, there are the following contributions to the exhaust plume: 1) the chamber combustion flow as it is ejected by the nozzle, 2) the interaction of the atmospheric flowfield with the nozzle effluent in the base region of the missile, and 3) the interaction of the atmospheric flowfield farther downstream with the effluent of the base region and the exhaust jet. This last interaction occurs over the entire length of the exhaust plume. These contributions to the exhaust plume must be considered with the earlier caveats of this section.

Computer programs currently in use to predict plume properties and effects are based on the assumption that plume flowfields can be described by steady-state solutions of the Navier–Stokes equation[1, 13–16] To accomplish this, the equation is solved by finite difference methods with assumptions governing 1) the introduction and effect of mixing between the effluent and atmospheric flowfields and 2) the paths along which finite rate chemical reactions occur. The results of computations by these methods are often in reasonable agreement with measured data for a number of plume properties, when coupled by appropriate sophisticated plume effects models. For example, the total IR radiation from an exhaust plume can usually be predicted to an accuracy of a factor of 2–4, as can the average peak values of IR station radiation. Radio frequency attenuation in the microwave region can usually be predicted with an accuracy of 3 dB (a factor of 2). Primary and secondary smoke spatial profiles and effects can usually be predicted quite accurately on the basis of steady-state plume flowfield assumptions.

The weaknesses of the steady-state assumptions first become apparent when one attempts to calculate plume properties or plume effects that depend on time dependent aspects of the exhaust plume flowfield. For example, scattering of microwave radiation by fairly sharp gradients in the free-electron concentration seems to be responsible for observed levels of rf-noise and RCS.[17] Computations of these phenomena cannot be made with the parameters generated by steady-state flowfield models. A turbulent, time-dependent structure must be superimposed on the computed steady-state flowfield. This can only be done with the current computer programs by artificially formulating the turbulence properties of the exhaust plume and superimposing them upon the steady-state profiles.[18] Measured instantaneous (rather than time averaged) values of exhaust plume station radiation tend to fluctuate strongly about the spatially localized average values, a clear indication that time-dependent plume phenomena are real.[19] Temporal fluctuations may tend to be reduced for full plume signatures because of spatial averaging.

If the time-dependent behavior of the exhaust plume is real, and if it is based on underlying turbulence, a number of questions emerge.

1) Does the turbulent fine structure influence the mixing and chemical reactions occurring between the atmospheric flowfield and the exhaust "source flow" in a manner more complex than steady-state formulations can solve?

2) Do the local heat and pressure fluctuations generated by the localized turbulent chemical reactions feed back to modify the instantaneous turbulent structure of the flowfield?

3) How do the time-averaged exhaust plume properties and effects, based on fluctuating phenomena, differ from the "average" values that are computed on the basis of steady-state flowfield assumptions? Are there important local differences in values? Averaged differences?

4) What experiments could be performed to determine the importance and nature of the time-dependent effects on exhaust plume properties?

5) How could time-dependent exhaust plume computer programs be formulated so they would give accurate results and be practical to use?

6) Is there important information in the fluctuating signatures of exhaust plumes that has military significance for the design of rocket motors and of sensor systems?

When the answers to any of these questions is yes, major weaknesses in the plume flowfield models become apparent, and may include 1) lack of turbulence-combustion coupling, 2) inadequate treatment of base flow recirculation with chemical reactions, 3) lack of particle combustion modeling, including multiphase flow interactions, 4) inadequate criteria for predicting ignition of afterburning in some plume flows, 5) lack of models for turbulence introduction into the flow, 6) lack of models for three-dimensional chemically reacting plume flowfields, and 7) failure of current models to account for time-dependent plume-signature effects. Elimination of these weaknesses will require major investigations of both experimental and theoretical nature.[20]

What follows, in this section, to the extent it is based on the results of current steady-state exhaust-plume computer programs, should be treated cautiously; however, the information is probably qualitatively correct.

The contributions of exhaust products to IR emission and microwave interference increase with increasing exhaust plume temperature. Therefore, these contributions tend to be stronger in the presence of afterburning and behind strong shock waves. Since low-altitude shock waves are rather small, and the afterburning plume region (should one exist) is quite large, the latter is by far the major factor in most tactical missile exhaust signatures. The increases in emissivity and free electrons are very sensitive to temperature increases. Carbon dioxide (CO_2) is a major source of IR radiation in tactical exhaust plumes and, in spite of a strong atmospheric CO_2 absorption band, the major source of detectable IR signature in some important wavelength regimes. Combustion of carbon monoxide (CO) to form CO_2 is one of the major reactions that contribute to exhaust plume afterburning (the other is the formation of water, H_2O, from various hydrogen and oxygen containing radicals) and the resulting increase in CO_2 concentration also increases the IR emission in and downstream of the afterburning region of the exhaust plume. Other molecular species (including H_2O, CO, and HCl) also contribute measurably to the IR emission signature. Some self-absorption of radiation occurs within the exhaust plume as emissions from hot inner regions pass through cooler regions into the atmosphere and, of course, additional signature losses occur by absorption and scattering along the atmospheric path from the exhaust plume to any sensor, as shown in Fig. 1.

Solid particles in the plume emit continuum radiation in approximate proportion to the fourth power of their surface temperatures and in proportion to particle concentration. Specific details of particle size distribution and optical properties of the particles significantly affect this emission. Plume particulates also scatter radiation, and through this effect may be observed what seem like local anomalies in the intensity of IR emissions that depend on the spatial distribution of the plume's emitting sources. One of the more dramatic examples of this behavior is the so-called search-light effect, in which "black-body" radiation from the nozzle throat (i.e., from the combustion chamber) is scattered by particulates in the plume and, thus, appears from some range of viewing angles to originate from the plume.

IR radiation results from quantum effects. The emission of radiation from energy-releasing transitions in molecular rotation-vibration levels is responsible for molecular radiation and, thus, should follow local gas temperature fluctuations.

Particle radiation, obeying the Planck function (also a quantum effect) will tend to lag local gas temperatures because of thermal lag effects within the particles (the amount of lag is affected by particle size and, of course, any major exothermic or endothermic transitions, for example, melting, that affect the particles). Particle scattering effects should be relatively independent of temperature, except to the extent that particle optical properties change with temperature and that the radiation being scattered fluctuates at its source or along the radiation path.

Visible and uv radiation should follow local temperature fluctuations very closely since the former are generally due to changes in atomic electronic states and the latter, generally, to radiation emitted during free radical reactions, which are part of the source of the afterburning temperature changes. There will also be some amounts of visible and uv radiation emitted by hot particles in the exhaust, and the hotter the particles the greater the intensity of the radiation. Also, the peak wavelength of particle radiation shifts toward shorter wavelengths (higher energy photons) as the particles get hotter, in accord with the Planck function.

Free electron production occurs as a result of reactions, some of them multistep, that accelerate at high temperatures and pressures. Free electrons will persist for some short time after they are formed, until recombination by collision with other species occurs (recombination will, therefore, be slower at higher altitudes). Therefore, we would expect some differences between the effects of turbulence on emission and its effects on rf-interference phenomena.

Downstream of the plume afterburning region, the emission and free electron effects decrease fairly rapidly, more rapidly at lower altitudes than at higher. In this downstream-wake region, which may extend for many kilometers, primary and secondary smoke effects have their greatest importance. The particles, which in the aggregate form primary smoke, are in general the same particles that radiated and scattered in the afterburning region of the exhaust plume. Additional species of particles may also form by condensation at the lower temperatures that follow the afterburning region. Particle mass fractions as low as 0.01 (i.e., 1%) of the nozzle effluent can affect plume visibility or transmission to a degree that may be of tactical or operational importance. Although the particles generally exist everywhere downstream of the nozzle, in practical terms they assume importance only because their spatial extent is so great that they can be perceived from large distances, and/or extreme attenuation of optical (typically laser) guidance beams occurs during transmission through these considerable lengths of exhaust plume smoke.[21,22]

Secondary smoke, which is operationally important for the same reasons as primary smoke, forms only in regions of the exhaust plume where the local vapor pressure of the condensable species (usually water, water and HCl, or water and some soluble salts) exceeds their saturation vapor pressure at the local temperature and pressure for the sizes of condensation nuclei present. Condensation nuclei are just primary smoke particles, however, they may include very small size particles that are not otherwise important to optical effects. Condensation will start preferentially at lower vapor pressures on larger particles (unless they are hotter, due to temperature lag) and condense on smaller particles only as the vapor pressure reaches the saturation vapor pressure for the smaller size. Condensation will also start preferentially at lower vapor pressures on particles that contain soluble salts (the thermodynamics of solution is important here).[23–25]

Secondary smoke is composed entirely or predominantly of water droplets. Secondary smoke in exhaust plumes that contain HCl and/or HF will start to form at higher temperatures and lower ambient moisture levels than for plumes without the acid vapors. The optical properties of the acid-containing droplets will be somewhat different than those of pure-water droplets. Generally, the effect of the particulate condensation nucleus is neglected in the calculation of droplet optical properties, although it can certainly be important with the larger nuclei. At any given atmospheric conditions, secondary smoke will be thicker (larger and, perhaps, more droplets) in the presence of acid vapors than in their absence.

Exhaust plume afterburning affects the formation of secondary smoke in two ways. 1) Afterburning causes the formation of additional water in the plume, this tends to enhance subsequent condensation and formation of secondary smoke. 2) Afterburning, by adding more heat to the plume, delays condensation to locations farther downstream and, in marginal situations, may prevent condensation altogether.

Because mixing and chemical reactions do not scale with plume size and missile velocity in the same way, afterburning may be significantly different for exhaust plumes of the same propellant fired under different conditions of these variables. Therefore, one must be careful when extrapolating the results of condensation measurements on small motors, fired statically, to flight conditions, even for a phenomenon as apparently simple and straightforward as secondary smoke formation.

Secondary smoke can form in plumes containing significant amounts of primary smoke. When this occurs, the total effect on visibility and obscuration is roughly the sum of the individual effects until the optical density (optical depth) exceeds some value at which nonlinear effects assume increasing importance.[22] One caveat here, the larger primary smoke particles, which are optically more important, would be the first, theoretically, on which condensation starts; this could have a major effect on changes to plume visibility and obscuration, compared to the simple summation of primary and secondary smoke effects. However, the larger particles cool more slowly than the smaller ones and, therefore, condensation may start on the smaller particles, in which case summation of the two effects is a reasonable approximation. With current exhaust plume computer programs that treat nonequilibrium two-phase flow, these tradeoffs can be calculated, albeit somewhat tediously because of the great lengths of plume involved.

C. Plume Effects

This section is distinguished from the previous one by its more in-depth, although still cursory, treatment of the specific phenomena believed to be responsible for tactically important plume effects. The effects examined are the following:

1) emitted (radiation) signature effects: a) infrared, b) visible, c) ultraviolet, and d) millimeter wavelengths; and 2) interaction signature effects: a) smoke signatures, and b) radio frequency (including microwave) interference.

1. Emitted (Radiation) Signature Effects

Table 1 summarizes the state of the art of rocket exhaust plume radiation signatures.

Infrared. Infrared radiation emitted by the low- and medium-altitude exhaust plumes of tactical missiles derives from two major sources; molecular emission and particle emission.[26–28] The molecular radiation originates with energy input to, or excitation of, a molecule. When the excited molecule returns to a lower energy

Table 1 Rocket exhaust plume emitted radiation state of the art[26]

Wavelength, μm	Phenomenological understanding	Observational base, data	Prediction capability
Ultraviolet, 0.2–0.4	Fair controversy over radiation mechanisms: CO+O OH (A) particles	Fair and increasing Small tactical missiles with data validity questions Some large missiles	Poor Errors up to a factor of 10^2
Visible, 0.04–0.8	Fair Atomic transitions of minor species, Na, K (thermal vs chemi-luminescence issue) Solar scattering from smoke Search light effect	Fair Largely photographic data Some quantitative spectra	Probably poor but relatively untested Errors up to a factor of 10^2 for trace species Errors probably considerably less than a factor of 10 for smoke
Near infrared 0.8–2.0	Good Atomic transitions with overtones of molecular vibration-rotation transition (N_2O) Particulates Blackbody peaks in NIR for most plumes	Poor Some quantitative spectra	Probably fair but relatively untested
Short wave infrared, 2.0–4.0	Very good Molecular vibration–rotation bands (CO_2, H_2O, HF, HCl) Particulates	Very good Field– and wind-tunnel data giving spatial and spectral details 10–500,000 lb thrust, many propellants, ground-to-space conditions	Good Molecular band predictions can be made with ± 30% errors, errors of up to a factor of 3 occur for some altitudes and propellants Particulate signatures are not reliably predicted
Middlewave infrared, 4.0–8.0	Very good Molecular vibration–rotation bands (CO_2, CO, NO, H_2O) Particulates	Very good Field– and wind-tunnel data giving spatial and spectral details 10–500,000 lb thrust, many propellants, ground-to-space conditions	Good Molecular band predictions can be made with ± 30% errors, errors of up to a factor of 3 occur for some altitudes and propellants Particulate signatures are not reliably predicted
Long wave infrared 8.0–100	Fair Particulate thermal emission plus scattering of solar and earthshine Molecular hydride rotation, HF, HCl, H_2O Molecular bending vibrations, CO_2, NH_3	Fair High-altitude database available in 8–25 μm region	Poor Particulate and hydride molecular rotational systems predictions can be made to a factor of 5 Moleculate vibrational predictions can be made up to a factor of 20

state, a photon is emitted. Variations in the excitation process result in emitted energy that may be described as phosphorescence, fluorescence, chemiluminescence, x-rays, millimeter waves, radio-frequency waves, etc. The emitted photons are characterized by a wavelength or frequency, which is directly related to the energy of an individual photon by the Planck constant (h). Often the term wave number, which is the reciprocal of wavelength in units of cm^{-1}, is used. Emitted and received spectra will typically be described by some measure of power (watts) vs one of these three measures of the photon energy.

The infrared region of the electromagnetic spectrum can be characterized by wavelengths between approximately 0.7 and 100 μm (see Fig. 3 for a survey of the entire electromagnetic spectrum). The emission of radiation in this spectral region is dominated by rotation-vibration transitions of molecules. However, the far infrared (wavelengths longer than approximately 10 μm) also includes radiation contributions from pure rotational transitions, and some electronic transitions occur in the near infrared.

When the energy state of a molecule undergoes a photon emitting transition, the change occurs between two very precisely defined states. The quantum theory accurately describes the energy states in which a molecule can exist and the rules for transitions between those states. The quantized energy differences between the rotational states are smaller than those between the vibrational states, and

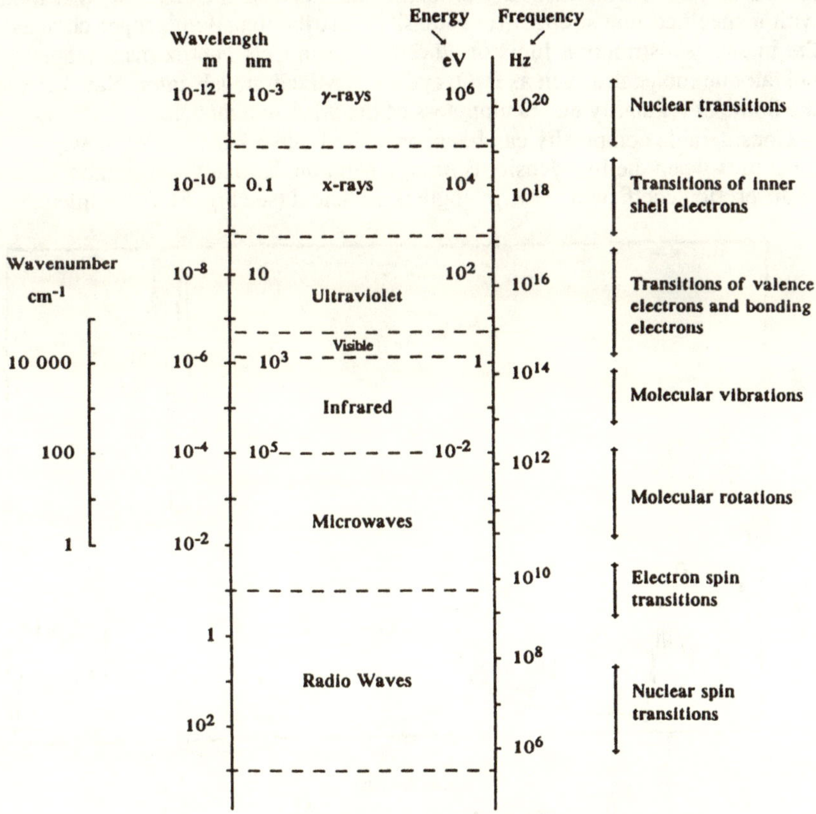

Fig. 3 Survey of the electromagnetic spectrum.[26]

according to the quantum rules (fully verified by observation) some transitions between vibrational states are also accompanied by a transition between rotational states. Therefore, the molecular emission spectrum of any molecule is defined with absolute accuracy. The photons emitted in these transitions define a line spectrum. However, in practice, each spectral line is broadened about the central frequency corresponding to the transition by three processes: natural broadening caused by unavoidable uncertainty in energy levels, collision broadening due to perturbation of energy levels by molecular collisions, and Doppler broadening due to thermally caused motion of molecules relative to the observer. As a result of such broadening, the emission spectra of some molecules appear to be continuous (to some degree of resolution) over a "band" of the spectrum. Figure 4 illustrates infrared emission (and absorption) data over a fairly wide spectral range. Bands for H_2O, CO_2, and CO can be clearly seen. Line structure on these bands is apparent as well.

A line-by-line computation of all of the molecular contributions to the infrared spectrum of an exhaust plume would clearly be difficult to set up and time consuming to perform. Band models have been developed to simplify the calculation of infrared spectra. Band models are commonly divided into three classes: the regular model, the statistical or random models, and the mixed models. The regular model considers absorption (or emission) by identical, equally spaced lines. This approximates the spectra of many diatomic molecules, such as HF or NO. Random models assume that all lines have the same line shape but are randomly distributed with a specified line strength (or intensity) distribution. With proper choices of the intensity distribution function, the random models approximate features of polyatomic molecules such as H_2O, and CO_2. Mixed models interpolate between the limits of regularity and randomness of the other two models.

Considerable complexity can be incorporated into a band model, however, for most uses when the line density is high, a band model that gives a smooth curve of absorption coefficient vs wavelength is sufficient (see Fig. 5). A complete band

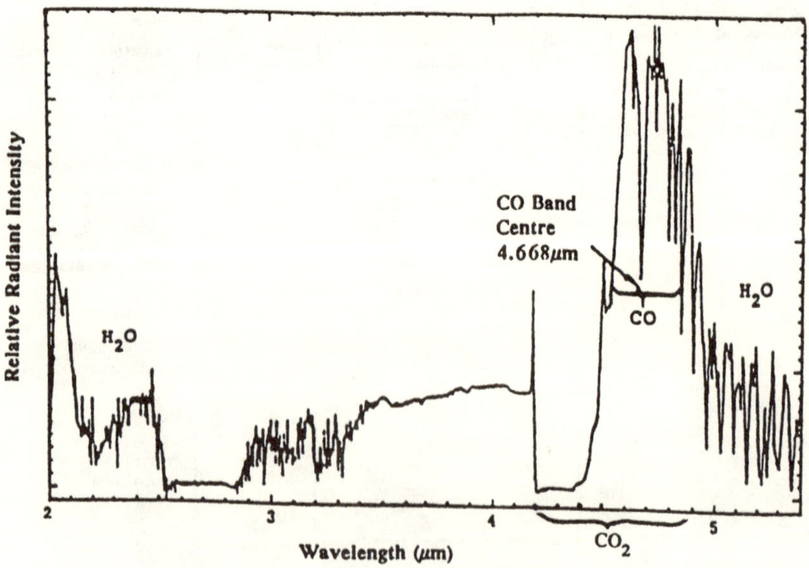

Fig. 4 Emission from a plume.

model will include the general effect of temperature (shown for the 4.3-μm CO_2 band in Fig. 6). This permits band models to be used for both emission and absorption calculations, with the reservations described later.

Band models are also used to calculate atmospheric absorption of radiation by molecular transitions, for example, in the widely used LOWTRAN[29] computer program. Difficulties may occur if band models are used to calculate both source-emitted radiation and absorption along the propagation path. These difficulties occur because the band model is just a model, a curve fit, based on fitting scientific data but not containing scientific fundamentals, and the failure of band models is due to their insensitivity to line-correlation phenomena. The band models are made by fitting discrete, though broadened, line spectra with a continuous curve. In actual emission–absorption situations, radiation will be absorbed by a molecule only if its photon energy (wavelength) corresponds exactly to an allowed energy transition of the molecule, that is, an exact correlation. However, in the band model calculation, a specific resolution is selected (usually in wave number units, cm^{-1}, for example, 25, 5, 1, etc.) and that interval of the "absorption" band is activated for either emission or absorption, as appropriate to the problem at hand. This is not a bad assumption if either the emitter or absorber is quantized. However, if both are quantized, it is easy to see that wavelengths that are not exactly the same may be calculated to interact, whereas in practice they would not. This line-correlation problem may occur when calculating exhaust plume self-absorption as well as atmospheric absorption. The problem can occur for absorption due to the same species of molecule as emitted the radiation or for a different species.

This kind of error can be avoided by calculating the contribution from each individual line. Such line-by-line calculations can be very cumbersome in spectral regions with many close or overlapping lines. However, when only a few lines

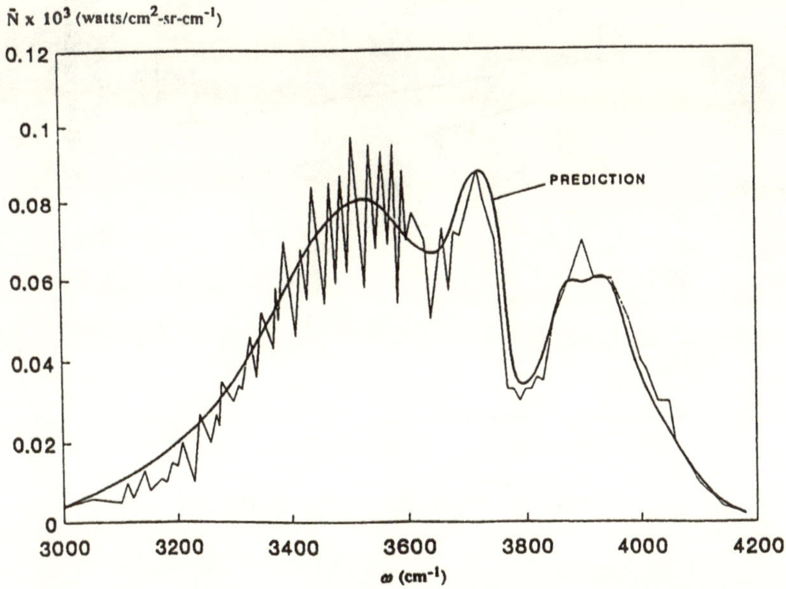

Fig. 5 Comparison of measurements and band model prediction for H_2O.

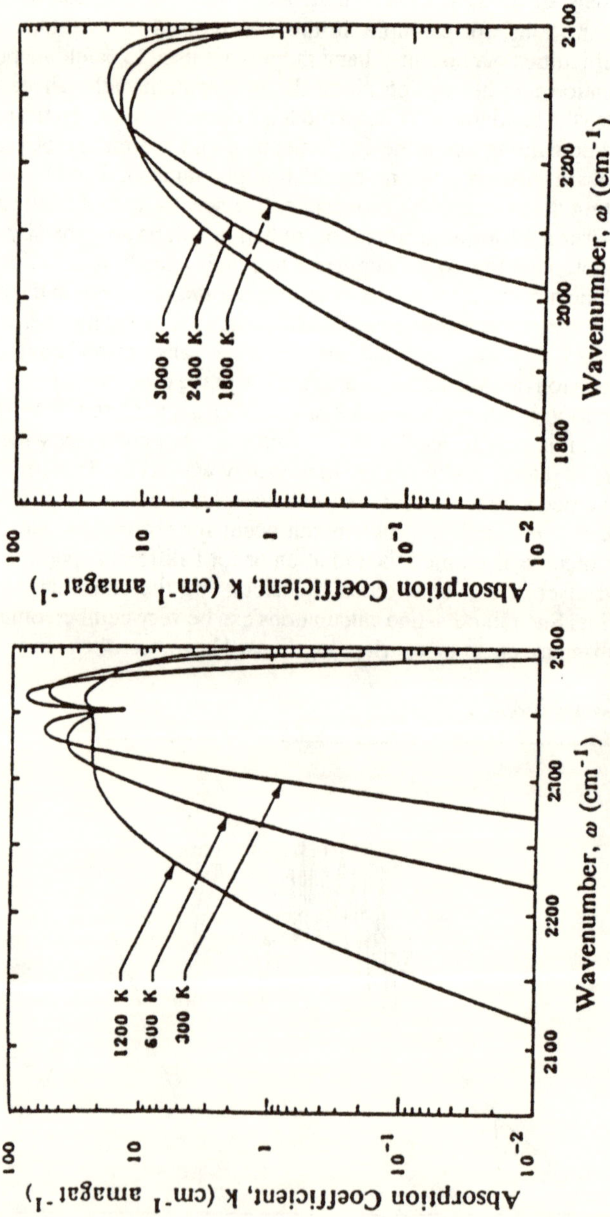

Fig. 6 Band model absorption coefficients for CO_2 4.3-μm band.

are present, line-by-line calculation can be tractable and, in fact, this procedure is used to calculate HCl spectra in some exhaust plume computer programs.

All radiation sources emit more strongly at high temperatures than at low. This is so because at high temperatures the higher quantum states are more heavily populated. This is even true of particulate radiation which follows the Planck function (shown as black-body radiation curves in Fig. 7). The difficulties with calculating particulate radiation, absorption, and scattering in exhaust plumes are that 1) we do not know the exact size distribution of the particles, 2) we do not know the optical properties as functions of temperature, and 3) even if we knew the other two important parameters, we do not know the particle shapes. Particles in exhaust plumes are usually not pure substances; for example, aluminum oxide particles in exhaust plumes are sufficiently contaminated that their absorption coefficients are much different than for the pure material. Therefore, even the most careful measurements of pure finely divided substances would not provide the necessary optical property data for use in calculating plume particle emission, absorption, and scattering effects. Furthermore, the optical property data are needed over the entire temperature range of interest in the plume.

Approximations to particulate scattering can be based on adjusting the black-body radiation curve (Fig. 7) for the emissivity of the radiating particles, for the particle size distribution (or effective particle size), and for the number density of the particles (particle-cloud density).[30] Although in practice this can be complicated, for the purposes of this overview it is clear that the result is continuum radiation with a spectral distribution that depends on the particle temperatures. One can judge, by examining Fig. 7, that for a typical plume afterburning region at 2500 K the continuum particle radiation (per μm) will peak at a wavelength of about 1.1 μm, will be an order of magnitude lower at 4 μm, and will drop another order of magnitude to 7 μm.

Determining the size distribution of plume particles is difficult and complicated by the problem of never knowing for sure if you have measured all the particles. Particle size distribution is further complicated because the important particles depend on what you are looking for. For example, for scattering of radiation, the important particle sizes are comparable to and larger than the wavelength. Therefore, in the visible, the important particle sizes are different than they are in the infrared. Exhaust plume particle number densities (at least for aluminum oxide) tend to be higher for smaller particles and decrease with increasing particle radius. In general, the largest 10% of the particles have 90% of the mass. Particle size distributions are often specified according to the reason for interest in them. One, therefore, must be cautious about using calculated particle size distributions (or sometimes, even distributions derived from measurements) for a purpose other than that for which they were intended. For example, in rocket motor performance calculations in which the effect of particles on delivered specific impulse is of interest, only the larger particles with most of the mass and drag are of interest, and, therefore, motor-performance models that give these[10] are accurate enough, without regard to the many smaller particles that may be present. Such a distribution would be totally inaccurate for calculating uv or visible light scattering[21] and might be inadequate for accurate determinations of infrared scattering as well.

In summary, infrared emission in plumes is comprised of lines emitted by rotation-vibration transitions in molecules and continuum emission from particles. The particles can also scatter radiation incident upon them or from other sources. Infrared radiation is attenuated during propagation through the atmosphere. This

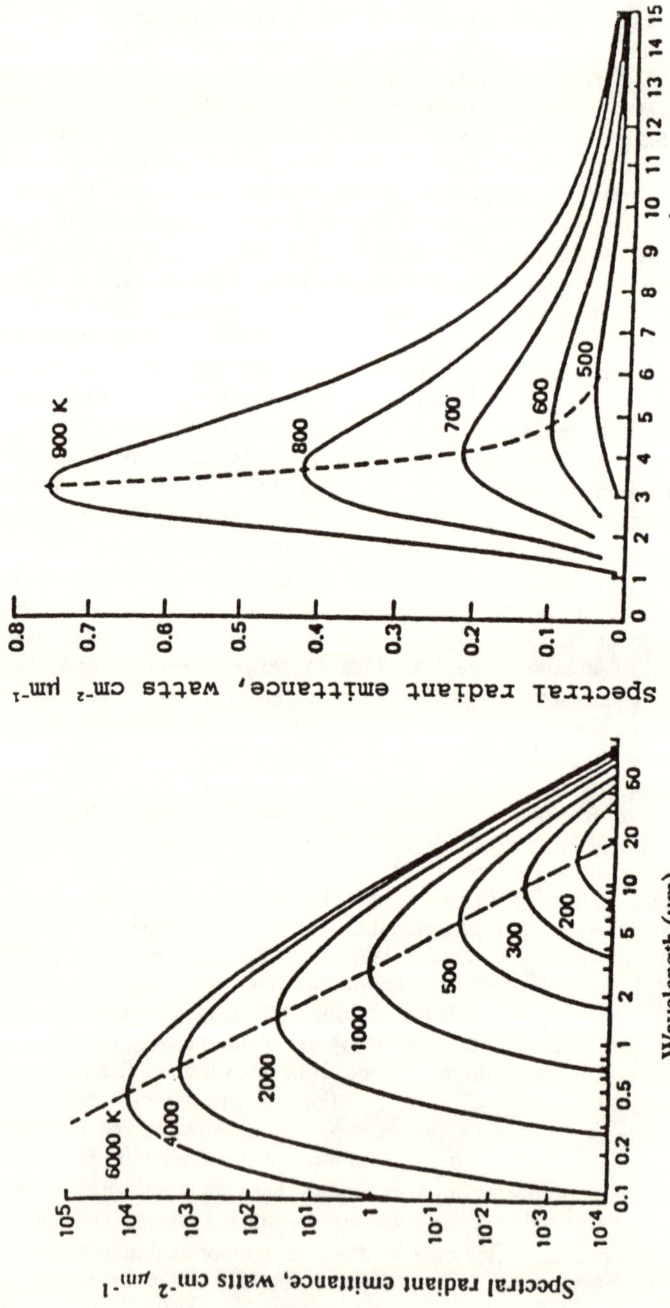

Fig. 7 Black-body spectral distribution.

attenuation can occur by continuum absorption and scattering by particles (for example, fog, clouds, or aerosols) or by molecular absorption by atmospheric species (such as CO_2, H_2O, or pollutants), as shown in Fig. 1.

Visible. Particle continuum radiation in the afterburning portion of a plume will tend to dominate visible radiation when particles are present. If the number density of particles is known, and temperatures of the particles can be estimated, an emissivity value between 0.05 and 0.2 may be used for typical plume afterburning temperatures.

Trace amounts of sodium atoms in the plume will produce atomic transition lines (yellow) at 0.589 and 0.5896 μm. Potassium lines at 0.7665 and 0.7699 μm will occur with trace amounts of that element, however, these lines are at the red end of the spectrum and verge on the near infrared, therefore, they would not be particularly visible.[31]

Frequently, exhaust plumes are judged to be not afterburning because of the absence of visible emission on film or video records. It should be clear that the exhaust plume of a minimum-smoke or reduced-smoke propellant, although afterburning, may well be invisible, because the emitted radiation is in wavelength bands not detectable with the medium used nor, in fact, sometimes by the human eye.

Atmospheric propagation attenuates visible radiation by absorption and scattering. The atmospheric daylight visibility range may vary from 0.5 km ($\sigma = 8$ km^{-1}) in moderate fog to 20 km ($\sigma = 0.2$ km^{-1}) for clear conditions to 50 km ($\sigma = 0.08$ km^{-1}) for exceptionally clear conditions. (Sigma represents the attenuation coefficient.) Even greater visibility ranges occur, particularly at flight altitudes, but on the Earth's surface as well. The Rayleigh scattering limit, due to scattering by air molecules, at sea level is 310 km ($\sigma = 0.013$ km^{-1}). These values for the attenuation coefficient may be applied to Beer–Lambert transmission–attenuation equation calculations of atmospheric losses of plume visible radiation, both for emission and for smoke signatures discussed later.[21]

Ultraviolet. In addition to continuum particulate radiation which predominates in heavily aluminized exhausts, ultraviolet emission also occurs due to some minor plume species. It is predicted by some that even in heavily aluminized plumes CO + O chemiluminescence is dominant at the middle ultraviolet and shorter wavelengths. A spectral band from about 0.25 to about 0.7 μm seems to be emitted by the combination of these two radicals to form excited CO_2 and the subsequent de-excitation of the CO_2 molecule. For typical afterburning plume temperatures, the radiation peak is at about 0.35 μm.

In hydrogen and hydrocarbon flames, studies have shown that a source of ultraviolet radiation is electronically excited OH radicals produced in chemiluminescent reactions. Both of the following reactions are thought able to produce excited OH radicals.

$$CH + O_2 \rightarrow CO + OH^*$$
$$H + OH + OH \rightarrow H_2O + OH^*$$

However, the latter reaction is thought to be more likely in afterburning solid rocket exhaust plumes. The OH spectrum contains a large number of lines, which in all but the highest resolution spectra will blend together to form a continuous band. Therefore, OH radiation in the ultraviolet may be thought of as band radiation for some calculations. Some of the OH line centers are 0.2609 μm, 0.2811 μm, and 0.3064 μm.

Ultraviolet radiation is very easily scattered by the atmosphere and absorbed by atmospheric ozone in the so-called solar-blind region of the spectrum which, between about 0.2 and 0.3 μm, peaks at about 0.25 μm. Aerosol and molecular scattering of ultraviolet radiation severely limits the useful range of ultraviolet radiation detectors for locating and tracking exhaust plumes.

All radiation is scattered by fog or clouds. As might be expected, the shorter wavelengths are more strongly affected than the longer. It is all a matter of the relationship between the droplet sizes and the radiation wavelengths.

Millimeter wavelengths. A number of workers have sought evidence of rf emissions from plumes. Many years ago, the author heard of over-the-horizon (OTH) radar systems detecting large plumes, but it is not clear that these were plume emissions rather than bistatic scattering of ambient radiation. Sume[32] reported comparisons of millimeter-wave emission measurements at 35 and 94 GHz which support the theory that the primary mechanism is based on free-free emission (bremsstrahlung) caused by collisions between free electrons and molecules in the plume. The results indicated no difference between the emissions from motor plumes containing aluminum oxide and those with no solid particles. Additional measurements of this phenomenon were under consideration in the United Kingdom (Royal Armament Research and Development Establishment) in the early 1990s. It is likely that other, unreported, investigations are or have been done in this area.

2. *Interaction Signature Effects*

The previous discussion in this section has dealt exclusively with radiation emitted by exhaust plumes. The following discussion covers interactions of plumes with radiation from other sources. The major interactions to be covered are the interactions of visible and infrared radiation with primary and secondary smoke particles, and the interactions of radio-frequency radiation, specifically microwaves, with free electrons and ions. The sources of the plume species were covered in a cursory manner in Sec. IV.B and will not be covered further here, although information is available.[17]

One effect that will not be covered is the scattering of radiation, particularly at laser wavelengths, by "clear gas" turbulence discontinuities of refractive index within the plume. This phenomenon has the effect of beam spreading the radiation (analogous to atmospheric turbulence effects) so that a small aperture detector may see apparent attenuation losses of up to 6 dB. There will also be "jitter" effects. Even the clear exhausts of liquid propellant rocket motors can cause this effect.[17]

Smoke signatures. Smoke particles in plumes can affect incident radiation by both scattering and absorption.[21–25] The magnitude of the effect of the interaction of radiation with a particle depends on the real and imaginary terms in the particle refractive index, the particle size, and the particle shape. All of these effects combine to give scattering and extinction cross sections (or coefficients) for each particle/wavelength combination. The total plume effect is the combination of all of the individual particle effects. When the optical depth of the plume is greater than about 0.5, multiple scattering effects become detectable.[22] The importance of multiple scattering will depend on the reason one is concerned about the smoke in the first place. (The optical depth is the exponential term in the Beer–Lambert transmission–attenuation equation.)

$$\text{transmission} = 1 - e^{(\sigma d)}$$

where σ is the attenuation coefficient of the medium and d the transmission path length in the medium. Multiple scattering occurs when enough photons are scattered more than once that the effect cannot be ignored and the single-scattering assumption fails.[22] The predominant contributor to plume primary smoke is aluminum oxide (Al_2O_3) particles that result from combustion of aluminum fuel in many solid propellants. Other metals or refractory ingredients may also cause primary smoke, however, these sources will predominate only in propellants that are completely or relatively free of aluminum fuel. Primary smoke exists at all locations downstream of the motor nozzle, although its importance as "smoke" begins downstream of the afterburning region of the exhaust plume. Since a flying missile lays out a trail of these particles during the entire time that the rocket motor is burning, the smoke plume may be many kilometers in length. The smoke trail spreads with time. However, the amount of spread depends on the missile velocity: the greater the velocity, the less the spread, but the longer the trail (for a given motor burn time). The smoke trail will be dissipated by wind. The smoke trail may be visible from as far away as 100 km or more, depending on the amount of aluminum in the propellant, the thrust of the motor, the sun-plume-observer angle (scattering angle), and the atmospheric visibility conditions.[21] At the lower extremes, as little as 0.5% aluminum in the propellant can cause visibility as great as 20 km for certain scattering angles, although for most scattering angles (between 60 and 170 degs) a visible range of 4 km or less is predicted. Both of these visible ranges are based on exceptionally clear atmospheric transmission conditions. If the atmospheric conditions are clear to very clear, the visible range of the plume will be halved.

If the sun is obscured, the visibility of the smoke trail is greatly reduced and, since under these conditions the background is also cloudy sky, the plume will in many cases appear slightly darker than the background; for all practical purposes, the smoke trail will not be visible, although it will have limited visibility with a dark Earth background. In fact, even if the smoke trail is illuminated by the sun it will be practically invisible against a cloud background.

In addition to increased visibility, the scattering of ambient light by the smoke plume may interfere with optical detection systems that operate to wavelengths as great as 5 μm. The degree of interference depends on the wavelength of the detector, the optical depth of the smoke along the propagation path to the detector, the effective field of view of the detector, and, of course, the solar illumination of the plume smoke. In such a scenario, although radiation from a target may indeed pass through the plume and reach the detector, the radiation scattered to the detector by the plume may be so great that it has the effect of an overwhelming background radiation.

Often, attenuation of radiation by the plume is a more important effect on detectors. In this situation an optical (laser) target illumination beam must pass through the missile smoke trail. If the scenario parameters are such that the beam must pass through a substantial length of the plume, attenuation losses of 2–3 orders of magnitude (20–30 dB) are entirely possible and may render the beam totally ineffective.

Composite propellants containing AP but no metal fuel (often called reduced-smoke propellants) defeat some of the problems of smoky plume trails. These propellants virtually eliminate primary smoke, down to the level of particulates necessary to maintain stable chamber combustion. However, the AP in these propellants causes HCl to form in the combustion process, which, as described earlier, can condense with plume and atmospheric water to form secondary smoke. It is generally considered that at typical atmospheric conditions, reduced-smoke contrails will form above 6-km altitude. Reduced-smoke contrails at sea level will form over a fairly wide range of cold or humid climate conditions.

Propellants with neither metal fuel nor AP oxidizer (often called minimum-smoke propellants[25]) eliminate this problem, however, even minimum-smoke propellants can have a condensation plume or trail of secondary smoke when the climate is cold enough and a sufficient amount of water vapor is present. It is generally considered that at typical atmospheric conditions, minimum smoke contrails will form above 8-km altitude. Minimum-smoke contrails at sea level will form only at the very coldest climate conditions even with fairly high ambient humidity and, since at these conditions there is very little atmospheric moisture, the effect may be unimportant.

All these remarks concerning visibility and transmissivity of primary smoke plumes apply also generally to secondary smoke plumes. The applicable caveat here is that secondary smoke particles (droplets) may be somewhat larger than primary smoke (although that depends on the number and size of the condensation nuclei and the amount of condensate) so the optical effects, when they occur, may be larger. If it were possible to get the water to condense on a very large number of very small condensation nuclei, of the order of 10^8 cm^{-3} in the condensation region of the plume, the ultimate droplet radii might be reduced to substantially less than 0.1 μm, and the optical effects would be substantially decreased.

Radio-frequency interactions. The free electrons produced in hot regions of exhaust plumes, primarily in the afterburning region and less importantly behind strong shock waves, can absorb energy from radio-frequency (microwave) radiation which they give up in collisions with molecular species in the plume. The two important parameters for calculating the microwave attenuation coefficient, unique to a given plume, are the electron density (or concentration, m^{-3}) and the collision frequency (s^{-1}), which is the number of collisions per second encountered by an average single electron, typically $\sim 2 \times 10^{11}$ multiplied by ambient pressure (in atmospheres). For radiation frequencies above 500 MHz, only the electrons need be considered. For lower frequency radiation, the ions present in the plume (typically there are 1,000–10,000 ions both positive and negative for every free electron present, however, because of their far greater mass the ions absorb much less microwave radiation) must be considered as well.

If absorption is the only process encountered by radiation in the plume the simple calculation of the attenuation coefficient of each region of the plume is sufficient to give a line-of-sight prediction of microwave attenuation. In general, this is not a bad first assumption for X-band radiation with composite-AP solid rocket propellants containing 5% aluminum or less.

However, for hotter propellants, with hotter afterburning, propagation through the plume is much more complicated. In general, the plume-microwave interference problem may be thought of in terms of the plume (a mathematical transfer function) modifying the electromagnetic field state. In other words, without the plume, a field exists in space; when the plume is interposed, the field is changed.

This concept is illustrated in Fig. 8. Although simple in concept, this approach is difficult in practice. It has been solved fairly successfully using the Fresnel–Kirchoff–Huygens formulation of the wave equation.[33] Computer codes using this formulation have been developed in the United States, France, and the United Kingdom.

It may be illustrative to describe the things that go on in the plume-microwave interaction in terms derived from elementary physics. We consider here a plume with arbitrarily large electron density (ED) values. The geometry we are considering involves a transmitting antenna on the missile body and a receiving antenna at a location such that the line of sight is at a very slight angle to the plume axis. The microwave radiation (MR) penetrates through the outer plume layers where the ED is low without any interaction other than slight absorption. As the MR penetrates into higher ED regions the electron density gradient causes refraction of the MR along every path and attenuation per unit length increases as well. As the MR penetrates into the overdense region of the plume (where the ED exceeds some critical value) attenuation losses are so high that for all practical purposes the MR is totally absorbed (actually attenuation losses of several hundred decibels are calculated to occur). While this is going on, MR propagating along the outer edges of the electrical plume behaves as though it were undergoing a combination of refraction and diffraction by the plume and sets up an interference pattern very similar to optical diffraction. The MR that reaches the receiver displays local intensity structure characteristic of diffraction, modified by some refraction, and with some evidence of line-of-sight attenuated or refracted direct radiation, depending on the locations of the antennas and the ED of the plume. Because of these paths of propagation, tactical missile microwave signal attenuation rarely exceeds 30 dB, no matter how high the ED. For much larger space missile exhaust plumes at higher altitude, attenuation as high as 60–70 dB has been measured, however,

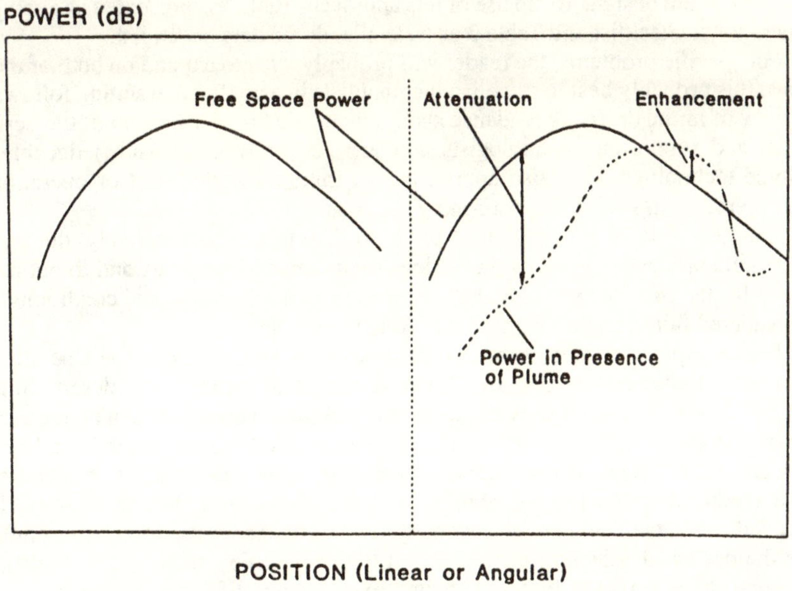

Fig. 8 Field concept of steady-state plume-rf interference.

this too is accounted for by the same propagation mechanisms. This effect can be modeled with fairly simple wave-diffraction assumptions.[17,34]

But this is not all that is happening. Some of the MR that reaches the receiver shows evidence of frequency shifts, clear evidence that the radiation has been scattered by elements with a range of velocities. This can only be accounted for if the MR is scattered by refractive index gradients much sharper than those required to explain the absorption, refraction, and diffraction results. It seems clear that there is a fine structure, probably due to turbulence, that causes substantial scattering with Doppler shifts corresponding, in general, to the mean flow velocities of the plume gases. Mathematical modeling of this concept of the process has given reasonable values of the spectrum and intensity of plume induced noise, including the frequency shifts seen. There also seems to be a strong surface type of scattering from the overdense boundary, when the exhaust plume ED is high enough for one to exist. The same approach has, at times, also given reasonable values of plume RCS, which is the same phenomenon, but with both transmitting and receiving antennas at the same location (monostatic) instead of at different locations (bistatic).[17,35,36]

V. Design Guidelines

A systems approach to determining guidelines for missile-plume tailoring that is still valid was considered for the specific problem of radar-guidance attenuation over 20 years ago.[37] Two check lists are needed. The first of these lists the missile operational requirements and operational goals related to plume signature; to the extent possible these should be stated as quantified numerical values. The second list contains all available information relating to the plume of interest or similar plumes; this list should include propellant and rocket motor information and all measured and calculated plume data, not just those for the plume effects of concern for the current design. To aid use of this approach, Table 2 is presented as a sample operational checklist and Table 3 as a sample plume data check list.

For specific problems, the reader will probably wish to expand on both of these lists. It is probably best to establish the final lists in a conference setting following review of initial drafts. Attendance should include a broad spectrum of the design team and supporting technologists, not just people who are knowledgeable in plume technology. It is also appropriate to initiate the draft list of operational requirements in similar brainstorming sessions.

The objective of this design effort is to determine, quantitatively, the available tradeoffs between all of the various requirements and goals and to optimize them for the missile system under consideration. If all operational conditions are considered fairly, some surprises are likely to emerge.

For example, consider further the fictitious anti-tank missile in the Operational Considerations section (Sec. III). Suppose a conflict scenario in a desert climate in which one combatant uses an all-climate missile (obtained from a major power arms source) containing a minimum-smoke propellant and the second combatant, realizing it will never use anti-tank missiles elsewhere, has chosen to produce and use a reduced-smoke propellant in the rocket motors of its anti-tank missiles. The second combatant will have an advantage in performance in the desert environment or the option of selecting a smaller and lighter missile design. He will also, in general, have a rocket motor with improved service life and greater handling, transport, and storage safety.

Table 2 Operational conditions checklist

1) Missile design	2) Operational conditions
Design requirements	Guidance mode(s)
Design goals	Targets
body signatures	Range
plume signatures	Velocity
Rocket motor	Altitude
Thrust/time	Maneuver capability
Propellant	Countermeasures
trace elements, K, Na, etc.	Climate/atmospheric conditions
Nonpropellant contribution	Threat conditions
Chamber conditions	Density
pressure/time	Range capability vs missile capability
chemical species	Guidance
Nozzle	Detectors active/passive
Missile flight conditions	range
External air flow	detection
velocity	identification
altitude	track
angle of attack	terminal
Body geometry/base	

Table 3 Plume data checklist

1) Rocket motor	3) Plume conditions (continued)
Thrust/time	unsteady flow
Propellant	afterburn ignition
trace elements, K, Na, etc.	shock structure
Nonpropellant contribution	particle flow
Chamber conditions	Flowfield calculations
pressure/time	signature calculations
chemical species	correlation with data
Nozzle	extrapolation to missile conditions
2) Data test conditions	uncertainties
External air flow	4) Visible signature
velocity	Radiation/flash
altitude	Smoke
angle of attack	Visibility
Body geometry/base	Guidance interference
3) Plume conditions	5) Passive sensor detected radition
Internal plume flow	Infrared signature
chemical species	Ultraviolet signature
reaction kinetics	Millimeter wave
temperatures	6) Radio-frequency interactions
pressures	Guidance attenuation
velocities	Guidance noise
ionization	Radar cross section
turbulent mixing	

In another situation, consider a goal involving reduced plume-IR signature. Suppose existing data indicate that with likely enemy detectors, the steady-state IR signature of a current motor capable of meeting the performance goals can be detected at ranges of 30–250 km depending on plume and detector altitudes and atmospheric and background conditions. Studies show that a motor might be designed and fabricated with only 10% of the IR signature of the current motor. However, calculations show that this will reduce detections to a range of 25–200 km for identical conditions, a trivial change and not an improvement in the operational context. Another option is a liquid-organic fueled ramjet engine, which after its rocket booster is exhausted, will have a signature only 1/40th that of the current rocket. Because the ramjet-powered missile will fly at different aspects to expected detector platforms, it is assumed that an IR signature only 1% of the current rocket plume can be achieved. With this condition, the IR detection range can be reduced to 19–100 km for identical conditions. If the operational goal had been identified as no detections beyond 50 km, that goal is now achievable in some significant percentage of encounters, and the value of these vs other operational parameters must be determined to optimize the system design.

As the final example, consider a plume-radar guidance problem. Everything possible has been done to reduce signal attenuation: afterburning is suppressed as much as possible considering the performance needed, the rocket motor nozzle lip (missile base) thickness has been minimized by boat-tailing the external missile-cylinder wall, and the two receiver antennas for the semiactive radar-guidance system are forward on the missile body. Signal attenuation of at least one of the antennas is always less than the 10 dB required for adequate signal-to-noise ratio at maximum thrusting range. Nonetheless, the missile sometimes has midcourse guidance problems that cause it to miss its target. After much head scratching and analysis of expensive telemetry data (which would not otherwise have been obtained), the project team comes to the awful conclusion that the antenna selector in the guidance system is selecting the wrong antenna for guidance. The system is programmed to select the antenna receiving the larger signal, which it is doing; but analysis of the telemetry data indicates that at the time of selection much of that signal is plume-induced noise on the carrier signal scattered into the antenna making it appear to be the larger signal. The problem is now severe, especially since the system has been deployed. Earlier in the design of the system it would have been possible to include noise-discrimination circuitry in the guidance system or an independent missile orientation sensor that could have eliminated or reduced this problem.

VI. Conclusion

This chapter has presented a brief overview of the plume technology areas applicable to tactical missile design. The reader is cautioned that the information presented here has been greatly simplified for the sake of brevity. Only trends are presented, hopefully with little loss of correctness. No attempt was made to show how to calculate or measure plume properties or effects. There are a number of phenomena associated with plumes at higher altitudes that were not even touched on in this chapter. Although this chapter is concerned with tactical missile exhaust plumes, which generally occur at low to medium altitudes, there are instances where higher altitude plume phenomena may emerge even in the tactical regime. Anyone desiring to explore this field more deeply will find, through the reference list, an entry into the extensive literature concerning the phenomena and theory of plume technology useful for solving specific practical problems.

References

[1]Zeller, B. M., "Terminology and Assessment Methods of Solid Propellant Rocket Exhaust Signature," NATO, Propulsion and Energetics Panel Working Group 21, AGARD-AR-287, 1993.

[2]Evans, G. I., and Facer, D., "Elastomer Modified Cast Double Base Propellants," AGARD CP 391, *Smokeless Propellants,* Vol. 198, pp. 13-1–13-17.

[3]Ramsay, D. A., "The Evolution of Radar Guidance," *GEC Journal of Research,* Vol. 3, No. 2, 1985, pp. 92–103.

[4]Balwanz, W. W., "Interaction Between Electromagnetic Waves and Flames, Part 6. Theoretical Plots of Absorption, Phase Shift and Reflection," Naval Research Lab., NRL Rept. 5388, Washington, DC, Sept. 1959.

[5]Piesik, E. T., "A Method to Define Low-Altitude Rocket Exhaust Characteristics and Impingement Effects," *Journal of Spacecraft and Rockets,* Vol. 7, No. 4, 1970, pp. 446–451.

[6]Piesik, E. T., "A Semi-Empirical Method Defining Low-Altitude Exhaust Flow Fields—Updated and With Data Comparisons," 14th JANNAF Plume Technology Subcommittee Meeting, Naval Weapons Center, China Lake, CA, 1983.

[7]Heimerl, J. M., "Suppression Workshop Summary," U.S. Army Ballistics Research Lab., BRL-MR-3580, Aberdeen, MD, April 7, 1987.

[8]Sellas, "High Impulse Density Propellants," Aerojet Techsystems Aerojet Rept. 8708-76F, Nov. 1966.

[9]Cruise, D. R., "Notes on the Rapid Computation of Chemical Equilibria," *Journal of Physical Chemistry,* Vol. 68, Dec. 1964, pp. 3797–3802.

[10]"Performance of Rocket Motors with Metallized Propellants," Propulsion and Energetics Panel Working Group 17, AGARD-AR-230, Sept. 1986.

[11]Victor, A. C., "Calculations of Rocket Plume Afterburning Coupled to Reacting Base Recirculation Regions," *Journal of Spacecraft and Rockets,* Vol. 14, No. 9, 1977, pp. 534–538.

[12]Bender, R. L., "Base Flow," *JANNAF Handbook, Rocket Exhaust Plume Technology,* Chemical Propulsion Information Agency, CPIA Pub. 263, July 1981, Chap. 5.

[13a]Dash, S. M., "Analysis of Exhaust Plumes and Their Interactions with Missile Airframes," *Tactical Missile Aerodynamics,* edited by M. J. Hemsch and J. N. Nielson, Vol. 104, Progress in Astronautics and Aeronautics, AIAA, New York, 1986, pp. 778–851.

[13b]Dash, S. M., and Sinha, N., "Exhaust Plumes and Their Interaction with Missile Airframes: A New Viewpoint," *Tactical Missile Aerodynamics: Prediction Methodology,* edited by M. R. Mendenhall, Vol. 142, Progress in Astronautics and Aeronautics, AIAA, Washington, DC, 1992, pp. 649–756.

[14]Mikatarian, R. R., Kan, C. J., and Pergament, H. S., "A Fast Computer Program for Non-Equilibrium Plume Predictions," Air Force Rocket Propulsion Lab., AFRPL-TR-72-94, Edwards AFB, CA, Aug. 1972.

[15]Jensen, D. E., Spalding, D. B., Tatchell, D. G., and Wilson, A. S., "Computation of Structures of Flames with Recirculating Flow and Radial Pressure Gradients," *Combustion and Flame,* Vol. 34, 1979, pp. 309–326.

[16]"Fluid Dynamic Flow Models," *JANNAF Handbook, Rocket Exhaust Plume Technology,* Chemical Propulsion Information Agency, CPIA Pub. 263, May 1975, Chap. 2.

[17]"Plume Electromagnetic Interactions," *JANNAF Handbook, Rocket Exhaust Plume Technology,* Chemical Propulsion Information Agency, CPIA Pub. 263, April 1977, Chap. 4.

[18]Cousins, J. M., and Jensen, D. E., "On the Computation of Ionization Levels in Rocket Exhaust Flames," *Combustion and Flame,* Vol. 52, 1983, pp. 111–125.

[19]Guernsey, C. S., and Luchik, T. S., "Analysis of High Frequency Image Data of Rocket Exhaust Plumes," 17th JANNAF Exhaust Plume Technology Meeting, April 1988.

[20]Victor, A. C., "Research Needs in Exhaust Plume Technology," 18th JANNAF Exhaust Plume Technology Subcommittee Meeting, Monterey, CA, Nov. 1989.

[21]Victor, A. C., and Breil, S. H., "A Simple Method for Predicting Rocket Exhaust Plume Smoke Visibility," *Journal of Spacecraft and Rockets,* Vol. 14, No. 9, 1977, pp. 526–533.

[22]Victor, A. C., "Effects of Multiple Scattering on Rocket Exhaust Plume Smoke Visibility," *Journal of Spacecraft and Rockets,* Vol. 26, No. 4, 1989, pp. 274–278.

[23]Victor, A. C., "Prediction of Rocket Exhaust Smoke Formation in Free Jets and Smoke Chambers," 1978 JANNAF Propulsion Meeting, CPIA Pub. 293, 1978.

[24a]Hoshizaki, H., et al., "Plume Visibility Detection Study," Air Force Rocket Propulsion Lab., AFRPL-TR-78-32, Lockheed, Nov. 1978.

[24b]Meyer, J. W., "Kinetic Model for Aerosol Formation in Rocket Contrails," *AIAA Journal,* Vol. 17, No. 2, 1979, pp. 135–144.

[25]Miller, E., "Smokeless Propellants," *Fundamentals of Solid-Propellant Combustion,* edited by K. K. Kuo, Vol. 90, Progress in Astronautics and Aeronautics, AIAA, New York, 1986, pp. 841–884.

[26]Victor, A. C. (ed.), "Rocket Exhaust Plume Radiation," *JANNAF Handbook, Rocket Exhaust Plume Technology,* Chemical Propulsion Information Agency, CPIA Pub. 263, May 1980, Chap. 3.

[27]Ludwig, C. B., Malkmus, W., Freeman, G. N., Slack, M., and Reed, R., "A Theoretical Model for Absorbing, Emitting, and Scattering Plume Radiation," *Spacecraft Radiative Transfer and Temperature Control,* edited by T. E. Horton, Vol. 83, Progress in Astronautics and Aeronautics, AIAA, New York, 1982, pp. 111–127.

[28]Spiro, I. J., and Schlessinger, M., *Infrared Technology Fundamentals,* Marcel Dekker, New York, 1989.

[29]Kneizys, F. X., "Atmospheric Transmittance/Radiance: Computer Code LOWTRAN 6," Air Force Geophysics Lab., AFGRL-TR-83-01187, Hanscom AFB, MA, 1983.

[30]Edwards, D. K., and Bobco, R. P., "Effect of Particle Size Distribution on the Radiosity of Solid-Propellant Rocket Motor Plumes," *Spacecraft Radiative Transfer and Temperature Control,* edited by T. E. Horton, Vol. 83, Progress in Astronautics and Aeronautics, AIAA, New York, 1982, pp. 169–188.

[31]Lyons, R. B., Wormhoudt, J., and Kolb, C. E., "Calculation of Visible Radiation From Missile Plumes," *Spacecraft Radiative Transfer and Temperature Control,* edited by T. E. Horton, Vol. 83, Progress in Astronautics and Aeronautics, AIAA, New York, 1982, pp. 128–148.

[32]Sume, A., "Millimeter Wavelength Emission from Solid Propellant Rocket Motor Plumes," Forsvarets Forskningsanstalt, FOA rapport, Sept. 1982.

[33]Senol, A. J., and Romine, G. L., "Three-Dimensional Refraction/Diffraction of Electromagnetic Waves Through Rocket Exhaust Plumes," *Journal of Spacecraft and Rockets,* Vol. 23, No. 1, 1986, pp. 39–46.

[34]Golden, K. E., Taylor, E. C., and Vicente, F. A., "Diffraction by Rocket Exhausts," *IEEE Transactions on Antennas and Propagation,* Sept. 1968, p. 614.

[35]Draper, J. S., Jarvinen, P. O., and Conley, T. D., "Analysis of Radar Return from Turbulent High-Altitude Rocket Exhaust Plumes," *AIAA Journal,* Vol. 8, No. 9, 1970, pp. 1568–1573.

[36]Rickman, J. D., et al., "Plume Attenuated Radar Cross Section Code—Users Manual," Aerodyne Research, Inc., AFRPL-TR-76-14, Billerica, MA, June 1976.

[37]Victor, A. C., "Recommended Guidelines for Determining Flight Radar Attenuation," JANNAF Plume Technology Working Group, Chemical Propulsion Information Agency, CPIA Pub. 205, Jan. 1971.

Chapter 9

Insensitive Munitions Technology for Tactical Rocket Motors

Andrew C. Victor*
Victor Technology, San Rafael, California 94901

I. Introduction

INSENSITIVE munitions (IM) are defined as munitions that reliably fulfill their performance, readiness, and operational requirements on demand but will minimize the violence of a reaction and subsequent collateral damage when subjected to unplanned stimuli. The objective of this chapter is to present scientific and technical information related to the sensitivity of munitions with particular emphasis on rocket motors and rocket propellants and to provide methods for designing munitions to meet the IM requirements as well as their other performance requirements.

Many catastrophic incidents occur because of the sensitivity of munitions to impact and thermal stimuli.[1] Such incidents are cataloged and reported by the armed services[2–4] and can result from accidents in transportation, "normal" handling, and routine operations, and from aggressive terrorist or battle action. The armed services of the United States have undertaken reduction of such incidents under their separate and joint IM programs.[5] The goal of these programs is to assure the development and deployment of IM. Because the IM problem is of equal importance in other nations, there are modes of international collaboration, particularly through NATO, but through other alliances as well, concerned with improving the science and technology of munition insensitivity and with interoperability, through common test and assessment methods.[6–9] The recently formed NATO Insensitive Munitions Information Center (NIMIC), housed at NATO headquarters in Brussels, Belgium, is an increasingly dynamic factor in international coordination.[10,11] The recent classification activities on "insensitive high explosives" for transportation safety requirements by the United Nations and its member nations have shown the need for hazard test protocols that can satisfy the needs of many different safety requirements.[12,13,19]

The U.S. armed services have a chain of documents defining procedures for achieving compliance of munitions with IM requirements.[5,14–18] These documents

*President.

are linked or are being linked to the NATO requirements.[19] The primary feature of these requirements is compliance with full-scale munition testing procedures and results, which are described elsewhere.[1,16,19] However, many additional tests are required at component and energetic material (EM) levels as well for "classification" and "qualification" of energetic materials.[19–23] In the U.S., all of these requirements fall under system safety program requirements.[24]

II. Background

The propellant in missile rocket motors generally constitutes up to 80% by weight of the explosive material in the missile. To meet the U.S. armed forces joint requirements for IM, one must ensure that safety hazards of the propulsion units of munitions are reduced to levels compatible with NAVSEAINST 8010.5B,[15] which (in combination with MIL-STD 2105B[16]) includes test descriptions and definitions of levels of reaction violence for insensitive munitions. Because the requirements of these documents did not apply to the design and development of propulsion units until 1985 and had no formal documentation until 1991, there has been no mature technology base from which to draw the improvements required.

All propellants have some degree of sensitivity. When reaction is initiated, all propellants respond with some level of violence. That is always true. The Joint Services Insensitive Munitions Program sets testable target levels for minimizing the sensitivity and reaction violence of the propellants and explosives used in munitions. The approach is still evolving. Propulsion performance is maximized by maximizing the energy delivered by the propellant in its rocket motor. Fortunately, sensitivity and explosive violence are not strictly monotonic functions of propellant energy, even though they generally are within any propellant family. Under research and development (R&D) programs currently proceeding in the U.S. and abroad, numerous approaches to achieving high-delivered propellant energy with reduced sensitivity and reaction violence are being tried.

Table 1 lists the required IM tests for U.S. munitions with requirements for performing and passing the tests.[16] Although the required IM hazard tests were described in an earlier volume of this series,[1] changes affected in 1994 have changed testing requirements significantly.[16] These changes strive to couple testing more closely to service use and logistical threats and to permit testing common to both IM and hazard classification requirements.[19] The U.S. services also require that a threat hazard assessment (THA) be used to evaluate the logistic and operational threats to munitions during their life cycles. The basic test requirements may be modified to more closely match the assessed threats.[16] The U.S. Army strives for very close coordination between in-service threats, system survivability, and IM test requirements.[25] The latest version of the controlling document for IM[16] permits some allowable munition responses to test stimuli to be modified from the passing criteria shown in Table 1 when such reponses are deemed acceptable on the basis of a THA. In terms of increasing hazard severity, munition reactions are ranked burning (type V), deflagration or propulsion (type IV), explosion (type III), partial detonation (type II), and detonation (type I).[16] It is this author's opinion that some propulsion reactions are sufficiently more hazardous than deflagrations (particularly for small, thin-walled rocket motors) that they deserve a separate category [as in the earlier MIL-STD-2105A(Navy) where they were ranked type VI].

Table 1 Insensitive munitions tests from MIL-STD-2105B[16]

IM test	Test parameters	Criteria for passing[a]
Fast cookoff	Test configuration from THA, fuel (JP-4,5,8 or JET A-1) or wood	No reaction more severe than burning
Bullet impact	1 to 3 type M2 0.50-cal AP bullets 850 ± 60 m/s, 80 ± 40 ms firing interval	No reaction more severe than burning
Sympathetic detonation	Applicability determined by THA, test configuration based on THA	No propagation of detonation
Slow cookoff	3.3°C/h heating rate or higher if determined by THA	No reaction more severe than burning
Fragment impact	12.7-mm mild-steel cube (2530 ± 90 m/s) (alt 1) 12.7-mm conical (140°) mild-steel (1830 ± 60 m/s) (alt 2) based on THA	No reaction more severe than burning[a]
Shaped charge jet	50-mm Rockeye-type SC donor or as determined by THA	No detonation[a]
Spall	81-mm precision SC impacting 25-mm RHA plate, impact by minimum of 4 spall fragments per 64.5 cm^2 of test item up to 40 fragments total	No sustained burning[a]

[a]Some passing criteria may be adjusted on the basis of a detailed THA that considers effects on munition environment and system vulnerability requirements.

The past focus of R&D programs in tactical rocket propulsion concentrated on improved performance (range, velocity, signature, service life, and cost) rather than reduced sensitivity. One exception to this has been the de facto requirement that propellants used in Navy rocket motors, particularly air-launched rocket motors, meet requirements for class 1.3 explosives.[19] (More details are given later in this chapter.) Other notable exceptions have been work by the U.S. Army and Navy and by other nations on low-vulnerability ammunition (LOVA) propellants for artillery gun propulsion, the U.S. Navy's R&D programs to reduce the output violence of munitions in response to fast cookoff (an area in which the Army continues to work productively), and development of plastic bonded explosives (PBX). The U.S. Navy's Insensitive Munitions Advanced Development (IMAD) program started in 1984 and Insensitive Munitions Technology Transition Program (IMTTP) started in 1986 are intended to provide a demonstrated technology base to support the design, development, demonstration, and production of insensitive missile components, including propulsion units.[26,27] The U.S. Army's IM program has since 1990 also achieved significant advances in developing less sensitive rocket propellants.[28,29] The U.S. Air Force IM program concentrates on bomb storage and transportation safety.[30] An insensitive munitions working group has recently been created in France.[31] The United Kingdom is also developing policy and technology for IM.[32,133] A reasonable idea of progress in this area can be obtained by tracking the symposia on insensitive munitions and on energetic materials sponsored by the American Defense Preparedness Association (ADPA).

Through work on explosives and strategic propellants and years of related basic and applied research, there are techniques and expertise that can be applied to the sensitivity problems of tactical propulsion.[33] However, more recent investigations have shown that tactical propellants present additional unique sensitivity problems.[26,34–37] Achieving necessary propellant energy and density levels to meet performance requirements (nonetheless still meeting IM requirements) is a special problem for controlled-smoke, nonmetallized propellants.

Bullet and fragment impact testing of rocket motors was introduced recently in response to IM requirements.[1] Bullet impact testing had been performed on warheads and bombs for many years using 20-mm armor-piercing projectiles at a standard velocity. More recently the joint U.S. armed services have agreed to a NATO-compatible bullet impact test using 0.50-caliber armor-piercing (AP) projectiles.[1,5,16] There is a substantial international database relating to 0.50-caliber bullet impact testing of rocket motors and propellants.[33,38–41] Smaller caliber bullet impact has been demonstrated to cause damage that results in explosion or detonation of some rocket motors upon subsequent bullet impact in the same vicinity (multiple-bullet impact).[181]

Fragment impact testing of rocket motors has been rather limited, and no standard method of generating the fragments has been defined. Multiple-fragment impact testing was standardized for the Navy under the IMAD program based on methods first demonstrated at the Naval Surface Warfare Center, Dahlgren, Virginia, in the early 1980s under what was then the U.S. Navy's Explosives Advanced Development Program. In 1990 the U.S. IM test requirement was changed from "multiple-fragment" to one that causes "two to five impacts on the ordnance."[16] The responses of rocket motors to the high-velocity impacts experienced in these tests [2800 fps (854 m/s) bullet impact and 8300 fps (2530 m/s) impact by steel cubes (fragments) each $\frac{1}{2}$ in. (12.7 mm) on an edge] varies from no reaction or mild burning through deflagration, propulsion, explosion, and detonation, depending on the propellant and a number of other design features of the motors as well as how they are struck by the projectiles.

Good and reproducible results have been reported using both powder and light-gas guns to propel single, well-aimed cylindrical or saboted-rectangular fragments at cased and uncased energetic materials.[42] Snyer[43] reports routinely achieving 8300 ± 300 fps with a 40-mm powder gun. With the multifragment generator that was used in U.S. Navy munitions testing through 1993, prediction and control of the exact number of impacts and the impact locations on a munition was not possible. The Army currently favors limiting the upper velocity of its conical-fragment impact test to 6000 fps because of the nature of the fragment threats normally encountered in operation. Thiokol Inc. has reported development of an explosively driven single-fragment launcher with good capability up to 6000 fps (1830 m/s).[35] The French use a spherical projectile in high-velocity impact studies.[44] There are reasons to prefer the perfect reproducibility of well-aimed spherical impact compared to the possible impact-aspect variations of flat-faced or tapered projectiles, in spite of the higher impact velocities required to initiate detonation with spheres.[45] There is evidence that the critical impact velocities of spheres can be accurately correlated analytically with data for other projectile shapes as shown later in Eqs. (42) and (43).[46,159]

Figure 1 illustrates the regimes of behavior that have been observed by a wide range of ordnance (warheads, rocket motors, and gun propellant charges) in response to projectile impact.[7] The mass coordinate in Fig. 1 actually corresponds

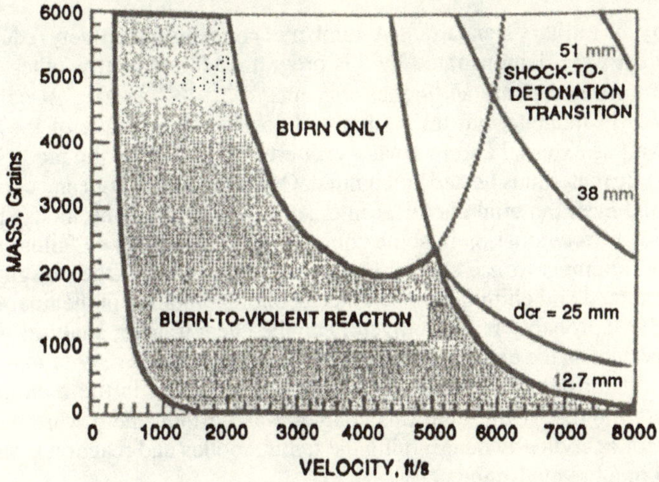

Fig. 1 Hazard plot showing typical regimes of ordnance responses to projectile impact.

to a mild-steel flat-faced cylindrical projectile with unity ratio of length to diameter. The calculated shock-to-detonation transition (SDT) boundaries for several different critical diameters are shown to the right. For at least some energetic materials, the burn-only region corresponds to impacts with sufficient energy to fully penetrate the munition and break up the case. This may prevent violent pressure burst reactions (violent deflagrations or explosions). The shaded region of Fig. 1 corresponds to a region of impact in which the projectile penetrates one side of the case, but lacks sufficient energy to pass entirely through the munition and leave through the opposite side. In such situations, ignition of the energetic material often occurs, and the possibility of growth to violent explosion, following rapid pressure buildup within the case, exists. The darkened region also represents potential paths to delayed transition to detonation [(XDT) X is unknown] responses following first penetration of the case. Fairly normal propulsive reactions may also result from reactions in this region. Propulsive reactions are extremely dangerous because they can spread the hazard of impact, burning propellant, and a live warhead to great distances. The region described as burn only in Fig. 1, which involves complete projectile penetration of the case, has indeed been observed to produce propulsion, violent deflagrations, and explosions for many rocket motors and other ordnance items. There are a number of reasons for this, including the presence of empty volume in the bore of a propellant grain, that are described later in this chapter. Instances of impact ignition by projectiles with velocities insufficient to pierce the case have been observed for some cased gun propellants. Figure 1 does not include effects of impact on previously damaged energetic material, which can greatly sensitize the material.

Successful reduction of the violence of munition reactions in fuel fires (fast cookoff) demonstrated by the U.S. Navy's Weapon Fast Cookoff Program[1,48] (which has subsequently become the IMTTP) has depended largely on reducing confinement (venting of the case) of the explosive material prior to ignition or so weakening the case that it bursts at a relatively low pressure on ignition. In this way it has been possible to modify warheads and rocket motors that exploded or detonated in fast cookoff so that the maximum reaction was only burning (as defined by the military standard[16]). A number of concepts for actively reducing case confinement were demonstrated by this program. More recently, other programs have demonstrated many additional concepts.[26,34–37,49,50,196] All fast cookoff test results clearly indicate that the design and construction details of the munition case as well as external and internal inert parts are critical to the mode by which the energetic material is heated and ignited. Other specific design and construction details (and even material chemical interactions) determine the modes by which ignition leads to detonation in some components or to explosive failure in others. Therefore, attempts to use standard subscale test articles in the design phase to help understand and eliminate violent fast cookoff responses in the final full-scale munition will probably fail. Intelligently applied heat transfer analyses with good laboratory data on the energetic material thermal properties behavior may be much more useful for determining the temperature distribution in the munition at the time of ignition.[46] At the time this chapter was written, there were no a priori methods for analytically determining the failure modes and reaction violence of a munition in a fire environment.

The standard slow cookoff test, used to demonstrate compliance with IM requirements, now adopted fairly universally, involves uniformly heating the munition at a rate of 3.3°C/h until a reaction occurs. It has been observed that slow cookoff reaction violence in some systems can be decreased by venting the munition case prior to the extremely rapid pressure buildup that often follows ignition of the energetic material. [Hydroxyl Terminated Poly Butadiene (HTPB) propellants are particularly notorious for showing this type of behavior once the pressure in the reacting zone reaches about 2000 psia (138 bars).] This decrease in reaction violence has been achieved with some sealed warheads by using stress risers in the case that crack under the pressure caused by an outgassing liner or the expanding main-charge explosive within. This approach cannot be used for rocket motors, which require the case strength for normal pressurized operation and which already have a venting nozzle that prevents any slow buildup of substantial pressure within the case. However, case or liner materials that soften when heated, or methods of venting the case, have been demonstrated to greatly reduce the violence of slow cookoff responses of some rocket propellants. Unfortunately, a few propellants currently in use are inherently detonable at slow cookoff heating conditions, even in small quantities. Polyurethane propellants with copper chromite burning rate catalysts are particularly susceptible, and some iron containing catalysts also greatly increase reaction violence. These materials may detonate even with no confinement; therefore, no venting mitigation system will provide the needed safety unless it results in ignition and burning of the material prior to autoignition. In the standard slow cookoff test, the entire propellant grain is often heated through, and additional heat due to decomposition of the propellant, which due to low thermal conductivity is unable to escape from the grain interior, may cause parts of the internal grain to rise to temperatures measurably more than 60°C higher than that of surrounding environment of air in the slow cookoff test oven.

There is a regime of heating rates between the current fast cookoff and slow cookoff test conditions[1] that is quite likely to occur in hazard scenarios.[51,197] This intermediate cookoff regime, which may cause extreme reaction violence similar to that experienced in slow cookoff, was studied by the IMAD program in FY 1988 to determine its range of applicability, its effect on reaction violence, and methods for mitigating its effects.[49] To do this, tests of full-size rocket motor reaction violence at 42°C/h (intermediate) heating rate were performed to compare the results with those obtained at 3.3°C/h (standard slow cookoff test heating rate) for rocket motors that detonated or exploded at the slower heating rate. The results indicate that reaction violence is greatly reduced for some propellants at the higher heating rate (which corresponds to 6–8 h heating duration for the motors tested).[26,52] Temperature probes and heat transfer computer modeling confirm that this effect occurs because at the higher heating rate only the outer regions of the propellant grain reach autoignition temperature.[46] However, some rocket motors that detonated at the 3.3°C/h heating rate also detonated at the higher heating rate. Small, unconfined samples of the propellants in these motors detonated in propellant slow cookoff visualization (SCV) tests at 14°C/h.

A related effect due to motor size has also been observed. For some propellants that react very violently in a slow cookoff test of an 8-in.-diam motor, the resulting reaction was relatively mild with a larger motor of about 20-in. diam. Computer modeling indicates that the reason for this difference is the same as that observed with the intermediate cookoff test; peripheral heating instead of complete temperature soaking has occurred. Another factor contributing to this difference in cookoff reaction violence is that the larger motors cook off at a lower environmental temperature at which less decomposition of the bulk of the propellant has occurred. The experimental evidence concurs with the theoretical calculations that indicate the importance of propellant thermal conductivity to the outcome of a slow cookoff test.

Lack of reproducibility in some slow cookoff tests has been observed, with successive identical tests resulting in a near-detonation-level explosion in one test and a relatively mild deflagration-level response[16] in another. No explanation for this disparity has been offered.

Another aspect of the sensitivity and reaction violence due to inadvertently initiated rocket propellants involves pressure buildup within the confining motor case. If a reacting energetic material is confined, internal pressure due to generation of decomposition and combustion gases can increase without relief. This chain of events may ultimately lead to a very violent explosion or a detonation. The time required for this progression and its ultimate violence depends on the stimulus, the explosive (and its physical and chemical condition), and the degree of confinement. Normal solid-rocket motor operation involves moderately rapid pressure buildup, as explained earlier in this book, to the level at which an equilibrium condition exists involving the propellant burning surface area and the propellant burning rate at the pressure maintained with sonic effluence at the nozzle throat. If the propellant is damaged by cracking or the presence of voids, it is possible for the propellant burning surface to be increased many times greater than the design surface area. When this happens, the pressure generated by the combustion gases may rise much faster than during design combustion to levels much greater than the case can endure; and an explosion may result. In some situations a detonation may be initiated in the propellant during this pressure buildup, giving rise to what is known as a deflagration-to-detonation transition (DDT).

In 1988, the IMAD Propulsion Project completed feasibility and proof-of-concept tests on a multihazard mitigation system that successfully vented rocket motor cases in fast cookoff, slow cookoff, and bullet impact tests.[49] Both thermite and linear-shaped charge case-cutting systems were demonstrated on steel rocket motor cases in these tests. IMAD work on active mitigation systems for rocket motors terminated at that point. It was determined that it is not feasible to develop a "generic" mitigation system since such a system must be tailored to the specific motor it is to be used on. It makes more sense to work with a specific weapon system, if active mitigation is needed. Many people believe that such mitigation, since it is intrusive to a rocket motor's primary function, should be considered only as a last resort. However, such mitigation systems are included in the current development of several rocket motors as part of their approved "insensitive munitions plans." The U.S. Navy requires that demonstration testing be used to prove that such mitigation systems will not inadvertently ignite the propellant grain. Recently, improved thermal sensors and activators for case-cutting mitigation systems have been developed.[196]

One category of currently available propulsion technology derives its insensitivity from separation of fuel and oxidizer. The IMAD program refers to such technology as alternate propulsion systems. Bipropellant gels, liquids, solids, and hybrids as well as airbreathing systems (ramjets, ducted rockets, and turbojets) comprise this category.[193] A bipropellant gel system developed by the Army (MICOM), TRW, and Talley was subjected to IM large-scale hazard tests (fast cookoff, slow cookoff, and bullet impact) under the Navy's IMAD program during FY 1989.[26,28,223] Pelletized separation of fuel and oxidizer within solid propellant motor grains is also possible, although it may cause difficulties with combustion stability. Recent work has shown that encapsulated or pelletized additives [such as glycidal azide polymer (GAP)] within the propellant grain can reduce reaction violence in IM tests, and despite local inhomogeneities, the average burning composition in the bore should remain rather constant.

Preliminary work on several new rocket motor case concepts (strip laminate, composites, and hybrids) shows promise for passive venting in response to some hazardous stimuli.[26,28,34–37,53–55] Some designs of these cases vent well and quickly in fires and in response to bullet and fragment impact, and some have the potential to relieve the violence of response to intermediate cookoff (and some slow cookoff) situations as well. There are, however, some concerns about the durability of cases fabricated with new technology, under long-lifetime, high-stress operational conditions. There are also concerns about production costs of the cases, although the most recent estimates of future production costs are not unreasonable for smart weapons.

The propulsion project of the IMAD program has defined technology goals for missile propulsion systems that meet the Navy's IM requirements.[26] The Army has defined similar goals for meeting their requirements.[25] These requirements include design principles for rocket propellants, cases, mitigation systems, and related test methods that form the basis for most technology development. Technology thrusts in the U.S. service propulsion R&D programs, in industrial research and development (IR&D), and in allied nations now show strong interdependence. Although most R&D efforts are focusing on new propellant formulations, including such areas as modified combustion characteristics, new binder systems, advanced energetic solids, advanced combustion modifiers, and methods for evaluating these new propellants for IM suitability, work on new case material and design concepts, case venting, and shielding of munitions from impact and thermal hazards also is in progress.

The initial goals of early programs that led to IM concentrated on reduction of munition hazards in response to thermal threats. Such hazards as detonations and explosions of munitions caused by fires on aircraft carrier decks, fires in transportation vehicles, and premature reactions in hot gun bores were reduced by the development of PBXs and LOVA gun propellants. Thus, reduced violence in response to thermal threats (corresponding to all heating rates) is the first requirement of an IM. This idea is universally accepted, and appears as the first step in the French MURAT protocol.[31,212] The MURAT protocol goes on to identify reduction of response violence to other threats as the next two levels of risk reduction.

Other nations have similar, if slightly different, test requirements from the United States. France's MURAT doctrine, for example, deletes the spall test and adds a heavy fragment (250 g) fragment impact test.[212] In addition, France recognizes the three levels of insensitivity as shown in Table 2. By allowing qualitative classification according to this range of levels, one can tailor maximum allowed responses to specific munition sizes and storage and use environments, thus permitting a wider range of design variables and yet still retaining control of safety.

III. Operational Considerations

A. Qualification of Munitions

Qualification is both the process by which energetic materials and munitions containing energetic materials are certified as safe and suitable for use and the

Table 2 MURAT tests from French DGA/IPE Instruction No. 260[212]

		Maximum reaction criteria for passing[a]		
IM test	Test parameters	MURAT*	MURAT**	MURAT***
Electrical	Static, EMR, lightning	NR	NR	NR
Drop	0–12 m on flat steel surface	NR	NR	NR
External fire	liquid hydrocarbon, no time limit	IV	V	V
Slow cookoff	Steady rate from 3 to 60°C/h	III	V	V
Bullet impact	1–3 AP 12.7-mm burst, 0–850 m/s	III	III	V
Sympathetic reaction	Identical munition, most vulnerable configuration	III	III	IV
Fragment impact (light)	3 simultaneous 20-g steel cubes, 0–2000 m/s	I	III	V
Fragment impact (heavy)	250-g steel parallelepiped, 0–1650 m/s	I	III	IV
Shaped-charge jet	Jet capable of penetrating 300-mm thick steel plate	I	I	III

[a] Allowable reaction level definitions: I complete detonation; II partial detonation; III violent case burst, projection of large fragments and unburned energetic material, shock wave (explosion); IV Nonviolent case burst, possible projection of energetic materials and end caps, no blast or fragmentation, possible propulsion of article is not permitted in order to pass the MURAT test for this level (deflagration); V Nonpropulsive combustion, nonviolent opening of case, no hazardous fragments beyond 15 m (burning); NR no reaction and safe to dispose.

status conferred upon them by such certification. Energetic materials must be qualified prior to use in development, product improvement, or other nonlaboratory R&D programs. Only "final (type) qualified" energetic materials may be used in weapons introduced for operational use. A qualified energetic material is certified on the basis of results of a number of laboratory tests and IM tests in generic hardware. A final (type) qualified energetic material will, in addition, have been demonstrated to be safe in its intended role or application. The fact that an energetic material has been final (type) qualified for a specific application does not confer that acceptance for other ordnance uses. Within the past few years the U.S. has appended its qualification requirements to include compliance with IM policy. Therefore, munitions that contain qualified energetic materials will also meet IM requirements or have well-documented, and supposedly justified, supporting waivers.

Qualification status is granted to an energetic material by a national authority of the developing nation, and in many cases, by the using nation as well. [For example, for the U.S. Navy this authority is the Weapons Systems Explosives Safety Review Board (WSESRB).] Equivalency of qualification status between nations is important when, through foreign military sales or joint operations, munitions produced in one nation are used by others. Achieving this equivalency is the purpose of NATO Group AC/310.[56] To help achieve this equivalency, NATO documents[8,18,23] form keystones to which the qualification requirements of individual nations[5,6,14–23] are linked. The links are complex, and only national requirements for the U.S. are referenced here. Many hundreds of pages of NATO documentation are used to describe the tests required for qualification in the individual nations. Determining equivalency requires expertise only alluded to in the Technical Considerations section of this chapter and its supporting references. NIMIC helps provide information at the working level that is needed to coordinate these requirements.[11]

B. Hazard Threats to Munitions

Qualification only certifies that a munition and its energetic material have given acceptable results in a number of tests. Although this, particularly since it now includes IM certification, confers a certain safety status, it provides no estimate of the probability of a munition's hazardous encounters in service use. The potential hazards to a munition throughout its life cycle are many and will vary not only with the type of munition and its typical use but also with unique deployment conditions. For this reason, the U.S. requires that an insensitive munitions threat hazard assessment (IMTHA or THA) accompany each munition's request for IM status (which precedes granting of qualification status).[15,16,58,197] The U.S. Army is now basing its approach to IM on a policy that requires an IMTHA to help define the munition development and test programs, and this is incorporated in the latest version of MIL-STD 2105.[16,25] The U.K.[133] and France[31] also consider hazards in operational situations.

An insensitive munition is never "totally safe." A properly done IMTHA should determine how much safety IM status achieves for a specific munition at all steps in the life cycle.[197,211] An IMTHA must include consideration of all energetic components of a munition and all foreseeable threats to the munition that may cause reaction of an energetic component.[4,52,59] The really important components are the warhead and rocket motor, which contain nearly all of the energetic material. However, other very small components must also be considered since reaction of one of these may initiate reaction in a large component.[1,52] Benign components, including containers and launchers, may also affect a munition's response to hazardous stimuli.

Specific areas the IMTHA should address include the following. The first area is the differences between test stimuli and threat stimuli. This includes cookoff heating rates, bullet calibers and velocities, fragment sizes, shapes, velocities, and spatial distribution, and donor-acceptor relationships including nonlike donor/acceptor munitions (in sympathetic detonation scenarios), storage arrangements, and barrier and container relationships to the scenarios. The second area is the differences between munition responses to test and threat stimuli. The third area is the differences between the amount of damage done by the threat alone and the combined effect of threat and reacting munition. This involves detailed knowledge of planned use environments for the munition. (For example, in one analysis of a very large warhead threat to a moderate-size combatant ship, the difference between the damage calculated to be done by the threat warhead alone and the threat warhead plus sympathetic detonation of the warheads of particular missiles stored aboard the ship was negligible.)

Figure 2 illustrates this approach to IMTHA. Figure 3 shows the various stimulus, munition, sensitivity, and output considerations that should be considered in an IMTHA and in the design of an IM.

Before an IMTHA can be performed one must have information on the various factors involved. This includes the use environment; potential threats, including intelligence information on the anticipated weapons of potential enemies; and the predicted responses of the munition to various threats. This last area, predicted munition responses, involves in-depth knowledge of the behavioral characteristics of the energetic material as well as munition design details involving case materials and configuration, case attachments, and storage and shipping configurations, and shielding.

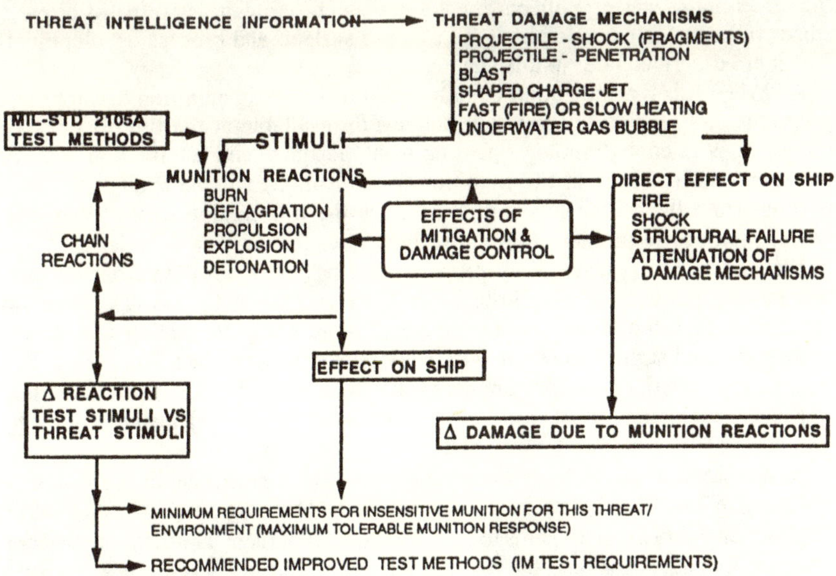

Fig. 2 THA approach (example of munitions on a ship); relationship of operational threats to IM tests.

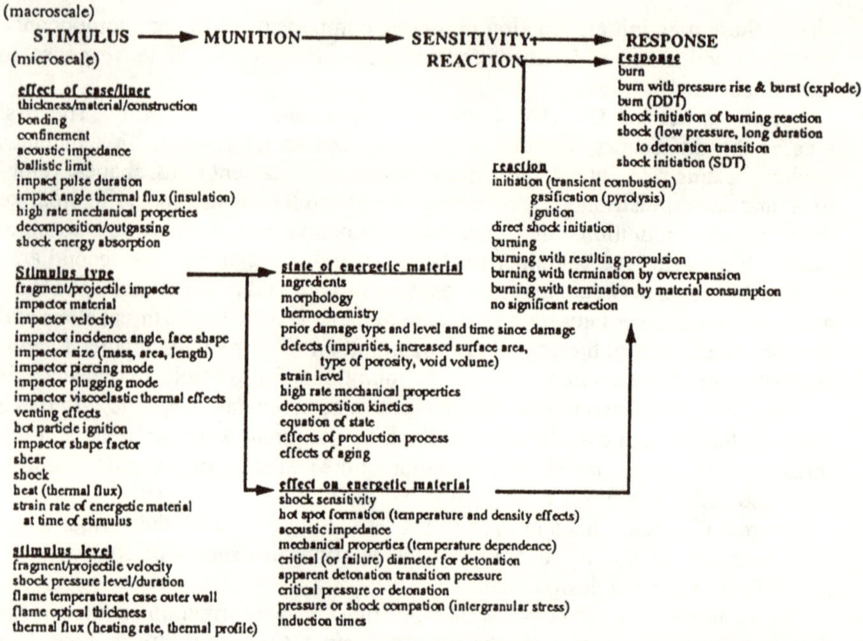

Fig. 3 Data components needed in considerations of munition sensitivity and response.

The various tests for qualification of energetic materials[20–22] include a number which give insights into this area. The U.S. Navy's IMAD propulsion program devoted considerable effort to devising tests and a protocol for their application for screening rocket propellants for the key safety issues related to IM at early stages of propellant development (i.e., before scaleup and process development) which are described later in this chapter.

An IMTHA is typically contractually scheduled early in munition development programs and requires information not readily available at the time it must be prepared. With enough money spent up front, adequate characterization data for the energetic materials in the munition (qualification) will be available, and it is often from these data that one must draw the information to use with analytical methods to predict munition responses to threats in the initial IMTHA. Subsequently, the system safety program will have the test data from the preliminary design tests to include in its IMTHA and IM test plans. The final IM assessment will then have the advantage of more complete testing and design refinement, and at that time there will be more system-level data. But even at that final stage, analytical methods are necessary to tie all of the information together to provide reasonable inputs for the IM assessment required by the system safety program.[24,46,197]

At the proposal phase, a contractor will naturally concentrate on the seven IM test areas of Table 1 only as minimally specified by Department of Defense (DOD) requirements.[16] (This also seems to be a predilection of many Government weapon program managers for the obvious reasons that it conserves resources for more traditional uses, it seems to meet the requirements, if not the intent of DOD IM policy, and better approaches although proposed have not yet been implemented.[60])

Under these conditions, development programs focus on creating IM designs that will pass the required tests.

Alternate or additional tests that simulate other possible threats will likely be suggested at the proposal phase usually only if there is serious doubt that the basic tests can be passed. However, to justify any alternate tests it is necessary to demonstrate threat feasibility (and, if possible, probability) with a preliminary IMTHA. An adequate IMTHA describes the nature and probability of the alternate threats, the design of alternate corresponding tests, and the expected responses (and their probabilities) of the munition design to both the basic tests and the alternate tests. It does not require much imagination to visualize the extensive scope of the prior test effort that would have been required to provide such an IMTHA with any quantitative degree of certainty and to realize that it would be prohibitively expensive and untimely in today's munitions-development environment.

Into this data vacuum it is prudent to introduce analytical methods for estimating the IM behavior of a proposed design in the preliminary IMTHA. Such methods are available,[46] and some are given later in this chapter. The reliance on such methods will be lessened, or at least inspire more confidence, if data from more preproposal research and development efforts are available on the proposed design concept or by analogy and extrapolation from similar design concepts. In the future, as appropriate scientific studies are completed and complementary analytical techniques evolve, engineering approaches to IM design will become more refined and systematized through experience. Even in that optimistically envisioned future, simple methods will have an important part to play for IMTHA, data correlation, preliminary design, and systems analysis studies.

IV. Technical Considerations

A. Detonation of Explosives and Propellants

A detonating energetic material produces chemical energy at the maximum rate for that material. In other words, the power output is a maximum during a detonation. For a tactical rocket motor the power output during a detonation (which might last about 0.1– 0.2 ms) would be between 10,000 and 100,000 times greater than during normal design combustion (which might be of the order of 1–10 s duration). By comparison, unconfined burning of composite propellants containing ammonium perchlorate (AP) typically lasts about 10 times longer than normal design combustion and, hence, has about 1/10 the power output. Under most conditions a detonation is generally considered to be more dangerous than any other reaction because it causes more damage, particularly through the supersonic projection of fragments and generation of blast overpressure and impulse. Whereas the total energy resulting from burning and deflagration-type reactions (including those classified as explosions) is usually greater than that from detonations, the power output, or rate of energy release, is much less. In point of fact, the ultimate damage to the environment caused by fires is often greater than from any other type of reaction. The following time scale to full development of each type of reaction is approximate; completion of each reaction takes longer: detonation, 1–5×10^{-6} s; XDT, $\sim 10^{-4}$ s (buildup time); severe explosion, $\sim 10^{-3}$ s; deflagration, 3–100×10^{-2} s; burn, >1 s.

A detonation is a combustion process in which the combustion products initially move toward the reacting surface, thereby building up pressure and maintaining a shock wave. In contrast, in a deflagration (technical term for any burning reaction, as opposed to IM use for assessing the reaction violence of munitions[1,16]) the combustion products move away from the reacting surface, as in a normally burning solid rocket propellant grain. The magnitude of the shock wave in a detonating explosive (detonation pressure) is of the order of several hundred thousand atmospheres (200–300 kbar or 20–30 GPa). The shock wave moves through a solid ideal explosive at a velocity which is typically several times greater than sonic velocity in the solid (detonation velocity D has a unique and simple relationship to the velocity of sound in the highly compressed gaseous products); thus yielding detonation velocities as high as 9–10 km/s in the highest density modern explosives.[61,62]

A detonation does damage to materials in contact with the explosive both by the direct effect of the shock wave and by the rapid expansion of the combustion product gases, which cause plastic rupture of encasing metal. Damage at a distance is caused by both the pressure wave from the explosion and by thrown fragments. Special effects of detonations may be used to accelerate metals to extremely high velocities. For example, in a shaped-charge jet warhead the detonation forms a jet of metal and accelerates it to velocities of the order of 10 km/s (known as the Munroe effect) in order to penetrate substantial thicknesses (of the order of up to a meter) of steel armor.

A number of international journals regularly report the latest research on detonation and combustion that has bearing on this chapter. There are also a number of excellent books available, several of which are referenced here.[63–72] The International Detonation Symposium, held at 4–5-year intervals,[73] is a superior source of recent advances and references to supporting literature. Computer programs are available for predicting shock wave propagation, effects, and initiation and resulting propagation and effects of detonation of energetic materials in one, two, and three dimensions.[69,73]

1. Initiation of Detonation

A detonation is initiated in an energetic material when enough energy is introduced to create a pressure wave which starts the process of growth of reaction to a detonation. Typically, the minimum initiating pressure for a detonation is many times less than the detonation pressure of a sustained detonation in the same explosive; the lower the minimum initiating pressure, the more sensitive the explosive. If the energy is introduced into the explosive as a shock wave, the critical energy criterion ($E_c = AP^k t$) often gives good results for values of k near 2 (Refs. 74–78). Detonations can also be initiated by energy sources other than shock waves. If insufficient energy is introduced, a detonation will not occur; instead, a deflagration (burn) may result, or ignition either may fail altogether or a weak combustion wave may extinguish before the energetic material is fully consumed.

Shock-to-Detonation Transition. The process just briefly described by which a shock wave initiates a detonation in an explosive is often called shock-to-detonation transition (SDT). Any energy introduction stronger than the critical energy will initiate a detonation provided the area over which it is introduced is equal to or greater than the area of a cylinder of diameter larger than the critical diameter d_{cr} of the explosive. A shock wave introduced over a smaller area of the explosive can still initiate a detonation if the critical energy criterion is equaled or exceeded when the shock wave has expanded to the size of the critical diameter. Typically, in an SDT the detonation will develop within a few microseconds after application of the shock.

For some insensitive explosives the critical diameter is very large (5 or more cm) and reliably initiating a detonation in these explosives can be quite a problem. For some composite propellants critical diameters may be of the order of 30 cm or more and detonation cannot occur in tactical size rocket motors. However, greatly reduced critical diameters (less than 6 cm) have been reported in high-burning rate HTPB composites that contain ferrocenic burning rate modifiers.[79] Heterogeneous explosives and propellants often behave anomalously in comparison with well-modeled ideal detonations.[198–200]

Typically, solid "minimum-smoke" propellants, both double-base formulations and composite formulations containing nitramine explosive molecules (such as HMX or RDX) have small critical diameters (1 cm or less) and exhibit typical SDT behavior. Larger critical diameters and greatly reduced shock sensitivity have been exhibited in minimum-smoke propellant formulations containing ammonium nitrate as the energetic solid constituent. Whereas reducing the concentration of high-explosive ingredients and using energetic binders to replace lost propellant energy and to substitute for typical inert binders will reduce shock sensitivity, energetic binders result in propellants that are more shock sensitive than propellants with inert binders. It has also been observed that when some, rather than all, of the nitramine is replaced with such crystalline oxidizers as ammonium nitrate or ammonium perchlorate, the SDT sensitivity is not commensurably reduced.

A detonation may occur even though the energy introduced does not meet the criteria for shock initiation. The alternate energy sources for this kind of initiation may be thermal or mechanical impact; however, they will not lead to a complete detonation unless they lead to the critical energy criterion being reached somewhere within the propellant. Several of these methods of introducing a detonation are described next.

Delayed-Detonation Transition (XDT). A delayed-detonation transition usually occurs by a mechanism that is not well known or understood (hence the X). Typically, XDT occurs about 150 μs after application of the initiating stimulus to the propellant case. The result described by Nouguez et al.[38] for tactical-size rocket motors appears to be an XDT type of reaction. In this situation the original initiating stimulus (an impacting 0.50-cal. projectile of approximately 1-km/s velocity) is not sufficiently energetic to cause an SDT reaction, but it starts a reaction that builds to a sufficiently energetic level. The web and bore dimensions of the rocket motor tested, the projectile velocity, and the propellant-case material all influence the reaction.[194]

Finnegan et al.[80a–80e] at the Naval Weapons Center (NWC, now NAWCWPNS) reported observations of the XDT effect in a one-dimensional apparatus in which two slabs of propellant with outer surfaces covered by typical rocket motor case materials are separated on their inner surfaces by an air gap. The researchers use a 1.9-cm-diam spherical projectile at approximately 1150-m/s velocity, although velocity is often varied. The high-speed photographs clearly show that propellant debris from the first slab ignites upon combined impact of debris and the projectile with the second slab and propagates a reaction back to the first slab. Finnegan et al.[80a–80e] studied both nondetonable AP/HTPB propellants and with a detonable propellants. With the AP/HTPB propellant, the ignition of the first slab occurs when recirculated, burning debris comes in contact with it near the edges of the test container; reaction does not propagate back through the debris cloud forming at the first slab after impact, probably because the combustion propagation rate is much slower than the debris velocity (approximately 750 m/s at the leading surface of the debris cloud). With the detonable propellant, the reaction that initiates at the second propellant surface appears to propagate back through the debris cloud forming at the first slab. When the debris velocity is taken into account, an estimated combustion (or detonation) propagation velocity of 1500 m/s or greater through the cloud is consistent with the high-speed photographic data.[80a–80e] They reported delay times between impact and detonation ranging from 108 to 221 μs. They obtained no detonation for air gaps of less than 0.6 cm or more than 7.6 cm. Finnegan et al.[80a–80e] observed detonation with an inert second slab and a detonable first slab. There is evidence that several mechanisms are involved in different regimes of their tests. It has been demonstrated in both their test and in cylindrical motors that a "bore mitigant" consisting of a low-density foam filler can prevent both the XDT reaction in susceptible propellants and relatively violent responses in nondetonable propellants. In some cases the bore mitigant has prevented impact-induced ignition. Tests have demonstrated that a bore mitigant can be designed so as not to interfere with normal propellant ignition. The test of Refs. 80a–80e has been dubbed the burn-to-violent reaction (BVR) test and is being used increasingly for routine propellant screening.

Much additional work remains to be done on both the SNPE and NAWCWPNS tests to explore the effects of dimensions, projectile shape and velocity, case materials, confinement, and temperature. The relationship between the French cylindrical tests and the planar tests should also be studied.

Deflagration-to-detonation transition (DDT). A deflagration-to-detonation transition occurs after some initiating stimulus causes a deflagration to occur in an energetic material and the deflagration grows to a detonation; such growth usually takes time on the order of milliseconds. A significant amount of experimental and theoretical research on DDT has been reported, and the papers reported at the Detonation Symposium[73] over the years are representative. Successful computer models of the DDT process involve pressure buildup resulting from a deflagration that causes compaction of porous regions of energetic material leading to growth of a shock wave strong enough to initiate a detonation. It has been observed that significant porosity in the energetic material is required for a DDT reaction to occur. Such porosity may be the result of faulty material production control or of excessive strain in the energetic material. Wachtell and McKnight[208] measured phenomena that might lead to this effect in early "explosiveness" tests. A fairly significant volume of porosity and fairly high confinement are needed to permit pressure buildup of a deflagration to a shock wave of sufficient strength to initiate a detonation. However, if the bulk of the energetic material has been sufficiently damaged (as evidenced by sufficient voids for the DDT to start growing) the required critical energy and the critical diameter are often significantly lower than for the undamaged material.[81]

2. *Concept of Explosiveness*

Several NATO nations rely heavily on explosiveness tests for assessing the sensitivity of energetic materials.[23,88,89] The concept of explosiveness involves igniting an energetic material confined in a metal tube and assessing explosiveness from measurements of time to tube rupture, the number of tube fragments created, and the amount of unconsumed energetic material. More explosive energetic materials rupture the tube in a shorter time, create more fragments, and leave less unconsumed energetic material. Typically, ignition is produced at one end of the tube by an electric match or by adiabatic compression. Tubes of different dimensions have been used, representing a many-fold range of volume of energetic material contained, with no apparent difference in the assessed degree of explosiveness. There is a lower limit to the tube wall thickness that must be used in order to provide required confinement during pressure buildup. It is also believed that there is an upper limit to tube wall thickness that makes a difference on the reaction level achieved. Some explosiveness tests have been run in the U.K. with confined propellants that were ignited by slow heating (much like the U.S. slow cookoff test) rather than by an electrical igniter, but this is not a standard test. It appears that the explosiveness test may be a measure of DDT.[208] Although explosiveness tests have particular relevance to the high-setback–pressure environment of artillery propellants and explosives, the phenomenon may be of importance to buildup to violent reaction and even detonation for all propellants and explosives under sufficient confinement. The new PBXs have generally behaved quite well in explosiveness tests.

3. *Thermal Initiation and Thermal Explosion*

Energetic materials are unstable and decompose ever more rapidly as their temperature is raised. Energetic materials can be characterized by an "ignition temperature"; however, the temperature at which ignition occurs more accurately depends on the size and shape of the energetic material and the rate at which it is heated. The current theoretical basis for analyzing slowly heated bulk quantities of energetic materials goes back over 60 years. Typical investigations involve studying the processes leading to ignition or "thermal explosion" of energetic material submerged in uniformly heated environments over a range of temperatures.[82] Under these conditions simple analytical relationships can be applied.[83]

With the more complex heating-rate relationships used in IM fast cookoff and slow cookoff testing, more complex analyses are required. These analyses are now possible, even with full three-dimensional geometry, using commercially available finite difference or finite element computer programs. For symmetrical geometries, spreadsheet calculations using linearized heat-transfer equations have been equally accurate.[46,214] Simple analytical expressions that have proved useful for cookoff calculations are given later in Eqs. (62) and (67). Although calculations have been quite successful in predicting time to a cookoff reaction at different temperatures and at different heating rates, they are not able to predict the violence of the resulting reaction.

The term thermal explosion is used to describe the process that occurs when the rate of heat evolution in an energetic material exceeds the rate of heat loss from its outer surface and a combustion reaction inevitably ensues. The term thermal explosion does not refer to the violence of the ensuing reaction, which may range anywhere from mild burning to a full detonation. The process by which the ignition grows to a violent reaction consuming the entire bulk of the energetic material is not treated in current analyses, although the methods used to analyze DDT reactions may be applicable.[73,81,84,85]

Detonation in some energetic materials can be initiated simply by heating them. This is typical of primary explosives, but it has also been noted to occur with some booster and main-charge explosives and with some rocket propellants. Often confinement of the energetic material is required, and some AP containing composite propellants with very large critical diameters (at normal temperatures) have reacted with vigor approaching a detonation (if not actually detonating) when slowly heated (slow cookoff test) while confined in their motor cases. Warhead and bomb booster explosives confined in typical metal booster housings have detonated, with subsequent initiation of the large main-explosive charge, when the munition is heated either rapidly in a fire or slowly. There is evidence that a DDT process is occurring in some confined energetic materials and that decomposition of the explosive or propellant prior to ignition forms gas filled voids, which in some propellants can double the apparent propellant volume. For a fully confined explosive charge, void growth will be limited by the free volume in the warhead case. Several composite (AP) propellants, normally having large critical diameters, detonate unconfined, even with sample dimensions of the order of centimeters, when slowly heated, indicating that the critical diameter of the heated material is much changed from that of the "class 1.3" cast propellant. Ingredient incompatibilities may play a major role in the thermally initiated detonations of some energetic materials. Standard test methods for assessing thermal behavior of propellants and explosives are used.[20,21,86,87]

To further complicate analysis of thermal ignition and subsequent reaction behavior, it is important to be aware that propellants and explosives soften when heated. Any complete analysis of the reaction would have to take account of the reduced strength of the energetic material matrix and its potential effect on subsequent combustion and mechanical behavior.

4. *Sympathetic Detonation*

When the detonation of one round of a munition initiates the detonation of one or more other rounds, a sympathetic detonation is said to occur. The initiating round is called the donor. The other rounds are called acceptors. When the time difference between detonation of the donor and the acceptors is so short that the result appears to be a single explosion both on high-speed camera and blast overpressure records, a mass detonation, mass explosion, or mass propagation has occurred. This is the most feared result of inadvertent initiation of stored munitions.

Distance between rounds can prevent propagation, or at least reduce its probability. For some small rounds, propagation will occur when the rounds are in close proximity to one another but not when they are either in direct contact with one another or when separated by substantial distances. Class A or class 1.1 (mass detonable) rounds are often stored close together but with large distances separating the individual storage sites. The minimum allowable distance is determined on the basis of quantity–distance (Q–D) criteria[90–92] to prevent propagation between sites.

The propagation of a detonation between munitions can be prevented by appropriate shielding. Sometimes a change in munition case material will prevent sympathetic detonation. For other munitions, changes in case design or installation of physical barriers may suffice.[93] As a simple example, for one type of tactical missile stored several within each container, storage of the fins between warheads was determined to provide sufficient shielding to prevent sympathetic detonation. Sometimes, head-to-tail arrangements of stored missiles that interpose rocket motors between warheads prevent sympathetic detonation. Some materials [synthetic scoria (an artificial material similar to volcanic rock) is one] are very effective in preventing sympathetic detonation caused by donor fragments. In addition to impact initiation, some people believe that the air-shock wave due to the donor detonation may cause sympathetic detonation. However, since the shock wave pressure drops very rapidly in air and expanding donor-case and fragment velocity is maintained for fairly long distances, fragment impact or case-on-case impact leading to a prompt detonation response (SDT) is a more likely mechanism.

For munitions stored in contact with each other, the energy from the detonation shock wave in the donor propagating through the cases may initiate acceptor munitions. Frey et al.[94] point out that propagation mechanisms for munitions within confining compartments can be considerably different than for unbuffered tests performed out in the open, and that the relative susceptibility of two explosives to sympathetic detonation may even reverse, depending on the conditions of the test.

Instances of mass propagation in bomb storage stacks have been reported (from tests) in which the propagation is between rounds stored diagonally and not between those stored adjacent to one another. Yet, for these rounds, single donor/acceptor sympathetic detonation tests demonstrate no propagation between rounds separated by either the nearest neighbor or diagonal distances. Workers in the field refer to "focusing" of donor energy in mass-propagating munition stacks and the effect of "confinement" by the stack. Recently, two-dimensional Eulerian-hydrocode calculations have successfully modeled the diagonal propagation effect seen in stack tests.[95,201] Simple calculations, based on a flyer-plate analogy, which have been shown to predict similar trends for diagonal propagation,[46] are presented later in this chapter. Recent field tests have substantiated the flyer-plate analogy, demonstrating that flyer plates impacting at the predicted velocity reproduce the diagonal propagation effect.[201]

Predictions of sympathetic detonation are usually based on an SDT initiation model that compares measured initiation pressure from a gap test for the acceptor explosive with the predicted shock pressure wave developed by the donor case, or fragment impact (based on calculated impactor dimensions and velocities, and donor case, acceptor case, explosive nonreactive Hugoniots, and shock wave pressure attenuation by the acceptor case). Because of other propagation effects that can occur, the assumptions of this model are still inadequate to use in lieu of testing; but of one thing you can be fairly certain: if this model predicts sympathetic detonation will occur, it most likely will.

Sympathetic detonation may also be propagated by XDT and DDT mechanisms. When this occurs there is some chance the result may be chain propagation rather than mass propagation because of time delays in each donor–acceptor propagation. The incidence of these other mechanisms is reduced by the use of PBXs and insensitive high explosives [(IHEs) also known as extremely insensitive detonating substances (EIDS)] recently developed.

Lesser sympathetic reactions may occur. For example, detonation of a missile warhead may ignite the attached rocket motor. The ensuing motor reaction may range in violence from burning to detonation. Even with a lesser response than detonation, some rocket motors have been observed to make significant contributions to the total measured blast overpressure from the reaction (for example, 20–50% TNT equivalency[96]).

Donor reactions less prompt than detonations can also cause propagation of reactions. An explosion (confined deflagration that bursts the case) can accelerate case debris to substantial velocities. A large, moderate-velocity fragment may initiate an acceptor detonation by SDT or a delayed detonation mechanism. Small fragments may penetrate the case and initiate XDT or DDT reactions. It is also possible to propagate reactions of lesser levels, for example, explosions, deflagrations or fires. Although such reactions are not as immediately devastating as detonations, the end result in a confined space, such as ship magazine, may be totally destructive. For these lesser reactions, the storage configuration, barriers, and containers, and other protective measures, such as sprinkler or water-deluge systems, may effectively reduce some slower propagating hazards. It is important to be aware that in confined spaces, any rapid combustion reaction can contribute to the blast overpressure in the enclosure, with overpressure behavior and effects similar to those from a confined detonation.[205]

B. Deflagration of Explosives and Propellants

In the technical sense described earlier, a deflagration is distinguished from a detonation by how the combustion gases move immediately following reaction. In the sense the term is used in the evaluation of insensitive munitions,[1] a deflagration is a combustion reaction sufficiently confined that munition case materials are thrown substantial distances. Such a deflagration is distinguished from an explosion in that in the latter, a significant air-blast overpressure may be measured. Thus, whereas detonation results in an explosion, a confined deflagration (in the technical sense) also may.

The distinction to bear in mind is that a supersonic combustion propagation velocity (detonation velocity) is unique to a detonation. A subsonic to sonic propagation velocity may occur for a deflagration. Thus, propagation modes for deflagration and detonation are fundamentally different. Deflagration is the design mode of combustion in a rocket motor with typical propagation velocities (burning rates) of the order of only centimeters per second. Lest this thought bring a sense of false security, note that deflagration is also the design mode for reaction of gun propellants.

A munition will burst when the internal pressure exceeds the strength of the munition case. If the rate of pressure increase is sufficiently high, the pressure may grow to many times the case strength before the case "senses" the pressure and bursts. This, in principle, is what happens during a detonation. It has also been observed to occur when the energetic material reaction is a very rapidly propagating deflagration. This has particularly been observed in some slow cookoff tests in which the case suffered brittle fracture, instead of the plastic fracture characteristic of a detonation, but other evidence, including metal fragment size, air-blast overpressure, and site damage, was much like that expected of a detonation. This type behavior is often observed in explosiveness tests.[88,89]

The point of this discussion is that although many tactical rocket motors are not capable of sustaining a detonation, pressure growth and subsequent explosion due to a deflagration reaction can cause extreme damage. Composite propellants containing ammonium perchlorate are very easy to ignite and very difficult, if not impossible, to extinguish. Extremely rapid pressure rise can occur in these propellants in the vicinity of projectile impact sites where the propellant has been damaged. In detonable propellants (for example, double-base or nitramine-filled-composite propellants) the deflagration can grow to a detonation by the XDT or DDT mechanisms already described. Therefore, the deflagration reaction of rocket motors is a serious concern for system designers tasked to develop insensitive munitions.

1. Initiation of Combustion, the Deflagration Hazard

The purpose of the igniter in a solid rocket motor is to bring the motor to design burning conditions as rapidly as possible without damaging the propellant grain. To do this, the igniter must create hot gases or particles that ignite the propellant grain uniformly over the surface that is designed to burn. If ignition occurs over only a small portion of the grain surface, propellant will be wasted burning in a nonpropulsive or less than optimum accelerative mode. If the pressure generated by the igniter is too great, grain cracking may occur that, by creating excessive burning surface, could cause the motor to burn at higher than design pressure and possibly to explode. Some propellants are more easily ignited than others and thus require smaller igniters. The igniter must be designed for the particular propellant and motor with which it will be used, with consideration of the complete thermal operating environment.

From the standpoint of IM, the hazard threat to a rocket motor can be considered to be an igniter. This igniter may be a bullet, fragment, or shaped-charge jet penetrating the motor; it may be an external fire; it may be a slowly heating environment, perhaps the neighborhood of a fire; it may be a devastating impact, perhaps due to a transportation crash, massive flying debris, or even the munition's own motion. The type of threat and the location of its effect in the rocket motor cannot be predicted. A threat that will act as an ignition stimulus for one motor may be insufficient to ignite another.

Generally, when projectiles (bullets or fragments) impact a solid rocket motor at high velocities (greater than the ballistic limit, i.e., penetration velocity of the case), they cause significant mechanical damage to the propellant. This is particularly true when the motor contains free volume, for example, the grain bore. The damaged propellant will typically be broken up to such a degree that even though only a small volume of the total propellant may be affected, a very substantial increase in exposed propellant surface may occur. If heat generated in the penetration process (probably by shear) ignites the propellant, a very rapid pressure rise will occur. This pressure rise sometimes occurs nearly as rapidly as passage of the projectile through the case. If the projectile perforates both sides of the case, it is possible that confinement will be reduced sufficiently to prevent a subsequent explosive release of pressure. It is possible, but by no means certain, since a substantial area of the case must be removed mechanically to reduce the internal pressure. However, when this occurs, some propellants, particularly minimum-smoke formulations, may fail to ignite or may extinguish rather than sustain burning. When propellant is ignited, there is always the potential hazard of a propulsive reaction that may thrust the munition significant distances and, thus, transport additional hazard to other areas.

Shock ignition. High-pressure shock waves of the magnitude associated with SDT in explosives will ignite rocket motors. These shock levels are generated by high-velocity fragment impacts. The ignition phenomenon is particularly interesting at very high-shock pressures. For example, an 8300-fps steel fragment impacting steel generates a shock pressure of about 650 kbar (with a primary shock duration of the order of twice the smallest fragment dimension divided by the velocity of sound in the fragment material) which will cause a pressure of about 160 kbar in a typical propellant either in contact with a steel case or directly impacted by a steel fragment. This pressure is near the level of detonation pressures in high explosives. Following impacts of this magnitude, even propellants that will not support a detonation in tactical size motors may behave locally as if a detonation had occurred and is dying out. The larger the impacting fragment, the larger the volume of propellant consumed in this explosive manner. Under impact conditions of this type, a rocket motor containing a nondetonable propellant will be ignited very rapidly and, depending on the case material and degree of confinement, give an initially explosive reaction that, combined with the impact momentum, throws case pieces and large amounts of propellant hundreds of feet. Depending on the composition of the propellant, the thrown pieces may continue to burn for some time.

Low-pressure shock waves with insufficient energy to initiate a prompt detonation[74] can initiate burning in explosives and propellants.[97] The energy fluence criterion for this phenomenon is similar in form to the critical energy criterion for SDT but involves constants of different magnitudes ($E = BP^n t$). For example, t for burning is on the order of 5–40 μs, whereas for SDT it is less than 1 μs; P is on the order of 1–10 kbar, whereas for SDT it is generally greater than 30 kbar for relatively small fluence areas.

Low-pressure shock ignition is not a likely occurrence in scenarios involving projectile impacts on tactical rocket motors because in these scenarios penetration of the case and subsequent ignition by shear heating and heat transfer from the projectile will occur, however, as noted earlier, instances of impact ignition with projectile velocities insufficient to pierce the case have been observed for some cased gun propellants. Translational impact hazards may also cause ignition by low-pressure shock waves acting over large areas of the munition case.

Impact ignition following penetration. When the shock wave generated in a propellant by an impacting projectile is insufficient to cause rapid ignition (i.e., SDT), the subsequent penetration of the case by the projectile is often followed by ignition. The mechanisms that cause ignition in this situation are complex. First of all, the projectile and any case debris are heated by the impact upon the case.[98] These materials then penetrate the propellant, continuing to be heated by friction as they decelerate, transferring heat into the propellant and generating additional heat as the propellant is deformed in shear and broken up. Most propellants are also cracked under these stresses.[99] Several methods that offer the promise of being adopted as standard tests have been used to study these phenomena.[100–105] The common link between these studies is that they all cause shear in impacted small propellant samples and they are all studied in the range from marginal ignition to complete sample combustion. These tests, although done quite differently, all seem to rank propellant ignitability and "deflagrability" in the same order and that appears to be the order of reaction violence in many projectile impact tests.[7,38–40] Therefore, these tests seem to be good predictors, based on current evidence, of rocket motor behavior in projectile impact tests. Both these tests and projectile impact tests show that propellant high-rate mechanical properties are important to deflagrability. Projectile impact test results show that propellant deflagrability increases drastically at temperatures below the dynamic glass-transition temperature (that is, the glass-transition temperature T_g adjusted for high strain rate, i.e., projectile velocity induced deformation and shear).[39,40]

The term deflagrability was coined by the IMAD program to represent a measurable characteristic of propellants that correlates with nondetonating violence in projectile impact tests on rocket motors. There is some evidence that laser ignition of propellants[106,206] leads to rankings for ignitability similar to those of the impact tests, but there have been no studies of subsequent reaction intensity of the laser ignited propellants and, of course, the laser ignited propellants will lack the mechanical damage experienced by impacted propellants, so mechanical property effects will not be accounted for.

Thermal ignition. As described earlier, the initial reaction resulting from a thermal stimulus (heating) to a propellant is ignition. The violence of the response of a thermally ignited rocket motor will depend on the nature, condition, and confinement of the propellant at the time of ignition. For example, in slow cookoff tests of some AP/HTPB rocket motors, substantial quantities of propellant have exuded from the nozzle prior to ignition. Ignition occurred in the exuded propellant and slowly burned toward the nozzle. Shortly after burning passed through the nozzle throat into the motor case, the motor exploded; all propellant was consumed in less than 15 ms (which was the measurement resolution); a strong air-blast overpressure wave was generated, and no fragments of the thin steel case were recoverable. The test site exhibited damage typical of a detonation or "partial detonation." Identifiable fragments of motor-case end closures were recovered many hundreds of feet from the cookoff site.[87]

A small-scale heating test was devised at NWC to measure the mechanical property changes (including volumetric expansion), temperature distribution, and conditions at ignition and reaction violence of small unconfined propellant samples.[87] The slow cookoff visualization (SCV) test (also known as the toaster oven test because a standard kitchen toaster oven was used as the original heat source) has provided major insights into the differences that may occur between slowly heated propellants with even only trace differences in composition. The test has also shown that although most propellants react relatively mildly to heating when unconfined, some will detonate. Correlation of SCV test results with full-scale motor slow cookoff test results and with other experiments indicates the importance of confinement to the reaction violence of a rocket motor in most cases. It must be noted that the applicable concept of confinement is dynamic rather than static because pressure rise in the rocket motor may be so rapid that inertial effects dominate those of case strength.

Under fast heating conditions typical of a rocket motor in a fire, or a fast cookoff test, ignition typically occurs at the outer surface of the propellant.[107] However, there have been internal ignitions of rocket motors in fast cookoff tests. This can happen when hot decomposition gases from liner or insulation materials accumulate in the bore, or when there are direct heat-conduction paths from the hot outer case to internal propellant surfaces, as might occur with inadequate bulkhead designs or insulation. Fast cookoff responses of rocket motors have been successfully modeled with finite difference heat conduction computer codes and spreadsheet models[46] that include global kinetic models for propellant decomposition and ignition (thermal explosion). When rocket motors are thermally shielded by internal or external insulation or by containment in launchers or shipping containers, internal ignition may be more likely to occur because of the longer time to reaction and the possibility of ignition through alternate heat paths.

In a strict sense, all munition reactions are the result of thermal ignition. Figure 4 shows in simple schematic form, the relationship between the various types of stimuli and munition reactions. For prompt detonation reactions (SDT) initiated by impulsive loading and for some delayed detonations (XDT), this heating is extremely local and occurs in a time so short that it is ignored in Fig. 4, although it is considered in first-principle modeling of the phenomena, usually in terms of shock-heated or shear-heated hot spots in the material matrix. For other stimuli and reactions, the heated volumes are usually larger and heating times are longer, from milliseconds to hours, and the major difference is determined by the size of the heated region. For penetrating projectiles, the initially heated region is relatively small, encompassing either the shear layer surrounding a hole in the energetic material or some volume of broken up material. For munitions directly exposed to fire, the heated region is usually just a thin layer at the outer surface. For slowly heated munitions, the entire munition may be heated to a nearly uniform temperature before runaway self-heating begins in well-insulated regions. Ignoring SDT reactions, the differences in material breakup and/or mechanical

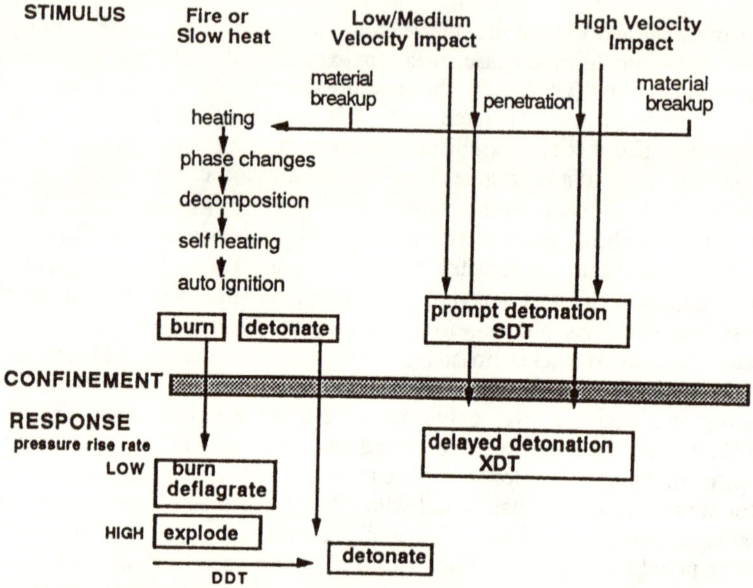

Fig. 4 Simple schematic chart of munition responses to threat stimuli.

properties and the amount of material heated prior to ignition account for the different responses of a munition to different stimuli. The munition case may protect enclosed energetic material from SDT, and it may prevent penetration by some projectiles; however, if it does not, the effect of case confinement on a reacting energetic material is generally to cause pressure to rise prior to case burst. The nature of the global energetic material following ignition and the pressure rise that can be contained by immediate confinement are generally responsible for the rate of pressure rise.[214] Hence, the pressure that will be reached before case burst depends on case strength, munition dimensions, and this pressure rise rate. Heavily confined energetic materials, finely divided (perhaps by material breakup) can escalate to an explosion or even transition from a burning reaction to a detonation (DDT). This property, routinely tested in heavily confining sealed vessels by a number of countries, is known as "explosiveness." Similar behavior ($dp/dt > 5$ GPa/ms) has also been observed in slowly-heated rocket motors in spite of the presence of a venting nozzle and case burst strength of only 0.02 GPa (3000 psi).

V. Design Guidelines

A. Rocket Motors

Lessons learned from large-scale IM testing of rocket motors (i.e., the tests in Table 1) indicate that some propellant failings can be compensated for by proper integration of all subcomponents of the motor (i.e., propellant grain, case, liner, insulation, case-venting mitigation devices, and even external attachments). To do this, one must understand the part the different components play in hazardous reactions. Typically, since funding for such important research is inadequate in the U.S., progress is slower than might otherwise be the case. The critical component, of course, is the propellant grain since that is what reacts and causes the hazard. It is important that screening tests be used on new propellant formulations early in development to optimize the selection of scaleup candidates for both performance and insensitivity characteristics. Early predictions of some IM test results may be made using simple techniques[46,108]; however, prior to incorporation into a developmental system motor, a propellant should have been tested according to the guidelines given in this chapter and elsewhere.[1,5,15,16,20–23,26,44,109] The U.S. Army recently identified IM issues and approaches for tactical rocket motors as shown in Table 3.[25]

1. Rocket Propellants

Since the beginning of IM technology programs, propellant formulation information has been restricted. Some relief was provided by papers presented at the 1994 Insensitive Munitions Technology Symposium sponsored by the ADPA.[216–227] From data released at that symposium it is clear that the commonly used AP/HTPB reduced-smoke and metallized propellants in rocket motors respond to many of the IM test stimuli with unacceptably violent reactions. Although these propellants are incapable of detonating in tactical-size munitions at normal storage or use temperatures, there is evidence that a transition to detonation may occur for some formulations of the propellants in slow cookoff tests without a vent in the case of sufficient area and proper location. Rocket motor cases incorporating polyolefin fibers have successfully reduced the reactions of some of these propellants to levels that pass all the IM tests in Table 1.[226]

Polyurethane-binder propellants containing AP and copper chromite burning-rate catalysts have been observed to detonate in slow cookoff and intermediate (higher rate) cookoff tests even with no confinement of the propellant.

There has also been concern about the detonability of minimum-smoke propellants, which are often as shock sensitive as PBXs and have exhibited undesirably violent slow cookoff reactions, and XDT responses in some bullet impact tests.

A number of apparently promising propellant candidates are being developed. Because motors with these propellants have generally been subjected to IM tests using graphite/epoxy cases, it is not possible to compare all of the results directly with results obtained on conventional propellants using steel or aluminum cases. Some of these candidates, recently reported, are given as follows.

a) The class 1.3 minimum smoke propellant based on an ammonium nitrate (AN) oxidizer in a ballistically modified smokeless casting powder, NOLLSGT sensitivity < 0.60 in., was IM tested in a 7-in. diam graphite/epoxy case and passed all tests with burning reaction or less; $I_{sp1000} \sim 235$ lb-s/lb (Ref. 220).

b) The class 1.3 minimum smoke propellant based on a crosslinked nitrocellulose binder system plasticized with nitrate esters in a castable doublebase (CDB) propellant, NOLLSGT sensitivity 0.45–0.50 in., was IM tested in small graphite/epoxy motors and passed slow cookoff, bullet impact, and fragment impact; $I_{sp1000} > 240$ lb-s/lb (Ref. 221).

c) The reduced smoke propellant based on replacing HTPB with hydroxy-terminated polyether (HTPE) binder, when 6-in.-diam card-gap tested for shock sensitivity, gave zero cards. In 10-in.-diam graphite/epoxy motor cases: fast cookoff, slow cookoff, bullet impact, and fragment impact tests were passed. I_{sp} is comparable to HTPB propellant.[217]

d) The minimum smoke propellants based on hydroxylammonium nitrate (HAN) replacing AP.[227]

e) Nitrate ester polyether (NEPE) propellants show promisingly mild slow cookoff responses, although some versions of this propellant type have given explosive responses.[129]

Table 3 Summary of tactical missile IM rocket motor issues and approaches

Issue	Approaches
No IM minimum signature propellants have high performance of current class 1.1 propellants.	Elastomer modified cast double base Ammonium nitrate propellants Non-energetic binders Energetic binders Castable double base Unfilled Filled Homogeneous propellant
High-performance composite smoky and reduced smoke propellants are deflagration hazards.	Extinguishable propellants Tough propellants Endothermic binders Motor cases that fail under thermal and impact stimuli to lessen reaction violence Passive mitigation devices to open motor case
Develop technology and demonstrate IM advantages of new case materials and fabrication techniques. Reduce costs for composite cases and other new case concepts and materials.	Graphite filament wound composite case Steel strip laminate case Braided case (Kevlar™ or graphite) Roll bonded case Hybrid (overwrapped/slotted metal) case Other (e.g., Polyolefin) composite case
IM testing is expensive and consumes valuable resources.	Develop and show small-scale testing that can predict full-scale item response Develop predictive models for response and behavior of proposed IM components and systems Build database of propellant/case material interaction using reduced scale test items
New concepts and approaches for future IM missile systems and smart missile system propulsion.	Gel propulsion Air-breather propulsion Develop new ingredients (oxidizers, energetic monomers/polymers, ballistic modifiers and stabilizers) Improve predictive modeling of systems response to IM stimuli Develop passive and active case opening devices Refine scaling factors from subscale testing to full-scale design
Igniters and igniter materials adequate for new insensitive propellants but that do not increase IM hazards.	High-temperature cookoff response Low-impact sensitivity Moderate-pressure rise rate on ignition
Shipping containers, launchers, and barriers that improve munition safety and survivability.	Nonflammable and thermal barrier Provide shielding to prevent/reduce round to round propagation of reaction Reduce impact hazard of handling, bullets/fragments Restrain debris and propulsion

f) Efforts are also underway to formulate propellants using energetic binders (such as GAP, polyglycidal nitrate, and others), energetic plasticisers, and AN as a solid oxidizer. Sensitivity, IM test, and performance data on these propellants have not been published.[222]

Many other propellants are under development and characterization to meet the IM requirements, although, as this is written, none has been introduced into an operational system.

2. *Rocket Motor Igniters*

If a hazardous external stimulus initiates reaction of a rocket motor igniter, the igniter will generally do what it is designed to do, that is, ignite the propellant grain. Igniter materials generally cook off at higher temperatures than propellants so they do not normally pose a thermal hazard. Impact ignition of an igniter causes an effect not much different than that due to impact ignition of propellant broken up (damaged) by a projectile. Since the igniter is typically at one end of a rocket motor, there will usually be a large unbroken portion of propellant grain to sustain burning caused by the igniter. Other factors involved will be propellant breakup in the vicinity of penetration and case damage. Ignition of the igniter does not increase the inadvertent detonation hazard of a rocket motor. Some increase in the likelihood of the rocket motor to react propulsively (i.e., fly about) results from projectile impacts on the igniter because of the significant length of unbroken grain remaining but, otherwise, the results are generally not unlike those of impacts elsewhere on the motor. A major exception to this may occur when impact and subsequent ignition occur for the igniter in a rocket motor with a tough propellant that resists impact damage or a propellant with a high-pressure deflagration limit (P_{dl}). For these situations the improvements designed into the propellant may not be as effective as they are for projectile impacts that miss the igniter. Another situation to be wary of is ignitor ignition within a propellant already softened by a slow-heating stimulus; this may severely damage the grain and expose and ignite excessive propellant surface, resulting in an explosive reaction.

3. *Guidelines*

Although rocket motor designers will make every attempt to select propellants that have the lowest sensitivity and reaction violence in both laboratory and motor-scale tests, there are still likely to be times when a propellant that is marginal or worse in the tests will be the only candidate to meet operational requirements. For such situations other motor subcomponents must be adjusted as much as possible to bring the entire weapon to IM requirements. NIMIC has recently published a guide to specific methods for achieving insensitive munitions that is available to NIMIC member nations.[230] The following guidelines should be helpful.

Detonability. A rocket motor may be considered nondetonable if any of the following conditions exist. 1) The propellant cannot detonate in the motor for reasons related to critical diameter; that is, the motor is not large enough to support detonation of that propellant. 2) The minimum initiation pressure for detonation combined with the area over which it must act exceeds any likely (or possible) stimulus to the motor. This must include XDT and DDT reactions. To assure passing of IM tests, this requirement is less stringent than it is to assure passing of all "possible" threats. 3) The run distance from the initiation site to fully developed detonation exceeds propellant grain dimensions.

There are complicating circumstances; for example, detonation is initiated more easily in damaged propellant.[81] The damage may occur before or as part of the initiating stimulus. Even initiation of damaged propellant will not cause a motor to detonate completely unless the remaining undamaged propellant can propagate the detonation, although the reaction may be completed as a low-velocity detonation or a very violent explosion.

Even if a propellant is found to be "detonable," the following system design modifications may reduce the operational hazards. 1) Design the warhead/motor interface to prevent propagation of warhead or motor detonation to the other component. 2) Design the motor case (including insulation and liner) to attenuate impact shocks. With tactical weight and volume constraints, this is easier said than done. For a propellant that is marginally initiable with IM test or threat fragments, a spherical motor will have lower probability of detonation than a cylindrical motor in any threat situation [including the IM fragment impact and sympathetic detonation (SD) tests] because of shape factors that reduce the vulnerable area. This principle also applies to warhead design and, in addition, a spherical warhead generates a larger, less dense fragment pattern with a higher kill probability in most encounters. 3) Design the motor case to be a "soft" fragment to reduce the SD hazard. 4) Design for rapid case venting upon high-velocity impact if a DDT hazard results. 5) Design the missile container to retard mass propagation (also retards impacting threat fragments).

Deflagrability. Even a propellant that passes the test protocol for deflagrability given later in this chapter may cause a violent reaction (particularly a propulsion reaction) if the case retains its integrity for any length of time following projectile impact. To prevent unacceptable levels of deflagration, propulsive, or explosive response, the following recommendations apply. 1) Design propellants to maximize the time between projectile impact and ignition of the propellant. 2) Minimize the pressure rise rate of impacted propellant (implies increased propellant toughness and minimum new surface created by impact, including a low glass-transition temperature, and consider using a bore mitigant such as a polyurethane foam[216]). 3) Develop propellants with high-pressure deflagration limit (P_{dl}) in air to quench or reduce ambient and low pressure combustion violence. 4) Maximize the extent of case failure caused by projectile impact to promote venting of propellant reaction products and prevent pressure buildup.

Cookoff. The following recommendations apply to design of rocket motors to avoid unacceptably violent cookoff reactions.

1) For fast cookoff, the case must fail before the propellant ignites (for aircraft carrier deck fuel fire stimulus) to prevent explosion, hazardous fragments, or a propulsive reaction. (For fast cookoff reactions in more confined areas, dispersal of the propellant may actually be technically preferable to a violent burning reaction, although this has never been demonstrated or stated officially. Small pieces of propellant burn up more quickly and are less hazardous as heat sources of ignition for other munitions.) A propellant that burns mildly at ambient or low pressures coupled with mild case failure (currently typified by composite or strip laminate cases) would be ideal. Rocket motors with active case mitigation devices to vent them may cause a "blow torch" effect that could readily propagate to contiguous munitions in storage environments, if these mitigation devices are activated.

Vetter[107] proposed using pyrolizable outgassing liners to build up decomposition gas pressure, collapse the grain (for star perforated propellant grains), and expose the case for a short time to direct flame heating with no radial backwall heat conduction path. Vetter demonstrated the feasibility of this approach with stress analyses. This model was further presumed to be correct on the basis of previous and subsequent test results and was subsequently proved by IMAD tests that used real time x-ray (side view) and video cameras (longitudinal view through the nozzle).[130] Pakulak[184] has reported the use of outgassing liners in warheads with violence reducing effects when combined with stress risers or other mechanical case venting mechanisms. Although these approaches do not yield easily to simple analyses, it is possible to calculate generated gas pressure at temperature from the liner material and apply the calculated pressure to heated case burst or explosive ejection calculations.

2) If unconfined propellant detonates in the SCV test,[87,105] it will not pass the large-scale slow cookoff test requirement in an actual motor unless it is ignited prior to autoignition. One way to prevent an extremely hazardous reaction with such a propellant over a large range of heating rates, if it must be used to meet rocket motor performance requirements, is to purposefully vent the case and ignite the propellant before heating brings it to the explosive or detonable state. This has been accomplished with active case-venting mitigation devices. Sometimes a similar effect, without prior venting, has occurred fortuitously in boost–sustain rocket motors in which only one of the two propellants is SCV detonable, but the other propellant autoignites earlier in slow cookoff tests leading to burning of all of the propellant in the rocket motor. Active venting of the motor case without igniting the propellant is difficult to achieve reliably, and when accomplished it will not prevent detonation of a propellant that detonates in the SCV test, although it has been demonstrated to greatly reduce reaction violence of rocket motors that, because of propellant foaming, appear to detonate or have low-velocity detonation (LVD) reactions in unvented metal cases. The general onus against active mitigation by service evaluators of munition design must not be forgotten (nor should the fact that there is less time before a reaction occurs) and should be explored with program management and appropriate safety offices for a specific system.

3) If "hard" case confinement leads to a detonation/explosion reaction in a slow cookoff test, an appropriate composite case or active case venting must be considered to permit case yielding or opening to prevent pressure buildup and ignition wave propagation. However, extremely violent slow cookoff reaction may only occur at very low-heating rates and for relatively small motor diameters. In that case the specific propellant may not constitute a real hazard in its full-size operational motor. If propellant thermal properties and decomposition kinetics are known, reasonably accurate predictions[46] of these phenomena can be made for a wide range of heating rates[57,197] related to hazard scenarios, including the standard 3.3°C/h heating rate of the IM slow cookoff test. If program management concurs, one might be able to take advantage of a higher slow-cookoff-test heating rate, based upon a THA.[16] Adiabatic (constant soak temperature) heating tests are more difficult to model because of the stiffness of the computer models and, thus, provide no advantage in energetic material assessment except for initial determinations of thermal parameters. In addition, the reactions that occur in energetic materials at relatively low-soak temperatures may differ substantially from those in most threat scenarios. Such low-soak temperatures are more applicable to long-term high-temperature storage of large, relatively unstable munitions.

The situation of thermal hazards in storage configurations is complex. A munition is a potential fire source in confined areas, particularly magazines. To avoid a propagation hazard, the exposure of munitions to potential fire sources (including each other) should be minimized. This is particularly true for rocket motors, some of which burn at extremely high temperatures (~3000°C) even when the propellant is unconfined and in the absence of atmospheric oxygen. Thermal shielding (insulation) will retard propellant heating by fire, however, one must be careful not to set up a situation which converts a fast or intermediate cookoff response into a far more hazardous slow cookoff type of response by overuse of insulation that can also prevent natural failure mechanisms of metal cases from operating.

4. *Mitigation Systems*

It should be clear from the earlier parts of this chapter that the reaction violence of munitions subjected to inadvertent impact or thermal stimuli, particularly the latter, can be reduced or "mitigated" by reducing confinement prior to explosive reaction. Devices designed to do this by venting the rocket motor or warhead case are commonly referred to as mitigation devices or mitigation systems.

An "active" mitigation system will contain a sensor to detect the hazardous stimulus, an initiator to start the chain of events leading to case venting, and a case opening device. Safety regulations often require that a safe/arm or out-of-line device be included in the initiation chain to prevent inadvertent activation of the mitigation system (for example, on the launch platform, or after launch, when the missile rocket motor is actually thrusting or in the post thrust phase when case integrity is needed for missile aerodynamics). Active mitigation systems have been designed and demonstrated to be effective in both fast and slow cookoff tests, and effectiveness (though not feasibility) has even been demonstrated against some bullet impact-induced reactions.[26,50] Both thermite and linear-shaped charge (LSC) case cutters have been used for venting in response to fast and slow cookoff tests. The faster LSC cutter is preferable because of its reproducibility and for the short heating times prior to ignition often experienced by munitions in fuel or energetic material fires. It is important that the sensor used survive the entire missile life cycle and be able to detect heating not only at the extreme rates used in fast and slow cookoff tests, but also at any intermediate rates that might occur in operational practice.

It has been demonstrated in both fast and slow cookoff tests that an energetic material that reacts to thermal stimulus before the main propellant in a rocket motor (or the explosive chain in a warhead) can be used as a case cutter (or as a "pre-ignitor").[26,49] There are many ways such case cutters can be designed into munitions, however, it is difficult to design devices based on this concept that also meet the safe/arm and out-of-line requirements. Eutectic metal mixtures for joints and seals have also demonstrated effectiveness in case venting,[34] and the use of epoxy joints has been considered as well. A recently developed intermetallic thermal cookoff sensor that meets safe/arm requirements can be tailored to initiate a case-cutting charge over a fairly wide temperature range tailored to specific threat and munition characteristics.[196]

Other mitigation concepts include motor cases which are actually not locked shut prior to arming the igniter for motor ignition. Such cases will vent readily through the liner upon propellant ignition, with very little confinement; however, any extra mechanical features will add to inert weight and cost. Because these mitigation concepts involve "natural" mitigation in that they are mitigated or vented in their benign state, they do not require separate heat sensors or safe/arm devices and may be considered "passive."[49] There is the hazard that such devices may still permit propulsive reactions to occur if not carefully designed.

Outgassing liners that decompose to "free-radical attaching" species have proved useful in reducing slow-cookoff violence of some warhead explosives. A similar approach might work for some rocket propellants.

Other passive mitigation concepts involve specific case materials and designs such as strip laminate and composite cases, which have been demonstrated to be very effective in avoiding violent rocket motor reactions in fuel fires.[26,34–37] Hybrid cases that involve a fiber-reinforced composite layer overwrapped on a thin metal cylinder case have been demonstrated in fuel-fire tests with careful pre-venting of the metal cylinder.[26,53,111,216] The overwrap provides the radial strength needed to contain motor operating pressure; the metal thickness is sized for stiffness. The metal in the hybrid case is an improvement over composite cases for attaching lugs and fins. When crack-growth criteria for the case metal material used have been applied to design of the small pre-vent openings, improved case-failure response to bullet impact has also been demonstrated. Designs for passive mitigation of responses to slow cookoff have generally not progressed very far, but some tests of full-scale rocket motors indicate that softening of some composite cases, or perhaps softening of the propellant-to-case bonding liner, can reduce the reaction violence of propellant grains subsequently ignited by thermal explosion. A major exception to this generallization involves the use of Polyolefin (Spectra®) fiber cases. Dhillon et al.[226] reported that with a composite case using alternate wraps of graphite (helical wrap) and Polyolefin (hoop wrap) fibers, AP/HTPB reduced-smoke propellant will pass the first six tests listed in Table 1 with no reaction greater than burning.[226] This is a milestone result. Major potential problems in the use of exotic new case materials and concepts, effective for mitigation, involve their ability to withstand performance requirements, handling, and long life cycles, as well as methods of mounting external attachments.

B. Packaging and Shielding

It is obvious that if a munition is isolated from its environment by enough shielding, it will never be exposed to an external threat capable of initiating a reaction. Containers used for transportation and storage of missiles or missile components are usually designed to prevent damage that may occur due to rough handling or accidents in handling. Packages providing sufficient protection to prevent inadvertent initiation of munitions in the IM tests would be very heavy, expensive, and generally impractical. But, fortunately, the packages do not have to be that good. Since most modern munitions, particularly those that are designed according to IM principles, pass many of the IM tests, or at worst, give only moderate pressure-burst deflagration reactions, the package protection may only have to provide a slight improvement in one area. For example, very reasonable package thicknesses can attenuate fragment and sympathetic detonation hazards to warheads, and possibly class 1.1 rocket motors, enough to prevent acceptor detonation.[47,117] However, the added confinement of containers or launchers can also increase the violence of deflagration reactions and contribute to propulsive reactions.

Barriers can be used in storage of packaged or unpackaged munitions to break up projectiles, reduce their velocity, and in other ways mitigate impact hazards. Barriers capable of attenuating fragment velocity and preventing the spread of fires due to burning propellant have been demonstrated too. Work related to such physical protection of munitions is progressing in many nations.

C. Propellant Selection

To determine whether or not the detonability, deflagrability, and cookoff characteristics of a developmental propellant are sufficiently benign to warrant subsequent scaleup, process development, and motor-scale testing, test procedures for laboratory-scale quantities (1 gal and less) are needed. The results of such tests should permit assessment of whether a motor containing the propellant will meet IM requirements unassisted and whether the design guidelines given earlier in this section might be used to meet IM requirements with that propellant. A recommended protocol for screening tests, to which some of the discussion refers, is given at the end of this section.

1. *Detonability*

Concerns about detonability of a rocket motor include the following:

1) Can the motor be detonated?

2) If it can be detonated, what are the minimum stimulus levels and types? Are they likely to be encountered in hazard situations, which situations?

3) How are the answers to the questions above affected by the condition of the propellant (i.e., damage, temperature, etc.)?

4) What are the hazards of mass propagation (sympathetic detonation) with this rocket motor?

5) Can the hazards of mass propagation be reduced or eliminated by motor design parameters? If so, how?

It is reasonable to deal with the first three questions by using screening tests involving relatively small amounts of propellant. Since the last two questions involve specifics of containment, case materials, and design, as well as stowage configuration, it is difficult to see how to address them with small-scale screening tests. However, if the propellant is detonable, analytical techniques are available for preliminary assessment of the sympathetic detonation hazard. Information on sympathetic detonation of tactical rocket motors is extremely limited. Hartman[118] has described SD tests using 5-in.-diam generic motor cases fabricated of both steel and graphite/epoxy composite in which a detonation propagated with a propellant (containing a low level of HMX) that had a 50% card-gap value of 65 cards in the Naval Ordnance Laboratory large-scale gap test (NOLLSGT). Propagation did not occur for another 65-card propellant that did not contain HMX; but it is not known if an actual detonation occurred in the donor for this second situation. Neither the exact propagation mechanism(s) nor the critical case and propellant parameters in these tests are known. Clearly much work remains to be done in this area.

Recently, reactions propagating through composite AP/HTPB propellants at sonic or slightly higher velocities (2–2.5 km/s) have been measured; but this reaction is not a full (or strong) detonation (velocity between 5 and 8 km/s).[199] What are the minimum stimulus levels and types for such reactions? How should such reactions be classified? (In this chapter they are referred to as low-velocity detonations.) Can these reactions propagate sympathetically, and if they can, what level of damage would result?

In the U.S., it has become common practice to distinguish between what are called class 1.1 (detonable) and class 1.3 (nondetonable) propellants by their behavior in the NOLLSGT.[21,119] The distinguishing criterion is at the 70-card configuration [0.70 in. of Plexiglas® (PMMA) attenuator between donor explosive and acceptor (propellant being tested)]. If a detonation does not occur at this condition, as evidenced by a punched hole in the steel witness plate located approximately 14 cm from the attenuator-acceptor interface, the propellant is classified as 1.3. Many arguments against this criterion have been expressed.[120] Although the test appears to be simple and one dimensional, it is neither. Involved are the complexities of the two- or three-dimensional physics of the card-gap test, the Hugoniot [or equation of state (EOS)] of the propellant, propellant heterogeneity, confinement of propellant by the test hardware and resulting reflected shock waves and rarefactions, the interactions of critical diameter, the run distance to detonation, and the brisance or detonation pressure of the reaction. Also, one must be concerned that the test sample is representative of the propellant in its normal production state (i.e., no difference in porosity, or composition, including local catalyst concentrations). Different size card-gap tests give different gap values for the same propellants, however the relationships between test diameter and gap thickness are understood.[37,89,121,122,202] Despite these qualifications, the NOLLSGT does not seem to have done a bad job as a crude discriminator for detonability assessment with respect to sympathetic detonation and fragment impact testing; although exhaustive assessments have not been made. There is some ambiguity between gap test results and rocket motor detonability to 23-mm high-explosive-incendiary (HEI) bullet impacts, for which a no. 8 blasting cap test is a better discriminator.[21,23]

Lundstrom's analytical assessment[123] of the relationship between initiating pressure and run distance for a large number of explosives and some propellants in the wedge test and correlation with shock wave levels from a number of stimuli including the NOLLSGT, fragment impact, and sympathetic detonation offers some explanation. Lundstrom defined a shock-sensitivity plane (SSP). The two coordinates of the SSP are 1) the slope of the linear portion of an explosive's POP plot[124] (plot of log initiating pressure P_i vs log run distance to detonation from wedge test) and 2) the value of the initiating shock pressure P_1 required to give a 1-cm run distance to detonation in the Los Alamos National Laboratory (LANL) wedge test for that explosive. The results of wedge tests on many explosives can be plotted on a single SSP display, as shown in Fig. 5. The results for any single ideal explosive correspond to a single point in the SSP. Lundstrom suggests that many explosive test results can be interrelated on the SSP, as shown in Fig. 5. For example, the 70-card NOLLSGT curve in the figure represents the locus of points for all explosives with a 50% point for detonation at 70 cards. The curves also represent a dividing line between explosives for which detonation will occur (below the curve but not above it) and those for which it will not. The "SD 1" Al (*barrier*) curve indicates that a 1-in.-thick aluminum barrier will prevent SD of >200 + card explosives with very small critical diameters. Comparison of the *Frag Impact bare* and *Frag Impact* (*0.1-in. case*) and *Frag Impact* (*0.5-in. case*) curves shows the protection provided by such cases in the standard IM fragment impact test. It must be recognized that the curves describe a wide range of explosives. There is a striking relationship between the shapes of NOLLSGT data curves and the curves representing calculated explosive responses to IM fragment impact (FI) and sympathetic detonation (SD) tests.[1,16] For heterogeneous explosives with nonideal behavior, the POP-plot slope may be nonconstant and prevent reasonable interpretation in the SSP.[125,126,198,199,202]

If all parameters on the SSP are correct, the point for an explosive or propellant material obtained from any two orthogonal data types would enable one to predict any other relationships shown on the plane. Any two types of variables that intersect on the SSP may be considered to have orthogonal components and, thus,

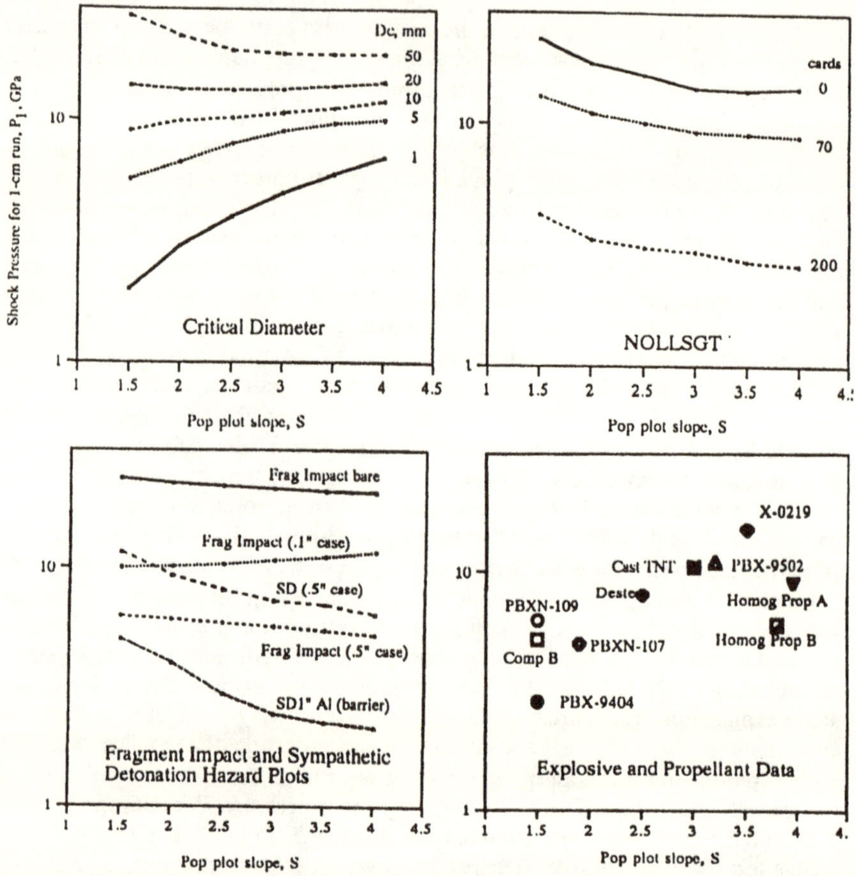

Fig. 5 Shock sensitivity plane showing comparison of calculated IM hazard test results with card gap and critical diameter results.[123]

not be directly relatable by measurement of only one of the variables. Thus, one should not expect to be able to determine card-gap sensitivity unambiguously by measuring critical diameter, although within a given propellant family, an analytical relationship between the two variables probably exists. There is one important caveat with respect to Fig. 5; the unreacted Hugoniot for all explosives plotted in the figure was assumed to be the same as that for the explosive Composition B. The uncertainty caused by this assumption is not expected to be greater than 1 GPa for any calculated value shown in the figure, as confirmed by hydrocode calculations.[151,202] These transformations between wedge test and gap/critical diameter data fail for insensitive heterogeneous energetic materials that tend to be located toward the upper right side of the SSP.[202]

For the plane shown in Fig. 5, the case-protected Frag Impact and baseline-SD curves are enclosed over a wide range of POP-plot slopes by NOLLSGT results corresponding to between about 50 to 100 cards (i.e., roughly 70 cards). The NOLLSGT at 70 cards inputs a pressure wave of approximately 5-μs duration at a pressure level of about 7 GPa (depending on the Hugoniot of the unreacted acceptor explosive) into the acceptor.

For a larger scale gap test, for example, the Naval Surface Warfare Center (NSWC) expanded large-scale gap test (ELSGT),[105,121] which is double the NOLLSGT scale, the FI and SD curves are in the vicinity of the 180-card curve for the same explosive EOS. This relationship between 70 cards in the NOLLSGT and 180 cards in the ELSGT is confirmed reasonably well by data reported by Graham et al.[37] in which an explosive with a 67/68 card 50% point in the NOLLSGT had a 50% point of 158/164 cards in the ELSGT. The Navy's IMAD propulsion program uses a modified ELSGT test of reduced acceptor length (to conserve material and reduce detonation output) that has agreed with results obtained in the unmodified tests. It must be noted, as shown clearly by Nouguez et al.,[110] that the critical initiation pressure in the ELSGT is lower than that in the NOLLSGT because of the greater shock energy fluence area and greater shock wave duration in the ELSGT. These factors are accounted for in the SSP, which is built up by Lundstrom's hydrocode results using the SMERF computer code with a Forest Fire burn model for initiation and growth of the detonation.

Clark[122] and Keefe[127] indicate that a gap test can be used to detect an XDT. Bernecker's ELSGT employing a dent plate as a detonation witness,[121] can be used to detect XDT provided time records of the test events are made. The plate dent evidence of a detonation can also be used to obtain some estimate of brisance and detonation velocity.[69] An XDT reaction, should one occur, is induced with a lower initiating shock pressure (thicker gap) than a prompt detonation (SDT) in the same explosive. Clark and Keefe observed XDTs induced in propellant in timed gap test measurements after passage of the initial shock wave. Presumably, the rarefactions following the initial wave cause sufficient propellant dilation and damage in tension from release of compression waves that the reflected wave can initiate a detonation. The ELSGT is large enough to give reliable results with explosives having critical diameters up to about 5 cm, which should be adequate to assess any explosive or propellant that might be initiated by even the largest and fastest threat-weapon fragments (see Fig. 1), although not necessarily large enough to assess sympathetic detonation of large munitions.

An even larger card-gap test has been developed for evaluation of insensitive high explosives, primarily those intended for use in bombs.[128] This super large-scale gap test (SLSGT) uses an acceptor explosive sample that is 20 cm in diameter and 40 cm long. In no way can this be considered a small-scale test and, in fact, it is larger than most tactical rocket motors or missile warheads.

The mandatory tests for solid rocket propellant qualification in the U.S. include detonability determinations by no. 8 blasting cap, NOLLSGT, and critical diameter measurement.[21–23] Detonation velocity and critical diameter can both be determined with sufficient accuracy in a single test using a pyramidally shaped propellant charge instrumented with electrical-resistance crush gauges to measure detonation velocity. This test, used for propellant samples of 4 in. maximum pyramid-base dimension, can also identify LVD phenomena. Initial screening of propellants should include these three tests. If detonation occurs, the propellant should either be rejected for scaleup and subsequent use or it must be further tested extensively as indicated by the protocol later in this section.

If the propellant does not detonate in the NOLLSGT or no. 8 blasting cap test, a larger card-gap test, operated at an appropriately scaled gap thickness is recommended as the pivotal test for screening the detonability of questionable propellant candidates. To overcome objections to the small size of the NOLLSGT,[119,120] a larger test, for example, the ELSGT,[121] should be used and, in fact, could be used as the initial screening gap test, if indicated by the critical diameter results. If a propellant with other admirable properties detonates in this test, additional investigation with a wedge test is required. [The United Nations test for detonability of insensitive high explosives or insensitive detonating substances, is similar in size to the ELSGT but uses a 70-mm PMMA gap (276 U.S. cards, corresponding to about 100 cards in the NOLLSGT, for explosives that can be adequately tested in the latter test).] XDT is another concern for detonable solid propellants. Even if the propellant appears nondetonable on the basis of the SDT tests already described, the "bore effect"[38] may initiate a detonation under IM fragment or bullet impact testing. Therefore, a propellant grain that could support a detonation based on the relationship between critical diameter and grain dimensions may need to be tested by the SNPE or NAWCWPNS methods (BVR test) described earlier.[38,80] The NAWCWPNS BVR test is also used for "deflagrability" screening.

2. *Deflagrability*

The term deflagrability applies to responses to bullet and fragment impact comprising ignition and combustion resulting in case-bursting deflagrations, explosions, and very high-velocity combustion (i.e., sonic velocity or, perhaps, LVD). The common factor in these responses seems to be high-pressure case burst and vigorous expulsion of its fragments (and usually reacting or nonreacting propellant debris) by the rapid expansion of combustion product gases. It is postulated that these deflagrability-related phenomena are caused by very rapid pressure rise before the mechanical action of the impacting projectile causes case breakup and relief of confinement (if, indeed, it can). Concerns about deflagrability of a rocket motor include the following questions one would like to see answered by propellant screening tests.

1) Does the propellant ignite rapidly and violently on impact?

2) Is the pressure rise rate acceptably low?

3) Is propellant breakup (free surface formation) at an acceptable level?

4) Will bullet/fragment impact test results on a full-scale motor pass the IM requirements?

5) What is the initiating stimulus (type and level)?

6) How is the resulting reaction classified as to promptness, pressure buildup rate, mechanical property effects, and relationships to case mechanical properties and confinement?

7) What is the reaction hazard as a propagation source?

8) What design parameters will reduce the reaction hazard (for example, propellant mechanical properties, ignitability, IM motor cases, etc.)?

9) Is the propellant pressure deflagration limit P_{dl} in air sufficiently high to provide low-pressure extinguishment or at least mild burning and thus prevent hazardous burning reactions?

The first six questions have been treated in the past mostly by projectile impact testing against subscale analog motors.[40] Such tests are expensive to execute and involve costly sample preparation. There is recent evidence that modified shotgun and Hopkinson bar tests developed and used in Australia[39,102,103] and a battery of tests based on drop-weight impact developed at the NSWC[100,101,105,132,229] give results that can be related to projectile impact test results. The questions regarding intrinsic propellant mechanical properties, P_{dl}, and ignitability can be answered using standard test techniques, including laser ignition.[106,206] Ignition by electron-beam heating[192] is different than laser ignition studies in that the electron beam provides volume heating instead of surface heating. When this chapter was written, the NSWC ballistic impact chamber (BIC) test was considered to be the most likely candidate for a US propellant-deflagrability-screening test and was being evaluated by IMAD propulsion project.[105] All IMAD developmental propellants were screened with that test and, for comparison, selected baseline propellants were also screened in the Australian tests.

If the propellant being screened for deflagrability has potential to detonate in a candidate rocket motor, it is desirable to screen for specific behaviors that can link deflagrability and detonability. Examples of such behaviors include XDT and DDT that may be studied in a burn-to-violent-reaction test. BVR behavior has been studied in analog motors[40] and in motor sections.[38] The planar BVR test fixture at NAWCWPNS may also be useful and has become a standard propellant assessment test in the U.S. Navy's IMAD propulsion program.[80,216] The NAWCWPNS BVR test, in use for several years to screen nondetonable propellants, has been able to distinguish between propellants that deflagrate violently and those that ignite and burn more mildly or not at all. Important information related to growth to violent reaction can also be obtained from shotgun impact tests (with associated combustion bomb studies and CBRED-II analysis[131]) and from split-Hopkinson bar tests which provide high-rate mechanical property data.

None of these methods can screen propellants for potential propulsive reactions that can result from any ignition of the propellant if case integrity is maintained.

3. *Cookoff*

Heat causes degradation, decomposition, and, ultimately, ignition and reaction of energetic materials. The phenomena involving application of heat to munitions and the ultimate result, are called cookoff. Standard IM cookoff tests on munitions are performed at high-heating rates (fast cookoff test or fuel-fire cookoff test) and at a low-heating rate (slow cookoff, 3.3°C/h), although higher (low) heating rates based on a THA of life-cycle threat stimuli are now permitted.[16] Test results indicate that a munition's response to fast cookoff is controlled largely by the design and material of the case and its attachments and the case–propellant interface. To achieve an acceptable reaction level in fast cookoff, the rocket motor case generally must fail structurally prior to propellant ignition.[26,33–37,107] There is no small-scale test that can be used to screen propellants for fast cookoff response, although Pakulak[108] reports some success with explosives.

The slow cookoff situation is more complicated. The responses of energetic materials to heating at various rates have been studied for many years, and numerous small-scale tests are available.[86] Pakulak's small-scale cookoff bomb (SCB) tests and analytical methods have been successfully used to predict large-scale responses of high-explosive warheads and bombs to various thermal environments.[108] Although these tests have not been as successful with most propellants, there is reason to believe that with proper instrumentation, heating rates, and loading methods, useful data for predicting slow-cookoff responses related to rocket motors may be obtained.

The SCV test provides some of the information necessary for cookoff screening of propellants[87] as do Butcher's unconfined and confined small-scale cookoff test methods.[129] Since the SCV test and Butcher's tests have been used on dozens of propellants, comparisons with full-scale slow cookoff test results can be made. Propellant material quantities necessary to perform these tests can be generated at the 1-pint mix levels. The unconfined tests are designed to provide the following data:

1) The bulk volume change of the propellant as a function of temperature.

2) Visible physical state changes that occur as a function of temperature. Most propellants undergo visually observable physical changes as a function of temperature. Some propellants soften, swell, and/or foam a great deal, whereas others show only small changes. Other propellants partially liquify due to binder depolymerization and/or melting of one or more ingredients. Sometimes the liquid or semiliquid phase foams and/or boils prior to autoignition. Significant color changes often occur as a function of temperature as well.

3) The radial thermal profile through the cylindrical propellant sample as a function of oven-air temperature and time right up to autoignition. Internal exothermic activities as well as endothermic decompositions and/or phase changes can be observed with thermocouple probes fixed in a three-dimensional spatial arrangement throughout the sample.

4) The location of autoignition. Autoignition can occur in the gas phase above the propellant sample, on a propellant surface exposed to air (and decomposition products), or within the volume of the propellant sample.

5) The composition and volume of gases given off by the heated propellant as a function of temperature and time can be determined in the SCV test, which has the capability for capturing effluent gas.

These data are useful for predicting whether sufficient physical property degradation and/or expansion will occur in a full-scale motor to cause propellant grain collapse and/or exudation of propellant through the nozzle. In addition, knowledge of the physical state of the propellant and the degree of propellant confinement at the time of autoignition provides important clues about how violently the propellant will react in a rocket motor exposed to slow cookoff. Data provided by the SCV test can also be used to estimate the time-to-reaction and cookoff temperature of full-scale motors through mathematical modeling. As a propellant formulation research tool, the SCV test can provide clues and insights regarding what is occurring within a propellant as a function of temperature. Knowledge of how propellant formulation variables affect propellant changes during slow cookoff can be obtained by conducting a matrix of tests in which one ingredient variation is made at a time. Most propellant companies now have tests that provide information similar to that obtained from the SCV test.[34–37, 129]

Several propellants have detonated during heating under the light (Pyrex glass) radial confinement of the SCV test. This is a clear indication that these propellants are unsatisfactory in rocket motors over a wide range of slow heating environments, and these propellants should not generally be further developed or considered for use in munitions. One exception to this rule is that such thermally detonable propellants may be used as booster propellants if fully enclosed by thermally nondetonable sustainer propellants, provided the latter reliably ignite, burn, open the motor case, and ignite the booster prior to thermal explosion of the booster propellant in all thermal scenarios.

Although the SCV test provides the visual and thermal data that it was designed to provide, it does not provide the data required to make an accurate determination of the violence to be expected from a full-scale slow cookoff test of a rocket motor (except for propellants that detonate in the SCV test, as already mentioned) because of 1) different confinement and 2) sample size affecting both the amount of propellant available for reaction growth and the autoignition temperature. The average temperature, just prior to autoignition, of the small propellant sample used in the SCV test usually does not match the lower average bulk temperature of the same propellant in a full-scale rocket motor at the time of slow cookoff (at 3.3°C/h heating rate). An HTPB/AP propellant that has expanded in volume (foamed) prior to autoignition seems to be much more sensitive to confinement than an HTPB/AP propellant that autoignites in a more consolidated state.[87] Some HTPB/AP propellants were observed to undergo expansion just prior to the autoignition temperature of the sample size used in the SCV test. The same propellant autoignited at a substantially lower bulk temperature in a full-scale motor configuration that was heated at 6°F/h. This means that, whereas expansion was observed in the SCV test, little or no bulk expansion and only localized expansion near the thermal runaway zone occurred in the full-scale motor. This illustrates that, in addition to the other guidelines just discussed, a subscale slow cookoff propellant screening test intended for use as a reaction violence predictive tool must be carefully designed so that the propellant being tested is at the relevant physical state at the time of autoignition, or erroneous interpretations of results are likely. These potential problems can be assessed with the spreadsheet method described earlier.

A variable confinement slow cookoff bomb (VCSCB) test devised at NWC to complement the SCV test and provide the additional data required for accurately predicting slow cookoff reaction violence of full-scale motors produced very important insights about the cookoff process.[26,87] In principle, the VCSCB test was an autoignition "explosiveness" test, similar to those used in the other nations for energetic materials at ambient temperature but that are held under fixed high confinement and ignited by a small propellant charge or adiabatic compressive heating.[18,88,89] To preserve reasonable similitude with tactical missile-scale heating rates and other critical dimensions, the VCSCB was about 10 cm in diameter and 23 cm long. The bomb was only half-filled with propellant to allow room for significant volumetric expansion to occur. Ultimately, NWC found the test too expensive for general propellant screening.

Butcher (Pt. II)[129] described a 2.5-in.-diam "pipe bomb" confined slow cookoff screening test that provides useful slow cookoff information, although critical case failure strength is not measured as it was in the VCSCB test. A remote pressure gauge connected to the internal volume of the bomb measures the pressure rise during reaction with a time resolution of a few microseconds. Butcher also described the burst failure and case fragmentation in terms similar to those used for evaluating explosiveness tests. Even though the remoteness of Butcher's pressure measuring gauge probably attenuated the actual rapidly increasing bomb internal pressure, he measured chamber pressures that were from 4 to 10 times greater than the bomb case-burst pressures for some AP/HTPB propellants. These peak pressures occurred very shortly after the initial pressure rise; in some cases the total rise time to 200 MPa (in a 20 or 55 MPa burst strength bomb) was no greater than 0.25 ms. For HTPB propellants, pressure rise rates of 1.8 to 5.0 GPa/ms were observed as the pressure rose past 20.7 MPa (3000 psi). Some polyether-binder propellants gave milder reactions, with the measured burst pressure not significantly exceeding the case-burst strength; pressure rise time was on the order of 5 ms, the pressure rise rate was only 0.019 GPa/ms at 20–55 MPa; from 95 to 260 times slower than with the HTPB propellants. There is obviously an analytical relationship between the pressure rise rate and the reaction violence in either Butcher's tests or full-scale motor tests.[214]

4. *Insensitive Propellant Screening Test Protocol*

The following preliminary protocol for insensitive propellant (IP) screening is suggested based on the considerations described earlier in this chapter and particularly in this section. The primary function of the insensitive propellant screening test protocol (IPSTP) is to save time and money by providing early information needed for a decision to reject a propellant from further development effort. The tests can be performed simultaneously in the three areas of detonability, deflagrability, and cookoff. The IPSTP described is shown schematically in Fig. 6.

Initial screening tests. Three tests, a gap test, the BIC test, and the SCV test, make up the recommended initial screening test protocol for insensitive propellants. If a propellant passes these tests as described next, scaleup is not contraindicated. It is not possible to state the case more positively.

For detonability, the ELSGT is recommended as the initial screening test. If the minimum initiating pressure for detonation is 70 kbar or greater (180 cards or less), the propellant meets the current requirement for propellant hazard class 1.3 and is considered suitable for scaleup and no further detonability testing is done. The 70-kbar criterion is tentative; it is based on an initiating pressure at 70 cards in the current NOLLSGT, and additional data on a number of propellants are needed to determine a final value for the criterion (see earlier discussion). The smaller NOLLSGT and the no. 8 blasting cap test, required for qualification anyway, could be used for initial propellant screening, in which case the ELSGT would be used only if the propellant passed the cap test and the NOLLSGT without detonating at 70 kbar. The NOLLSGT alone might provide adequate screening for some materials, to determine if the IM hazard tests can be passed, but it would be ambiguous for materials that pass, with respect to large-fragment threats related to threat/hazard assessment scenarios (see Fig. 1) and would not have a realistic size relationship to tactical motor grain dimensions. It should be clear from Fig. 5 that although the 70-kbar

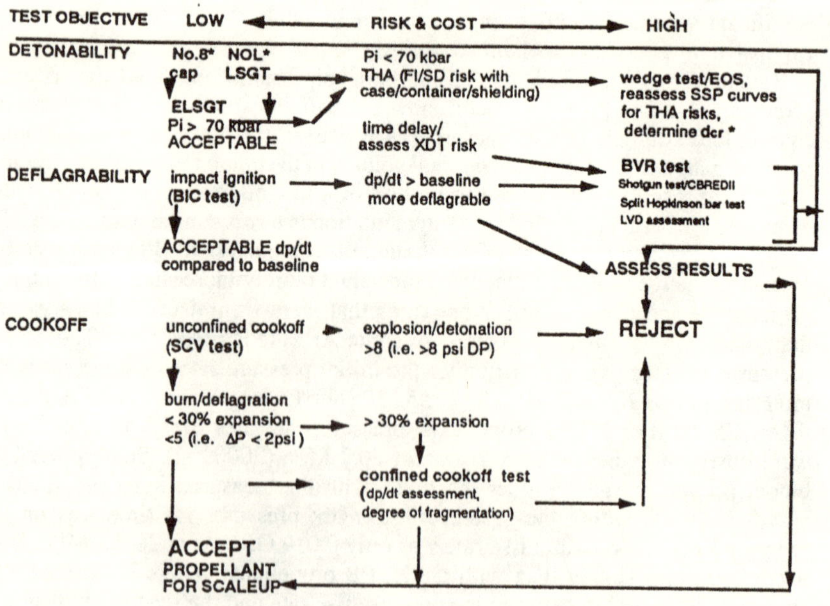

* Standard tests for explosive qualification in U.S.

Fig. 6 Schematic diagram of a proposed IPSTP.

criterion will not assure that a motor will avoid detonation in the standard IM fragment impact test,[16] it is close to the curve for a well-protected, detonable propellant grain.

The initial assessment of deflagrability in the IPSTP is based on simple tests capable of causing propellant breakup and ignition but not capable of demonstrating a run-to-detonation. The IMAD propellant program has evaluated the BIC test[26, 101, 105] (including comparisons with laser ignitability tests) as a deflagrability screening test; and passing criteria are only qualitative, namely, improvement over the propellant to be replaced or over some baseline propellant(s). Values of critical impact energy, ignitability, pressure-rise rate, and total energy and power output as functions of impact are measured and compared with values from acceptable and unacceptable propellants. The pressure rise rate in the first 100–200 μs appears to be the most relevant variable for correlating BIC tests with motor-scale bullet impact test results.[101] The standard drop-weight impact test that is mandatory for propellant qualification is not an effective substitute for the BIC test. The NAWCWPNS BVR test was used as a deflagrability screening tool following an explosives accident at NSWC/WO that halted all laboratory activity, including use of the BIC test, for two years. BVR appears to be a very useful test for interpreting the mechanisms by which projectile impact leads to propellant grain reactions of different violence.

The initial slow cookoff test in the protocol is the SCV test or an equivalent. It is risky to base propellant acceptance for slow cookoff on the SCV test alone, however, accumulated data[87] indicate that if the volume expansion of the propellant prior to autoignition is slight (less than 30%) and the reaction is fairly mild (a ranking of 5 or less, which corresponds to less than 2-psia overpressure pulse measured in the NAWCWPNS heating oven), the propellant has a good chance of passing the IM slow cookoff test in a full-scale motor. These test results should be accompanied by computational thermal analysis of a full-scale motor, as described later in this chapter, to indicate whether the phenomena observed at SCV test temperatures would be expected to occur at full-scale motor test temperatures. The standard self-heating tests that are mandatory for propellant qualification are not effective substitutes for the SCV test; however, the self-heating test data can be analyzed to obtain useful information on decomposition kinetics that is related to time to cookoff in both the SCV test and in full-scale motors. These standard data have been successfully used to calculate the measured SCV cookoff time and temperature and to predict rocket motor slow cookoff times and temperatures as described later in this chapter.[46]

Follow-on screening tests. If a propellant fails to pass any one of the initial IPSTP tests, it should either be rejected or investigated further. Further investigation has all the uncertainties and costs of a major energetic-material research project, but follow-on propellant screening tests may still be justified on the basis that they provide information necessary for performing a credible IMTHA and for predicting munition sensitivity and reaction violence of operational munitions in scenario situations that are both within and beyond the stimulus range of the IM hazard tests.

If the minimum initiating pressure for detonation is less than 70 kbar, the propellant is either rejected or tested further. A THA should be done to determine if the rocket motor in which the propellant will be used can pass the IM requirements for FI and SD without detonating by an SDT mechanism. This assessment should be made with full consideration of any protective layers that may stand between the candidate propellant and impacting projectiles. This determination can be made crudely, by examining Fig. 5, by the methods given later in this chapter, or with an appropriate hydrocode analysis, provided wedge test data are available. The propellant's unreacted Hugoniot and critical diameter, which are important parameters for making this determination, can be obtained by several methods, including appropriate analysis of wedge test data. Comparable data can also be obtained from instrumented flat-face impact, high-velocity projectile, or slapper-plate tests. For nitramine containing propellants, Eq. (35) in the next section of this chapter can be used to estimate critical diameter if no materials of greater sensitivity are present.

Next, it may be desirable to investigate the potential for XDT, as described earlier. With a gap test, initiation thresholds for both SDT and XDT can be detected. If a threshold for XDT is found, this indicates a complex susceptibility that requires further testing, and a BVR test is appropriate. It is clear from the results of the work at SNPE[38] and NAWCWPNS[80] described earlier, that size factors are important to the test results. Therefore, close approximation to actual-use motor radial dimensions may be essential for reliable test results. Because of the complex interactions involved, it may be possible to modify a rocket motor design to avoid XDT. It currently appears that the NWC BVR test, which is being further evaluated, may be an important tool for correlating the relationships of projectile size, shape, and velocity with motor bore size and web thickness, and with propellant detonability, deflagrability, and high-rate mechanical properties in terms of tendency to undergo a delayed transition to detonation. Thiokol has instrumented a similar test with external blast gauges to assess reaction violence. It is not known at this time how susceptible propellants that pass the 70-kbar card-gap test initiation criterion, as determined with the ELSGT, are to XDT reactions tested with the French or NAWCWPNS BVR tests.

As indicated earlier, the BVR test is also used to study, at realistic size and impact conditions, the impact ignition of burning, deflagration, and explosion (less than detonation) reactions and subsequent reaction growth. The author believes that a BVR test configuration as close to the final motor dimensions and shape as possible is most appropriate as a propellant screening tool.

Shotgun/CBRED II assessment and Hopkinson bar tests can provide useful information on propellant breakup and mechanical properties at high-deformation rates, for correlation with BIC and BVR test results.[131,132] The results thus obtained have bearing on deflagration reactions as well as the DDT mechanisms they were originally designed to explore.

Tests for XDT and LVD reactions give aspects of deflagrability that are related to detonability. LVD behavior in a shock-initiated propellant is a concern in relation to IMTHA scenarios. This behavior, if it can occur, should be observable in the mandatory critical diameter measurement for qualification, particularly if a pyramidal test specimen is used. Otherwise, it is not likely that one would include a LVD measurement in a screening test. LVD behavior is a concern for propellants that do not detonate by an SDT mechanism. If a propellant detonates by SDT, LVD is a much lesser concern.

If a propellant fails to pass the SCV cookoff test by reason of reaction violence or volumetric expansion, a slow heating test in a container of known confinement behavior must be performed. A test of Butcher's type[214] is a reasonably inexpensive way to obtain critical design data. Such a test could be performed in an instrumented SCB. Even if the propellant passes the SCV test, it is wise to

INSENSITIV

o determine the explosiveness llage to allow foaming to occur) -venting active mitigation system ase, may be needed to alleviate the st may demonstrate cookoff behaviors

perform a confined cook and analysis of the candidate motor size of the confined, heate ficative of potential to pass the IM require- and whether some Some variation of Eqs. (71) and (61) may be or a thermally silence in a rocket motor. It should be noted that reaction violents soften at elevated temperatures and mechanical that, based rising pressure. Either by blocking flow paths or by and othe rface, such deformation may lead to violent pressure ments cenarios.

use ening investigation moves to the right in this group of tests, e great additional expense is incurred. The decision to incur such be made with consideration of the risks of ultimately rejecting anyway [perhaps as late as the final (type) qualification phase] and value of the propellant for other reasons (for example, performance, signature, or excellence in other insensitive propellant screening tests).

margin, inves- head ic-

D. Warheads

Almost all missiles used in tactical warfare have explosive-containing warheads. The few exceptions are those missiles used for countermeasures, reconnaissance, or kinetic energy (KE) penetrators. A missile's explosive-containing warhead is always a potential hazard threat to the rocket motor. It must be remembered that warheads are designed to detonate. The U.S. Navy's IMAD program has developed and scaled up PBXs that pass the IM tests in tactical missile-size warheads.[27] A few normal design features of warheads and the explosives aid that success. For completeness, a brief discussion of insensitive missile warheads is included. More detailed information has been published in a companion volume to this book.[210]

1. *Insensitive High Explosives*

Generally explosives burn rather slowly. Unless a PBX detonates by a prompt shock initiation mechanism (SDT), is an inferior material as demonstrated by an explosiveness test (and should thus be rejected from further use), or is ignited in a sealed, heavily confining case, ignition of the explosive often leads to very mild burning.

Considerable effort has been devoted to developing insensitive high explosives.[27–32, 36, 37, 89, 95, 110, 112–114, 210] Some modern insensitive high explosives being developed have critical diameters of 5 cm or more. These large critical diameters are required to prevent sympathetic detonation in stack configurations. These explosives are very difficult to initiate reliably. In fact, developing reliable boosters and initiation systems for these explosives, so they can be used in warheads or bombs, is quite a difficult technical problem. This class of IHE is now known universally as extremely insensitive detonating substances (EIDS), and ranked hazard class/division 1.5 for shipping. Articles containing EIDS, if they pass full-scale cookoff, impact, and propagation tests, are ranked hazard class/division 1.6 and known as Articles EEI (explosives, extremely insensitive).

2. *Warhead Case Design*

Warhead case walls are generally fairly thic
some hard targets before initiation or so they ca
significant mass. The thick case attenuates the shoc
to such an extent that shock initiation in the standa
can be prevented for a number of explosives. For exa
head case 1 cm thick will increase the critical impact vel ation of
1-cm-diam impactor acting on the explosive within, by 6 nts of
pact on the bare explosive; whereas a typical steel rocket ats
thick will increase the critical impact velocity by only 15%. Lin
between the explosive and the case also assist in reducing the in
to the explosive, but many are less effective for a given thickness
Because of the thickness of the warhead case, other forms of safet
not available in rocket motors, are available. Warhead IM programs are
tigating possible improvements to the shock attenuating properties of war
cases and liners by looking at new materials, including composites and ret
ulated structures. Dual-explosive warheads, in which a less sensitive outer explosive in the warhead cylinder protects the more sensitive, more energetic inner explosive have been demonstrated, with some explosives, to be effective in reducing reaction violence in the IM fragment impact test and in sympathetic detonation tests, with no loss in the warhead's effectiveness upon purposeful detonation.[47,115]

Warhead cases, unlike rocket motor cases, can be completely sealed. In heating situations, thermal expansion and decomposition of the warhead liner and explosive can cause cracking of the completely filled, sealed case prior to thermal ignition. For some warheads this behavior has been enhanced by designing a longitudinal stress riser into the case to promote venting. The case generally bursts at the stress riser prior to ignition of the explosive. For some warheads loaded with shaped-charge jet submunitions, or grenades, a longitudinal stress riser in the outer case has also worked, presumably after ignition of the explosive (since this type of warhead has ullage volume) due to rather slow initial pressure rise in the combustion (~5 ms rise time or greater) and relatively low-burst pressure at the stress riser.

3. *Fuse and Booster Design*

To meet safety requirements[24] warhead initiation systems are usually designed with an out-of-line lead between the initiator and the booster.[1] The booster is the high-explosive charge that initiates the main-charge explosive. The initiator is usually a primary explosive or electric squib. Sometimes inadvertent initiation of the fuze-booster chain causes initiation of the main-explosive charge.

Improvements to fuse-booster systems are being sought in IM R&D to develop design methodologies and in development programs for improvements to specific existing munitions and for new munitions. There are two driving forces for these improvements; first, to minimize or eliminate inadvertent initiation of a main-charge explosive by its initiation chain and, second, to reliably initiate the new insensitive high explosives being developed. Among the approaches being tried are 1) slapper (flying plate) detonators in which the booster explosive is not in contact with the main charge but instead accelerates a plate (projectile) that initiates the main charge by impact only if the booster is initiated in its design mode; 2) multipoint initiators composed of several boosters, each too weak to initiate the main charge if initiated separately or nonsimultaneously, but which when initiated simultaneously create a Mach stem of sufficient energy to initiate the main charge; and 3) Munroe effect initiators, using the shaped-charge jet principle and effective only when fired in design mode. One technique that may be effective involves the use of steel plates, cylinders, or cones embedded in the main-charge explosive to reflect and focus the booster shock wave.[116] Other initiation concepts, including some based on sophisticated matching of acoustic impedances in the initiation chain, are also being studied. One bit of noteworthy arcane information: in fast cookoff tests, booster-explosive detonations have occurred in aluminum booster cases but not in steel cases, indicating the continuing need to be concerned with chemical compatibility.

VI. Simple Analytical Methods for Munition Hazard Assessment

This section presents a wide range of simple calculation methods for determining energetic material and munition behavior important to insensitive munitions issues, assembled or presented here for the first time. Methods for calculating both impact and thermal initiation of reactions are included. Methods are also included for calculating explosive and warhead behavior (as a source of threat fragments). With these methods, one can rapidly perform preliminary design and threat hazard assessments for insensitive munitions. These methods are abstracted from recent publications.[46,197,202,204] The methods are generally algebraic and, except for fast and slow cookoff problems, can be solved with a pocket calculator, although a desktop computer spreadsheet greatly speeds solution. More sophisticated two- and three-dimensional numerical computer techniques are available, and those who follow the references will be led to them. For the range of problems solved by the simple methods given here, the more sophisticated models do no better and require orders of magnitude more investment and time. For detailed design calculations of complex shapes, the more complex models are essential.

A. Impact Initiation of Reactions

This section presents calculation methods for initiation of detonation and ignition in munitions subject to impact by bullets, warhead fragments, and shaped-charge jets. Also provided are methods for estimating the velocities, size distributions, and spatial distributions of fragments from threat warheads. The term explosive is often used generically to denote energetic material that may be either high explosive or propellant.

1. Detonation Behavior of Explosives

Simple methods for calculating the detonation behavior of explosives are available. With these methods, one can obtain results comparable in quality to those obtained with such codes as Tiger BKW or CHEQ. The methods are useful in the absence of laboratory data.[209]

Detonation pressure. The Kamlet–Jacobs formula[134] gives the detonation pressure P_d in an explosive as

$$P_d = 1.558\,\rho_o^2 N M^{\frac{1}{2}} Q^{\frac{1}{2}}, \quad \text{GPa} \tag{1}$$

where N is the number of moles of the detonation product per gram of explosive, ~0.03; M is the average molecular weight of detonation product gas, ~30; and Q is the chemical energy of the detonation reaction, ~1000 cal/g.

If one has trouble estimating values for N, M, or Q needed in Eq. (1), the detonation pressure can be estimated from the calculated propellant specific impulse (I_{sp}) of the explosive formulation by the method of Gill, Asaoka, and Baroody (GAB).[135] The GAB method uses the Navy's Propellant Evaluation (PEP) Code,[136] although any similar equilibrium thermochemical code may be used.

$$P_d = 4.44\rho_o^2(0.009807[I_{sp}^{1000}]_{14.7}) - 2.1, \quad \text{GPa} \tag{2}$$

Here I_{sp} is in lb-s/lb as obtained from PEP or NASA/Lewis codes with superscript and subscript indicating chamber and exit pressures, respectively, in psia.

For a wide but reduced variety of explosives the GAB formula may be simplified to

$$P_d = 4.44\rho_o^2(0.009807[I_{sp}^{400}]_{14.7}), \quad \text{GPa} \tag{3}$$

which is in the same form as Eq. (1).

Detonation velocity. The detonation velocity D can be approximated by Jacobs formula found in Ref. 134 as

$$D = (0.809/\rho_o + 1.052)P_d^{\frac{1}{2}}, \quad \text{km/s} \tag{4}$$

Walker[137] has described a method for estimating detonation velocity on the basis of the Hugoniots of the explosive's constituent elements and the measured or calculated detonation pressure.

$$D = \Sigma_e(U_{s_e}[W_e/\Sigma_e W_e])[1 + f(P_d)], \quad \text{km/s} \tag{5}$$

where W_e is the formula weight of the element equal to $M_e \times n_e$, for example, for $C_6H_6N_6O_6$, $n = 6$ for each element and

$$f(P_d) = 2.0286(-3) + 2.231(-3)P_d + 9.6429(-6)P_d^2 - 4.1667(-7)P_d^3$$

$$Us(\mathrm{C}) = 4.5319 + 0.11651P_d + 1.0717(-4)P_d^2 - 1.5162(-5)P_d^3$$

$$Us(\mathrm{H}) = 5.976 + 0.35362P_d + 1.6859(-3)P_d^2 - 5.0439(-6)P_d^3$$

$$Us(\mathrm{N}) = 1.2364 + 0.51667P_d + 1.5555(-2)P_d^2 - 1.9072(-4)P_d^3$$

$$Us(\mathrm{O}) = 2.7904 + 0.18343P_d + 1.9501(-2)P_d^2 - 1.045(-5)P_d^3$$

It is conceivable that other elements could also be used in the Walker formulation; for those elements that do not react promptly to form gaseous products in the detonation some approach involving partition of energy might be used to obtain good calculated values of detonation velocity.

Heats of detonation and explosion. Baroody and Peters[138] published a method based on rocket-motor specific-impulse (I_{sp}) calculations for estimating heats of explosion and heats of detonation. This method is summarized in Fig. 7, which also shows the calculated results obtained for two densities of HMX, an unmetallized composite rocket propellant, and several common explosives. This method can be used to obtain reasonable values of Q for use in Eq. (1).

Section summary. With the data generated by the preceding equations, one can calculate the detonation pressure P_d and the detonation velocity D for an explosive if its chemical formulation and density are known. With this information, one can proceed to estimating warhead performance.

Baroody & Peters calculation of Heats of Detonation and Explosion, IHTR 1340, 7 May 1990.

A Density	B Enthalpy (ch)	C Enthalpy (ex)	D Heat Content	E Enthalpy (ch-ex)	F Hdet cal/g	G Hdet cal/cc	H Enth exh, 14.7psi	I Heat Content	J Enthalpy (ch-ex)	K Hexpl cal/g	L Hexpl cal/cc	M Enth(298) ch14.7psi	N Hcomb cal/g
HMX													
1.9002	6.054	-132.32	140.779	-1383.74	1524.52	2896.89	-74.7351	529.234	-807.891	1337.13	2540.8		
1.8988	6.1	-132.3	140.788	-1384	1524.79	2895.27	-74.66	530	-807.6	1337.6	2539.83	-155.09	1611.9
RDX													
1.816	6.6	-132.23	140.61	-1388.3	1528.91	2776.5	-74.42	532.4	-810.2	1342.6	2438.16	-155.09	1616.9
AP/HTPB (86/14) Reduced Smoke Propellant													
1.806	-51.59	-171.17	189.78	-1195.8	1385.58	2502.36	-121.41	524.42	-698.2	1222.62	2208.05	-206.01	1544.2
TNT													
1.63	-4.8	-108.34	113.21	-1035.4	1148.61	1872.23	-56.31	266.36	-515.1	781.46	1273.78	-147.17	1423.7
HMX/Al, 80/20													
2.019	4.88	-166.55	132.841	-1714.3	1847.14	3729.38	-81.93	827.775	-868.1	1695.88	3423.97	-220	2248.8
PBXN-109													
1.68	4.43	-148.24	172.12	-1526.7	1698.82	2854.02	-65.81	650.68	-702.4	1353.08	2273.17	-200.67	2051
AFX-931													
1.67	-19.95	-160.98	127.99	-1410.3	1538.29	2568.94	-94.01	559.27	-740.6	1299.87	2170.78	-223.02	2030.7
H-6													
1.76	-1.69	-152.73	222.73	-1510.4	1733.13	3050.31	-70.31	619.365	-686.2	1305.57	2297.79	-213.7	2120.1
PBXN-107													
1.64	5.86	-111.01	147.565	-1168.7	1316.27	2158.67	-53.48	288.94	-593.4	882.34	1447.04	-200.67	2065.3

B=chamber enthalpy at 1000 psi
C= exhaust enthalpy at 0.0017 psi
D=heat content for exhaust condition (0.0017 psi)
E=(C-B)*1000/gfw, cal/g
F=-E+D, cal/g
G=F*A, cal/cc
H=exhaust enthalpy at 14.7 psi
I=heat content at exhaust condition (14.7 psi)
J= (H-B)*1000/gfw, cal/g
K=-J+I, cal/g M=enthalpy run with PEP option 8 for T=298K, P=14.7 psia
L=K*A, cal/cc N=heat of combustion=10*(B-M) approx, since B should be comb at 1 atm.

Fig. 7 Calculating heats of detonation and explosion with PEP code.[136, 138]

2. *Warhead Behavior of Explosives for Insensitive Munitions Considerations*

The warhead behavior of explosives involves the acceleration of metal, as in case fragmentation and shaped-charge jet generation, and the generation of blast shock waves and overpressures and impulse in air and water, where gas bubble energy is also of interest. Our primary interest, from the IM standpoint, is the calculation of fragment velocities and sizes that may be important in fragment impact and sympathetic detonation scenarios. Fragment velocities, size distribution, and spatial distribution can be estimated with the equations given in this section, and the explosive behaviors calculated in the preceding section. With this information, one can generate threat parameters relevant to IMTHA fragment impact and sympathetic detonation scenarios.

Fragment velocity. The maximum velocity of metal fragments in the detonation of a cased explosive is approximated by the Gurney formulas. The simplest expression of the Gurney formula for symmetrical configurations is

$$V_{\mathrm{Gurney}} = \surd\{2E/(\mu + n/[n+2])\} \tag{6}$$

where $\mu = M/C$, and M is the mass of the metal in warhead case and C the mass of the explosive charge. The Gurney constant, in meters or feet per second is $\surd 2E$; values of n are 1 for a flat sandwich of explosive between two equivalent flat metal plates, 2 for a cylinder, and 3 for a sphere.

In addition, formulas for unsymmetrical sandwiches[142] are useful for flyer-plate warhead-booster performance calculations. Equation (7) may be used for an "open-faced sandwich," with metal on only one face, although other formulas have been proposed as well.[143]

$$V_{\mathrm{Gurney}} = \surd\{2E/\{\mu + ([1+2\mu]^3 + 1)/(6[1+\mu])\}\} \tag{7}$$

For an unsymmetrical sandwich with metal mass of N on one face and M on the other

$$\begin{gathered} V_M = \surd\{2E/(1+A^3)/(3[1+A]) + A^2N/C + MC\} \\ V_N = AV_M \end{gathered} \tag{8}$$

where

$$A = (1 + 2[M/C])/(1 + 2[N/C])$$

The Gurney constant $\surd 2E$ can be approximated by the simple expression

$$\surd 2E = 0.338D, \text{ km/s} \tag{9}$$

or by the equation of Kamlet and Finger[144]:

$$\begin{aligned} \surd 2E &= 233\rho_o^{-0.6}P_d^{\frac{1}{2}}, \text{ m/s} \\ &= 887\rho_o^{0.4}(NM^{\frac{1}{2}}Q^{\frac{1}{2}})^{0.5}, \text{ m/s} \end{aligned} \tag{10}$$

with P_d is kilobars and one kbar equal to 0.1 GPa.

For a cylindrical warhead, the appropriate Gurney formula is applicable only for the cylindrical portion, and the values of M and C used must be adjusted to eliminate end effects. Odintsov[145] recently published expressions applicable to the ends of cylindrical warheads. Equation (11) is derived from that work:

$$V_{\mathrm{end}} = \surd(dMV_{\mathrm{Gurney}}/4Lm) \tag{11}$$

where d is the warhead diameter, M the mass of warhead cylinder case section, L the warhead cylinder length, and m the mass of warhead case end section.

It is also possible to estimate the velocity of the expanding warhead case at different amounts of radial expansion. For cylindrical warheads, the initial elastic-plastic expansion of the case occurs as it expands from its original radius to about 1.2 times that radius. At the end of this phase the case radial velocity is about 60% of the calculated Gurney velocity for explosives that release all their energy in the detonation. The maximum velocity (as calculated by the Gurney formula) is that achieved at the end of fragment acceleration. About 95% of this velocity is achieved with the fragments at a radius of about 1.6 to 1.8 times the initial warhead radius (again for explosives that release all their energy in the detonation).[139] Explosives that do not release all their energy in the detonation are frequently used in bombs and underwater warheads. For these, a significant fraction of the combustion is released in "afterburning" that occurs subsequent to the detonation. For such afterburning explosives, the general shape of the velocity vs case expansion curve is the same but the 60 and 95% velocity radii just described are generally significantly larger.[110] Dehn[140] has published equations based on Taylor's method that calculate fragment velocity as it increases with case expansion. These are approximated in Eq.(12):

$$\begin{aligned} V_R/V_{\text{Gurney}} &= \max\left(V_0/V_{\text{Gurney}}, \left[K\left(1-(R_D/R_E)^2\right)\right]^{\frac{1}{2}}\right) < 1.0 \\ V_0 &\sim \left\{\left[\left(c_{oc^2}+4s_c P_c/\rho_{oc}\right)^{\frac{1}{2}} - c_{oc}\right]/2s_c\right\}(R_D - t_c)/R_D \end{aligned} \tag{12}$$

which is the particle velocity in the donor case times a radial expansion factor[204]

$$P_c \sim \rho_{oc} P_{cj}/(\rho_{oc} + \rho_{oe})$$

which is the approximate pressure induced in the donor case by a grazing detonation wave. (The pressure in a case of material c due to a normally incident detonation wave is given by $2P_c$.)

In Eq. (12) R_D is the original case radius, t_c the case wall thickness, R_E the expanded radius, and K a constant representing the degree of combustion that occurs during detonation. For explosives that release virtually all their energy in the detonation, the value of K is of the order of 1.35. When the value obtained with Eq. 12 is less than V_0 or greater than unity, V_R/V_{Gurney} should be set equal to the terminal values indicated. P_{cj} is the detonation pressure or Chapman–Jouget pressure of the explosive. This formulation provides some accounting for initial expanding velocity V_0 of the donor case due to a sweeping detonation wave needed to calculate shock transfer for very closely spaced donor-acceptor combinations in sympathetic detonation scenarios. A logarithmic expression may replace Eq. (12) for explosives that release energy more slowly.[204]

For an exploding cylindrical warhead with partial additional circumferential confinement, such as a bomb stored in a stack (typical of the sympathetic detonation stack test), it is reasonable to substitute a reduced value for the effective case mass M_i for at least part of the case. This will result in a higher calculated Gurney velocity for that part of the case. For example, this appears to be consistent with Lundstrom's calculations[126] of the effects of stack confinement and fragment focusing in sympathetic detonation stack tests in which the expanding donor case appears to be focused into a planar shape, so that the problem appears as a large flyer plate impacting on a cylindrical acceptor. Glenn and Gunger's work[201] is consistent with this hypothesis.

Fragment size distribution. The Mott equation is used to estimate the size distribution of fragments from a warhead

$$N(m) = N_o \exp(-m/a)^{\frac{1}{2}} \tag{13}$$

equal to the total number of fragments of mass greater than m where a is one-half the average fragment mass in grams, $N_o = M/(2a)$ is the total number of fragments (M is total mass of fragments)

$$a = B\left(t_o[D_i + t_o]^{\frac{3}{2}}/D_i\right)(1 + \mu/2)^{\frac{1}{2}}$$

where B is a constant $\sim 338.1/P_d$, (kbar), t_o is casing thickness, (in.), and D_i is the internal diameter of the cylindrical case, (in.).

The fragment velocity and size formulas in this section are in common use and are frequently modified to extend their useful range.[146,147,205] For sympathetic detonation predictions, the calculated fragment size distribution is irrelevant for an unconfined donor-acceptor surface separation distance x of one munition diameter d or less.

Fragment spatial distribution. The spatial distribution of fragments about a detonating cylindrical warhead is not uniform.[139,148–150] Naturally fragmenting metal warhead cylinders typically fracture into 20 to 23 initial radial bands: therefore, the typical bandwidth (peak to peak or valley to valley) about the cylinder axis varies between 15 and 18 deg. These bands break up further during subsequent expansion into the ultimate fragment size distribution given by the Mott distribution. However, the number of fragments per circumferential angle increment may vary by as much as a factor of 4 or 5 between the peaks and valleys caused by the initial fracture. Sewell[149] gives a rule of thumb that the number of initial fracture sites is given by Eq. (14):

$$F = V_c/(2u_{pc}) \tag{14}$$

where the number of fracture sites is equal to the number of axial fragment bands. In Eq. 14 V_c is the initial circumferential velocity of inner wall and is equal to $2\pi \times$ (radial velocity). The radial velocity is approximated by the sweeping wave pressure divided by the wall acoustic impedance. For steel and a typical explosive this radial velocity would be about (20 GPa)/(45.2 GPa/mm/μs) = 0.442 mm/μs. Also in Eq. (14), u_{pc} is the critical particle velocity. For a typical warhead-case steel, u_{pc} for shear is 200 ft/s or 0.061 mm/μs. With these input values, the number of circumferential fracture sites $F = 22.8$.

The azimuthal (polar) distribution of fragments from a detonating cylindrical warhead is limited to a fan with small angular dispersion. For single-end initiation, the peak angular fragment density is angled from the normal (90 deg) by an amount that can be approximated by one-half the Taylor angle,[142] or about 5 deg.

$$(\sin^{-1}[\theta]) = V_{\mathrm{Gurney}}/2D \tag{15}$$

The band or fan of fragments from a cylindrical warhead, typically about 15–20 deg wide, is not as narrow as Eq. (15) implies. The differential (percent of total fragments per degree) patterns are approximated by Eqs. (16) and (17) for end- and center-initiated cylindrical warheads. The cumulative patterns are obtained by integrating or summing these equations. Here ϕ is the angle the fragment trajectory makes to the cylinder axis of symmetry.

End initiated,

$$\%/\mathrm{deg} = 11\exp(-0.04(95-\phi)^2) \tag{16}$$

Center initiated,

$$\%/\mathrm{deg} = 1.1694\sqrt{\{25-[(90-\phi)/2.2]^2\}} \tag{17}$$

B. Shock Initiation of Detonation

This section is divided into two parts. The first part provides solutions to fragment-impact initiation problems. The second part provides solutions to sympathetic detonation problems. The minimum fragment impact velocity, or critical-impact velocity V_i that will initiate SDT reaction of an explosive, can be estimated by the equations given in this section for a number of different situations. With these equations one may estimate probabilities of SDT given impact by the fragments generated in the calculations of the preceding section. Other factors that decrease fragment velocity and spread the fragment pattern and otherwise reduce the chances of an impacting fragment detonating a munition were given by Wagenhals et al.[170] and include divergence of the fragment pattern with distance, air drag, cylindrical munition shape, obliquity of impact, and random orientation of fragment face. Drag data on warhead fragment shapes are available for application to problems of this type.[171,172] The equations given apply only to "chunky" fragments. [The shorter duration shocks caused by flyer plate impacts (generally of radius more than six times thickness[155]) require higher shock pressures (higher velocity) to cause SDT.]

In the absence of a hydrocode capability, or with insufficient time to use it adequately, there are still many simple, yet useful, calculations that can be made. The shock sensitivity plane, described earlier, displays both wedge test results (POP plots) and the results of hydrocode calculations by Lundstrom[151] to provide a very accessible framework for such analyses. The ordinate of the SSP, shown in Fig. 5, is the shock pressure P_1 entering an explosive in the wedge test that exactly results in a 1-cm ($X = 10$ mm) run distance to detonation. The abscissa (S) is the slope of the POP plot of log P vs log X as shown in Eq. (18):

$$X = 10(P_1/P_e)^S \tag{18}$$

in millimeters, with P in GPa.

Each explosive is assumed to have an exactly linear POP plot, and this results in a single point for each explosive in the SSP. Lundstrom obtained the curves in Fig. 5 that correspond to the various test results by using reactive-hydrocode calculations for specific explosive properties as defined by their points in the SSP. Any explosive point that lies above the line corresponding to a particular test will not detonate in that test, whereas, any explosive that lies below the test curve will detonate. Some explosive and propellant values are shown in Fig. 5 to aid in practical use of the figure as it stands.

1. Fragment Impact Calculations

To predict whether an acceptor explosive will detonate due to an impacting projectile, it is necessary to follow the following steps.

1) Determine the impactor dimension d_e at the case-explosive interface that results from the impactor diameter d_i (for equivalent flat-faced cylindrical impactor)

on the outer case surface and determine the matched shock pressure condition at the explosive P_e that results from the impact velocity V_i.

2) Determine the depth X in the explosive to which a shock of strength P_e must propagate to transition to a detonation, based on one-dimensional wedge test results.

3) Determine that the diameter of the shock at depth X is greater than or equal to the acceptor explosive critical or failure diameter d_{cr}.

4) If the width of the shock at depth X is less than the acceptor explosive failure diameter, it is possible to use an expanding wave approximation, as described later [Eqs. (37–39)] to get a solution.

The only difficulty with this approach is that many calculations are required to identify the critical conditions for propagation. It is easiest to start with the explosive; specify a list of values for P_e, and determine the corresponding particle velocity u_{pe} and shock velocity u_{se} in the explosive for each value of P_e from the momentum conservation equation:

$$P_e = \rho_o u_{pe} u_{se} = \rho_o u_{pe}(c_{oe} + s_e u_{pe}) \tag{19}$$

where the subscript 0 refers to the unshocked material, subscript e to the explosive, c to the case material, and i to the impactor. In Eq. (19) ρ is the material density, u_s the shock velocity, u_p the particle velocity (material velocity) in the direction of shock wave, P the pressure, V the specific volume, and c_o the sonic velocity in the unshocked material. Values of the parameters of the conservation equations for some materials are shown in Table 4.

The run distance X to detonation at each value of shock pressure P_e is obtained from the sensitivity relationship in Eq. (18).

Table 4 Shock properties of selected materials

Material	ρ, g/cm^3	c_o, km/s	s
Tungsten	19.224	4.029	1.237
Titanium	4.527	5.037	0.955
Titanium	4.527	4.937	1.019[a]
Steel	7.89	4.58	1.49
Steel	7.9	4.5	2.6[a]
Aluminum	2.785	5.328	1.338
Copper	8.93	3.94	1.489
PMMA (Plexiglas)	1.186	2.654	1.488
Water	0.998	1.647	1.921
Polyurethane	1.265	2.486	1.577
Polyrubber	1.01	0.852	1.865
Graphite/epoxy	1.53	3.3	2.2[a]
Composition B	1.715	3.03	1.73
Nitroguanidine	1.71	2.72	1.5
PBXN-109	1.66	1.75	2.78
PBXN-110	1.657	3.70	1.905
PBXN-107	1.63	2.449	2.019
PBX 9404	1.84	2.43	2.57
Pressed TNT	1.54	2.08	2.44
Destex	1.694	2.31	1.83
Adv. min smoke propel.	1.62	2.2	2.0[a]

[a]From Ref. 228.

The velocity of sound in the explosive (c_{se}) is calculated with Eq.(20).

$$c_{se} = (u_{se} - u_{pe})(u_{se} + s_e u_{pe})/u_{se} \tag{20}$$

Equation (21) gives the smallest impactor diameter that will initiate detonation (SDT) in the bare explosive for each specific value of P_e, where d_{cr} is the failure diameter or critical diameter for detonation of the explosive, and is discussed with respect to Eqs. (32–35).

$$d_e = 2X\tan(\theta) + d_{cr} \tag{21}$$

where

$$\theta = \cos^{-1}(u_{se}/c_{se})$$

and θ is the assumed angle of the lateral rarefaction that intrudes from the edge of the impactor on the explosive surface into the shocked, unreacted region of the explosive, decreasing the diameter of the highest pressure region in the explosive. If this diameter becomes less than the critical diameter before X is reached, failure of initiation is predicted. In his work, Green[154] used $\theta = 45$ deg, regardless of explosive properties.

The relationship between impact velocity and shock pressure in the explosive for a projectile of material i with density ρ_i directly on the explosive is calculated with Eq.(22).

$$V_e = u_{pe} + P_c/(\rho_i u_{si}) \tag{22}$$

where

$$u_{si} = c_{oi} + s_i u_{pe}$$

To simplify the large number of calculations needed to obtain results, Eq. (22a) may be used as an approximation in combination with inert material attenuation Eqs. (28–31). Equation (22) was derived by curve fitting a large number of solutions to Eqs. (19–22) over the range of P_1 from 3 to 20 GPa and the range of S from 1.5 to 6.0 for an explosive with the unreacted Hugoniot of composition B.

$$V_e = \left[(20S/d_e)(P_1/10.68)^S\right]^{(1.3/S)} \tag{22a}$$

Equation (22a) was obtained from the curve fit relationship $V_e = (2Sd_{cr}/d_e)^{(1.3/S)}$ using the analytical expression for d_{cr} given later in Eq. (33). Equation (22b) gives the comparable expression for an explosive with the unreacted Hugoniot of explosive PBXN-109.

$$V_e = \left[(15.89S/d_e^{0.9})(P_1/10.68)^{0.9S}\right]^{(1.25/S)} \tag{22b}$$

For the cased (c) explosive, the conservation shock-jump conditions give

$$\rho_e u_{se} u_{pe} = \rho_c u_{sc}(2u_{pc} - u_{pe}) = P_e \tag{23}$$

$$\rho_i u_{si}(V_i - u_{pc}) = \rho_c u_{sc} u_{pc} = P_i \tag{24}$$

Arranging terms, one obtains the easily solved quadratic for u_{pc}

$$2s_c u_{pc}^2 + (2c_{sc} - s_c u_{pe})u_{pc} - (c_{sc}u_{pe} + P_e/\rho_c) = 0 \tag{25}$$

and u_{sc} and u_{si} are easily obtained across the projectile-case matched conditions

$$u_{sc} = c_c + s_c u_{pc} \qquad u_{si} = c_i + s_i u_{pc}$$

The impactor velocity on the case is given by Eq. (26).

$$V_i = u_{pc} + P_i/(\rho_i u_{si}) \tag{26}$$

The impactor diameter required at the outer case wall to give the conditions already calculated in the explosive is given by Eq. (27).

$$d_i = d_e + 2t_c \tan\left(\cos^{-1}[u_{se}/(u_{sc} + s_c u_{pc})]\right) \tag{27}$$

where t_c is the case thickness, mm.

Equation (27) is applicable only for case thickness no greater than about one-half the impactor diameter. For thicker cases, the link between Eqs. (21) and (27) forces the interaction zone diameter at depth X in the explosive to become less than d_{cr}. For case thicknesses up to the impactor diameter, and possibly greater (but data are lacking), the author has found that the exponential attenuation calculated by Eqs. (28–31) can replace the more laborious approach in Eqs. (23–27) for the inert cover or liner materials shown.[46]

Steel

$$V_i/V_e = 0.949\cdot \exp[1.035(t_c/d_i)] \tag{28}$$

Aluminum

$$V_i/V_e = 0.814\cdot \exp[0.9051(t_c/d_i)] \tag{29}$$

Titanium

$$V_i/V_e = 0.902\cdot \exp(0.869[t_c/d_i)] \tag{30}$$

PMMA

$$V_i/V_e = 0.929\cdot \exp[0.953(t_c/d_i)] \tag{31}$$

The constants in Eqs. (28–31) include corrections for shock pressure matching and the effect of release wave angle in the case calculated with Eq. (27). The curves of V_i vs d_i generated by Eqs. (28–31) give results for $d_i = d_e$. The equations as shown apply for steel impactors only, upon the four different inert materials shown. The equations for steel and titanium are based on hydrocode calculations.[153] The equation for PMMA (Plexiglas) is based on fitting data from James et al.[160] for $t_c/d_i \leq 0.7$. The equation for aluminum was obtained by curve fitting to values calculated using Eqs. (18–27) for $t_c/d_i \leq 0.4$ and PBX-9404 explosive and further checked by comparison with data of Refs. 75, 76, and 160 out to $t_c/d_i = 1.0$, as in Ref. 46.

The calculation procedure for either sympathetic detonation or fragment impact initiation of detonation is so greatly simplified by use of these exponential equations that it is worthwhile to develop such equations by generating data with Eqs. (18–27) for the specific impactor, case, liner, and explosive materials of the problem at hand and fitting them to the exponential form, before proceeding to the detonation predictions.

As shown by Eq. (21), the value of the explosive's failure diameter is a critical datum for these calculations. The experimentally determined failure diameter of an explosive is the diameter of the smallest unconfined circular cylinder that will sustain a steady detonation. Our interest here is somewhat different, since what we are using in the calculation is the smallest diameter of a shock wave that can initiate a steady detonation confined within a volume of the explosive. For explosives that behave ideally, agreement of the simple calculation methods of the preceding section with experimental data demonstrate comparability of the two concepts of failure diameter. The same comparability exists when Green's method[154] is used to predict critical velocities of impactors that are smaller than the failure diameter (for example, shaped-charge jet impact). This is still no guarantee that the same degree of comparability will hold for nonideal explosives and propellants.

The failure diameter of an explosive is best obtained from direct measurements. A general predictive capability for shock initiated detonation by projectile impact or sympathetic detonation followed in this chapter requires a predictive method for failure diameter. Lundstrom[123] has presented hydrocode calculations that show a fairly consistent relationship to failure diameter for a number of explosives. Lundstrom's results are fit well by Eq. (32), which is taken from Ref. 46.

$$d_{cr} = 10P_1^{(9.025 \log S)}/S^{(8.2335+0.3766S)}, \quad \text{mm} \tag{32}$$

Lundstrom's method and Eq. (32) give results in good agreement with measured failure diameter for many conventional high explosives, such as PBX-9404 and Composition B. Even insensitive TATB-containing explosives, such as PBX-9502 and X-0219, are fit quite well. Unfortunately, Eq. (32) predicts failure diameter values that are too low by as much as a factor of 100 for some of new insensitive high explosives and composite insensitive propellants with apparent high values of POP plot slope ($S > 4$).

Price[203] examined a simpler linear relationship (failure diameter vs run distance at 8.3-GPa shock pressure) that can be expressed, in terms of Eq. (19), by Eq. (33).

$$d_{cr} = 6.85(P_1/8.3)^S \tag{33a}$$

modified here to be

$$d_{cr} = 10(P_1/10.68)^S \tag{33b}$$

Equations (33) can be combined with Eq. (22a) to give Eq. (34), which is very useful for plotting results. A similar relationship can be derived from Eq. (22b) as well.

$$P_1 = \left[22V_e/(20S/d_e)^{(1.3/S)}\right]^{0.77} \tag{34}$$

The two forms of Eqs. (33) give identical results only for $S = 1.5$. Price's relationship is based on a linear empirical fit for ideal explosives with low values of slope. Price showed that her linear relation did not fit data for such insensitive TATB-containing explosives as PBX-9502 or X-0219. However, as expressed here Eq. (33b) fits those explosives far better, and in fact, almost as well as does Eq. (32). Although predicted values of failure diameter with Eqs. (33) for many insensitive explosives with $S > 4$ are lower than measured data, the discrepancy is at least an order of magnitude less than with Eq. (32). Because of these factors and its comparative simplicity, the author prefers to use Eqs. (33) in initial calculations. However, for many insensitive, composite propellants and explosives, these relationships between wedge test results and measured critical diameter do not hold.

Equation (35) presents two expressions for failure diameter obtained by curve fits to a large number of published values for nitramine (RDX or HMX) containing inert-plastic-binder explosives and propellants.[202] Most of the materials used in the curve fit contained aluminum and ammonium perchlorate. However, the second expression extrapolates well to high-nitramine explosives that contain neither AP nor Al. The results of Eqs. (35) can be used in Eqs. (21), (22a), (22b) and (33). However, it should be noted that although values of P_1 (perhaps we should call it P_1') thus obtained by convolving Eq. (35) through Eq. (34) are not necessarily constant and are not appropriate for use in Eq. (18), they may be used in the SSP to represent the effective sensitivity of insensitive explosives to fragment impact and sympathetic detonation stimuli.

$$\begin{aligned} d_{cr} &= 1500\,(N)^{-1.181}, \quad \text{mm; for N = 5–66\% nitramine} \\ d_{cr} &= 126\exp(-0.048N), \quad \text{mm;} \quad \text{for } N = 10\text{–}86\% \text{ nitramine} \end{aligned} \tag{35}$$

For impactors of diameter less than the explosive's critical diameter, SDT can apparently occur with very high-impact velocities [such as the IM test fragment (2530 m/s) or a shaped-charge jet], although there are speculations that a shear heating and ignition mechanism could be operating[165] (see next section). Chick cited an empirical observation[166] that prompt surface impact (SDT) initiation of bare explosives can occur with shaped-charge jet impact for ratios of $d_{cr}/d_e < 5$. For this situation, Eq. (36) can be used to obtain the critical jet impact velocity. Held[195] provides sufficient information to obtain values for the constant in the equation, which agrees quite well with the results from Eqs. (37–39):

$$V_{ji}^2 \rho d_e = K_1 \tag{36}$$

with K_1 a constant, where ρ is jet density.

For impactors that are either smaller than the critical diameter or that will not create shock width greater than the critical diameter at the appropriate run distance to detonation, a different approach must be used. Instead of tracking the highest pressure region with the angle of the lateral rarefaction, the shock wave is assumed to expand with commensurably decreasing pressure with distance into the explosive.[202] The best fit to a range of data has been obtained assuming that the diameter of the expanding shock follows a curved channel given by Eq. (37):

$$\begin{aligned} D_{eqv}/d_e &= [1 + (2X_i'/bd_e)^2]^{\frac{1}{2}} \\ b &= 1/\left[\tan(\cos^{-1}(u_{se}/c_{se})\right]^{\frac{1}{2}} \end{aligned} \tag{37}$$

The scale factor, b governs the distance scale for the flow channel divergence. X_i' is defined as in Green's method[154] as the axial distance corresponding to the equivalent diameter D_{eqv}.

The equivalent pressure P_{eqv} is defined by Eq. (38):

$$P_{eqv} D_{eqv} = P_e d_e \tag{38}$$

The solution method of Eq. (21) is applied as shown in Eq. (39), where u_{se} and c_{se} are the values at the equivalent pressure P_{eqv}, and X_b is the longitudinal distance from the position where the channel width equals D_{eqv} to the position where the lateral rarefactions starting there have reduced the shock width to the critical diameter d_{cr} at an appropriate distance from the impact, $X_i' + X_b$:

$$X_b = (D_{eqv} - d_{cr})/\left[2\tan(\cos^{-1}(u_{se}/c_{se})\right] \tag{39}$$

Equations (37–39) are solved iteratively. The solution is obtained from the smallest run distance, X_i', that will give a solution to Eq. (39). The solution is determined in terms of the impactor diameter d_e and the impactor velocity V_e as given in Eq. (22).

These solutions for bare explosive or propellant are combined with case or case and liner effects that can be calculated by Eqs. (23–26) or by the exponential expressions of Eqs. (28–31). The method of Eqs. (37–39) gives results that agree well with data for small very high velocity flat-faced impactors, as described by the constant K_1 in Eq. (36). For shaped-charge-jet impact, values of K_1 from data are generally 3–4 times larger than for flat-faced impactors, and results of calculations should be adjusted accordingly to obtain approximate predictions using Eqs. (37–39).

Chick ascribes initiation occurring when $d_{cr}/d_e > 5$ to the bow wave of the penetrating shock, which cannot be modeled by the method just given. It is also possible for shear heating effects to operate in this situation,[166] for which Eq. (40) can be used to obtain the critical jet velocity for bow wave initiation. Data reviews can be used to obtain values of the constants for particular materials.

$$V_{jb}^2 d_e \rho^{\frac{1}{2}} = K_2 \tag{40}$$

with K_2 a constant.

For high-velocity impacts (greater than sonic velocity in the impacted material) by spherically tipped (instead of flat tipped) projectiles with sphere-tip radius equal to rod radius, a rough rule of thumb obtained from the work of James and Hewitt[156] is that about a 50–100% higher impact velocity is required to initiate detonation than with flat-tipped projectiles

$$V_{i(\mathrm{sph})} = G V_{i(\mathrm{flat})} \tag{41}$$

where G ranges between about 1.4 (tungsten) and 1.8 (steel) and depends on the impactor material as well as on the explosive material, impactor and case dimensions, and impact velocity. The experimentally based work of Liddiard and Roslund[157] indicates the factor in Eq. (41) is in the same range as their results. Their work is recommended as an alternate method of calculating. [Sewell's relationship[158] that can be expressed as $d_{\mathrm{sph}} = d_{\mathrm{flat}}(0.4393 V_i - 0.060814 V_i^2 - 0.00029359 V_i^3)$ seems to give values of d_{sph} for low V_i that are lower than reported data.]

For impact velocities lower than sonic velocity in the explosive, Ferm and Ramsay[159] give

$$d_{\mathrm{sph}} = (2X + d_{cr})[1 + (c_{oe}/V_e)^2]^{\frac{1}{2}} \tag{42}$$

where c_o is the bulk sonic velocity in the explosive, described later, and X is the run distance to detonation. For impacts upon cased explosives, the impact velocity upon the case V_i, and the sonic velocity in the case c_{oc} should replace corresponding values[202] for the explosive in Eq. (42).

For rods with the spherical tip radius greater than the rod radius, $V_{i(\mathrm{sph})}$ is lower and approaches, with increasing sphere radius, the value for a flat-tipped rod. The relationship of $V_{i(\mathrm{sph})}$ to the relative radii of the rod and sphere is given by Eq. (43), which is also applicable with some amount of protective case on the explosive, provided the value of $V_{i(\mathrm{flat})}$ also applies to the cased explosive:

$$V_{i(\mathrm{sph})} = V_{i(\mathrm{flat})} \times 10^{(0.2785\, r_{\mathrm{rod}}/r_{\mathrm{sphere}})} \tag{43}$$

For cone-tipped projectiles, James et al.[160] recent work shows a dependence of V_i on cone angle, θ. As a rough rule of thumb, Eq. (44) gives the relationship of increasing V_i as a function of decreasing cone angle (where a flat surface has a 180-deg cone angle):

$$V_{i(\mathrm{cone})} = V_{i(\mathrm{flat})}(1 + 0.0183[180 - \theta]) \tag{44}$$

for θ between 180 and 120 deg. Equation (44) is also applicable with some amount of protective case on the explosive.

Johansson and Persson[64] show a relatively small effect of obliquity in impacts upon bare explosive for angles up to 8-deg. At larger angles of obliquity an abrupt increase in critical impact velocity commences. For oblique impacts of flat-faced projectiles, a rough rule of thumb used by Sewell[161] is that the critical impact velocity V_i, is increased by about 61 m/s for every 1-deg increase of obliquity:

$$V_{i(\mathrm{obl})} = V_{i(\mathrm{flat})} + 61\alpha, \mathrm{m/s} \tag{45}$$

where α is the angle of obliquity in degrees.

For side-on impacts against cylindrical munitions, the effect of this trend alone can be expressed as a probability less than unity of SDT given a randomly located impact of a specific size projectile at a specific velocity. This is because at most fragment impact velocities (for example, the 2530 m/s "unaimed" IM fragment impact test) there may be a large invulnerable area that can result in hits that do not cause SDT. For example, consider a fragment impact scenario against a cylindrical munition for which the fragment velocity is equal to $V_{i(\mathrm{obl})}$ as calculated by Eq. (45) for $\alpha = 20$ deg. For impacts at all angles greater than 20 deg off the cylinder normal (assuming flat projectile face normal to direction of travel), the target will not detonate by an SDT mechanism; since sin 20 deg = 0.34, this represents a 66% probability that a single randomly located impact on the cylinder will not cause a detonation by SDT. This is not as good as it sounds, since with only three such impacts on the cylinder, the detonation probability rises to 96%. Sewell and Graham[162] suggest that if the angle of obliquity [or for pointed impactors, the cone angle of Eq. (44)] exceeds the minimum jetting angle for the impacting materials, the impulsive load follows dynamic pressure relations [of Eq. (56)] rather than shock wave equations. The critical angle range for jet formation for steel-upon-steel impacts is given by Eq. (46)[163,164]:

$$\tan^{-1}(5.945/V_i) > \theta_j > \sin^{-1}(0.864/V_i), \mathrm{deg} \tag{46}$$

where V_i is impact velocity in km/s and θ_j is the angle the projectile surface tangent makes with the target surface.

2. *Sympathetic Detonation Calculations*

The method of Ferm and Ramsay,[159] as exemplified by Eq. (42), has been extended to explore sympathetic detonation (SD) problems involving cylindrical donors and acceptors.[204] The key assumption in developing this approach is that a shock wave is introduced into the cylindrical acceptor case by impact of a large cylindrical or planar impactor. Release waves are assumed to originate on the surface of the acceptor case at the location where the phase velocity of the leading surface contact point is equal to the bulk velocity of sound, c_{oc}, on the acceptor case surface. For one-on-one SD configurations, the donor case is assumed to expand as a growing cylinder (of radius $R_D + \Delta R$) with velocity V_R given by Eq. (12). For stack geometries causing confinement of the expanding donor case metal, the impacting case is assumed to distort to a planar shape due to confinement effects of nearest neighbor munitions on the expanding detonation product gas.[126] For simplification in this analysis, and consistent with some hydrocode calculations,[169,201] the velocity of the effective plane impactor is assumed to be V_{Gurney}, Eq. (6).

For the situation of a planar impactor on a cylindrical case, Eq. (47) applies:

$$d_i/D_A = V_R/c_{oc} \tag{47}$$

where d_i is the effective width of the planar impactor (shock width on acceptor case), D_A the outer diameter of the acceptor cylinder, D_D the outer diameter of donor case, V_R the impacting velocity of the donor case upon the acceptor as given by Eq. (12), and c_{oc} the bulk sonic velocity in the acceptor case material as in Table 3.

For the situation of a cylindrical impactor upon a cylindrical case, Eqs. (48) apply:

$$d_i/D_A = V_R/c_{oc} * (\Delta R + D_D/2)/(L + D_A/2 + D_D/2) \tag{48a}$$

exact, ΔR is case expansion at impact and

$$d_i/D_A = V_R/c_{oc} * (L + D_D/2)/(L + D_A/2 + D_D/2) \tag{48b}$$

approximate, L is intercase distance.

The calculated values of d_i obtained by Eqs. (47) and (48) may be worked backwards through the earlier equations in this section [specifically, Eq. (34) to identify pertinent energetic materials for avoiding sympathetic detonation. Figure 8 shows the results of such a calculation displayed on the SSP for Mk 83 (1000-lb, 30-cm-diam, 9.525-mm steel case thickness) bomb donor and acceptors containing AFX-1100 explosive. Note the difference between P_1 from wedge test and from Eqs. (33). The calculation indicates that a critical diameter d_{cr} between 70 and

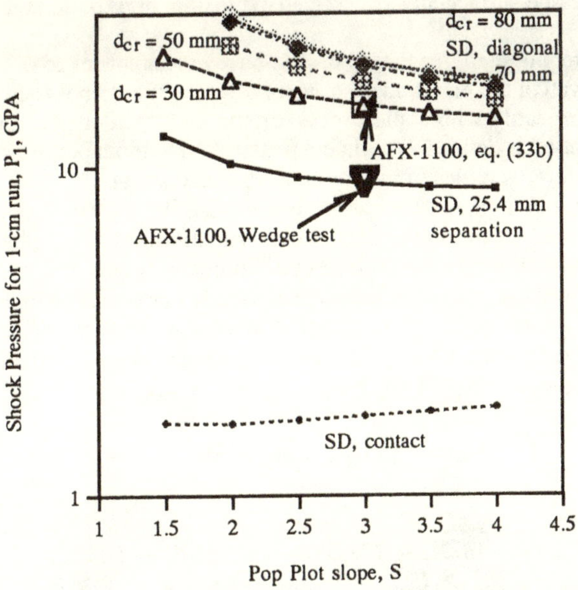

Fig. 8 SSP showing threshold line for sympathetic detonation with Mk 83 bomb containing AFX-1100, ($\sqrt{2}E$ = 2 km/s), d_{cr} = 30, 50, 70, and 80 mm and Eq. (33b).

80 mm is required to pass a sympathetic detonation test in the stack-diagonal position. More complex effects involving wave interactions, which can be calculated with hydrocodes, are beyond the capability of this simple approach.

C. Other Reaction Modes Initiated by Impact

Projectiles with insufficient energy to initiate prompt detonation of a munition can still cause reactions that range in violence from burning through deflagration, explosion, and detonation. From a simple analytical standpoint, it is possible to do little more than estimate the perforation of the case and ignition by shear heating. Parametric studies can be done involving as variables, burning rate vs pressure, case vent area, and burning surface area. Such studies can be used to estimate the case vent and energetic material damage conditions that will permit pressure buildup to levels that cause hazardous case bursts; in fact, such studies have been done.[181] One must always be aware that stimuli or environmental conditions like confinement may be outside the range of available data. A protocol for assessment of the many effects related to impact initiation is available.[166]

SDT initiation modes for cased explosives may be envisioned as occurring in the time frame prior to projectile perforation of the case. Other impact initiation modes than SDT are too complicated for simple analysis. Specifically, at this time a priori prediction of reaction violence is not possible. The mechanisms by which these modes occur are not known with certainty, and analytical and experimental research in these areas is ongoing. The basic mechanisms, as currently envisioned, are XDT and DDT, as described earlier and other mechanisms of impact-induced reactions that lead initially to propellant burning. Then, depending on a number of factors, the burning reaction may 1) continue to completion as burning, 2) increase in violence to cause a case burst with some projection of debris (what is referred to as deflagration[16]), 3) consume explosive at a nearly sonic rate or burn over a greatly increased surface area causing an explosion, or 4) build to a detonation as in a DDT.

The first requirement for these burning-based mechanisms to occur is penetration of one wall of the munition case. More often, perforation is required since this leaves the projectile with some excess kinetic energy after penetrating the case wall. Perforation is defined as compete penetration without plugging. (There have been only a few examples of initiation of serious reactions by projectiles that did not have sufficient energy for case penetration.) The tip shape of the projectile has only a small effect on the perforation-ballistic limit for normal impact. For complete perforation, the ballistic limit of conically tipped projectiles is slightly larger than for flat-faced projectiles. Also, rounded or conically tipped projectiles will have a greater tendency to ricochet at oblique impact angles, particularly when the impact obliquity angle exceeds the tip half-angle.[176] The ballistic limit of a steel case can be calculated approximately by Eq. (49)[177]:

$$V_{50} = C\{4\rho t A/(\pi m \cos\theta)\}^{0.61}, \quad \text{m/s} \tag{49}$$

where

$$C = 332.43 + 171.06B - 14.286B^2 - 3.1111B^3$$

$B = \text{BHN}/100$, where BHN is the Brinell hardness number. ρ the projectile density; t the case thickness, cm; A the projected frontal area of projectile, cm^2; m the projectile mass, g; and θ the angle of obliquity of impact. For a cylindrical target θ is the tangent angle at the impact point. This equation applies for a blunt projectile and a steel case that is thick enough that the bracketed term in Eq. (49) is always greater than 0.125. For other case materials than steel the equation may become more complicated and Ref. 177 should be used.

For harder aluminum alloy case materials, the ballistic limit equation is given approximately by Eq. (50).

$$V_{50} = \{190[4\rho t A/(\pi m \cos\theta)]^{1.75} + 120\}, \text{m/s} \tag{50}$$

If the ballistic limit is exceeded, the projectile moves into the case material with an initial residual velocity given by Eq. (51) as

$$V_r = \left(V_i^2 - V_{50}^2\right)^{\frac{1}{2}}(1.0 + \rho_c t/\rho_p L) \tag{51}$$

where V_i is the initial projectile velocity, L the length of the impactor, t the thickness of the case, ρ_c the density of case material, and ρ_p the density of projectile material.

An alternate approach is to use the THOR equations, which can be applied both to barrier and case perforation. The THOR equations have the added advantage that they account for changes that happen to projectiles during perforation, although they do not account for spall from barrier or case materials. The THOR equations provide a simple method for preliminary calculations of the limiting velocity V_o, for perforation of specific target materials of specific thickness, as well as the residual velocity V_r following perforation, and the residual mass m_r of the perforating fragment. All of these are calculated for steel fragments as a function of the mass m_s, velocity V_s, and obliquity to the target surface θ, of the impacting fragment. The THOR equations do not generate secondary fragments from spalling or fracture of target materials. The general form of the THOR equations is as follows:

$$V_r = V_s - 10^{c_1} \cdot (t_0 A)^{\alpha_1} m_s^{\beta_1} (\sec\theta)^{\gamma_1} \cdot V_s^{\lambda_1} \tag{52}$$

$$m_r = m_s - 10^{c_2} (t_0 A)^{\alpha_2} m_s^{\beta_2} (\sec\theta)^{\gamma_2} \cdot V_s^{\lambda_2} \tag{53}$$

where t_o is the case thickness in centimeters, masses are in grams, velocities are in meters per second, and subscripted parameters (c, α, β, γ, and λ) are characteristics of target materials. Also A is the presented area of the fragment at impact in square centimeters and equal to $K m_s^{2/3}$. $K = 0.3079$ for spherical fragments, $K = 0.3799$ for cubic fragments, and $K = 0.5199$ for random fragment shapes can be used in lieu of more specific data.

The limiting velocity for perforation is given by Eq. (54); $\theta_{\max}$ [Eq. (55)] is the maximum obliquity angle that will give the desired residual velocity V_{rd}.

$$V_o = \left\{10^{c_1} \cdot (t_o A)^{\alpha_1} m_s^{\beta_1} (\sec q)^{\gamma_1}\right\}^{1/(1-\gamma_1)} \tag{54}$$

$$q_{\max} = \sec^{-1}\left\{\left[(V_s - V_{rd})/\left(10^{c_1} \cdot (t_0 A)^{\alpha_1} m_s^{\beta_1} V_s^{\lambda_1}\right)\right]\right\}^{1/\gamma_1} \tag{55}$$

Figure 9 shows results of spreadsheet calculations using Eqs. (52–55). The calculations are for a 16-g fragment of random shape impacting a 1/4-in. thick target sheet of hard homogeneous steel at 2000 m/s. Following perforation of the sheet, the target thickness that will prevent perforation by the residual mass

THOR Fragment Penetration Spreadsheet

Material	Residual velocity equation constants c1	α1	β1	γ1	λ1	Residual mass equation constants c2	α2	β2	γ2	λ2
Mild steel	3.6901	0.889	-0.945	1.262	0.019	-2.478	0.138	0.835	0.143	0.761
hard homo steel	3.7661	0.889	-0.945	1.262	0.019	-2.6711	0.346	0.629	0.327	0.88
face-hard st	2.3053	0.674	-0.791	0.989	0.434	-1.5342	0.234	0.744	0.469	0.483
Cast iron	2.0793	1.042	-1.051	1.028	0.523	-8.89	0.162	0.673	2.091	2.71
2024T-3 Al	3.9356	1.029	-1.072	1.251	-0.139	-6.3215	0.227	0.694	-0.361	1.901

Layer 1

Vs, m/s	2000		Assume random fragment (A=K*ms^.667), K random =0.5199	
ms, g =	16 A=	3.30116 sq cm	Other fragments: K cube=	0.3799
targ thick,cm=	0.635		K sphere=	0.3079
obliq, ang, deg	0			

TARGET MATERIAL: Hard homogeneous steel (row 2)

Residual vel=	1052.27 m/s
Residual mass	3.34051 grams
limit velocity	934.124 m/s for perforation
max obl	56.9066 deg for perforation

Layer 2

targ thick,cm=	0.38924 A=	1.16179 sq cm
obliq, ang, deg	0	
Residual vel=	0.01407 m/s	
Residual mass	1.76072 grams	

Fig. 9 Spreadsheet results for THOR penetration equations; 16-g random-shape steel fragment, 1/4-in.-thick, 2000-m/s impact velocity, hard homogeneous steel target.

and residual velocity is calculated. Also shown are values of the 10 subscripted parameters for five typical target materials.

Ignition of the energetic material will occur by a combination of heat gained by the projectile in the gun barrel, heat from the case debris and the projectile after their violent impact, and heat generated by shear heating as these items move through the energetic material. The least damage that can be caused to the energetic material is the hole through which the metallic intruders have moved. It is more common to have severe breakup of the energetic material due to these intruders. This generates increased surface area that can burn following ignition and result in a very rapid pressure rise.

Sewell and Graham[162] use Eq. (56) to calculate the dynamic pressure generated as a projectile penetrates explosive:

$$P_{\text{dyn}} = \tfrac{1}{2}\rho_e V_r^2 C_D \tag{56}$$

where P_{dyn} is in GPa if the explosive density ρ_e is in g/cm^3, and V_r is in m/s. The drag coefficient $C_D = 2.68$ for $V_r > 100$ m/s or $C_D = 1.8$ for $V_r < 100$ m/s (Ref. 162).

This dynamic pressure is then combined with the work of Frey[165] to obtain a temperature in the shear-heated explosive based on an assumed shear heating mechanism. If one assumes the shear velocity is equal to the projectile residual velocity, Eq. (57) approximates temperature rise as a function of residual velocity. This may be used, if the ignition temperature of the energetic material is known, to make a first approximation to the lowest residual velocity that might cause ignition:

$$T = 25 + 0.4V_r + 0.003V_{r^2},\ ^\circ\text{C} \tag{57}$$

with V_r in meters per second. Any temperature rise due to shear heating competes with cooling thermal conductivity and endothermic phase change effects. Nevertheless, Eq. (57) gives as a rough approximation 350 m/s as an approximate minimum residual velocity required to ignite more sensitive energetic materials. It must be recognized that the concept of an ignition temperature, being a dimensional balance between a source temperature, heat loss by conduction and phase change, and exothermic decomposition must be used cautiously. The methods used in this chapter's section on slow cookoff have been used to estimate an ignition temperature for appropriate shear-layer thicknesses t_c, as shown by Eqs. (58) and (59). Equation (58) gives the estimated minimum shear-layer temperature, T°C, and Eq. (59), the corresponding residual velocity V_r, m/s, calculated as a function of the assumed shear-layer thickness t_c (Ref. 202). As the shear-layer thickness approaches 1 μm, a value suggested by Frey, Eqs. (58) and (59) give values of 600°C and 400 m/s, respectively:

$$T = -123.6 \log t_c + 372,\ ^\circ\text{C} \tag{58}$$

for reduced smoke AP/HTPB propellant and

$$V_r = -56.0 \log t_c + 280,\ \text{m/s} \tag{59}$$

Melting of the energetic material will absorb heat and increase the velocity required for ignition. Heating of the projectile during case penetration and hot spall will tend to reduce the velocity required for ignition.

Sewell and Graham[162] proposed calculations of case confinement effects, including projectile induced venting, on reaction buildup (specifically, by the buildup of internal pressure). To make such calculations in a meaningful way one must also know the burning rate slope vs pressure for the contained energetic material and the surface area of the burning portion as a function of time. For very rapid reactions, as suggested earlier, the dynamic confinement conditions of the reacting region, rather than the static conditions given by measurable case burst pressure are critical. One must be careful to design vents of sufficient area to prevent "choking" of the gas flow from the munition; flow choking (with mechanically or thermally damaged propellant) can occur for small-diameter rocket motors even with both ends removed. Once the combustion-gas flow from the openings in a munition case is choked (e.g., reaches sonic velocity at its temperature and pressure) the internal pressure will continue to rise with some increase in mass efflux. If sufficient volume burning rate is achieved, violent case burst can occur even though the case is vented. A bullet hole presents little resistance to the pressure rise. The critical flow conditions are given by Eq. (60), the well-known nozzle-flow relationship.

$$\left(\frac{dm}{dt}\right)_v = A_t Pk\sqrt{\{[2/(k+1)]^{(k+1)/(k-1)}\}}/\sqrt{(kRT)} \tag{60}$$

where $(dm/dt)_v$ is the mass flow rate from vents; A_t the vent area; P the internal pressure in the case; k the ratio of specific heats of product gas, typically about 1.2; T the temperature of combustion products within case; and R the gas constant.

Assuming ideal gas behavior, the pressure buildup rate within a case can be estimated with equation (61).

$$dP/dt = ((dm/dt)_b - (dm/dt)_v)RT/MV \tag{61}$$

where M is the average molecular weight of the product gas; V the gas volume within case, which increases with time as $dV/dt = dm/dt_{(\text{produced})}/\rho_p$; ρ_p the propellant density; and $(dm/dt)_b$ the mass production rate at burning surface $(A_b) \sim \text{constant} \cdot A_b \rho_p P^n$.

Recent work[178] reported a relationship between response of munitions to the bullet impact test (0.50 cal.) and response of the energetic materials to friability (shotgun and relative quickness - pressure rise rate) tests and hot ball ignition (800°C).[23] The results show absence of reaction or only mild combustion in bullet impacts up to 1140 m/s for explosives whose laboratory tests give 0.5 MPa/ms quickness and no ignition with the hot ball test. Combustion and deflagration reactions in bullet impact were reported corresponding to pressure rise rates of 3–7 MPa/ms and ignition occurring with the hot ball. With hot ball ignition, materials with quickness of about 8–20 MPa/ms showed mostly deflagrations with one reaction ranked an explosion [HMX/polyurethane (PU) at a projectile impact velocity of 1140 m/s] in the bullet impact test, and for quickness above 24 MPa/ms a mix of explosions and detonations occurred in bullet impact tests. Composition B explosive detonated in all its bullet impact tests at velocities above 740 m/s and had a quickness of 114 MPa/ms.

Ammonium perchlorate (AP) containing materials have shock-to-ignition thresholds considerably lower than their thresholds for detonation. Some iron-containing burning-rate catalysts can reduce the ignition temperature as well as the friction sensitivity threshold of AP.[187] For some other energetic materials, the thresholds for shock ignition and SDT seem to be quite similar.[155,179]

Results from Liddiard and Forbes[180] on low-level shock-induced ignition in water-attenuated shock-ignition tests[162] tend to follow a $P^n t$ = const relationship, with t between 5 and 50 μs and n between 0.7 and 2.6. In general, the percentage standard deviation of the constant for a specific explosive is between 10 and 25% over the entire range of data. For TNT-based explosives, n tends to be about 2 (1.7–2.6), indicative of a critical energy relationship; for plastic-bonded explosives, n tends to be about 1 (0.7–1.6), indicative of a critical impulse relationship. However, there are exceptions.[180] The one fairly stable factor in the water-attenuated shock tests is that the minimum-shock pressure that will cause ignition in a large number of explosives is about 5 ± 2 kbar (there are some exceptions apparent in the data).

The effect of multiple-bullet impacts on a cased energetic material can be significant, as shown by Milton and Thorn[181] (Fig. 10). All of the propellants tested by Milton and Thorn contained RDX or HMX and were shock detonable if struck hard enough (harder than with the bullets tested). Also, Milton and Thorn found 0.223-cal. bullets more effective in causing violent reactions than 0.50-cal. bullets. A possible reason for this is that the hyperdamaged zone for the 0.50-cal. bullets may have been larger than the propellant samples themselves. When the first bullet grazed the edge of the grain bore, more propellant was damaged and the probability of a closely placed second bullet causing a very violent reaction was increased. Although it is premature to use this work to generate analytical relationships, it is clear that if a bullet impacts a region damaged (but not destroyed) by a previous bullet, the reaction will often increase in violence, presumably because of the increased sensitivity and surface area of the damaged energetic material.

D. Thermal Threats

Exposure to high temperatures will initiate reaction in energetic materials. Two heating rate regimes are of concern in IM testing, 1) rapid heating by flame

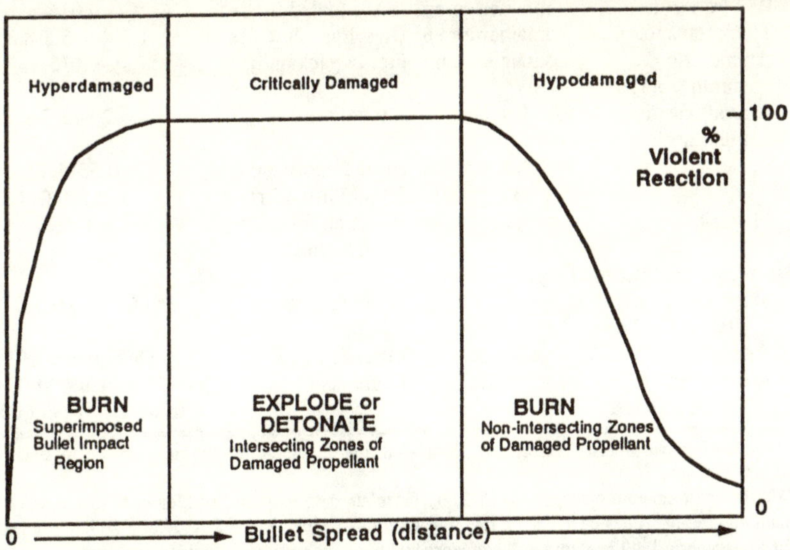

Fig. 10 Effect of multiple bullet impact spatial distribution on reaction violence.[181]

impingement directly on munitions and 2) slower heating as might occur in an area subject to heating, but without direct flame impingement on munitions. The first of these conditions is exemplified by the IM fuel-fire fast cookoff test (FCO)[1,16] or the wood-based bonfire test.[19] At such heating conditions, bare munitions are brought to ignition in about 30 s–5 min, depending on design details and specific test-fire conditions. The second condition is tested, in extremis, by the IM slow cookoff test (SCO) at a heating rate of 6°F/h (3.3 K/h). Assessment of threat conditions indicates a wide range of possible intermediate heating rates as indicated by Table 5, with 50°F/h (27.8 K/h) being a reasonable estimated likely minimum heating rate under shipping or storage conditions.[51,183,197] In actual threat situations, the entire range of heating rates, covering five orders of magnitude, is of concern. In this section, simple analytical methods for calculating cookoff behavior for all heating rates are presented.

Simple one-dimensional quasistatic heat transfer calculations can be used to estimate fast cookoff times of cylindrical ordnance with metal cases. However, modern, low-conductivity composite cases cannot be analyzed as accurately in the same simple way. The most critical parameter for the analysis is the heat flux to the ordnance from the fire. One must remain aware that heating will be different for transparent flames of propane test fires than they are for the opaque smoky flames of wood and jet-fuel fires. Good values for material thermal conductivities and energetic material self-accelerating decomposition and ignition behavior are also needed. Reaction violence cannot be calculated in a simple way with the present state of the art.

Table 5 Typical estimated energetic material surface heating rates for thermal threat scenarios

Logistic cycle	Configuration	Thermal environment	Heating rate, K/s
Fire scenarios			
Rail transportation	container/comp	Wood fueled fire	0.016
Truck transportation	container/comp	Diesel/gasoline fire	1.6-2.8–5.5-8.3[a]
Ammunition dump storage	container/comp	Fire in packaging	0.015
Aircraft carrier flight deck	AUR/launcher	Jet fuel fire	1.6-2.8–5.5-8.3[a]
	AUR/launcher	Heated debris pile	0.55–1.7[a]
	AUR or comp	Jet fuel fire debris	0.055–0.011
Magazine	container/comp	Munition fire (mass fire)	3.3–16.7
Slow/intermediate heating scenarios			
Shipboard Fire (Below Deck)	AUR	Heated magazine walls	(50°F/h) 0.0077
Steam Leak	AUR	Magazine	(6°F/h) 0.00093[b]
Aircraft Jet or Huffer Exhaust	AUR	Flight deck	(~30 min at 600–1000°F followed by cool down)

[a]Typical of range of fast cookoff test heating rates as temperature rise on energetic material outer surface.

[b]Maximum temperature reached (437 K by calculation[51]) may be insufficient to cause reaction in many munitions. It has been shown that a heating rate as high as 22°F (0.0034 K/s) is more justified for this scenario.[197] The steam will condense and flood the magazine, moving the temperature toward 373 K. This is the slow cookoff test heating rate.

Simple hand analysis of slow cookoff problems is generally limited to simple symmetric shapes of the type that were published over 30 years ago involving constant boundary temperatures.[82,83] Modern computers, using finite difference or finite element codes can be used to readily solve problems with varying boundary temperatures to predict time to cookoff. Time-dependent quasistatic solutions of symmetric shapes (slabs, cylinders, and spheres) obtained with desktop computer spreadsheets agree with data and give the same results as the finite difference codes.[214] There are no a priori methods for predicting the violence of the slow cookoff reaction, although a simple equation for this purpose, based on small-scale test measurements of pressure-rise rates is presented in this section. These same comments apply as well to intermediate cookoff heating rates (greater than 6°F/h but less than direct exposure to flame).

There are two aspects to heating munitions to reaction of IM concern that can be considered separately: 1) heating of inert materials placed between the heat source and the energetic material and 2) heating and ignition of the energetic material. Inert materials will include all thermal barriers such as container walls, launchers, munition case, and its inner and outer insulation. Most thermal threat scenarios can be represented by a heating rate of the energetic material surface β, which to a first approximation is considered to be constant. The ignition temperature of an energetic material is a monotonic function of the heating rate at its surface that can be calculated with Eq. (62), which was derived to give the same results as use of the numerical-analysis cookoff spreadsheet (which has accurately predicted fast and slow cookoff ignition temperatures and times to ignition for a number of munitions).[213,214] The cited references explain the derivation of the equation and the time or heating rate dependent variables, $a(t)$, $\delta(t)$, and $E(\beta)$; otherwise, Eq. (62) bears a strong resemblance to Eq. (69) for critical temperature, T_{cr}. The solution to Eq. (62) occurs when $T_{ign} = T_s$:

$$T_{ign} = E(\beta)/\{R\{\ln(p'QZ/C_p) + 1 - \ln[(p'\beta)^2/(1-(1+p'\beta)e^{-p'\beta})]\}\} \quad (62)$$

where

$$p' = E(\beta)\left[a(t)\right]^2/R\delta(t)\alpha T_s^2]$$

and $a(t) = \sqrt{(\alpha t)}$ if $\sqrt{(\alpha t)} < a_0$, otherwise $a(t) = a_0$, the characteristic dimension (radius of the cylinder) and

$$\delta(t) = \delta_{slab} + (\delta_{cyl} - \delta_{slab})a(t)/a_0 = 0.88 + (2 - 0.88)a(t)/a_0$$

$$E(\beta) = E_0 - 0.032\log(\beta) + 0.236[\log(\beta)]^2$$

where E_0 is the constant value of activation energy for $\beta \leq 0.001$, otherwise $E(\beta) = E(0.001)$. The rest of the terms are defined as follows.

a_0 = characteristic dimension, slab half-thickness or cylinder or sphere radius
C_p = specific heat, cal/g-K
E_0 = constant value of activation energy, cal/mole
Q = heat of reaction, cal/g
R = gas constant, 1.987 cal/mole-K
T_s = estimated surface temperature, K
t = heating duration time, s, $(T_s - T_0)/\beta$

Z = collision number, s^{-1}
α = thermal diffusivity = $\lambda/\rho C_p$
δ = shapefactor, 0.88 for slabs, 2 for solid cylinders, and 3.32 for spheres
λ = thermal conductivity, cal/s-cm-K
ρ = density, g/cm^3

For slow heating scenarios the heating rate of the energetic material surface β can often be approximated well by the heating rate of the environmental medium. For higher intermediate heating rates (40-K/h air temperature rise rate, for example), a somewhat slower surface temperature rise may be appropriate, as indicated by data.

There are good physical reasons for the given corrections to the characteristic dimension a_0 and the shape factor δ as a simple means to include transient effects in Eq. (62). The correction to activation energy is based only on its success with two energetic materials, HMX explosive and AP/HTPB[b] propellant from Table 6. Therefore, one should be cautious before extending this method blindly to other energetic materials. However, by comparing this method with the spreadsheet method described earlier for a number of materials, a more extensive database can be built to explore the generality of the assumption. Because the solution to Eq. (62) occurs for $T_{\text{ign}} = T_s$, the iterative calculation is most easily obtained with a spreadsheet because of the visibility of the calculation thereon or by using iterative logic in a memory-containing pocket calculator. Figure 11 compares calculations using Eq. (62) with results from spreadsheet calculations and slow- and intermediate-cookoff measurements on two sizes of rocket motors containing AP/HTPB[b] propellant.

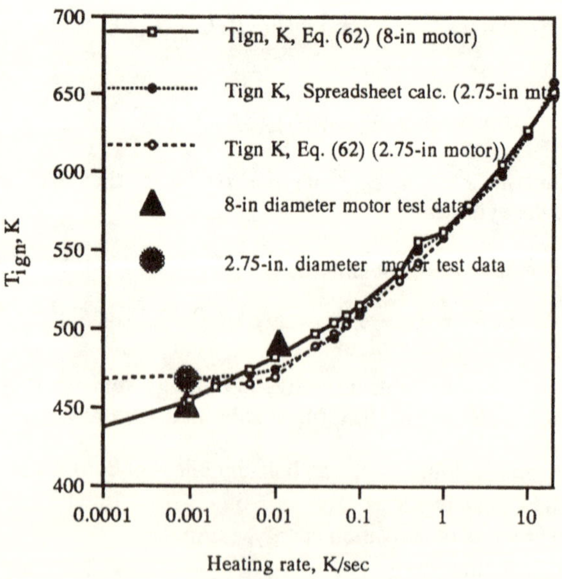

Fig. 11 Comparison of Eq. (62) with cookoff data and spreadsheet calculations.

1. Fast Cookoff

The most important thing one can do to analyze the fast cookoff problem of a munition exposed to direct flame impingement is a heat transfer analysis. For heat transfer analyses, simple one-dimensional cylindrical analysis will be adequate (if applicable to the munition geometry) for metal cased munitions, if there are no other heat paths from the munition case to the interior of the propellant (for example, metal bulkheads) and no important end heating effects. Nonmetal cases, such as fiber/epoxy composites, will experience charring, burning, and other physical changes that obviate the use of a simple heat transfer calculation.

The most critical factor in a heat transfer calculation simulating a fast cookoff heating scenario is the selection of the proper heat flux to the munition. Measurements at Sandia National Laboratories[185] determined a maximum heat flux in large hydrocarbon pool fires of about 160 KW/m^2. These values are consistent with those measured at the Naval Weapons Center (unpublished) in the late 1970s (6–10 Btu/ft^2-s), which indicate that radiation is the predominant heat transfer mechanism. The heat flux, including changes as the surface temperature of the heated cylinder rises, can be approximated by Eq. (63):

$$Q = 60\left(\varepsilon_f \varepsilon_s (T_f/1000)^4 - \varepsilon_s (T_s/1000)^4\right) + 0.006(T_f - T_s), \text{KW/m}^2 \qquad (63)$$

where T_s is the munition surface temperature in Kelvins, T_f the flame temperature in Kelvins, and ε_f and ε_s the flame and surface emissivities, respectively, nominally chosen equal to unity for wood or fuel fires.

The flame emissivity in Eq. (63) can be varied for degree of fire soot. The heat flux is many times greater than would be obtained by assuming an initial heat flux from an outer wall at the nominal flame temperature (1144 K) flowing to an inner wall at ambient temperature (for example, 300 K) and that is why this factor is so critical. In open propane burner tests, the measured heat flux values are about one-half the magnitude of those in liquid hydrocarbon fuel fires and are more strongly dependent on orientation and consistent with a predominant convection heat transfer mechanism.

Because the thermal conductivity of the metal case is at least 100 times higher than that of the liner or energetic material immediately in contact with its inner surface, as a first approximation, the case wall temperature rise can be approximated by the ignoring heat flux from its backwall (Q_{liner}, which ranged from <2 to 7% of the influx). Ignoring the backwall heat flux in Eq. (63) reduces calculated time to cookoff by about 3%:

$$T_{\text{ave}} = T_{\text{ave}}(t - \Delta t) + (Q' - Q_{\text{liner}})\Delta t / \left[C_{p\text{case}} \pi \rho \left(R_o^2 - R_i^2\right)\right] \qquad (64)$$

where T_{ave} is the average case wall temperature. The subscript $(t - \Delta t)$ represents the condition at the previous time step. $T_s = T_{\text{ave}} + \Delta T/2$, see Eqs. (62) and (63). Also, R_o and R_i are the outer and inner case wall radius, respectively, and $Q' = 2\pi R_o Q$, assuming a unit length cylinder. $C_{p\text{case}}$ is the specific heat of case material, and Δt is the time step used in the step-by-step calculation.

The temperature gradient through the case wall is given by the usual cylindrical heat flow equation

$$\Delta T = Q' \ln(R_o/R_i)/2\pi \lambda_{\text{case}} \qquad (65)$$

where λ_{case} is the thermal conductivity of case.

The heat flux through the liner/insulator is calculated by

$$Q_{\text{liner}} = 2\pi\lambda_{\text{ins}}\Delta t[T_{\text{ave}} - \Delta T/2 - T_{\text{liner}(t-\Delta t)}]/\ell n(R_i/R_p) \tag{66}$$

where λ_{ins} is the thermal conductivity of liner, $T_{\text{liner}(t-\Delta t)}$ is the temperature of inner surface of liner or outer propellant/explosive surface, and R_p is the radius of the inner liner surface or outer propellant/explosive surface.

The ignition temperature and time to ignition of the propellant/explosive can be calculated either by using a spreadsheet [including the exponential self-heating term in Eq. (68)] to calculate its temperature rise and ignition or by using Eq. (62) with the case and liner temperature rise rate as calculated.

Figure 12 shows the results of some fast cookoff calculations with the spreadsheet. The results show calculated energetic material surface temperature vs time. Five of these calculations were for 8-in.-diam cylindrical munition cases including AP/HTPB[b] composite propellant (Table 6) in cases with 0.1-in. aluminum, 0.1-in. steel, and 0.5-in. steel wall thicknesses, each with 1/8 and 1/4 in. of insulation/liner, and PBXN-109 explosive in a 0.5-in.-thick case with 1/8-in. liner. Two of the calculations were for 2.75-in.-diam rocket motors. The calculated cookoff times are consistent with similar data for actual munitions and propellant ignition phenomena.[186, 187] The method has been used to calculate cookoff of ordnance in launchers or containers, including the effect of melting the aluminum launcher as shown by Fig. 13.[214]

Table 6 Thermal phenomena: critical temperatures[c] and parameter values

Explosive	T_{cr}, °C	ρ, g/cm^3	Q, cal/g	Z/s^{-1}	E, kcal/m	$\lambda \times 10^4$, cal/cm/s/K	C, cal/g/K	T_d, °C[d]
HMX	253	1.81	500	5 (19)[d]	52.7	7.0	0.27	287
RDX	215	1.72	500	2.015 (18)	47.1	2.5	0.27	204
TNT	287	1.57	300	2.51 (11)	34.4	5.0	0.36	300
PETN	200	1.74	300	6.3 (19)	47.0	6.0	0.25	202
TATB	331	1.84	600	3.18 (19)	59.9	10.0		384
DATB	320	1.74	300	1.17 (15)	46.3	6.0	0.23	
NQ	200	1.63	500	2.84 (7)	20.9	5.0	0.3	
HNS	320	1.65	500	1.53 (9)	30.3	5.0	0.4	
N-109		1.68	525	1.023 (14)	36.5	13.0	0.34	
NC		1.5	500	8.46 (18)	48.5	3.0	0.31	
AP/HTPB[a]		1.806	500	1.29 (10)	32.8	12.7	0.31	
AP/HTPB[b]		1.715	300	1.35 (8)	27.0	7.5	0.29	
Aluminum		2.79				5300	0.21	
Steel		7.89				1000	0.11	
Insulation (typical)		1.45				4.0	0.2	
Graphite composite (est)		2.0				100–1000	0.4 (strong T dependence)	
Glass		2.17				124	0.2	

[a,b]Typical values for two different (AP/HTPB) reduced smoke composite propellants.
[c]Lowest experimental values for "a" between 0.003 and 0.039 slab cm thickness.
[d]T_d represents deflagration point or ignition temperature (shown for comparative information only).
[e]Numbers in parentheses are powers of ten.

When a similar calculation was attempted with a composite case, such a large temperature gradient built up through the case wall that it was obvious the case would be destroyed by the fire before conducted heat could ignite the energetic material, however, the spreadsheet method can be used for case conductivites as low as 0.001 cal/cm/s K and lower if the case is divided in thinner shells. It is clear, however, that a more complicated procedure is needed for composite case fast-cookoff analysis. For munitions with insulating thermal coatings, the temperature rise rate for a given thermal environment can be included to slow the temperature rise at the energetic material surface by using data from back-face heating data for such coatings.[215] For example, the effects of intumescent, charring, subliming, reflective, or simple insulating coatings on surfaces are more difficult to calculate accurately, but they can be treated by coupling graphically to spreadsheet calculations, based on back-face heating data for such coatings. The temperature–time

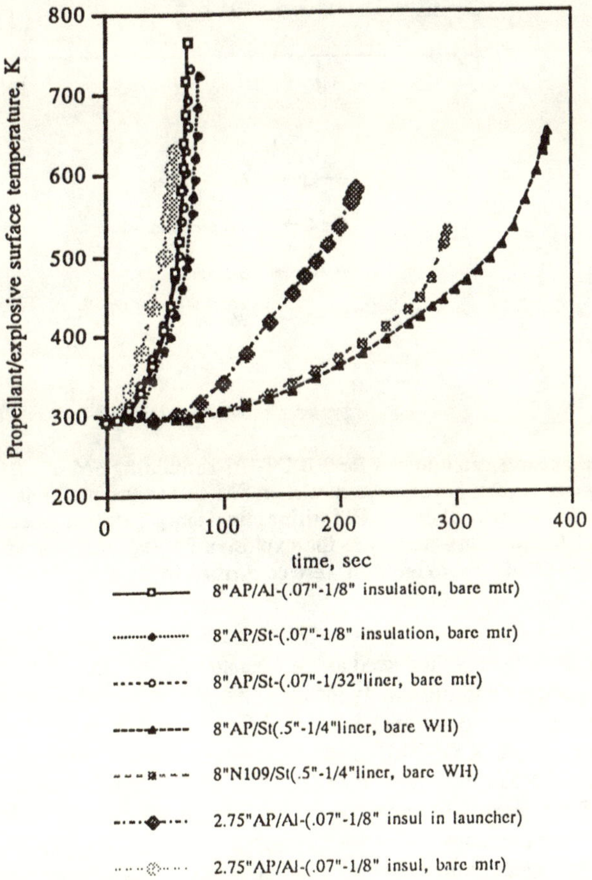

Fig. 12 Propellant/explosive surface temperature time histories from fast cookoff spreadsheet calculations for AP/HTPB propellant and PBXN-109 explosive in cases of aluminum and steel with 0.07-in. (0.178-cm) and 0.5-in. (12.7-cm) wall thicknesses and several thicknesses of insulation or liner, and one cookoff calculation for a launcher-protected 2.75-in. motor; curves stop at calculated cookoff temperatures for each condition.

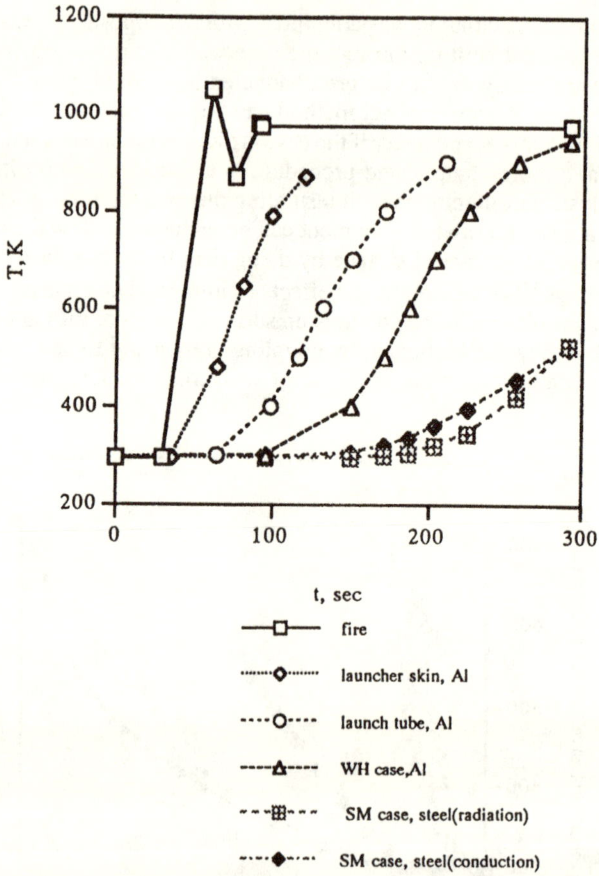

Fig. 13 Fast cookoff calculations for PBXN-107-containing SM of 2.75-in.- (7-cm-) diam warhead in launch tube within launcher, SM curves two methods of calculating heat transfer from warhead case to SM within; final temperature is calculated interior SM case wall temperature at which the explosive ignited. Calculated launch tube temperature rise and time to ignition were confirmed by data.

curve for the back face is then used as the heat source for calculating heat transfer to inner components and ultimately into the energetic material. External fire retardant coatings on munition cases behave in different ways to retard heat transfer to the case. For example, RX-2390, an intumescent material, gives a backwall heating rate of about 1.25 K/s, relatively independent of the original insulation thickness (between 1.5 and 6 mm), however, an initial delay in significant backwall heating of about 75 s/mm precedes the backwall temperature rise. Photon diffusive coating (PDC), a heat reflective material, is characterized by a backwall heating rate of 5 K-mm/s; to estimate the backwall heating rate in an 1144-K flame-temperature fire, divide 5 by the coating thickness in millimeters.

For cased munitions (case and liner thicknesses, t_{case} and t_{liner}, cm) directly exposed to flame temperature T_f K, as given by Eqs. (67) β closely approximates the heating rate of the energetic material surface for a liquid fuel or wood fire with T_f between 1033 and 1366 K (Ref. 214). Once β has been determined with

Eqs. (67), Eq. (62) can be used to estimate time to cookoff in a fast-cookoff scenario at the proper ignition temperature. For cases thicker than about 0.5 cm and liners thicker than 0.4 cm, the exponential term in Eq. (67) becomes too large, and the linearized representation of Eq. (67a) is recommended.[214]

$$\beta = 1/\left[0.589(t_{\text{case}}\rho_c C_{\text{pc}})^{0.9}(1 + 0.0005/\lambda_{\text{case}})(1000/T_f)^4 \times \exp\left(0.00189 T_f t_{\text{liner}}/(t_{\text{case}})^{0.4}\right)\right] \tag{67}$$

$$\beta = 1/\{(-0.0469 + 0.425 t_{\text{case}}(\rho_c C_{\text{pc}}) + [0.846 + 0.921 t_{\text{case}}(\rho_c C_{\text{pc}})^{0.79}] t_{\text{liner}})(1 + 0.0005/\lambda_{\text{case}})(1033/T_f)^3\} \tag{67a}$$

There is currently no a priori method for calculating the violence level to be expected in a fast cookoff reaction. If all pertinent parameters (material condition, vent area to burning surface area, thermal parameters, burning rate and slope, case strength at all locations, etc.,) are known, or can be reasonably estimated, such a calculation could be attempted, but it would involve a fairly large effort.

2. *Slow Cookoff*

From the perspective of preparing IMTHAs and IM test plans, there are two pertinent questions concerning slow cookoff analysis. First, how will the munition react in the standard IM SCO test (heating rate, 6°F/h), and, second, what other heating rates are of operational importance, and how will the munition react at those heating rates?

From the perspective of IM preliminary design, the question of how design variables affect the heat flow, total cookoff time, and the ignition site are important. Of course, the ideal solution method would also predict reaction violence and its dependence on design variables, but that is currently beyond even the most sophisticated analytical methods.[84,85] Simple methods for predicting slow cookoff reaction violence will be unavailable until more complex methods have succeeded and the important physical parameters are identified and routinely measured and reported.

The primary equation of thermal decomposition and heat transfer applicable to predicting thermal explosion (increase of temperature to the ignition point due to self-heating) is Eq. (68), the Franck–Kaminetskii differential equation. For materials that undergo physical changes, it may be necessary to include appropriate changes in density and thermal conductivity (and even Z and E) during the calculation. For materials that melt, it may be necessary to include a convection term in the heat transfer calculation.

$$-\lambda\nabla^2 T + \rho C_p(\mathrm{d}T/\mathrm{d}t) = \rho Q Z w \exp(-E/RT) \tag{68}$$

[The Laplacian operator ∇^2, in the special cases of spheres, infinitely long cylinders, and infinite slabs, reduces to $\nabla^2 T = (\partial^2 T/\partial x^2) + (m\partial T/x\partial x)$] where m is the shape factor and is 0 for slabs, 1 for cylinders, and 2 for spheres.

When the reaction heating term (the right side of Eq. (68)) is zero, the equation is the well known heat flow equation. Because Eq. (68) is not solvable in closed form, it is common practice to solve it for the limiting adiabatic boundary condition,

$\partial T/\partial t = 0$. This defines the critical temperature T_{cr} in Eq. (69). If the exposure temperature is less than T_{cr}, self-heating ignition will never occur:

$$T_{cr} = E/(R\ell n[(a^2 \rho QZEw)/(T_{cr}^2 \lambda \delta R)], \quad \text{K} \tag{69}$$

where a is the slab half-thickness or cylinder or sphere radius and δ is the shape factor, 0.88 for slabs, 2 for cylinders, and 3.32 for spheres.

Note that the unknown variable T_{cr} appears on both sides of the Eq. (69), which can be quickly solved iteratively on a pocket calculator. It is helpful to note that the left side of the equation is relatively insensitive to the guessed value of T_{cr} on the right side. A 20-K error in T_{cr} on the right side leads to an error of only about 1 K on the left. If an energetic material is exposed to a temperature greater than T_{cr}, it will eventually cook off. The time to cookoff can be calculated from Eq. (70):

$$t_{co} = (\rho C_p a^2/\lambda) Fn \tag{70}$$

where $Fn = 10^{lc}$ and

$$lc = -0.008511 - 0.0173v - 0.0061754v^2 + 4.0756 \times 10^{-5} v^3$$

for a cylinder geometry

$$v = E[(1/T_{cr}) - (1/T)]$$

where T is the environmental temperature. For a sphere, the value of Fn is about 0.5 as large, and for a slab about 2.5 times larger.

Equations (69) and (70) are readily set up in the memory of an inexpensive pocket calculator for solutions within a minute of problem definition.

For slow- and intermediate cookoff problems with a known heating rate β, Eq. (62) gives excellent results as shown in Fig. 11. For sufficiently low-heating rates, the insulating effect of the munition case can often be ignored, and the surface heating rate of the energetic material β, can be set equal to that of the surrounding air. A spreadsheet method[214] or finite difference numerical methods may also be used. Calculated ignition temperatures within 2 K of measured values, and times to cookoff within 2–4% of test data are typical.

There are no currently available simple methods for calculating, a priori, the violence of slow cookoff reactions. Recent research indicates the importance of energetic material type, condition, and dynamic confinement at the instant of ignition. When sophisticated calculation methods become available they will operate in the microscale regime of time steps comparable to those required in reactive hydrocode calculations of SDT and in DDT analyses.[84,85]

To a rough approximation, the effective maximum internal pressure at case burst appears to be estimated by Eq. (71). This gives results in reasonable agreement with Butcher's data[129] (assuming $c = 1000$ m/s), and consistent with experience in full-scale slow cookoff tests[87]:

$$P(\text{burst}) = (a/c)\mathrm{d}p/\mathrm{d}t + P_{\text{static burst}} \tag{71}$$

where a is a critical dimension (i.e., radius of test item), c is the velocity of sound in the material within the test item; $\mathrm{d}p/\mathrm{d}t$ is the pressure rise rate due to combustion at the case static-burst pressure, $P_{\text{static burst}}$. The pressure rise rates may be very high, as described earlier. Measured pressure rise rates and time delays from ignition to rapid pressure rise (0.2–5 ms, depending on propellant) correlate well with one-dimensional burn rates for propellants and exponential increases in burning surface area with time.[214] Equation (61) has successfully calculated the very high-pressure rise rates and delay times to onset of accelerated pressure rise rates measured for AP/HTPB propellant and some others following autoignition at slow cookoff heating rates when exponential increases in burning surface area are assumed to initiate at a particular pressure (between about 1.4 and 7 MPa).[132,214] The calculations also indicate the dimensions of a munition that might transition to detonation by a DDT mechanism.

When subjected to very rapid pressure rises, the effective dynamic response of the case may be estimated from its dynamic yield strength [approximately 140,000 psi (0.965 GPa) for aluminum and 350,000–700,000 psi (2.4–4.8 Gpa) for steels] rather than its static strength,[205] although that is not necessary to explain what has been observed here.

VII. Summary and Conclusions

This chapter has given an overview of the ideas of insensitive munitions, the specific hazards of rocket motors, the relationships of these hazards to the physics and chemistry of reactions in energetic materials, design approaches that are useful for making munitions insensitive, and simple calculation methods for predicting hazardous reactions. Under the pressure of proposal preparation and for use in preliminary hazard assessments, systems analyses, and test plans, methods of this simplicity can be invaluable.

Tradeoffs are available; and even if they are not yet very clear, the underlying philosophy is: the higher the risks involved in using a particular propellant, the more thoroughly it must be studied and understood before it is scaled up, loaded into rocket motors, and tested for performance and full-scale IM requirements.[133] As indicated earlier, the full-scale IM tests are not a guarantee of operational insensitivity (for example, a munition that passes the IM fragment impact tests is not guaranteed to pass all operational impacts—nor all possible repetitions of the same test). The data obtained from screening tests on riskier propellants can be applied to threat/hazard assessments to permit assignment of risks and include them in tradeoff considerations.[197]

Finally, some words of caution are appropriate. Insensitive munitions is a system problem, and the entire system and its life-cycle environments should be considered when attempting to reduce the sensitivity or reaction level of a single component. From reading this chapter and its companion chapter,[1] one might get the opinion that the insensitive munitions problem is tied together fairly neatly. This is far from true. Perusal of Fig. 3 shows only some of the areas that have not been covered in detail in the text; there are many others.

As stated earlier, an IM is never totally safe. Does this mean that the guidelines and requirements for achieving IM status are inadequate? Yes! It also takes into account the instability of energetic materials; they degrade from the time they are first produced until they are either used or disposed of, and the sensitivity of many increases throughout this time. One never knows when in its lifetime a munition may encounter an inadvertent stimulus, and its response will be affected by that timing and the munition's prior history. It also reflects the unfortunate fact that no matter how much testing of an energetic material is done, at least to a practical degree, there are always potential behaviors, statistically rare, that from time to time cause devastating catastrophes involving munitions. There is no reason to believe that such events will be totally eliminated by current IM requirements, but they should be reduced. The U.N. requirements for extremely insensitive detonating substances, also known as insensitive high explosives, may help provide even less hazardous munitions.[12,19]

It cannot be emphasized too strongly that the key to meeting both the published requirements and the intended purposes of insensitive munitions is in-depth understanding of the reactive behavior of their energetic materials.

References

[1]McQuaide, P. B.,"Test and Evaluation of Insensitive Munitions," *Test and Evaluation of the Tactical Missile,* edited by E. J. Eichblatt Jr., Vol. 119, Progress in Astronautics and Aeronautics, AIAA, Washington, DC, 1989, pp. 203–232.

[2]"Accident Incident Data Bank," Naval Surface Warfare Center, Dahlgren, VA.

[3]"Explosive Incident Summaries," Naval Weapons Station, Yorktown, VA, Naval Sea Systems Command, June 1984.

[4]Bentley, R.,"Peacetime Stimuli Potentially Hazardous to Air Force Munitions," 1990 Joint Government/Industry Symposium on Insensitive Munitions, American Defense Preparedness Association, White Oak, MD, March 13–14, 1990.

[5]"Memorandum of Agreement on Establishment of a Joint Requirement for Insensitive Munitions," Washington, DC, Sept. 3, 1987.

[6]*Pilot NATO Insensitive Munitions Information Center Newsletter* (PNIMIC), Applied Ordnance Technology, Columbia, MD, April 1991 (see Ref. 11).

[7]"Hazard Studies for Solid Propellant Rocket Motors," AGARDograph 316, AGARD, Neuilly sur Seine, France, 1990.

[8]"NATO Standardization Agreement, Principles and Methodology for the Qualification of Explosive Materials for Military Use," NATO AC/310 Working Group, STANAG 4170.

[9]*Insensitive Munitions (Les Munitions a Risque Atténué) Conference* (Bonn, Germany), Oct. 21–23, 1991, AGARD-CCP-511, Oct. 1991.

[10]Blue, D., Daugherty, E. A., Defourneaux, M., and Stokes, B.,"NIMIC—An Evolution to a Revolution," American Defense Preparedness Association, Insensitive Munitions Technology Symposium, Williamsburg, VA, June 15–18, 1992.

[11]*NIMIC Newsletter*, NATO Headquarters, B-110 Brussels, Belgium, Jan. 1992.

[12]Ward, J. M.,"Hazard Class/Division 1.6 Test Protocol," 1990 JANNAF Propulsion Systems Hazards Subcommittee Meeting, Laurel, MD, April 1990.

[13]Swisdak, M. M., Jr.,"Hazard Class/Division 1.6: Articles Containing Extremely Insensitive Detonating Substances (EIDS)," Naval Surface Warfare Center, Silver Spring, MD, NSWC-TR-89-356, Dec. 1, 1989.

[14]"U.S. Navy Insensitive Munitions Policy," U.S. Navy, OPNAVINST, 8010.13B, June 27, 1989.

[15]"U.S. Navy Insensitive Munitions Requirements," Naval Sea Systems Command, NAVSEAINST 8010.5B, Dec. 5, 1989.

[16]"Hazard Assessment Tests for Non-Nuclear Ordnance," Military Standard, MIL-STD-2105B, 1994.

[17]"Department of Defense—Ammunition and Explosive Safety Standards," Department of Defense, DoD-6055.9 ASD(M, I, and L), July1984.

[18]"Guidance on the Assessment of the Safety and Suitability for Service of Munitions for NATO Armed Forces," NATO-AOP-15, March 1985.

[19a]"Department of Defense—Explosives Hazard Classification Procedures," Department of Defense, U.S. Army, TB 700-2, Navy NAVSEAINST 8020.8, Air Force TO 11A-1-47, Defense Logistics Agency DLAR 8220.1, 1994, rev.

[19b]*Recommendations on the Transportation of Dangerous Goods*, 6th ed., ST/SG/ AC.10/ 1/Rev.6, United Nations Pub., New York, 1989.

[20]"Qualification of Energetic Materials," Naval Sea Systems Command, NAVSE- AINST 8020.5B, May 16, 1988.

[21]"Qualification Procedures for Explosives (High Explosives, Propellants, and Pyrotechnics)," Military Standard, MIL-STD-1751(A), Final Draft, Aug. 31, 1993 (pending approval).

[22]"Safety and Performance Tests for Qualification of Explosives," Naval Ordnance Systems Command, NAVORD OD 44811, Jan. 1, 1972.

[23]"Manual of Tests for the Qualification of Explosive Materials for Military Use," NATO-AOP-7, Feb. 1988, also AOP-7, Annex-1, July 1989.

[24]"Military Standard, System Safety Program Requirements," Department of Defense, DoD-MIL-STD-882B, March 1984.

[25]Murfree, J. A., Ayers, O. E., Hodges, J. C., and Wright, J. W.,"Army Insensitive Munitions (IM) Programs and Objectives for Army Missile Systems, Update," American Defense Preparedness Association, Insensitive Munitions Technology Symposium, Williamsburg, VA, June 15–18, 1992.

[26]DeMay, S.,"Navy IM Propulsion," 1990 Joint Government/Industry Symposium on Insensitive Munitions, American Defense Preparedness Association, White Oak, MD, March 13–14, 1990.

[27]Porada, D.,"US Navy Insensitive Munitions Overview," American Defense Preparedness Association, Insensitive Munitions Technology Symposium, Williamsburg, VA, June 15–18, 1992.

[28]Ayers, O.,"Insensitive Munitions Propulsion," 1990 Joint Government/Industry Symposium on Insensitive Munitions, American Defense Preparedness Association, White Oak, MD, March 13–14, 1990.

[29]Serao, P.,"US Army Insensitive Munitions Overview," American Defense Preparedness Association, Insensitive Munitions Technology Symposium, Williamsburg, VA, June 15–18, 1992.

[30]Jenus, J.,"US Air Force Insensitive Munitions Overview," American Defense Preparedness Association, Insensitive Munitions Technology Symposium, Williamsburg, VA, June 15–18, 1992.

[31]Chizallet, M., et al.,"French Companies Create IM Working Group The Club MURAT," American Defense Preparedness Association, Insensitive Munitions Technology Symposium, Williamsburg, VA, June 15–18, 1992.

[32]Cumming, A. S.,"Insensitive High Explosives and Propellants—The United Kingdom Approach," American Defense Preparedness Association, Insensitive Munitions Technology Symposium, Williamsburg, VA, June 15–18, 1992.

[33]"Hazard Studies for Solid Propellant Rocket Motors," AGARD, Neuilly sur Seine, France, AGARD-CP-367, May 1984.

[34]Hartman, K. O.,"Hercules IM Program," 1990 Joint Government/Industry Symposium on Insensitive Munitions, American Defense Preparedness Association, White Oak, MD, March 13–14, 1990.

[35]Thomas, W. B., and Hightower, J. O.,"Thiokol Approaches to Meeting Insensitive Munitions Challenges," 1990 Joint Government/Industry Symposium on Insensitive Munitions, American Defense Preparedness Association, White Oak, MD, March 13–14, 1990.

[36]Taylor, R. J.,"Aerojet IM Activities," 1990 Joint Government/Industry Symposium on Insensitive Munitions, American Defense Preparedness Association, White Oak, MD, March 13–14, 1990.

[37a]Graham, K. J., Spear, G., and Lynch, R. D.,"Insensitive Munitions, Atlantic Research Corporation Capabilities and Approach Towards the Solution of a System Problem," 1990 Joint Government/Industry Symposium on Insensitive Munitions, American Defense Preparedness Association, White Oak, MD, March 13–14, 1990.

[37b]Lynch, R.,"Development of Insensitive High Explosives Using Propellant Technology," AIAA/SAE/ASME/ASEE 26th Joint Propulsion Conf., AIAA Paper 90-2457, Orlando, FL, July 1990.

[38]Nouguez, B., Berger, H., Gondouin, B., and Brunet, J.,"An Odd Bore Effect on Bullet Induced Detonation of High Energy Propellant Grains," *Proceedings of the Joint International Symposium on Compatibility of Plastics and Other Materials with Explosives, Propellants, Pyrotechnics and Processing of Explosives, Propellants and Ingredients,* Virginia Beach, VA, Oct. 23–25, 1989, American Defense Preparedness Association, Alexandria, VA, 1989.

[39]Ho, S. Y., Fong, C. W., and Hamshere, B. L.,"Assessment of the Response of Rocket Propellants to High-Velocity Projectile Impact Using Small-Scale Laboratory Tests," *Combustion and Flame,* Vol. 77, 1989, pp. 395–404.

[40]Manners, D. J.,"Model Rocket Motor Studies for Reduced Vulnerability," *Combustion and Detonation Phenomena,* 19th International Annual Conf. of ICT 1988, Karlsruhe, Germany, June 29–July 1, 1988, Fraunhofer-Institut fur Chemische Technologie, 1988, pp. 29-1–29-14.

[41]"Smokeless Propellants," AGARD, Neuilly sur Seine, France, AGARD-CP-391, 1986.

[42]DeFourneaux, M.,"Survey of Recent Works on the Mechanisms of Ballistic Impacts on Munitions or Energetic Materials," NATO Insensitive Munitions Information Center, Brussels, Belgium, NIMIC-MD-109-92, April 8, 1992.

[43]Snyer, W. H.,"Ballistic Delivery of Projectiles for the IM Fragment Impact Test Requirement," American Defense Preparedness Association, Insensitive Munitions Technology Symposium, Williamsburg, VA, June 15–18, 1992.

[44]Isler, J., and Gimenez, R.,"Experimental Assessing of Energetic Materials for Insensitive Munitions Applications," *Proceedings of the Joint International Symposium on Compatibility of Plastics and Other Materials with Explosives, Propellants, Pyrotechnics and Processing of Explosives, Propellants and Ingredients* (San Diego, CA), American Defense Preparedness Association, Alexandria, VA, April 1991.

[45]DeFourneaux, M.,"Development of a Methodology for Evaluating the Vulnerability of Munitions," NATO Insensitive Munitions Information Center, Brussels, Belgium, NIMIC-MD-048-92, Feb. 17, 1992.

[46]Victor, A. C.,"Simple Analytical Relationships for Munitions Hazard Assessment," DDESB Explosives Safety Seminar, Anaheim, CA, Aug. 18–20, 1992.

[47]Swierk, T.,"U.S. Navy Ordnance IM Technology," 1990 Joint Government/Industry Symposium on Insensitive Munitions, American Defense Preparedness Association, White Oak, MD, March 13–14, 1990.

[48]Bowen, R., and Bates, K. S.,"U.S. Navy Insensitive Munitions Program," 1990 Joint Government/Industry Symposium on Insensitive Munitions, American Defense Preparedness Association, White Oak, MD, March 13–14, 1990.

[49]Diede, A.,"Summary of Rocket Motor Hazard Miti-gation Concepts Investigated Under the IMAD Program Prior to October 1988," Naval Weapons Center, China Lake, CA, NWC TM 6686, 1990.

[50]Diede, A.,"Initial Development and Evaluation of a Retrofittable Multi-hazard Mitigation System for Rocket Motors," Naval Weapons Center, China Lake, CA, NWC TP 6849, 1990.

[51]Fontenot, J. S., and Jacobson, M.,"Analysis of Heating Rates for the Insensitive Munitions Slow Cookoff Test," Naval Weapons Center, China Lake, CA, NWC TM 6278, July 1988.

[52]Victor, A. C.,"Insensitive Munitions Seminar," Victor Technology, Ridgecrest, CA, 1992.

[53]Hicks, T. A., and Victor, A. C.,"Designing Tactical Composite Motor Cases for Hazard Conditions," TTCP Workshop on Damage to Composite Pressure Vessels, RARDE, Waltham Abbey, UK, Sept. 6–7, 1988.

[54]Mason, A. C.,"The Design Features of Rocket Motors Relating to Insensitive Munition Response to Thermo-Mechanical Stimuli," Insensitive Munitions (Les Munitions a Risque Atténué) Conf., Bonn, Germany, AGARD-CCP-511, Oct. 1991.

[55]Hartman, K. O.,"Hazards Reduction for Tactical Missiles," Insensitive Munitions (Les Munitions a Risque Atténué) Conf., Bonn, Germany, AGARD-CCP-511, Oct. 1991.

[56]Brace, G. G. W.,"Aims and Requirements of NATO Group AC/310," Advisory Group for Aerospace Research and Development, Hazard Studies for Solid Propellant Rocket Motors, AGARD, Neuilly sur Seine, France, AGARD-CP-367, May 1984.

[57]Daugherty, E.,"Pilot NIMIC," 1990 Joint Government/Industry Symposium on Insensitive Munitions, American Defense Preparedness Association, White Oak, MD, March 13–14, 1990.

[58]Mathre, J. K.,"Insensitive Munitions Threat Hazard Assessment Methodology," NAWCWPNS, China Lake, CA, NWC TP 7093, 1993.

[59]Victor, A. C.,"Insensitive Munitions Threat Hazard Assessment, a System Safety Approach," JANNAF Safety and Hazard Classification Panel Meeting, Pt. Mugu, CA, June 2, 1992.

[60]Propulsion Systems Hazards Subcommittee, JANNAF, Safety and Hazard Classification Panel, Nov. 13–14, 1991.

[61]Davis, W. C.,"High Explosives," *Los Alamos Science,* Vol. 2, No. 1, 1981, pp. 48–75.

[62]Davis, W. C.,"The Detonation of Explosives," *Scientific American,* May 1987, pp. 106–112.

[63]Taylor, J., *Detonation in Condensed Explosives,* Clarendon, Oxford, England, UK, 1952.

[64]Johansson, C. H., and Persson, P. A., *Detonics of High Explosives,* Academic, London, 1970, 1981.

[65]Meyer, R., *Explosives,* 3rd ed., VCH Verlagsgstesellschaft mbH, Weinheim, Germany, 1987.

[66]Baker, W. E., Cox, P. A., Westine, P. S., Kulesz, J. J., and Strehlow, R. A., *Explosion Hazards and Evaluation,* Elsevier, Amsterdam, 1983.

[67]Kinney, G. E., and Graham, K. J., *Explosive Shocks in Air,* Springer–Verlag, New York, 1985.

[68]Henrych, J., *The Dynamics of Explosion,* Elsevier, Amsterdam, 1979.

[69]Mader, C. L., *Numerical Modeling of Detonations,* Univ. of California Press, Berkeley, CA, 1979.

[70]Fickett, W., and Davis, W. C., *Detonation,* Univ. of California Press, Berkeley, CA, 1979.

[71]Fickett, W., *Introduction to Detonation Theory,* Univ. of California Press, Berkeley, CA, 1985.

[72]Price, D., and Jacobs, S. J.,"Notes from Lectures on Detonation Physics," transcribed and edited by F. J. Zerilli, Naval Surface Weapons Center, Silver Spring, MD, NSWC MP 81-399, Oct. 1981.

[73]Tasker, D., and Short, J. M. (ed.), *Tenth Symposium (International) on Detonation* (Boston, MA), Naval Surface Warfare Center, White Oak, MD, July 12–16, 1993 (see for examples of state-of-the-art papers and references to previous papers).

[74]Walker, F. E., and Wasley, R. J.,"A General Model for the Shock Initiation of Explosives," *Propellants and Explosives,* Vol. 1, 1976, pp. 71–80.

[75]James, H. R.,"Critical Energy Criterion for the Shock Initiation of Explosives by Projectile Impact," *Propellants, Explosives and Pyrotechnics,* Vol. 13, 1988, pp. 35–41.

[76]James, H., and Hewitt, D. B.,"Critical Energy Criterion for the Initiation of Explosives by Spherical Projectiles," *Propellants, Explosives and Pyrotechnics,* Vol. 14, 1989, pp. 223–233.

[77]Andersen, W. H.,"Approximate Method of Calculating Critical Shock Initiation Conditions and Run Distance to Detonation," *Propellants, Explosives and Pyrotechnics,* Vol. 9, 1984, pp. 39–44.

[78]Lee, P. R.,"Critical Power Density: A Universal Quantitative Initiation Criterion," Royal Ordnance plc, Westcott, Buckinghamshire, UK, 1987.

[79]Brunet, J., and Salvetat, B.,"Detonation Critical Diameter of Advanced Solid Rocket Propellants," *Proceedings of the Joint International Symposium on Compatibility of Plastics and Other Materials with Explosives, Propellants, Pyrotechnics and Processing of Explosives, Propellants and Ingredients,* New Orleans, LA, American Defense Preparedness Association, April 18–20, 1988.

[80a]Finnegan, S. A., Schultz, J. C., Heimdahl, O. E. R., and Lindfors, A. J.,"Backed Plate Impact Fragmentation Behavior," Eleventh International Symposium on Ballistics, Brussels, Belgium, May 9–11, 1989.

[80b]Finnegan, S. A., Schultz, J. C., Heimdahl, O. E. R., and Lindfors, A. J.,"A Study of Impact-Induced Propellant Reactions Using a Planar Rocket Motor Model," American Defense Preparedness Association, Insensitive Munitions Technology Symposium, Williamsburg, VA, June 15–18, 1992.

[80c]Finnegan, S. A., Schultz, J. C., Heimdahl, O. E. R., and Lindfors, A. J., Tenth Symposium (International) on Detonation 1993.

[80d]Finnegan, S. A., Pringle, J. K., Schultz, J. C., Heimdahl, O. E. R., and Lindfors, A. J.,"Impact-Induced Delayed Detonation in an Energetic Material Debris Bubble Formed at an Air Gap," *International Journal of Impact Engineering,* Vol. 14, 1993, pp. 241–254.

[80e]Finnegan, S. A., Gehris, A. P., and Pringle, J. K.,"Comparative Study of Composite Case Materials for Mitigation of Impact Induced Violent Reactions in Solid Rocket Motors," Insensitive Munitions Technology Symposium, American Defense Preparedness Association, Williamsburg, VA, June 6–9, 1994.

[81]Boggs, T. L., Price, C. F., Richter, H. P., Atwood, A. I., and Lepie, A. H.,"Detonation of Undamaged and Damaged Energetic Materials," *Combustion and Detonation Phenomena,* 19th International Annual Conf. of ICT, Karlsruhe, Germany, June 29–July 1, 1988, Fraunhofer-Institut fur Chemische Technologie, 1988, pp. 30-1–30-1.3.

[82]Rogers, R. N.,"Thermochemistry of Explosives," *Thermochimica Acta,* Vol. 11, 1975, pp. 131–139.

[83]Zinn, J., and Mader, C. L.,"Thermal Initiation of Explosives," *Journal of Applied Physics,* Vol. 31, No. 2, 1960, pp. 323–328.

[84]Skocypec, R. D.,"An Evaluation of Cookoff, Status and Direction," 1991 JANNAF Propulsion Systems Hazards Subcommittee Meeting, Albuquerque, NM, March 18–22, 1991.

[85]Raun, R. L., Butcher, A. G., Caldwell, D. J., and Becksteadt, M. W.,"An Approach for Predicting Cookoff Reaction Time and Reaction Severity," 1992 JANNAF Propulsion Systems Hazards Subcommittee Meeting, Silver Spring, MD, April 1992.

[86a]Pakulak, J. M, Jr., and Anderson, C. M.,"NWC Standard Methods for Determining Thermal Properties of Propellants and Explosives," Naval Weapons Center, China Lake, CA, NWC TP 6118, March 1980.

[86b]Pakulak, J. M.,"Prediction and Application of Small-scale Techniques to Cookoff of Full-Scale Motors," 1990 JANNAF Propulsion Systems Hazards Subcommittee Meeting, Laurel, MD, April 1990.

[87]Diede, A., and Victor, A.,"Propellant and Rocket Motor Behavior in Low Heating Rate Thermal Environments," Technical Cooperation Program (TTCP) Subgroup W Action

Group (WAG-21) on The Hazards of Energetic Materials and Their Relation to Munitions Survivability, Australia, March 13–17, 1989.

[88]Dyer, A. S., Haskins, P. J., Hubbard, P. J., and Hutchinson, C. D.,"The Growth and Decay of Explosive Deflagrations in Munitions in Simulated Factory Accident Scenarios," *8th Symposium (International) on Detonation* (Albuquerque, NM), Naval Surface Warfare Center, White Oak, MD, July 1985, pp. 211–227.

[89]Belanger, C., and Hooton, I.,"Low Vulnerability Characteristics of Cast-Cured PBXs," 1990 Joint Government/Industry Symposium on Insensitive Munitions, American Defense Preparedness Association, White Oak, MD, March 13–14, 1990.

[90]Swisdak, M. M., Jr.,"Navy Explosive Safety Improvement Program," *Minutes of the 20th Explosives Safety Seminar,* Vol. I, Norfolk, VA, Aug. 24–26, 1982, pp. 285–302.

[91]Shopher, K. R., and Jacobs, E. M.,"Suppression of Propagation Between Stacks of Bombs," *Minutes of the 22nd Explosives Safety Seminar,* Vol. II, Anaheim, CA, Aug. 26–28, 1986.

[92]Bascombe, K. N., and Wyatt, R. M. H.,"The Quantity/Distance Category of Gun and Rocket Propellants," *Propellants and Explosives,* Vol. 1, 1976, pp. 15–19.

[93]Collis, D. L.,"New Techniques to Reduce Explosion and Fragment Severity of Mass Detonable Munitions," *Minutes of the 20th Explosives Safety Seminar,* Vol. II, Norfolk, VA, Aug. 24–26, 1982, pp. 1143-1165.

[94]Frey, R., Watson, J., Gibbons, G., Boyle, V., and Lyman, O.,"The Response of Compartmentalized Ammunition," 1990 Joint Government/Industry Symposium on Insensitive Munitions, American Defense Preparedness Association, White Oak, MD, March 13–14, 1990.

[95]Corley, J.,"Insensitive Munitions for GP Bombs," 1990 Joint Government/Industry Symposium on Insensitive Munitions, American Defense Preparedness Association, White Oak, MD, March 13–14, 1990.

[96]Held, M.,"TNT-Equivalent," *Propellants, Explosives and Pyrotechnics,* Vol. 8, 1983, pp. 158–167.

[97]Liddiard, T. P., Jr.,"The Initiation of Burning in High Explosives by Shock Waves," *4th Symposium (International) on Detonation,* U.S. Naval Ordnance Lab., Oct. 12–15, 1965, ACR-126, Office of Naval Research, White Oak, MD, 1965, pp. 487–495.

[98]Sun, J., and Chen, M. M.,"A Theoretical Analysis of Heat Transfer Due to Particle Impact," *International Journal of Heat and Mass Transfer,* Vol. 31, 1988, pp. 969–975.

[99]Fong, C. W.,"Crack Initiation in Perforated Propellants Under High Strain Rate Impact Conditions," *Propellants, Explosives and Pyrotechnics,* Vol. 10, 1985, pp. 91–96.

[100]Coffey, C. S., DeVost, V. F., and Woody, D. L.,"Towards Developing the Capability to Predict the Hazard Response of Energetic Materials Subjected to Impact," *Ninth Symposium (International) on Detonation* (Portland, OR), Aug. 28, 1989, Naval Surface Warfare Center, White Oak, Silver Spring, MD, 1989, pp. 965–974.

[101]Coffey, C. S., Woody, D. L., and Davis, J.,"The Ballistic Impact Chamber Test, Some Results and Interpretations," 1990 JANNAF Propulsion Hazards Subcommittee Meeting, JHU/APL, Laurel, MD, April 3–5, 1990; also"Impact Testing of Explosives and Propellants," Naval Surface Warfare Center, NSWCDD/TR-92/280, June 1992.

[102]Ho, S. Y.,"The Mechanism of Impact Ignition of Energetic Materials," *Ninth Symposium (International) on Detonation* (Portland, OR), Aug. 28, 1989, Naval Surface Warfare Center, White Oak, Silver Spring, MD, 1989.

[103]Ho, S. Y., and Fong, C. W.,"Relationship Between Impact Ignition Sensitivity and Kinetics of the Thermal Decomposition of Solid Propellants," *Combustion and Flame,* Vol. 75, 1989, pp. 139–151.

[104]Weiss, R. R.,"Review of USAF Treatment of Solid Propellant Rocket Motor Hazards," *Hazard Studies for Solid Propellant Rocket Motors,* AGARD-CP-367, Lisse, The Netherlands, May 28–30, 1984, pp. 2-1–2-15.

[105]DeMay, S., Coffey, S., and Bernecker, R.,"An Overview of IM Requirements and Subscale Hazards Screening Tests," AIAA/SAE/ASME/ASEE 26th Joint Propulsion Conf., AIAA Paper 90-2455, Orlando, FL, July 1990.

[106]Atwood, A. I., Price, C. F., Boggs, T. L., and Richter, H. P.,"Transient Combustion Analysis of Energetic Materials," *Combustion and Detonation Phenomena,* 19th International Annual Conference of ICT, Karlsruhe, Germany, June 29–July 1, 1988, Fraunhofer-Institut fur Chemische Technologie, pp. 1-1–1-14.

[107]Vetter, R. F.,"Reduction of Fuel Fire Cook-off Hazard of Rocket Motors," Naval Weapons Center, China Lake, CA, NWC-TP 5921, June 1977.

[108]Pakulak, J. M., Jr.,"Simple Techniques for Predicting Sympathetic Detonation and Fast and Slow Cookoff Reactions of Munitions," Naval Weapons Center, China Lake, CA, NWC TP 6660, June 1988.

[109]Davenas, A., et al., *Technologie des Propergols Solides,* Masson, Paris, 1989, pp. 323–347; English ed., *Solid Rocket Propulsion Technology,* Pergamon, Oxford, England, UK, 1993.

[110]Nouguez, B., Derrien, J., Andre, M., and Donze., G.,"Insensitive High Explosives and Munitions at SNPE," 1990 Joint Government/Industry Symposium on Insensitive Munitions, American Defense Preparedness Association, White Oak, MD, March 13–14, 1990.

[111]Gehris, A., Yehle, A. W., DeMay, S., Hicks, T. A., and Bernard, J. P.,"Hybrid SPARROW/SHRIKE IM Motor Case Development," American Defense Preparedness Association, Insensitive Munitions Technology Symposium, Williamsburg, VA, June 15–18, 1992.

[112]Kendall, E.,"The Use of Polymer Bonded Explosives to Meet IM Requirements: The Challenges to Industry—A UK Industry Viewpoint," 1990 Joint Government/Industry Symposium on Insensitive Munitions, American Defense Preparedness Association, White Oak, MD, March 13–14, 1990.

[113]Rutkowski, J., Orosz, J., and Mezg,"Development of Insensitive Propellants and Explosives at ARDEC," 1990 Joint Government/Industry Symposium on Insensitive Munitions, American Defense Preparedness Association, White Oak, MD, March 13–14, 1990.

[114]Lynch, R.,"Development of Insensitive High Explosives Using Propellant Technology," AIAA/SAE/ASME/ASEE 26th Joint Propulsion Conf., AIAA Paper 90-2457, Orlando, FL, July 1990.

[115]Nouguez, B., and Isler, J.,"Insensitive Warhead Concepts: SNPE Progress," Insensitive Munitions Technology Symposium, American Defense Preparedness Association, Williamsburg, VA, June 15–18, 1992.

[116]Ladd, P., Demers, J., and Spahn, P.,"Advanced Initiation Concepts for Boostering Large Critical Diameter Explosives," 1990 Joint Government/Industry Symposium on Insensitive Munitions, American Defense Preparedness Association, White Oak, MD, March 13–14, 1990.

[117]Wagenhals, M., Heimdahl, O. E. R., and Lundstrom, E. A.,"Methodology for Conducting Fragment Impact Analyses," 1990 JANNAF Propulsion Systems Hazards Subcommittee Meeting, Laurel, MD, April 1990.

[118]Hartman, K. O., and Hercules, Inc., private communication, March 1990.

[119]Amster, A. B., Noonan, E. C., and Bryan, G. J.,"Solid Propellant Detonability," *ARS Journal,* Oct. 1960, pp. 960–964.

[120]Brown, B., "What's Wrong with the NOL Card Gap Test?" 1980 JANNAF Propulsion Systems Hazards Meeting, Monterey, CA, Oct. 1980.

[121]Bernecker, R. R.,"Shock Sensitivity Tests, Insensitive Munitions, and Underwater Explosives," 1988 JANNAF Propulsion Systems Hazards Subcommittee Meeting, Los Angeles, CA, March 1988.

[122]Clark, D.,"Some Do's and Don'ts of Hazards Testing—A Summary of the Missile Propellant Safety Evaluation Program," 1987 JANNAF Propulsion Meeting, Propulsion Hazards Specialist Session, San Diego, CA, Dec. 1987, available from the author, Thiokol, Inc. Wasatch Div.

[123]Lundstrom, E. A.,"A New Approach to Shock Sensitivity Testing of Energetic Materials," 1989 JANNAF Propulsion Systems Hazards Subcommittee Meeting, San Antonio, TX, Feb. 1989. (Fig. 4 contains modifications made by Lundstrom subsequent to publication of this reference.)

[124]Ramsay, J. B., and Popolato, A.,"Analysis of Shock Wave and Initiation Data for Solid Explosives," *Fourth Symposium (International) on Detonation,* Naval Ordnance Lab., White Oak, MD, Oct. 12–15, 1965, ACR-126, Office of Naval Research, pp. 233–238.

[125]Kim, K.,"Development of a Model of Reaction Rates in Shocked Composite Explosives," *9th Symposium (International) on Detonation* (Portland, OR), Aug. 28–Sept. 1., 1989, pp. 946–957.

[126]Lundstrom, E.,"Advanced Bomb Family Sympathetic Detonation Analysis," Naval Weapons Center, China Lake, CA, NWC TP 7120, March 1991.

[127]Keefe, R. L.,"Delayed Detonation in Card Gap Tests," *7th Symposium (International) on Detonation,* Annapolis, MD, June 16–19, 1981, pp. 265–272.

[128]Glenn, J. G., and Gunger, M. E.,"Super Large Scale Gap Test," Insensitive Munitions Technology Symposium, American Defense Preparedness Association, Williamsburg, VA, June 15–18, 1992.

[129]Butcher, G. A.,"Propellant Response to Cookoff as Influenced by Binder Type," AIAA/SAE/ASME/ASEE 26th Joint Propulsion Conf., AIAA Paper 90-2524, Orlando, FL, July 1990; also,"Propellant Response to Cookoff as Influenced by Binder Type, Part II Effects of Confinement," 1991 JANNAF Propulsion Systems Hazards Subcommittee Meeting, Albuquerque, NM, March 18–22, 1991.

[130]Farmer, J., et al.,"A Study of Rocket Motors and Large-Scale Hazards Testing for the Insensitive Munitions Advanced Development (IMAD) Propulsion Program," Naval Weapons Center, China Lake, CA, NWC TP 6840, 1988.

[131a]Price, C. F., and Atwood, A. I.,"CBRED II A Versatile Tool for the Characterization of Damaged Propellants," JANNAF Propulsion Systems Hazards Subcommittee Meeting, Albuquerque, NM, March 18–22, 1991.

[131b]Atwood, A. I., et al.,"A Measurement of Propellant Damage Due to Fragment Impact," JANNAF Propulsion Systems Hazards Subcommittee Meeting, Albuquerque, NM, March 18–22, 1991.

[132]So, W., and Francis, E.,"Impact Test Analysis," AIAA/SAE/ASME/ASEE 26th Joint Propulsion Conf., AIAA Paper 90-2450, AIAA Paper 90-2524, Orlando, FL, July 1990.

[133]Mawbey, M. W. S.,"Assessing the Response of Naval Armament Stores in Credible Accident Environments," *Journal of Energetic Materials,* Vol. 7, Nos. 4-5, 1989, pp. 231–242.

[134]LLNL Explosives Handbook, Lawrence Livermore National Lab., Livermore, CA, UCRL-52996 Change B, Jan. 1985.

[135]Gill, R., Asaoka, L., and Baroody, E.,"On Underwater Detonations, 1. A New Method for Predicting the CJ Detonation Pressure of Explosives," *Journal of Energetic Materials,* Vol. 5, 1987, pp. 287–307.

[136]Cruise, D. R.,"Theoretical Computations of Equilibrium Composition, Thermodynamic Properties, and Performance Characteristics of Propellant Systems," Naval Weapons Center, China Lake, CA, NWC TP 6037, Rev. 1, Nov. 1991.

[137]Walker, F. E.,"Calculation of Detonation Velocities from Hugoniot Data," *Propellants, Explosives, Pyrotechnics,* Vol. 15, 1990, pp. 157–160.

[138]Baroody, E., and Peters, S.,"Heat of Explosion, Heat of Detonation, and Reaction Products: Their Estimation and Relation to the First Law of Thermodynamics," Naval Ocean Systems Center, Indian Head, MD, NOS, IHTR 1340, May 1990.

[139]Pearson, J.,"A Fragmentation Model for Cylindrical Warheads," Naval Weapons Center, China Lake, CA, NWC TP 7124, Dec. 1990.

[140]Dehn, J. T.,"Models of Explosively Driven Metal," *Eighth Symposium (International) on Detonation,* Albuquerque, NM, July 15–19, 1985, pp. 602–612.

[141]Glen, G., and Gunger, M. E.,"Sympathetic Detonation Predictive Methods for Mk-82 General Purpose Bombs," Insensitive Munitions Technology Symposium, American Defense Preparedness Association, Williamsburg, VA, June 15–18, 1992.

[142]Walters, W. P., and Zukas, J. A., *Fundamentals of Shaped Charges,* Wiley, New York, 1989.

[143]DeFourneaux, M.,"Energy Transfers in Explosive Propulsion," *Sciences et Techniques de l'Armement,* Vol. 47, No. 3, 1973, UCRL-Trans-10778, Oct. 1974, pp. 723–930.

[144]Kamlet, M., and Finger, M.,"An Alternate Method for Calculating Gurney Velocities," *Combustion and Flame,* Vol. 34, 1979, pp. 213–214.

[145]Odintsov, V. A.,"Expansion of A Cylinder with Bottoms Under the Effect of Detonation Products," *Combustion, Explosion, and Shock Waves,* Vol. 27, 1991, pp. 94–97.

[146]Hirsch, Held, M., et al., *Propellants, Explosives, Pyrotechnics,* recent issues.

[147]Zulkowski, T.,"Development of Optimum Theoretical Warhead Design Criteria," Naval Weapons Center, China Lake, CA, NWC TP 5892, Dec. 1976.

[148]Rinehart, J. S., and Pearson, J.,"Behavior of Metals Under Impulsive Load," American Society for Metals, Cleveland, OH, 1954.

[149]Sewell, R. G. S.,"Fragmentation of Uncontrolled Cylinders," COMARCO, Ridgecrest, CA, Sept. 1987.

[150]Pearson, J.,"A Fragmentation Model Applied to Shear-Control Warheads," Naval Weapons Center, China Lake, CA, NWC TP 7146, May 1991.

[151]Lundstrom, E. A.,"Shock Sensitivity Testing and Analysis for a Minimum Smoke Propellant," 1990 JANNAF Propulsion Systems Hazards Subcommittee Meeting, Albuquerque, NM, March 1991.

[152]Lundstrom, E. A., private communication, Naval Air Warfare Center, China Lake, CA, Jan. 1992.

[153]Heimdahl, O. E. R., and Dimaranan, L. F.,"Study of Impact Induced Detonation for Steel and Titanium Covered PBXN-107," 1992 JANNAF Propulsion Systems Hazards Subcommittee Meeting, Naval Surface Weapons Center, Silver Spring, MD, April 27–May 1, 1992.

[154]Green, L.,"Shock Initiation of Explosives by Impact of Small Diameter Cylindrical Projectiles," *Seventh Symposium (International) on Detonation,* Naval Surface Weapons Center, White Oak, MD, NSWC, June 1981, pp. 273–277.

[155]James, H. R.,"Critical Energy Criterion for the Shock Initiation of Explosives by Projectile Impact," *Propellants, Explosives, Pyrotechnics,* Vol. 13, 1988, pp. 35–41.

[156]James, H. R., and Hewitt, D. B.,"Critical Energy Criterion for the Initiation of Explosives by Spherical Projectiles," *Propellants, Explosives, Pyrotechnics,* Vol. 14, 1989, pp. 223–233.

[157]Liddiard, T., and Roslund, L.,"Fragment Impact Sensitivity of Explosives," Naval Surface Weapons Center, White Oak, MD, NSWC TR89-184, Sept. 1991.

[158]Sewell, R. G. S.,"Fragment Impact Response of Warheads," JTCG Surface Target Survivability Meeting, Eglin AFB, FL, Jan. 1975.

[159]Ferm, E. N., and Ramsay, J. B.,"Spherical Projectile Impact on Explosives," *9th Symposium on Detonation,* Paper 41, U.S. Government Printing Office, Aug. 1989, pp. 662–665.

[160]James, H. R., Haskins, P. J., and Cook, M. D.,"Effect of Case Thickness and Projectile Geometry on the Shock Initiation Threshold for a Given Explosive," *Insensitive Munitions,* AGARD, Neuilly sur Seine, France, AGARD, Conf. Preprint 511, Oct. 1991.

[161]Sewell, R. G. S., private communication, Consultant Physicist, Ridgecrest, CA.

[162]Sewell, R. G. S., and Graham, K. J.,"Fragment Initiation of Cased Explosives," *Air Weaponry Technology Program for Strike Warfare, FY 1983,* Second Quarterly Rept.; Vol. 4, Warheads, Naval Weapons Center, China Lake, CA, NWC, TP 6350-6, 1983.

[163]Carpenter, S.,"High Energy Forming," Denver Research Institute, Denver, Co., Rept. AMMRC CTR 74-69, Nov. 1974.

[164]Sewell, R. G. S.,"Effects of Velocity and Material Properties on Design Limits for Linear Shaped Charges," Naval Ordnance Test Station, China Lake, CA, NOTS TP 3894.

[165]Frey, R. B.,"The Initiation of Explosive Charges by Rapid Shear," *Seventh Symposium (International) on Detonation,* Naval Surface Weapons Center, White Oak, MD, June 1981, pp. 36–42.

[166]Boggs, T. L., and Dickinson, C. W. (eds.), *The Hazards of Energetic Materials and Their Relation to Munitions Survivability,* Technical Cooperation Program (TTCP) Subgroup W Action Group(WAG)-11, Summary Rept. Workshop Naval Weapons Center, China Lake, CA, March 1990, now NIMIC.

[167]Marsh, S. P., *LASL Shock Hugoniot Data,* Univ. of California Press, Berkeley, CA, 1980.

[168]Hancock, P., O'Connor, J., Spahn, P., and Wilson, W.,"Fragment Impact Studies for Explosives, Cases, and Liners," Insensitive Munitions Technology Symposium, American Defense Preparedness Association, Williamsburg, VA, June 15–18, 1992.

[169]Glenn, J. G., McCormick, M., and Gunger, M. E.,"Sympathetic Detonation Predictive Methods," Insensitive Munitions Technology Symposium, American Defense Preparedness Association, Williamsburg, VA, June 15–18, 1992.

[170]Wagenhals, M., Lundstrom, E., Heimdahl, R., Randolph, R., and Boggs, T.,"Which Threat? Which Response? or Determining Vulnerability of Weapons to Real World Ballistic Impact Hazards," 1988 JANNAF Propulsion Systems Hazards Subcommittee Meeting, Los Angeles, CA, March 1988.

[171]Daniels, P., McDonald, J. W., and Morgan, V. K.,"Subsonic Transonic, and Supersonic Drag Categories of Warhead Fragments," Naval Surface Weapons Center, Dahlgren, VA, NSWC TR 81-112, May 1981.

[172]*Engineering Design Handbook, Design for Control of Projectile Flight Characteristics*, U.S. Army Materiel Command, AMCP 706-242, Sept. 1966, pp. 4–6.

[173]Brunet, J., Hamaide, S., Nouguez, B., and Pitiot, F.,"Bullet Impact Behavior of Solid Propellant Grains," *Insensitive Munitions,* AGARD, Neuilly sur Seine, France, AGARD, Conf. Preprint 511, Oct. 1991.

[174]Finnegan, S. A., Schultz, J. C., Pringle, J. K., and Lindfors, A. J.,"The Relationship Between Ballistic Impact Damage and Violent Reaction in Cased Propellant," 1991 JANNAF Propulsion Systems Hazards Subcommittee Meeting, Albuquerque, NM, March 1991.

[175]Backman M. E., and Goldsmith, W.,"The Mechanics of Penetration of Projectiles into Targets," *International Journal of Engineering Science,* Vol. 16, 1978, pp. 1–99.

[176]Ipson, T., Recht, R., and Schmeling, W.,"Effect of Projectile Nose Shape on Ballistic Limit Velocity, Residual Velocity, and Ricochet Obliquity," Denver Research Institute for Naval Weapons Center, China Lake, CA, NWC TP 5607, Dec. 1973.

[177]"Penetration Equations Handbook for Kinetic Energy Penetrators," 61 JTCG/ ME-77-16, 1977.

[178]Isler, J., Gimenez, P., and Hamaide, S.,"Experimental Addressing of Energetic Materials for IM Applications," Joint International Symposium on Compatibility of Plastics and Other Materials with Explosives, Propellants, Pyrotechnics and Processing of Explosives, Propellants and Ingredients, American Defense Preparedness Association, San Diego, CA, April 1991.

[179]Andersen, W. H., and Louie, N. A.,"Projectile Impact Ignition Characteristics of Propellants, I. Deflagrating Composite Propellants," *Combustion Science and Technology*, Vol. 20, 1979, pp. 153–160.

[180]Liddiard, T. P., and Forbes, J. W.,"A Summary Report of the Modified Gap Test and the Underwater Sensitivity Test," Naval Surface Warfare Center, Silver Spring, MD, NSWC TR 86-350, March 1987.

[181]Milton, R. W., and Thorn L. B.,"Rocket Motor Design Considerations for Bullet and Fragment Impact Impact," Insensitive Munitions Technology Symposium, American Defense Preparedness Association, Williamsburg, VA, June 15–18, 1992.

[182]Cantey, D. E.,"Insensitive High Explosive Munition Redesign Study *Technical Report*," Contract F08635-87-C-0225, Sunnyvale, CA, LMSC-F230369, Jan. 1990.

[183]Victor, A. C.,"IM Threat Hazard Assessment Reports on the ABF, AIWS, and ARS," Victor Technology, Ridgecrest, CA, 1991.

[184]Pakulak, J. M.,"ABF Cookoff," ABF Technology Transfer Conf., San Diego, CA, April 2–4, 1991.

[185]Gregory, J. J., Keltner, N. R., and Mata, R.,"Thermal Measurements in Large Pool Fires," *Transactions of the American Society of Mechanical Engineers*, Vol. 111, May 1989, pp. 446–454.

[186]Fontenot, J. S.,"Summary Report of Insensitive Munitions Testing of Bombs, Rockets, and Missiles," Naval Weapons Center, China Lake, CA, NWC TP 7077, Jan. 1991.

[187]Bazaki, H., and Kubota, N.,"Friction Sensitivity Mechanism of Ammonium Perchlorate Composite Propellants," *Propellants, Explosives, Pyrotechnics*, Vol. 16, 1991, pp. 41–47.

[188]*A Manual for the Prediction of Blast and Fragment Loading on Structures*, U.S. Dept. of Energy, Albuquerque Operations Office, Amarillo, TX, DOE/TIC-11268, Change 2–1, April 1982, pp. 6–54.

[189]Kornhauser, M.,"Engineering Methods of Calculating Munition Sensitivity to Impact by Bullets and Fragments," 3C Systems, Inc., Wynnewood, PA, SBIR Contract N00024-87-C-5163, July 1987.

[190]Kornhauser, M.,"Engineering Methods of Calculating Sympathetic Detonation of Shielded and Unshielded Munitions," 3C Systems, Inc., Wynnewood, PA, SBIR Contract N00024-87-C-5162, July 1987.

[191]Howe, P. M., *The Response of Munitions to Impact*, Army Ballistic Research Lab., Aberdeen, MD, ARBRL-TR-02169, June 1979.

[192]Stolovy, A., et al,"Thermal Initiation of High Explosives by Electron Beam Heating," *Journal of Energetic Materials*, Vol. 5, 1987, pp. 181–238.

[193]Bonner, B. H.,"The Role of Liquid Propellants in Insensitive Munitions Policy," *Propellants, Explosives, Pyrotechnics*, Vol. 16, 1991, pp. 194–196.

[194]Hamaide, S., Quidot, M., and Brunet, J.,"Tactical Solid Rocket Motors Response to Bullet Impact," *Propellants, Explosives, Pyrotechnics*, Vol. 17, 1992, pp. 120–125.

[195]Held, M.,"Initiation Phenomena with Shaped Charge Jets," Ninth Symposium (International) on Detonation, Portland, OR, Naval Surface Warfare Center, White Oak, MD, Paper 211, Aug. 28, 1989.

[196]The Rocketeer,"Intermetallic Thermal Sensor," Naval Air Weapons Station, China Lake, CA, Patent applied for, Navy Case No. 74073, July 29, 1993, p. 17.

[197]Victor, A. C.,"Insensitive Munitions Threat Hazard Assessment: Methodology and Examples," 1993 JANNAF Propulsion Systems Hazards Subcommittee Meeting, Fort Lewis, WA, May 1993; also ADPA *1994 Insensitive Munitions Technology Symposium Proceedings,* American Defense Preparedness Association, Williamsburg, VA, June 6–9, 1994 pp. 85–106.

[198]Dick, J. J.,"Nonideal Detonation and Initiation Behavior of a Composite Solid Rocket Propellant," *Seventh Symposium (International) on Detonation,* U.S. Naval Academy, Annapolis, MD, U.S. Government Printing Office, Washington, DC, NSWC MP 82-334, 1981, pp. 620–623.

[199]Lindfors, A. J., and Heimdahl, O. E. R.,"An Energy Transport Model for the Shock Initiation of Composite Explosives and Propellants," *Tenth Symposium (International) on Detonation* (Boston, MA), Naval Surface Warfare Center, White Oak, MD, July 12–16, 1993.

[200]Kennedy, D. L., and Jones, D. A.,"Modelling Shock Initiation and Detonation in the Non-Ideal Explosive PBXW-115," *Tenth Symposium (International) on Detonation* (Boston, MA), Naval Surface Warfare Center, White Oak, MD, July 12–16, 1993.

[201]Glenn, J. G., and Gunger, M.,"Simulating Sympathetic Detonation Effects," *Tenth Symposium (International) on Detonation* (Boston, MA), Naval Surface Warfare Center, White Oak, MD, July 12–16, 1993.

[202]Victor, A. C.,"A Simple Method for Calculating Shock Initiation of Explosives by Projectile Impact," 1993 JANNAF Propulsion Systems Hazards Subcommittee Meeting, Fort Lewis, WA, May 1993.

[203]Price, D.,"Examination of Some Proposed Relations Among HE Sensitivity Data," *Journal of Energetic Materials,* Vol. 3, 1985, pp. 239–254.

[204]Victor, A. C.,"A Simple Method for Calculating Sympathetic Detonation of Munitions," 1993 JANNAF Propulsion Systems Hazards Subcommittee Meeting, Fort Lewis, WA, May 1993.

[205]Victor, A. C.,"Warhead Performance Calculations," Victor Technology, Ridgecrest, CA, July 1993.

[206]Finnegan, S. A., and Atwood, A. I., et al.,"A Study of Impact-Induced Violent Reactions in Cased Propellant Using Planar Impact Model and Radiant Ignition/Burn Rate Measurements," 1993 JANNAF Propulsion Systems Hazards Meeting, Fort Lewis, WA, May 1993.

[207]Roux, M., Marlin, F., Brassy, C., and Gillard, P.,"Numerical Determination of the Thermal Diffusivity and Kinetic Parameters of Solid Explosives," *Propellants, Explosives, Pyrotechnics,* Vol. 18, 1993, pp. 188–194.

[208]Wachtell, S., and McKnight, C. E.,"A Method for Determining the Detonability of Propellants and Explosives," *Third Symposium on Detonation,* (Princeton Univ. Princeton, NJ), Office of Naval Research, ACR-52, Sept. 1960, pp. 635–658.

[209]Souers, P. C., and Kury, J. W.,"Comparison of Cylinder Data and Code Calculations for Homogeneous Explosives," *Propellants, Explosives, Pyrotechnics,* Vol. 18, 1993, pp. 175–183.

[210]Carleone, J. (ed.), *Tactical Missile Warheads,* Progress in Astronautics and Aeronautics, Vol. 155, AIAA, Washington, DC, 1993.

[211]Pizzo, J. T., Spear, G. B., Graham, K. J., and Lynch, R. D.,"Quantitative Threat Hazard Assessment Methodology," 1993 JANNAF Propulsion Meeting, Monterey, CA, Nov. 15–19, 1993.

[212]"French National Doctrine with Regards to Less Sensitive Munitions," (Munitions á Risques Atténués), Director for Armaments, DGA/IPE Instruction No. 260, France, July 1993.

[213]Creighton, J. R.,"The Variation of the Ignition Temperature of Solid Explosives as a Function of Heating Rate," 1993 JANNAF Propulsion Systems Hazards Subcommittee Meeting, April 1993.

[214]Victor, A. C.,"Exploring Cookoff Mysteries," 1994 JANNAF Propulsion Systems Hazards Subcommittee Meeting, San Diego, CA, August 1994.

[215]Koo, J. H., Miller, M. J., and Kneer, M. J.,"The Effect of Hydrocarbon Flames to Fire Retardant Materials," Combustion Fundamentals and Applications, Joint Technical Meeting, Central and Eastern States Sections of the Combustion Institute, Pittsburg, PA, March 15–17, 1993.

[216]DeMay, S. C.,"Recent Advances in the Navy's Insensitive Munitions Advanced Development Propulsion Program," *ADPA 1994 Insensitive Munitions Technology Symposium,* American Defense Preparedness Association, Williamsburg, VA, June 6–9, 1994 (not printed in initial Proceedings).

[217]Comfort, T. F., et al.,"Insensitive HTPE Propellants," *ADPA 1994 Insensitive Munitions Technology Symposium,* American Defense Preparedness Association, Williamsburg, VA, June 6–9, 1994 (not printed in initial Proceedings).

[218]Avnon, I., and Peretz, A.,"Slow Cookoff Research of AP Based Composite Propellants," *ADPA 1994 Insensitive Munitions Technology Symposium,* American Defense Preparedness Association, Williamsburg, VA, June 6–9, 1994, pp. 170–178.

[219]Magnum, M. G., Cherry, C. C., and Wiechering, R. E.,"Combustion Behavior of Reduced Smoke Propellant Under Cookoff Conditions," *ADPA 1994 Insensitive Munitions Technology Symposium,* American Defense Preparedness Association, Williamsburg, VA, June 6–9, 1994, pp. 191–199.

[220]DeFusco, A., et al.,"Development and IM Testing of a Class of 1.3 Minimum Signature Propellants," *ADPA 1994 Insensitive Munitions Technology Symposium,* American Defense Preparedness Association, Williamsburg, VA, July 6–9, 1994, pp. 207–220.

[221]Harrod, C. E.,"An Insensitive Nitrocellulose Based High Performance Minimum Smoke Propellant," *ADPA 1994 Insensitive Munitions Technology Symposium,* American Defense Preparedness Association, Williamsburg, VA, July 6–9, 1994, pp. 221–228.

[222]Campbell, D., Marshall, E. J., and Cumming, A. S.,"Development of Insensitive Rocket Propellants Based on Ammonium Nitrate & PolyNIMMO," *ADPA 1994 Insensitive Munitions Technology Symposium,* American Defense Preparedness Association, Williamsburg, VA, July 6–9, 1994, pp. 229–239.

[223]Allen, B. D.,"Gels—A "Smart" Insensitive Munitions Preparedness Solution," *ADPA 1994 Insensitive Munitions Technology Symposium,* American Defense Preparedness Association, Williamsburg, VA, July 6–9, 1994, pp. 240–246.

[224]LeRoy, M.,"SNPE Methodology for Insensitive Rocket Motors," *ADPA 1994 Insensitive Munitions Technology Symposium,* American Defense Preparedness Association, Williamsburg, VA, July 6–9, 1994, pp. 439–448.

[225]Allen, C. A.,"JAVELIN Insensitive Munitions Results," *ADPA 1994 Insensitive Munitions Technology Symposium,* American Defense Preparedness Association, Williamsburg, VA, July 6–9, 1994, pp. 449–461.

[226]Dhillon, M., Weyland, H., and Miller, R.,"Insensitive Munitions Rocket Motor," *ADPA 1994 Insensitive Munitions Technology Symposium,* American Defense Preparedness Association, Williamsburg, VA, July 6–9, 1994, pp. 462–472.

[227]Rothgery, E. F.,"Safety and Hazards Evaluation of Hydroxylammonium Nitrate Solutions," *ADPA 1994 Insensitive Munitions Technology Symposium,* American Defense Preparedness Association, Williamsburg, VA, July 6–9, 1994, pp. 431–438.

[228]Lundstrom, E.,"The Design of Ordnance to Survive Fragment Impact," *ADPA 1994 Insensitive Munitions Technology Symposium,* American Defense Preparedness Association, Williamsburg, VA, July 6–9, 1994 (not printed in initial Proceedings).

[229]Baker, P. J., and Mellor, A. M.,"Critical Initiation Energy Tests on AP Composite Propellants," *ADPA 1994 Insensitive Munitions Technology Symposium,* American Defense Preparedness Association, Williamsburg, VA, July 6–9, 1994, pp. 179–190.

[230]Kernen, P., and the NIMIC Staff,"Ways and Methods to Insensitive Munitions — IM Recipes," NATO Insensitive Munitions Information Center, Brussels, Belgium, NIMIC-PK-425-93, 31 Oct. 1993.

Chapter 10

Gas Turbines

Jeffrey M. Stricker*

Wright Laboratory, Wright–Patterson Air Force Base, Ohio 45433

Nomenclature

A	= area
Aa	= annulus area
a	= speed of sound
BHI	= blade height
BPR	= bypass ratio
C	= absolute velocity
C_f	= nozzle friction coefficient
CFG	= gross thrust coefficient
$CNPR$	= nozzle choking pressure ratio
Cp	= specific heat at constant pressure
C_v	= nozzle velocity coefficient
DOR	= degree of reaction
ER	= expansion ratio
FAR	= fuel-air ratio
FG	= gross thrust
FHV	= fuel heating value
FN	= net thrust
FR	= ram drag
g	= gravitational constant
H	= enthalpy
L/H	= burner-length-to-height ratio
M	= Mach number
N	= slip factor
NPR	= nozzle pressure ratio
P	= pressure
PR	= pressure ratio
R	= universal gas constant
RT	= residence time

*Assessment Group Leader, Engine Integration and Assessment Branch.

r	=	radius
SM	=	stall margin
SR	=	space heat release rate
T	=	temperature
$TSFC$	=	thrust specific fuel consumption
U	=	wheel speed
V	=	velocity
Vol	=	volume
W	=	relative velocity
W_f	=	fuel flow
Wa	=	airflow
Wg	=	gas flow
α	=	absolute angle
β	=	relative angle
δ	=	pressure correction
η	=	efficiency
θ	=	temperature correction
ρ	=	density

Subscripts

amb	=	ambient
burn	=	burner
bypass	=	bypass stream
comp	=	compressor
core	=	core stream
corr	=	corrected condition
diff	=	diffuser
dome	=	burner dome
eff	=	efficiency
f	=	friction
hub	=	component hub
jet	=	nozzle exhaust
noz	=	nozzle
oper. pt.	=	compressor operating point
ov	=	overall
pass	=	burner passage
poly	=	polytropic
prop	=	propulsive
ram	=	inlet ram
ref	=	reference
spec	=	specific
stall	=	compressor stall point
T	=	stagnation condition
ther	=	thermal
tip	=	component tip
turb	=	turbine
v	=	velocity

I. Introduction

THE expendable gas turbine engine used for unmanned weapon systems is the subject of this chapter. Turbomachinery design methods will be discussed which relate in general to this type propulsion device. Many factors including cost, starting, and dimensional constraints make expendable engines unique compared to man-rated engines. These will also be described.

The turbine engine has inherent qualities which make it the preferred propulsion device for many applications. Put in terms of specific impulse, the turbojet will deliver greater than 2400 s. The turbofan has even better fuel consumption characteristics with a specific impulse of about 3500 s or better. The rocket engine, on the other hand, is below 500 s. This means that the turbine engine is best suited when range and/or powered flight to the target are required.

To gain a better understanding of the factors which have led to the use of gas turbines in expendable vehicles, it is useful to take a historical look at their development. The following section describes the historical appearance of turbine engines.

II. Background

The use of gas turbine engines for unmanned applications is a fairly recent development. Gas turbines first appeared in the late 1950s in several target drones such as the Firebee (J-33 turbojet) and QF-80 (F-80 fighters used as drones, J-33). These expendable systems used off-the-shelf man-rated engines. Missiles such as the Hound Dog (J-52 turbojet) and Snark (J-33 turbojet) appeared in the mid 1950s. These used man-rated engines as well. Consideration for gas turbine engines specifically designed for expendable applications did not occur in the United States until the late 1960s.

The U.S. Navy first considered an expendable turbine engine for a subsonic antiship missile. This system became known as the Harpoon, which is powered by the Teledyne J-402 turbojet (see Fig. 1). The Harpoon had its beginnings in 1968 as

Fig. 1 J-402 turbojet engine.

a study examining the design requirements for a tactical air-to-surface vehicle. In 1971, the program was combined with another program which provided surface-to-surface capability. For the surface launch variant, a rocket motor accelerated the vehicle to the J-402's takeover point. The first prototype flight tests (air, submarine, and ship launched) occurred in 1974–1975, and the Harpoon entered service in 1976. Its combined long-range, low-altitude cruise, and 500-lb warhead have made it a mainstay of the U.S. Navy. In fact, because of Harpoon's success, a land attack version [stand-off land attack missile (SLAM)] was developed which used the Harpoon propulsion and airframe with the Maverick imaging infrared (IIR) seeker. The Harpoon and its variants have been used by the United States in a number of conflicts going back to the U.S. attack against Libya in the Gulf of Sidra in 1986.

Around the same time the Navy first considered the Harpoon's development, feasibility studies were performed by the Air Force to examine ways to improve the bomber fleet's survivability and effectiveness. The subsonic cruise armed decoy (SCAD) became the favored approach in 1970. The intent of SCAD was to have an unmanned system which could act as a decoy. Because the SCAD was armed, the enemy would have to defend against it. This, in effect, reduced the amount of defensive resources which could be used against the bombers.

The SCAD program became the ALCM (designated AGM-86A) in 1973. ALCM was originally a nuclear-armed weapon system which cruised to the target at low altitude to avoid detection. ALCM has a unique trapezoidal cross sectional shape which allows it to be carried internally in B-1 and B-52 rotary racks. The original AGM-86A was intended to fit on the SRAM racks. To enhance range, the vehicle length was extended to make room for more fuel. This new vehicle was designated the AGM-86B. All known versions of ALCM have the extended length. It was necessary for the B-52s and B-1s to have a new rack to fit the

Fig. 2 F-107 turbofan engine.

AGM-86B internally. The result, however, is that ALCM has a reported range in excess of 1500 n mile.

ALCM has had a fairly long gestation period. Its first flight occurred in 1976, and the vehicle entered service in 1981. The ALCM is powered by the Williams International F-107 turbofan (see Fig. 2). A turbofan was selected because of the long-range requirements of ALCM.

The Navy had unique requirements which made ALCM unsuitable. The Navy strategic subsonic missile system had to be able to be launched from the air, surface, and ocean depths. This forced the Navy's long-range missile to have a round body with limited diameter and length to fit in a wide variety of launchers. The Tomahawk, designated BGM-109, was developed as a result. The Williams F-107 was tailored to meet Navy specification requirements including salt water exposure, starting, etc. For submarine operation, a steel capsule was designed for torpedo launch. Rocket boost was required for all submarine and ship-borne launches, and cruise under turbofan power was set at about 0.7 Mach, 200 ft.

Key milestones for Tomahawk occurred within a year or two of ALCM. The Tomahawk program began in 1972 as a competitive effort with ALCM. The development contract was awarded in 1976, and first year of service occurred in 1983. The Tomahawk was designed to carry both conventional and nuclear warheads. The conventionally armed variant was used extensively in Desert Storm, playing a key role in the early days of the conflict to destroy command and control facilities.

Basic design characteristics for the Harpoon, ALCM, and Tomahawk are described in Table 1. Since the development of these three systems, technological advancements have continued to improve the state of the art of expendable missile turbine engines. Cost, reliability, and performance continue to be major design challenges. Advancements in technology resulted in the development of many

Table 1 Design performance characteristics of several operational systems, Refs. 5 and 7

	Harpoon (AGM-84A)	ALCM (AGM-86B)	Tomahawk (BGM-109)
Manufacturer	McDonnell Douglas	Boeing	General Dynamics
Propulsion	J-402-400	F-107-WR-101	F-107-WR-400
Length, ft, in.	12, 7	20, 9	18, 3
Diameter, in.	13.5 round	24.5 trap.	20.4 round
Launch, weight, lb	1160	2825	2500
Cruise speed, Mn/SL	0.75	0.65	0.7
Range, mile	> 57	1550	
Warhead(s)	500-lb conv. (pene./blast)	W-80 (nuclear)	Nuclear 1000-lb conv.
Users	Air Force, Navy, Marine Corps	Air Force	Navy
Carriers	Ship, Sub, P-3C, A-6E, S-3B, A-7E F/A-18, B-52	B-52, B-1B	Ship, Sub
First year service	1976	1981	1983

new expendable engines including the Williams F-112 [which powers the advanced cruise missile (ACM)], Williams P8300, and the Teledyne CAE 382. In addition to missile propulsion, turbine engines are finding their way into a number of unmanned air vehicles used as target drones and reconnaissance vehicles. Future use of turbomachinery in unmanned systems appears to be growing. In the area of subsonic missile propulsion, the unique capabilities of turbine engines in terms of range, low-altitude cruise, and powered flight to the target make it the most favorable choice for a variety of applications.

This short historical background from the expendable turbojet's first major appearance to its current state hopefully has shown the reader how the gas turbine engine became a major player in the expendable vehicle arena. The next step is to define the components that make up a turbine engine and how they work.

III. Turbine Engine Cycle

There are many different variations to the gas turbine engine operating on the Brayton cycle. All, however, incorporate the same basic elements: inlet, compressor, combustor, turbine, and nozzle. Figure 3 illustrates a single-spool turbojet configuration. Because of the complexity of the turbine engine, the figure shows a standardized station numbering system which is used across the industry.

A very useful way to understand the thermodynamic characteristics of the Brayton cycle is to look at the entropy vs enthalpy, or H–S diagram. Valuable insight is gained by considering the ideal case first and then examining real world effects. For the analysis described, the thermodynamic relations γ and specific heat at constant pressure Cp will be used (see Appendix A). This is not an altogether accurate method because these parameters vary as a function of temperature and pressure. For turbine engine components such as the compressor, the pressure and temperature, and, therefore, γ and Cp, change continuously throughout the process. In addition, at extremely high temperatures (greater than 2000°F), gas dissociation occurs. For greater accuracy, cycle performance predictions are normally done using relative pressures to determine the enthalpy changes. However, since expendable turbine engines typically operate below 2000°F, the procedure outlined next, using average values for Cp and γ across each component, will provide quite accurate results.

A. Ideal Turbojet Cycle

A generalized H–S diagram for the ideal turbojet is shown in Fig. 4. A prime is used to denote ideal conditions. The analysis to predict component performance uses the stagnation conditions throughout, except for the exhaust nozzle. In this

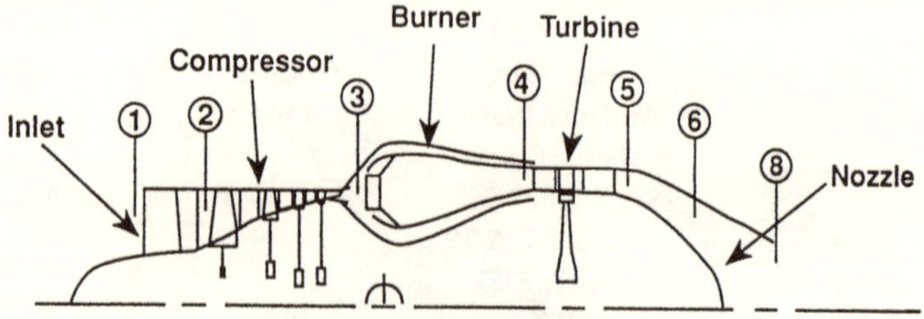

Fig. 3 Turbojet configuration.

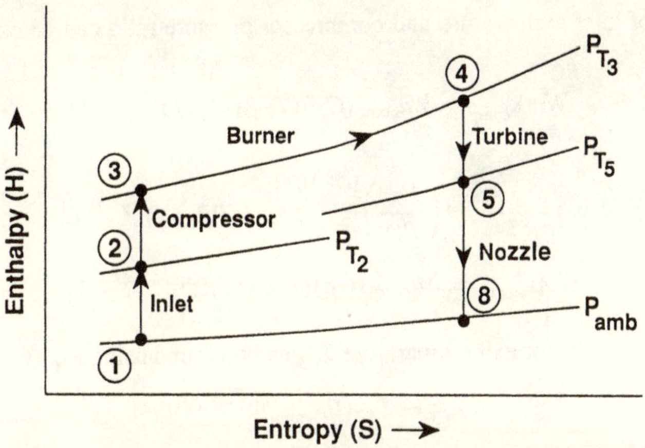

Fig. 4 Enthalpy vs entropy H–S diagram for ideal turbojet.

way, the kinetic energy can be taken into account. The nozzle is a special case because the production of thrust is dependent on exit velocity and static pressure. The following sections describe the thermodynamic relations used to determine component characteristics.

B. Ideal Inlet

Station 1–2 represents the ideal inlet in Fig. 4. The ideal inlet is defined as adiabatic and reversible, i.e., isentropic (entropy is constant). The inlet collects freestream air and slows it to speeds which the compression system can absorb. As the air is slowed down to low speed there is an increase in static pressure up to approximately stagnation levels. This pressure rise is referred to as the ram pressure rise. The inlet ideal stagnation temperature and pressure are determined using the following gas dynamics relations

$$M_0 \frac{V_0}{a_0} = \frac{V_0}{\sqrt{\gamma g R T_{\text{amb}}}} \tag{1}$$

$$T_{T1} T_{T2} = T_{\text{amb}} \left(1 + \frac{\gamma - 1}{2} M_0^2 \right) \tag{2}$$

$$P_{T1} P'_{T2} = P_{\text{amb}} (T_{T2} / T_{\text{amb}})^{\gamma/\gamma-1} \tag{3}$$

C. Ideal Compressor

As shown in Fig. 4, the compression process occurs from station 2 to 3. The ideal compressor analysis is assumed to be isentropic. To get a pressure rise across the compressor, work input is required. The amount of work is determined using the first law of thermodynamics.

$$\text{Work}'_{\text{comp}} = Wa_{\text{comp}}(h'_3 - h_2) = Wa_{\text{comp}}(Cp)(T'_{T3} - T_{T2}) \tag{4}$$

Normally, the compressor exit stagnation temperature T_{T3} is unknown. Compressor pressure ratio is assumed, and $T_{T3}$is calculated. The compressor work in terms of inlet temperature and compressor pressure ratio can be expressed as follows:

$$\text{Work}'_{\text{comp}} = Wa_{\text{comp}}(Cp)(T_{T2})[(T'_{T3}/T_{T2}) - 1] \tag{5}$$

Since

$$\frac{T'_{T3}}{T_{T2}} = \left(\frac{P_{T3}}{P_{T2}}\right)^{[(\gamma-1)/\gamma]} = (PR_{\text{comp}})^{[(\gamma-1)/\gamma]} \tag{6}$$

$$\text{Work}'_{\text{comp}} = Wa_{\text{comp}}(Cp)(T_{T2})\left[(PR)^{[\gamma-1/\gamma]} - 1\right] \tag{7}$$

The ideal compressor exit temperature T'_{T3} can be found using Eq. (6).

$$T'_{T3} = T_{T2}(PR_{\text{comp}})^{[(\gamma-1)/\gamma]} \tag{8}$$

D. Ideal Burner

An ideal burner is assumed to be adiabatic and have no pressure losses, and the combustion of the fuel is complete. In Fig. 4, the burner is represented as going from stations 3 to 4. Normally, the combustor exit temperature T_{T4} is assumed. The temperature rise across the combustor is used to determine the fuel/air ratio (*FAR*) required. To calculate the fuel/air ratio, an enthalpy balance between the reactants (air and fuel) and the products (combustion products at the combustor exit temperature) is performed. Figure 5 shows the ideal *FAR* required to achieve a given temperature rise ($T_{T4} - T_{T3}$) for several inlet air temperatures [T_{T3}]. With this chart fuel flow can be calculated.

$$W_f = Wa_{\text{comp}}(FAR') \tag{9}$$

Note that this figure is for JP-10, a common missile engine fuel. The fuel heating value for JP-10 is 18,100 Btu/lbm. To correct for alternative fuels, multiply the *FAR* derived from Fig. 5 by the ratio of heating values. Table 2 shows characteristics for several common gas turbine fuels.

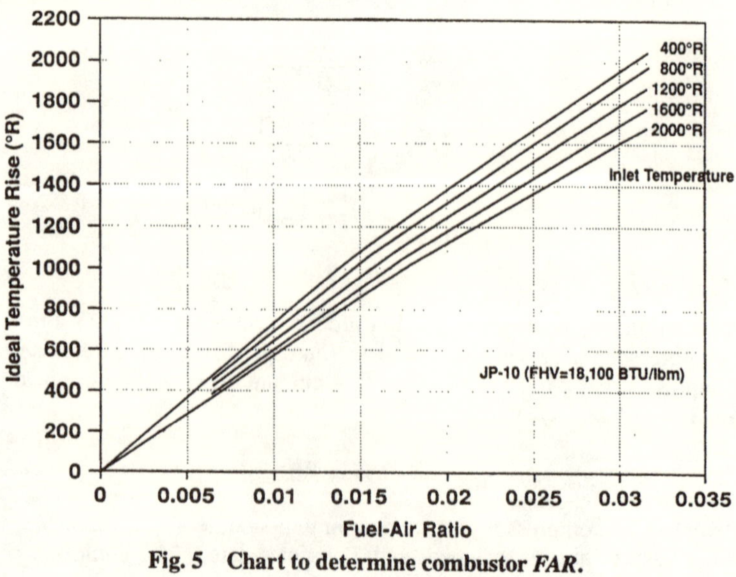

Fig. 5 Chart to determine combustor *FAR*.

Table 2 Properties for typical gas turbine fuels

	JP-4	RJ-4	RJ-6	JP-10
Formula	CH_2	$C_{12}H_{20}$	$C_{700}H_{999}$	$C_{10}H_{16}$
Mean molecular weight	120	164.3	9414.7	136
Heating value, Btu/lbm	18,400	18,250	18,320	18,100
Heating value, Btu/gal	117,090	140,660	153,570	142,000
Specific gravity at 68°F	0.776	0.925	1.016	0.941
Freezing point, °F	−76	< −40	< −54	−130
Flash point, °F	——	>140	>151	128
Stoichiometric fuel/air ratio	0.0676	0.0699	0.0717	0.0703

E. Ideal Turbine

The turbine provides the necessary power to drive the compressor. Station 4 to 5 represents the turbine process on the H–S diagram. For an engine operating in equilibrium, the work produced by the turbine is equal to the work required by the compressor, $\text{Work}'_{\text{turb}} = \text{Work}'_{\text{comp}}$.

Turbine work is defined as

$$\text{Work}'_{\text{turb}} = (Wa_{\text{comp}} + W'_f)(h'_4 - h'_5) = (Wg_{\text{turb}})Cp_{\text{turb}}(T'_{T4} - T'_{T5}) \tag{10}$$

As in the compressor, this expression can be put in terms of turbine expansion ratio ER_{turb}

$$\begin{aligned}\text{Work}'_{\text{turb}} &= Wg_{\text{turb}}(Cp)(T'_{T4})[1 - (T'_{T5}/T'_{T4})] \\ &= Wg_{\text{turb}}(Cp)(T'_{T4})\left[1 - (ER_{\text{turb}})^{[(\gamma-1)/\gamma]}\right]\end{aligned} \tag{11}$$

Since the turbine work is known, this expression can be solved to determine the turbine expansion ratio:

$$ER_{\text{turb}} = \left[1 - \frac{\text{Work}'_{\text{comp}}}{(Wg_{\text{turb}})(Cp)(T'_{T4})}\right]^{[\gamma/(\gamma-1)]} \tag{12}$$

Ideal turbine exit pressure and temperature are then determined using the turbine expansion ratio:

$$P_{T5} = P_{T4}(ER_{\text{turb}}) \tag{13}$$

$$T'_{T5} = T'_{T4}(ER_{\text{turb}})^{[(\gamma-1)/\gamma]} \tag{14}$$

F. Ideal Nozzle

Since the constant pressure lines on the H–S diagram diverge with increasing entropy, there is excess enthalpy and pressure after expanding through the turbine. It is this basic principle which allows the turbine engine to produce thrust. The excess enthalpy and pressure are available to expand through the nozzle (Fig. 4, stations 5–8).

For the vast majority of expendable turbine engines, a simple converging nozzle is used. For this reason, the converging nozzle will be considered first. The ideal nozzle assumes, once again, that the process is isentropic. Hence, there are no total pressure losses, i.e., $P'_{T8} = P_{T5}$. The first step is to determine whether the engine has sufficient exhaust pressure to choke the nozzle. This is done by comparing the nozzle pressure ratio to the choking pressure ratio.

Nozzle pressure ratio (NPR) is expressed as

$$NPR = (P_{T5}/P_{\text{amb}}) = (P'_{T8}/P_{\text{amb}}) \tag{15}$$

Nozzle choking pressure ratio ($NCPR$), or critical pressure ratio as it is sometimes called, is calculated using the following expression:

$$NCPR = \left(1 + \frac{\gamma - 1}{2} M_8^2\right)^{[\gamma/(\gamma-1)]} \tag{16}$$

where choking occurs when Mach number is equal to one. For this special case, Eq. (16) can be simplified to

$$NCPR = [(\gamma + 1)/2]^{[\gamma/(\gamma-1)]} \tag{17}$$

If NPR is greater than or equal to $NCPR$, the converging nozzle is choked. If the nozzle is choked, the nozzle exit static pressure and temperature are determined as follows:

$$T_8 = T_{T8} \Big/ \left(\frac{\gamma + 1}{2}\right) \tag{18}$$

$$P_8 = P_{T8} \Big/ \left(\frac{\gamma + 1}{2}\right)^{[\gamma/(\gamma-1)]} \tag{19}$$

The exit velocity for the choked nozzle is

$$V_{\text{jet}} = a(M_8) = \sqrt{\gamma g R T_8} \tag{20}$$

If the converging nozzle is unchoked, the exit static pressure is equal to ambient pressure. For this case, the exit Mach number must be determined. The nozzle pressure ratio is expressed as a function of exit Mach number as follows:

$$NPR = \frac{P_{T8}}{P_{\text{amb}}} = \left(1 + \frac{\gamma - 1}{2} M_8^2\right)^{[\gamma/(\gamma-1)]} \tag{21}$$

Rearranging to solve for exit Mach number

$$M_8 = \left\{\left[\left(\frac{P_{T8}}{P_{\text{amb}}}\right)^{[(\gamma-1)/\gamma]} - 1\right] \frac{2}{\gamma - 1}\right\}^{\frac{1}{2}} \tag{22}$$

Exit static temperature and pressure are found using the following expressions:

$$T_8 = \frac{T_{T8}}{1 + [(\gamma - 1)/2] M_8^2} \tag{23}$$

$$P_8 = \frac{P_{T8}}{(T_{T8}/T_8)^{[\gamma/(\gamma-1)]}} \tag{24}$$

As before, the exit velocity is given by

$$V_{\text{jet}} = M_8\sqrt{\gamma g R T_8} \tag{25}$$

The nozzle exit area is determined using the mass flow equation

$$A_8 = \frac{W g_8}{\rho_8 V_{\text{jet}}} = \frac{W g_8}{(P_8/RT_8)(V_{\text{jet}})} \tag{26}$$

For the ideal converging-diverging nozzle, the throat is choked, and the exit static pressure is expanded to ambient. The exit to the diverging section is defined as station 9. The equations describing the unchoked nozzle are used to determine nozzle performance except the exit Mach number is supersonic.

With the methodology just described, performance characteristics can be found for the ideal turbojet cycle. Specific thrust and thrust specific fuel consumption are the principle performance parameters. Before these parameters are derived, however, corrections to account for real-world effects will be discussed.

G. Nonideal Turbojet Cycle

Real-world considerations introduce inefficiencies into the gas turbine engine. These included friction, shock waves, incomplete combustion, secondary flow effects, separation, and leakages. One factor which is normally not accounted for is heat transfer out of the system. It is standard practice to assume the gas turbine operates adiabatically, i.e., no heat transfer (except in the combustor, of course). This assumption vastly simplifies the calculations and is reasonable because of the large energy flow rates relative to the heat losses through the case.

Once again, the H–S diagram will be used to help explain the gas turbine engine cycle. Figure 6 illustrates the turbojet cycle, taking into account real-world effects. Comparing this figure to the ideal H–S diagram (Fig. 4), the reader can see the impact of inefficiencies to the cycle. As in the ideal component descriptions, the reader may find it helpful to refer back to Fig. 6 as the actual component performance calculations are described.

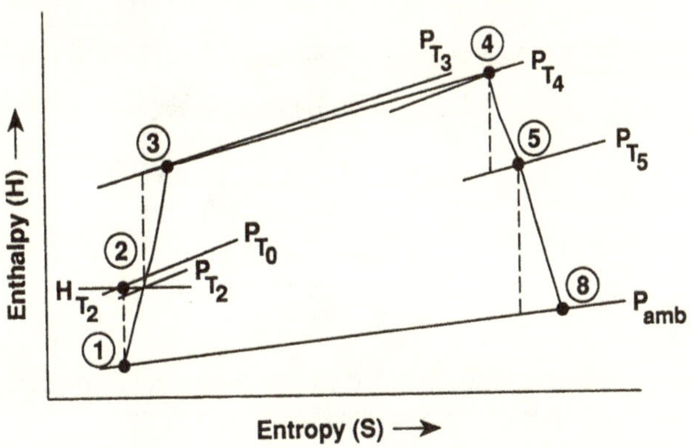

Fig. 6 Enthalpy vs entropy H–S diagram for real turbojet.

H. Actual Inlet

To account for inefficiencies in the inlet, the approach is to assume an adiabatic process with a total pressure loss. Ram recovery (sometimes referred to as ram efficiency) is used to account for the pressure loss and is defined as

$$\eta_{\text{ram}} = P_{T2}/P_{T0} \tag{27}$$

Since no work or heat transfer occurs in the inlet, the stagnation temperature is the same as that used for the ideal turbojet [Eq. (2)]. Note in Fig. 6 that the entropy increase due to the real (irreversible) process results in a total pressure loss, with ideal and actual exit stagnation temperatures being equal.

I. Actual Compressor

For the compressor, efficiency is quantified by comparing the ideal work to the actual work required for the same pressure ratio. This efficiency, commonly called compressor adiabatic efficiency, is defined as

$$\eta_{\text{comp}} = \frac{\text{ideal isentropic work}}{\text{actual work}} = \frac{h'_{T3} - h_{T2}}{h_{T3} - h_{T2}} \tag{28}$$

Using Eq. (7) and solving for the actual work

$$\text{Work}_{\text{comp}} = [Wa_{\text{comp}}(Cp)(T_{T2})]/\eta_{\text{comp}}\left[(PR)^{[(\gamma-1)/\gamma]} - 1\right] \tag{29}$$

In the H–S diagram (Fig. 6), the ideal and actual pressure for the compressor are equal. Losses associated with a compressor are book kept as higher exit temperature. This, in turn, results in a higher horsepower required to run the actual compressor. The exit temperature is found by

$$\eta_{\text{comp}} = \frac{\Delta H'}{\Delta H} = \frac{Wa_{\text{comp}}(Cp)(T'_{T3} - T_{T2})}{Wa_{\text{comp}}(Cp)(T_{T3} - T_{T2})} \tag{30}$$

Since airflow and Cp are considered equal in the ideal and actual case, this expression can be simplified to

$$\eta_{\text{comp}} = [(T'_{T3} - T_{T2})/(T_{T3} - T_{T2})] \tag{31}$$

Solving for actual compressor exit temperature

$$\begin{aligned} T_{T3} &= [(T'_{T3} - T_{T2})/\eta_{\text{comp}}] + T_{T2} \\ &= T_{T2}\left\{\frac{[(T'_{T3}/T_{T2}) - 1]}{\eta_{\text{comp}}} + 1\right\} \\ &= T_{T2}\left[\frac{(PR^{[(\gamma-1/\gamma)]} - 1)}{\eta_{\text{comp}}} + 1\right] \end{aligned} \tag{32}$$

An alternative efficiency which is sometimes used is called polytropic, or stage efficiency. Polytropic efficiency is defined by the expression

$$T_{T3}/T_{T2} = (PR_{\text{comp}})^{[(\gamma-1)/\eta_{\text{poly}}(\gamma)]} \tag{33}$$

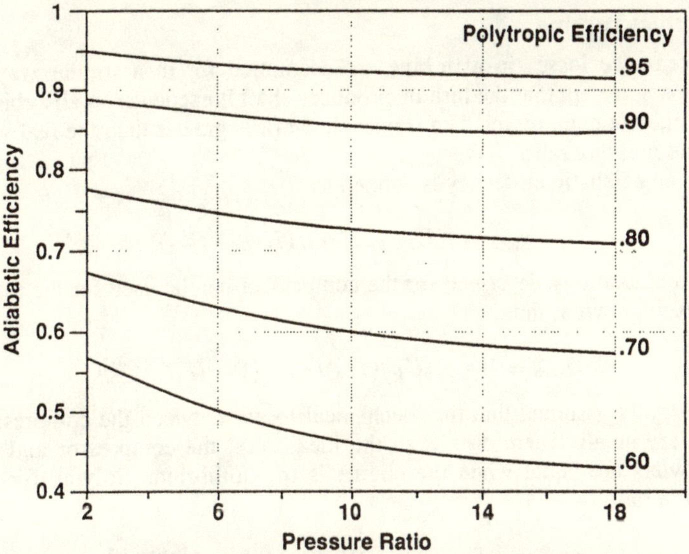

Fig. 7 Impact of polytropic efficiency on adiabatic efficiency.

In physical terms, the polytropic efficiency is equal to the efficiency for each incremental(differential) pressure rise. Since adiabatic efficiency normally falls with increasing pressure ratio, the use of polytropic efficiency to determine compressor exit temperature is sometimes the preferred approach. For a cycle study where a range of pressure ratios is being considered, it can be difficult to determine the adiabatic efficiency. Polytropic efficiency, on the other hand, can be assumed to remain constant with increasing pressure ratio. Hence, polytropic efficiency better represents the real-world effect of pressure ratio on adiabatic efficiency.

Adiabatic efficiency is related to polytropic efficiency by the following relation:

$$\eta_{\text{comp}} = \frac{(PR)^{[(\gamma-1)/\gamma]} - 1}{(PR)^{[(\gamma-1)/\eta_{\text{poly}}(\gamma)]} - 1} \tag{34}$$

Figure 7 shows the variation of adiabatic efficiency as a function of pressure ratio for various lines of constant polytropic efficiency. Note how adiabatic efficiency approaches the polytropic efficiency as pressure ratio approaches 1.0 (i.e., when the differential pressure rise becomes infinitely small).

J. Actual Burner

The primary loss mechanisms associated with a combustor include incomplete combustion and pressure losses. Incomplete combustion implies that a portion of the fuel's energy does not go into a temperature rise. A higher fuel flow rate is required to get a given temperature rise. To account for this, the burner efficiency is defined as

$$\eta_{\text{burn}} = Wf'/Wf \tag{35}$$

where Wf' is determined using Fig. 5. Note in Fig. 6 the drop in total pressure across the combustor. The total pressure loss is due to both friction and heat addition in a flow with nonzero Mach number.

K. Actual Turbine

Performance losses in a turbine are accounted for in a similar way to the compressor, except that the turbine produces shaft horsepower vs absorbing shaft power like the compressor. The isentropic work is greater than the real work for the same pressure ratio.

Turbine adiabatic efficiency is defined as

$$\eta_{\text{turb}} = [(H_{T4} - H_{T5})/(H_{T4} - H'_{T5})] \tag{36}$$

The methodology described for the compressor can be used for the turbine to define turbine work, thus,

$$\text{Work}_{\text{turb}} = Wg_{\text{turb}}(Cp)(T_{T4})(\eta_{\text{turb}})[1 - (ER)^{[(\gamma-1)/\gamma]}] \tag{37}$$

It is normally assumed that the mechanical losses between the compressor and turbine are small. Therefore, as in the ideal case, the compressor and turbine horsepowers are equal when the engine is in equilibrium. Solving for turbine expansion ratio

$$ER_{\text{turb}} = \left[1 - \frac{\text{Work}_{\text{comp}}}{Wg_{\text{turb}}(Cp)(T_{T4})\eta_{\text{turb}}}\right]^{[\gamma/(\gamma-1)]} \tag{38}$$

As in the ideal case, the exit total pressure is found using Eq. (13). The actual turbine exit temperature is given by

$$\eta_{\text{turb}} = \frac{Wg_{\text{turb}}(Cp)(T_{T4} - T_{T5})}{Wg_{\text{turb}}(Cp)(T_{T4})\left[1 - (ER)^{[(\gamma-1)/\gamma]}\right]} \tag{39}$$

As in the compressor, the flow and Cp can be assumed to be equal between the ideal and actual case. Solving for actual turbine exit temperature

$$T_{T5} = T_{T4}\left\{1 - \eta_{\text{turb}}[1 - (ER)^{[(\gamma-1)/\gamma]}]\right\} \tag{40}$$

Figure 6 shows both the ideal and actual turbine expanding to the same pressure line. The real turbine compared to the ideal has a higher exit temperature and lower work output.

The concept of polytropic efficiency can also be used for the turbine. It is represented by

$$T_{T4}/T_{T5} = (1/ER)^{\{[\eta_{\text{poly}}(\gamma-1)]/\gamma\}} \tag{41}$$

Relating turbine polytropic efficiency to adiabatic efficiency

$$\eta_{\text{turb}} = \frac{1 - (ER)^{\{[\eta_{\text{poly}}(\gamma-1)]\gamma\}}}{1 - (ER)^{[(\gamma-1)/\gamma]}} \tag{42}$$

L. Actual Nozzle

Internal losses associated with the nozzle can be book kept in a number of ways. In this section, an isentropic efficiency approach as described for the compressor and turbine will be used. However, a more common method is to calculate performance for an ideal nozzle, then correct the thrust to account for nozzle losses. This will be described in detail in the Sec. III.M.

Nozzle isentropic efficiency is defined as

$$\eta_{\text{nozzle}} = \frac{H_{T5} - H_8}{H_{T5} - H_8'} = \frac{T_{T5} - T_8}{T_{T5} - T_8'} \tag{43}$$

The methodology used to determine the choking pressure ratio for the ideal nozzle applies equally well to the real nozzle. Assuming a choked converging nozzle, the nozzle exit static pressure is expressed as

$$P_8 = P_{T8}\left[1 - \frac{1}{\eta_{\text{noz}}}\left(1 - \frac{T_8}{T_{T8}}\right)\right]^{[\gamma/(\gamma-1)]} \tag{44}$$

where Eq. (18) is used to determine T_8.

Figure 6 illustrates the difference between the ideal and actual nozzle. It is the residual enthalpy after the turbine has extracted the necessary work to drive the compressor which provides excess energy for thrust production. Clearly, the more efficient the nozzle, the more energy is available for thrust work.

M. Performance Indicators

Thrust for the gas turbine engine is determined by applying the momentum equation along the engine's axial direction. The net thrust is equal to the sum of the internal and external forces acting on the engine. Figure 8 illustrates the gas turbine as a control volume with the various forces acting on it. Summing the internal and external forces acting on the engine yields

$$FN = \frac{Wg_{\text{jet}}V_{\text{jet}}}{g} - \frac{Wa(V_0)}{g} + A_{\text{jet}}(P_{\text{jet}} - P_{\text{amb}}) \tag{45}$$

The first two terms are commonly referred to as the momentum thrust. The third is called the pressure thrust. An alternate way to express the net thrust using parameters derived from the engine analysis is

$$FN = Wa/g[(1 + FAR)V_{\text{jet}} - V_0] + A_8(P_8 - P_{\text{amb}}) \tag{46}$$

Note that when the exhaust static pressure P_8 is fully expanded to ambient pressure, the pressure thrust term is zero. For the ideal case, a fully expanded nozzle always produces peak thrust. However, friction losses, part power performance

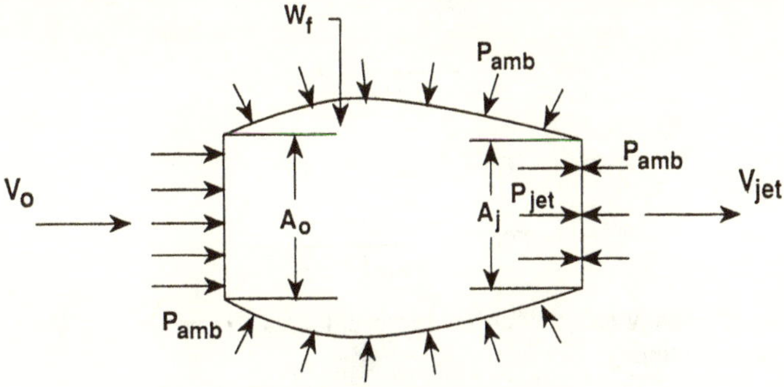

Fig. 8 Turbojet engine control volume.

and length considerations make the decision to use a converging or converging-diverging nozzle more difficult. A good rule of thumb is that if the nozzle pressure ratio is over 3.5, a converging-diverging nozzle should be considered.

Net thrust is often expressed in terms of gross thrust and ram drag, which are defined as

$$\begin{aligned} FN &= \text{gross thrust } (FG) - \text{ram drag } (FR) \\ &= \left[\frac{Wa(1+FAR)V_{\text{jet}}}{g} + A_8(P_8 - P_{\text{amb}})\right] - \left[\frac{Wa(V_0)}{g}\right] \end{aligned} \tag{47}$$

This way of subdividing the terms in the net thrust equation is advantageous because nozzle performance losses can be quantified using the nozzle gross thrust coefficient (*CFG*). *CFG* has the advantage over nozzle efficiency in that it is easy to measure in a test cell. Nozzle losses are book kept by modifying Eq. (47) as follows:

$$FN = CFG(FG) - FR \tag{48}$$

A useful way to analyze an engine's thrust production potential is net thrust per pound of airflow, or specific thrust. Specific thrust is useful because it is an indicator of thrust per unit size. It is defined as

$$\frac{FN}{Wa} = \frac{1}{g}[(1+FAR)V_8 - V_0] + \frac{A_8}{Wa}(P_8 - P_{\text{amb}}) \tag{49}$$

Another performance parameter commonly used which is an indicator of fuel consumption efficiency is thrust specific fuel consumption (*TSFC*) defined as

$$TSFC = \frac{Wf}{FN}(3600) \tag{50}$$

The units for *TSFC* are in lbm/h-lbf. *TSFC* is sometimes described in terms of overall efficiency. Overall efficiency is defined as

$$\begin{aligned} \eta_{\text{ov}} &= \frac{\text{thrust power}}{\text{chemical energy input rate}} \\ &= \frac{FNV_0}{Wf(FHV)}\left[\frac{1 \text{ Btu}}{778 \text{ ft-lbf}}\right] \\ &= \frac{V_0}{TSFC(FHV)(778)} \end{aligned} \tag{51}$$

Solving for *TSFC* yields

$$TSFC = \frac{V_0}{\eta_{\text{ov}}(FHV)(778)} \tag{52}$$

Overall efficiency can be further represented as the product of the propulsive and thermal efficiency.

$$\eta_{\text{ov}} = \eta_{\text{prop}}\eta_{\text{ther}} \tag{53}$$

The gas turbine engine produces thrust by increasing the velocity and/or pressure of the working fluid. The propulsive thrust is an indicator of how efficiently the kinetic energy is used

$$\eta_{prop} = \frac{\text{thrust power output}}{\text{available propulsive power}} \tag{54}$$

Net thrust can be defined using a term called effective jet velocity.

$$FN = Wa/g(V_{eff} - V_0) \tag{55}$$

Note that when the nozzle is fully expanded, $V_{jet} = V_{eff}$. Substituting for thrust power output and available propulsive power, propulsive efficiency is given by

$$\eta_{prop} = \frac{FN(V_0)}{FN(V_0) + (Wa/2g)(V_{eff} - V_0)^2} \tag{56}$$

Combining this with Eq. (55)

$$\eta_{prop} = \frac{Wa/g(V_{eff} - V_0)V_0}{Wa/2g\left(V_{eff}^2 - V_0^2\right)}$$

$$= \frac{2V_0}{V_{eff} + V_0} \tag{57}$$

Hence, the propulsive efficiency approaches 100% when the effective jet velocity approaches the freestream velocity. However, as propulsive efficiency goes to 100%, the thrust goes to zero. Finally, Eqs. (51) and (57) can be combined to determine thermal efficiency.

$$\eta_{ther} = \frac{\eta_{ov}}{\eta_{prop}} = \frac{(Wa/2g)\left(V_{eff}^2 - V_0^2\right)}{Wf(FHV)(778)} \tag{58}$$

It is the turbine engine designer's challenge to maximize FN/Wa while minimizing the *TSFC*. This will minimize the propulsion system size as well as fuel consumption. Unfortunately, these two parameters usually go in opposite directions from each other. Depending on the configurational requirements, a balance must be maintained between the performance indicators. If long range is the principle requirement, *TSFC* will be critical, with FN/WA a secondary consideration. If minimum propulsion system size is the driver, as is often times the case in missile applications, FN/WA is more important.

N. Turbofan Cycle Considerations

An alternative cycle often used for long-range vehicles is the turbofan. The advantage of the turbofan cycle compared to the turbojet is that for a given overall pressure ratio and turbine inlet temperature the turbofan has a higher propulsive efficiency. This is due to the fact that the turbofan moves a larger mass of air at a lower velocity. Hence, although its specific thrust is lower, its *TSFC* is considerably better than a turbojet with the same thrust.

Typical turbofan cycles with additional station numbers are illustrated in Fig. 9. A distinction is made between "mixed" and "separate" flow turbofans. The mixed

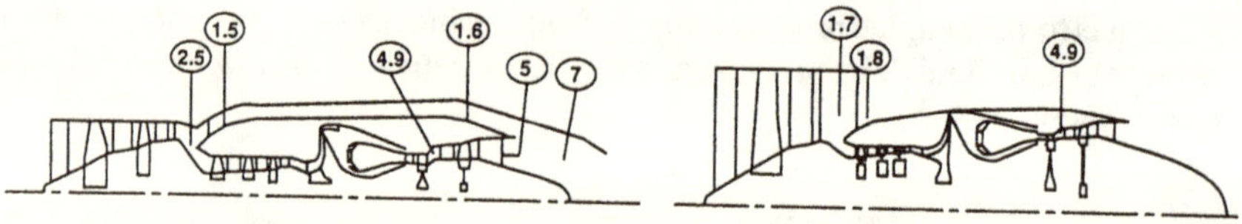

Mixed Flow Turbofan   Separate Flow Turbofan

Fig. 9 Common turbofan configurations.

flow turbofan has a single nozzle, whereas the separate flow design has a nozzle for both the core and fan streams.

The turbofan's airflow breaks into two streams after the fan. The flow split between the core and bypass is defined as bypass ratio.

$$BPR = \frac{Wa_{\text{bypass}}}{Wa_{\text{core}}} = \frac{Wa_{\text{fan}}}{Wa_{\text{core}}} - 1 \tag{59}$$

The thermodynamic equations derived for the turbojet components apply equally well to the turbofan. An additional power balance is required between the fan and low-pressure turbine. The total net thrust for the separate flow turbofan is the sum of the net thrust for the fan and core nozzles.

$$FN(\text{separate flow}) = FN(\text{fan nozzle}) + FN(\text{core nozzle}) \tag{60}$$

The mixed flow turbofan puts an additional constraint on the cycle. For the two streams to mix, the static pressures must be balanced. It is undesirable to have a large difference in total pressures between the core and mixer because this can result in a large total pressure loss due to shear effects. Therefore, fan and core pressure ratio should be set such that the total pressure in the bypass duct and the turbine exit are within 5% of each other.

There are a variety of options to determine mixer inlet and exit conditions. It is necessary to make an assumption which can be used to define the static pressure at one of the mixer stations. To illustrate the methodology, the core stream (station 5) Mach number will be assumed. The core static pressure becomes

$$P_5 = \frac{P_{T5}}{\{1 + [(\gamma - 1)/2]M_5^2\}^{[\gamma/(\gamma-1)]}} \tag{61}$$

The static pressures at the point where the two streams meet subsonically are equal ($P_5 = P_{16}$). This analysis will assume a constant area mixer, that is, the sum of the incoming areas equal the exit area ($A_5 + A_{16} = A_7$) (see Ref. 14). The bypass stream Mach number is found as follows:

$$M_{16} = \sqrt{\frac{2}{\gamma - 1}\left[\left(\frac{P_{T16}}{P_5}\right)^{[(\gamma-1)/\gamma]} - 1\right]} \tag{62}$$

By conserving the total enthalpy, mixed flow total temperature T_{T7} is given by

$$T_{T7}(Wg_7) = T_{T16}Wa_{16} + T_{T5}Wg_5 \tag{63}$$

Expanding and solving for T_{T7}

$$T_{T7} = \frac{T_{T16}(Wa_{16}) + T_5(Wa_5 + Wf)}{Wa_{16} + Wa_5 + Wf} \tag{64}$$

Dividing by Wa_5, and substituting BPR and FAR

$$T_{T5} = \frac{T_{T16}(BPR) + T_{T5}(1 + FAR)}{BPR + 1 + FAR} \tag{65}$$

Since FAR is small, this expression can be simplified with little loss in accuracy to

$$T_{T7} = \frac{T_{T16}(BPR) + T_{T5}}{BPR + 1} \tag{66}$$

The bypass and core areas are found by using the mass flow continuity equation to derive

$$\text{area}_i = \left[\frac{Wa_i}{P_{Ti}M_i}\right]\sqrt{\frac{RT_{Ti}}{\gamma g\{1 + [(\gamma - 1)/2]M_i^2\}^{-(\gamma+1)/(\gamma-1)}}} \tag{67}$$

where $i = 5$ or 16.

The momentum equation can be written as

$$A_5P_5(1 + \gamma M_5^2) + A_{16}P_{16}(1 + \gamma M_{16}^2) = A_7P_7(1 + \gamma M_7^2) \tag{68}$$

This expression can be used to determine the mixer exit Mach number[14]

$$M_7 = \left[\frac{2F}{1 - 2\gamma F + \sqrt{1 - 2(\gamma + 1)F}}\right]^{\frac{1}{2}} \tag{69}$$

where

$$F \equiv (1 + BPR)\left[1 + (BPR)\left(\frac{T_{T16}}{T_{T5}}\right)\right] \times \left[\frac{1}{\sqrt{f(M_5)}} + BPR\sqrt{\frac{T_{T16}/T_{T5}}{f(M_{16})}}\right]^{-2} \tag{70}$$

$$f(M) \equiv M^2\{1 + [(\gamma - 1)/2]M^2\}(1 + \gamma M^2)^{-2} \tag{71}$$

P_7 can be found using Eq. (68).

$$P_7 = \frac{A_5P_5(1 + \gamma M_5^2) + A_{16}P_{16}(1 + \gamma M_{16}^2)}{A_7(1 + \gamma M_7^2)} \tag{72}$$

With the exit Mach number and static pressure known, the exit total pressure is found using

$$\frac{P_{T7}}{P_7} = \left(1 + \frac{\gamma - 1}{2}M_7^2\right)^{[\gamma/(\gamma-1)]} \tag{73}$$

The mixer analysis just described represents performance for a fully mixed exit flow. In reality, the flow exiting the core and bypass streams does not mix completely. A commonly used parameter to address incomplete mixing is called mixing gain (MG). The gross thrust is adjusted to account for this effect as follows:

$$FG(\text{actual}) = MG[FG(\text{mixed}) - FG(\text{separate})] + FG(\text{separate}) \tag{74}$$

The mixing gain is dependent on several factors including bypass ratio, core and bypass total pressure difference, mixing length, etc. For a moderate bypass ratio engine ($BPR \leq 1.0$) with matched core and bypass total pressures, the mixing gain is usually around 0.7–0.8. For higher bypass ratios, the mixing gain usually diminishes.

IV. Off-Design Performance Prediction

Missiles propelled by gas turbines typically are launched, accelerate, and cruise at a variety of altitudes and Mach numbers. Consider the Harpoon, for example. It is launched from aircraft traveling at moderate subsonic speeds as well as ships which are essentially stationary. It accelerates to and cruises at relatively high subsonic velocities. It must operate all over the globe, from arctic to tropic latitudes. Because of this, gas turbine engines must be able to operate across a wide range of power settings and flight conditions. The methodology used to predict off-design performance is described in this section.

To illustrate the impact of off-design operation on performance, a generic expendable class turbojet will be considered. The cycle is consistent with typical existing expendable turbojets. The design point (0.6 Mn/sea level) compressor pressure ratio is 6, and turbine rotor inlet temperature is 1900°F. Figure 10 shows the maximum power thrust, $TSFC$, and airflow as a function of flight condition. As altitude increases, net thrust decreases. This is due primarily to the decrease in airflow, which is a function of density. With increasing flight speed, $TSFC$ gets worse. Compressor pressure ratio decreases with increasing flight speed and, as a result, the cycle fuel consumption increases.

Another interesting characteristic is engine part power performance. Figure 11 shows the variation of several performance parameters as a function of engine rpm. Thrust falls at a significantly higher rate than speed. This effect is even more substantial for turbofans. The $TSFC$ remains essentially flat to around 50% thrust

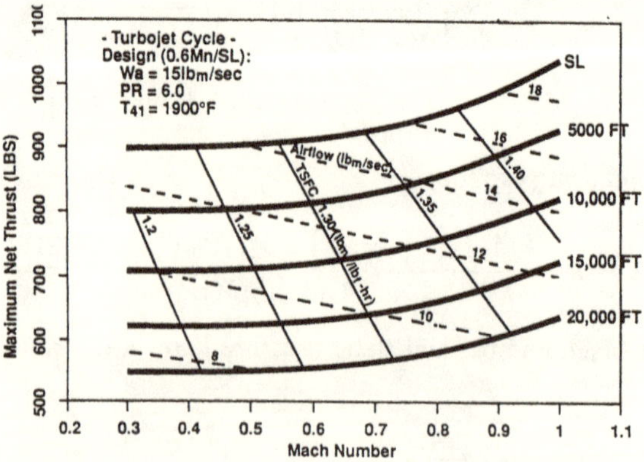

Fig. 10 Impact of flight condition on performance.

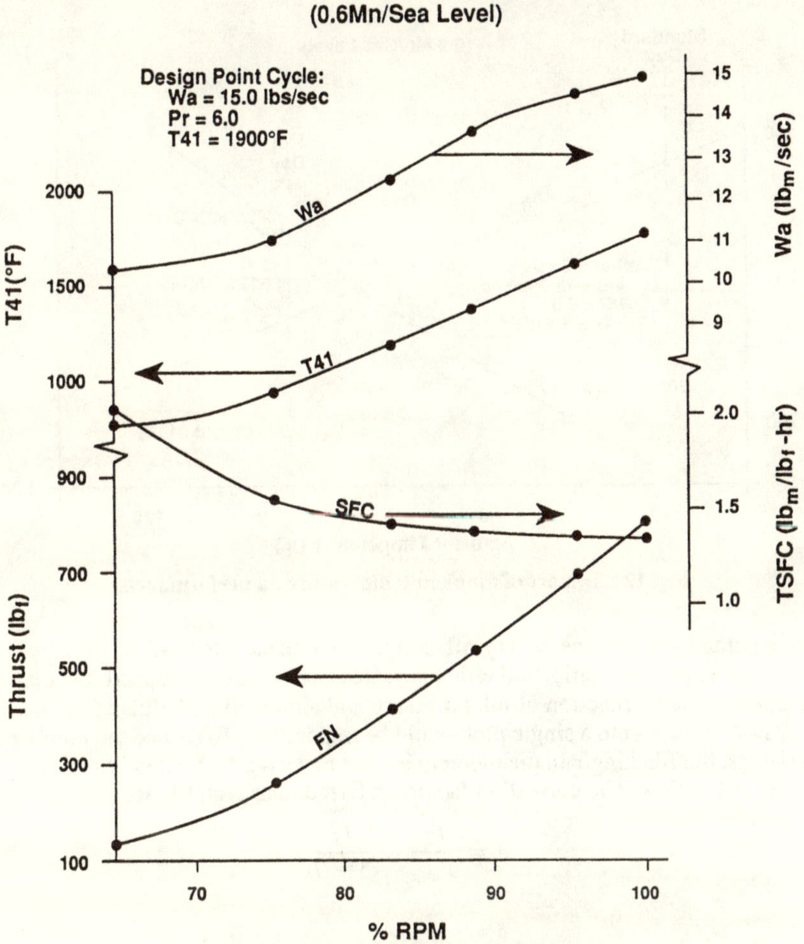

Fig. 11 Turbojet throttling characteristics.

and then climbs rapidly as thrust decreases further. This characteristic is very much cycle dependent. Often times, minimum *TSFC* will occur at maximum thrust. The turbine inlet temperature falls at a much faster rate than airflow. Both these factors contribute to the rate of thrust rolloff.

Ambient air conditions also have a profound effect on engine performance. Figure 12 shows the variation of ambient temperature on thrust and *TSFC*. The military specifications for standard day, tropical day, and hot day ambient temperatures are indicated on this figure. Clearly, thrust is very sensitive to inlet temperature. *TSFC* is also negatively influenced by increasing ambient temperature. It is important to be able to carry out an expendable vehicle's mission when ambient temperature is greater than 59°F. Because of this, expendable engine size is normally set by hot day performance.

In general, off-design performance varies with design-point cycle characteristics such as pressure ratio, bypass ratio, efficiencies, etc. To determine how components perform off-design, a methodology is needed to represent the critical

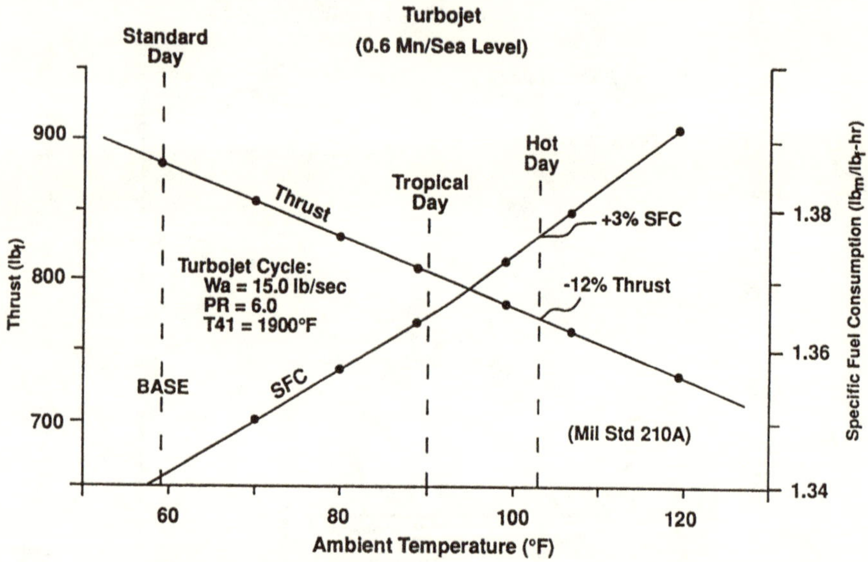

Fig. 12 Impact of ambient temperature on performance.

cycle parameters. The necessary off-design component information includes airflow, speed, pressure ratio, and efficiency. Because airflow and speed for rotating components are a function of inlet pressure and temperature, reducing the necessary information onto a single plot would be impossible. To reduce the number of variables, the Buckingham pie theorem is used to derive "corrected" functions for speed and airflow. The correction factors, referred to as θ and δ, are

$$\theta = \frac{T_T}{T_{\text{amb}}} = \frac{T_T}{518.7} \tag{75}$$

$$\delta = \frac{P_T}{P_{\text{amb}}} = \frac{P_T}{14.696} \tag{76}$$

The corrected speed and flow, respectively, are

$$\%N_{\text{corr}} = \%\text{rpm}/\sqrt{\theta} \tag{77}$$

$$Wa_{\text{corr}} = Wa\sqrt{\theta}/\delta \tag{78}$$

Figure 13 shows the typical format used to represent fan and compressor performance. This chart is referred to as a compressor map. It is a plot of pressure ratio vs corrected flow for constant corrected speed. Efficiency contours are also shown. Moving along a constant corrected speed line, the bottommost point (lowest pressure ratio) represents no back pressure on the compressor. The line connecting the uppermost pressure ratio points for each corrected speed line is referred to as the stall line. This line represents the point where the compressor will stall. Stall in the compressor may or may not trigger surge in the engine system. The line just below the stall line is the operating line. The operating line corresponds to the locus of points at which the compressor, within a given engine, will operate. For compressor stability, the operating line must be below the stall line by a certain

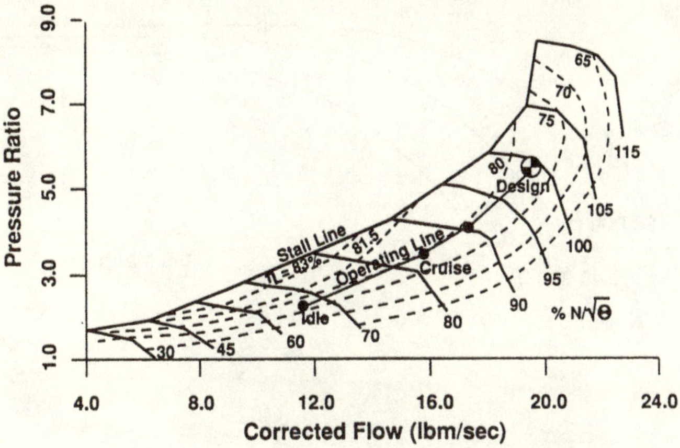

Fig. 13 Typical compressor map.

amount throughout its whole operating line. The difference between the operating line and stall line is the stall margin. Stall margin is defined as

$$SM = (PR_{\text{stall}} - PR_{\text{oper. pt.}})/PR_{\text{oper. pt.}} \tag{79}$$

Stall margin is normally taken at a constant corrected flow but sometimes is calculated at constant corrected speed. Stall margin is needed for a variety of reasons including inlet distortion and hysteresis during transition from one power setting to another. For conceptual design purposes, a commonly used value for minimum stall margin across the operating envelope is 20%.

A properly designed combustor will deliver nearly 100% efficiency at its design point. Modern combustors perform well over a large throttling range. Typical combustor performance is represented in Fig. 14. Note that in the example engine's normal operating range, combustor efficiency remains high. Only near idle does it fall to any great extent. Combustor efficiency gets slightly worse with altitude because the pressure goes down with increasing altitude. The implication is that a wide range, efficient combustor is easy to design. This is not the case. In fact, the combustor was one of the major problems in the first turbine engines which were developed. Hans Van Ohain, one of the co-inventors of the gas turbine engine, had to use hydrogen for fuel in his experimental models because of its favorable light-off and burning characteristics. Over the years, a great deal of experimental work in the field of combustor design has been performed to determine desirable design features. The design methodology used for combustors continues to be largely empirically based.

There are a number of alternative ways to define a turbine's performance characteristics. For this discussion, turbine cycle characteristics are illustrated using the map shown in Fig. 15. The turbine map contains similar parameters to the compressor map. The format is turbine pressure ratio (the inverse of expansion ratio) vs corrected flow and efficiency for lines of constant corrected speed. Corrected flow increases steeply along a constant corrected speed line, and then levels off. The leveling off occurs when the turbine chokes. Pressure ratio increases as the back

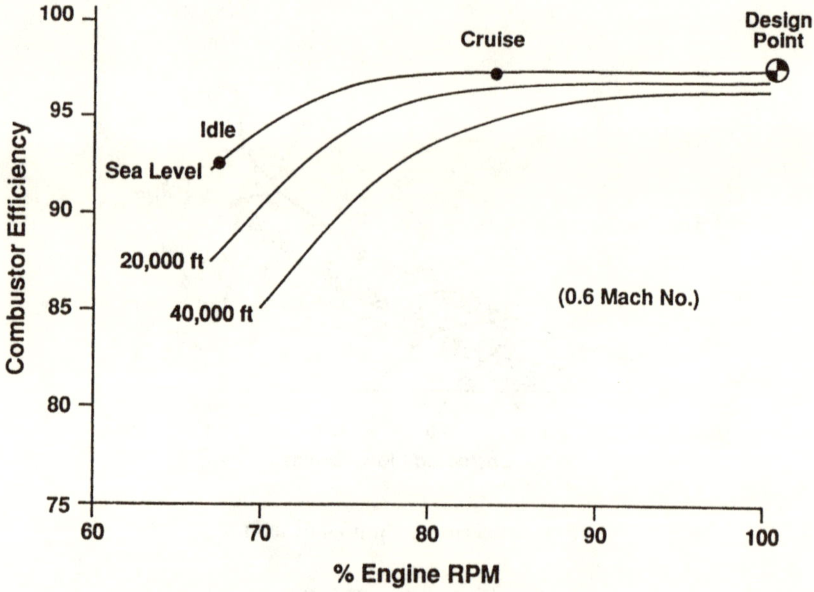

Fig. 14 Typical combustor performance.

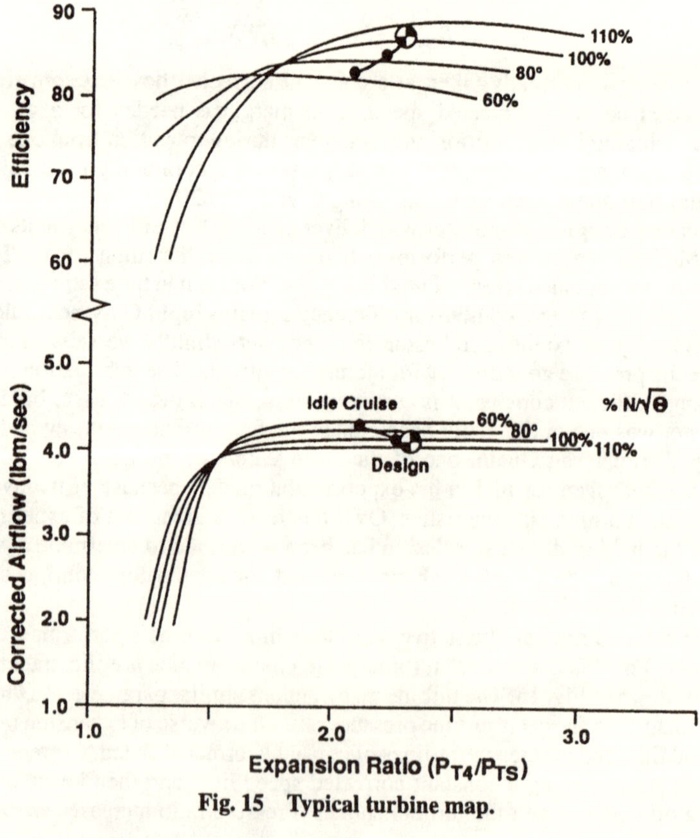

Fig. 15 Typical turbine map.

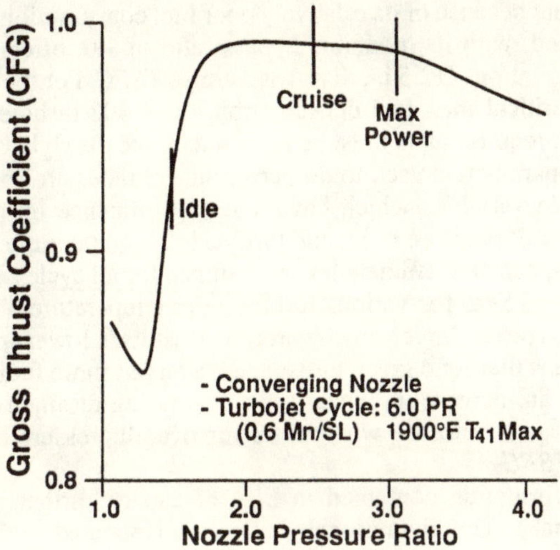

Fig. 16 Nozzle performance.

pressure goes down. Since increasing pressure ratio implies higher horsepower output, one might think that the horsepower would continue to increase. In fact, efficiency peaks and then rolls off with increasing pressure ratio. The net effect is that any turbine has a peak horsepower output. The peak horsepower is commonly referred to as the limit loading for a turbine.

The principle nozzle performance parameter, nozzle gross thrust coefficient (*CFG*), is shown in Fig. 16 as a function of nozzle pressure ratio. At high nozzle pressure ratios, the nozzle will perform quite well. As the engine is throttled back, the nozzle begins to operate less efficiently because of nonideal expansion levels.

The off-design characteristics of individual components have been described, but there are additional criteria associated with linking components to form a complete engine. Mass flow continuity requires that the flow exiting one component be equal to the flow entering the next. In addition, the compressor and turbine must have matched horsepower and physical speed. For multispooled systems like turbofans, the fan must be matched to the low-pressure turbine and the compressor must be matched to the high-pressure turbine. The methodology used to satisfy these criteria simultaneously is to apply numerical techniques, creating error functions which are dependent on the speed, flow, and work. An iterative approach is used to drive the errors to zero. Because of the large number of variables involved, the computer has become a critical element of off-design performance prediction. It has allowed the gas turbine industry to make off-design performance one of the aspects used to select a cycle. Because of the computer, it has become feasible to do a much more rigorous tradeoff to determine the "optimum" cycle for a given set of mission requirements.

V. Cycle Selection

The choice of the appropriate engine cycle is strongly driven by mission requirements. The ALCM, for example, must have long-range capability (~1500 n mile), yet fit in the B-52 and B-1 rotary rack. A turbojet such as the J-402 would not be appropriate because of its relatively poor fuel consumption rate. The F-107, on the other hand, with its moderate bypass ratio of 1.0, offers a good balance between size limitations (12.5 in. diam) and cruise *TSFC*. For the Harpoon, where thrust is more critical than fuel consumption, the J-402 turbojet is appropriate. Clearly, mission requirements must be known to make the right cycle selection.

The primary parameters used to do performance trades are specific thrust and *TSFC*. Key cycle variables which have a strong influence on performance are bypass ratio, overall pressure ratio, and turbine inlet temperature. For this type of tradeoff analysis, constant efficiencies are assumed for all cycles. Figure 17 shows specific thrust vs *TSFC* for various turbine inlet temperatures, overall pressure ratios, and bypass ratios. Increasing bypass ratio results in lower specific thrust and *TSFC*. This means that for a given thrust level, a larger, more fuel efficient engine will result from an increase in bypass ratio. Increasing turbine inlet temperature provides higher specific thrust, whereas raising overall pressure ratio will usually result in lower *TSFC*.

The cycle information contained in Fig. 17 can be further expanded to include dimensional and configurational constraints. Uncooled, nickel-based metallic turbines are limited to around 1900°F. A single-stage, uncooled, metallic high-pressure turbine is desired to minimize cost. Turbine corrected work can be used to define the maximum work capability of a single stage turbine:

$$\text{turbine corrected work} = \Delta H_{T\,(\text{turb})}/\theta \qquad (80)$$

A limit is suggested of somewhere around 32 btu/lbm for a single-stage turbine. Although higher loaded single-stage turbines have been designed and produced, this is a good first estimate for preliminary analyses. Engine front face diameter, although not usually the limiting diameter, is a good indicator of engine size. A simple method to calculate the front face diameter is to assume a specific flow and

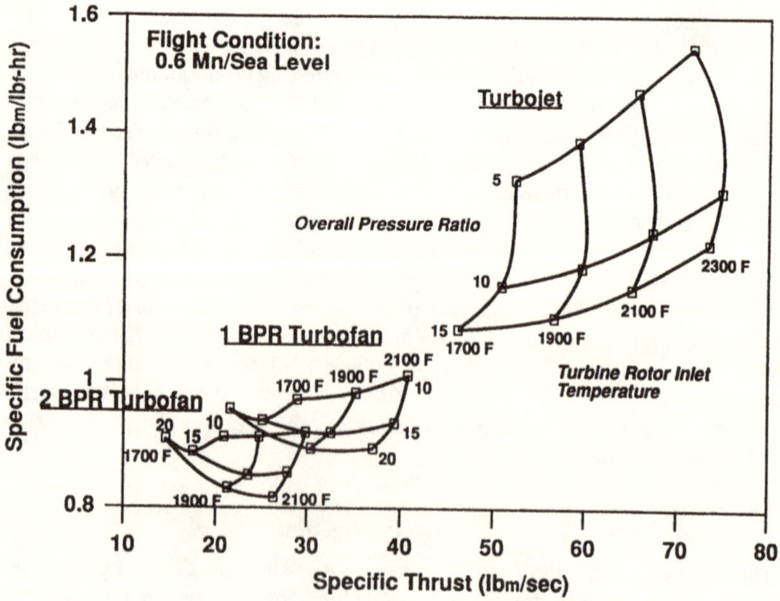

Fig. 17 Cycle performance comparison.

hub-to-tip ratio and calculate the diameter as a function of specific thrust. Specific flow is defined as

$$Wa_{\text{spec}} = \frac{Wa_{\text{corr}}}{Aa} = \frac{Wa\sqrt{\theta}}{\delta Aa} \tag{81}$$

Specific flow has the interesting characteristic of being dependent on Mach number. It peaks at Mach 1.0 and drops off at speeds above and below the choking velocity. Most compressors have a specific flow of around 36 to 40 lbm/s-ft^2. Annulus area is calculated using the following algebraic expression

$$\begin{aligned} Aa &= \pi\left(r_{\text{tip}}^2 - r_{\text{hub}}^2\right) \\ &= \pi\left(r_{\text{tip}}^2\right)\left[1 - (r_{\text{hub}}/r_{\text{tip}})^2\right] \end{aligned} \tag{82}$$

Substituting into Eq. (81),

$$Wa_{\text{spec}} = \frac{Wa\sqrt{\theta}}{\delta(\pi)(r_{\text{tip}})^2\left[1 - (r_{\text{hub}}/r_{\text{tip}})^2\right]} \tag{83}$$

Solving for flow

$$Wa = \frac{Wa_{\text{spec}}(\delta)(\pi)(r_{\text{tip}})^2\left[1 - (r_{\text{hub}}/r_{\text{tip}})^2\right]}{\sqrt{\theta}} \tag{84}$$

So, specific thrust is

$$\frac{FN}{Wa} = \frac{FN\sqrt{\theta}}{Wa_{\text{spec}}(\delta)\pi(r_{\text{tip}})^2\left[1 - (r_{\text{hub}}/r_{\text{tip}})^2\right]} \tag{85}$$

As an example, assume a weapon system requires 600-lb thrust at 0.6 Mach, 200-ft altitude. The missile diameter is 16 in., so maximum engine diameter should not exceed 14 in. The following represents a good first guess for engine design parameters.

$$Wa_{\text{spec}} = 36 \text{ lbm/s-ft}^2$$

$$r_{\text{tip}} = 6 \text{ in.}$$

$$r_{\text{hub}}/r_{\text{tip}} = 0.35$$

$$T_{41} = 1900^\circ\text{F (materials limit)}$$

Figure 18 expands on the Fig. 17 by adding the limiting factors. The 6-in. inlet radius limit requires the specific thrust to be greater than 19.8 lbf/lbm/s. Based on this constraint, the maximum bypass ratio allowable is around 2.0. The single-stage high-pressure turbine restricts the overall pressure ratio to around 10. Hence, the selected cycle for this example is a 10 pressure ratio, 1900°F T41, 2.0 bypass ratio turbofan. To make the final selection, part power performance for a few designs in the general vicinity of the selected point should be examined.

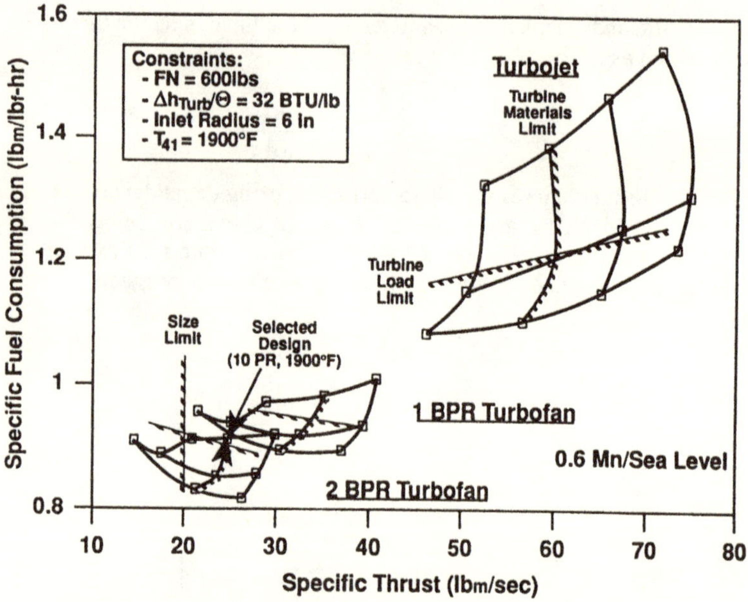

Fig. 18 Cycle performance comparison with design constraints.

VI. Component Flowpath Design

In the following sections, the process used to define preliminary component design features will be briefly covered. The analysis method provided gives the designer a simple technique to determine general characteristics including such things as tip speed, stage count, rough component dimensions, etc. Components to be described include the centrifugal compressor, axial compressor, combustor, turbine, and nozzle. A more comprehensive computational analysis that describes complex three-dimensional aero/thermodynamic behavior is far beyond the scope of this chapter, and realistically is not applied in the early stages of the design process. At this point it is most desirable to be able to determine a given cycle's general flowpath. Results from this simplified technique can be used to accurately predict key configurational characteristics.

A. Centrifugal Compressor Design

Centrifugal compressors are commonly used in expendable turbine engines. Centrifugal compressors offer the advantage of high-pressure ratio in a single stage, which is desirable in expendable applications since cost is a major driver. Fewer stages means fewer parts, which provides a potential cost savings compared to an equivalent pressure ratio, multistage axial compressor. Centrifugal compressors are easier to manufacture in small sizes. Often times they are used as the last stage in a combined axi/centrif compressor. Engines incorporating centrifugal compressors include the J-402, F-107, F-112 (advanced cruise missile) and Teledyne CAE 373 (U.S. Army MQM-107D Streaker).

There are a variety of nomenclatures used throughout the industry to represent centrifugal compressor aerodynamics. The centrifugal compressor nomenclature used here is shown in Fig. 19. Vector diagrams can be drawn at various streamlines

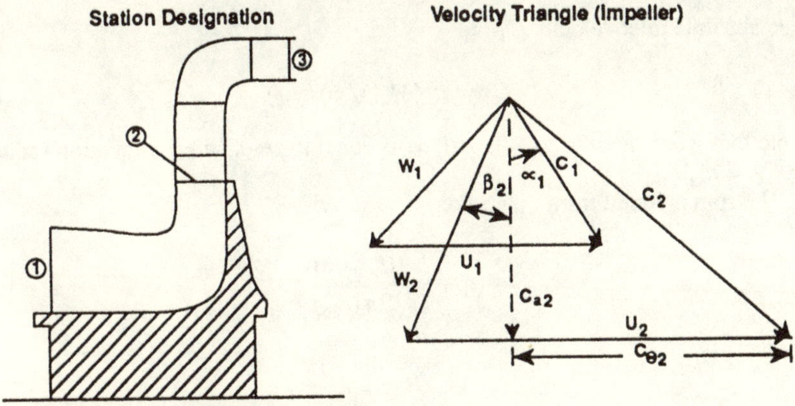

Definitions:
C - Absolute Velocity (ft/sec)
W - Relative Velocity (ft/sec)
U - Wheel Speed (ft/sec)
β - Relative Angle (Degrees)
∝ - Absolute Angle (Degrees)
C_θ - Absolute Velocity Component in the U direction (ft/sec) - Swirl Velocity
C_a - Absolute Velocity Component normal to the wheel speed

Fig. 19 Centrifugal compressor nomenclature.

along the span of the blade. For the purposes of this discussion, tip conditions are described. Often times the hub and mean (r_h and r_m) streamlines are examined to get a feel for blade twist. The vector diagrams represent relative (W) and absolute (C) velocities at the impeller inlet and exit stations. Relative refers to the velocity the rotating blade sees in its frame of reference, and absolute is the actual velocity.

The first step in determining the compressor velocity diagram is to calculate inlet specific flow. The inlet specific flow is determined using the inlet axial Mach number and swirl angle. The specific flow expressed as a function of Mach number is

$$Wa_{\text{spec}} = \frac{Wa\sqrt{\theta}}{\delta(Aa)} = \sqrt{\frac{\gamma g}{RT_{\text{std}}}}\frac{144(P_{\text{std}})M_{a1}/\cos\alpha_1}{\left\{1+[(\gamma-1)/2](M_{a1}/\cos\alpha_1)^2\right\}^{\{(\gamma+1)/[2(\gamma-1)]\}}} \tag{86}$$

The minimum inlet hub radius is set by shaft diameter, bearing support, or shaft mounted alternator size. For first guess purposes, 1.5–2.5 in. radius can be assumed for the minimum hub radius. With hub radius assumed, tip radius can be calculated.

$$r_{\text{tip}} = \sqrt{Aa/\pi + r_{\text{hub}}^2} \tag{87}$$

The inlet speed of sound is

$$a_1 = \sqrt{\gamma g R T_1} = 49.05\sqrt{\left\{T_{T_1}\Big/\left[1+\frac{\gamma-1}{2}\left(\frac{M_{a1}}{\cos\alpha_1}\right)^2\right]\right\}} \tag{88}$$

The absolute inlet velocity C_1 is

$$C_1 = a_1(M_{a1})/\cos(\alpha_1) \tag{89}$$

Note that when there is no preswirl, α_1 is equal to zero, and the equation reduces to $C_1 = a_1 M_{a1}$.

The rpm is found from tip speed

$$\text{rpm} = \frac{U_{\text{tip1}}(60)}{2\pi(r_{\text{tip}})} \tag{90}$$

The $C_{\theta 1}$ is

$$C_{\theta_1} = C_1 \sin(\alpha_1) \tag{91}$$

Inlet relative velocity is found using the Pythagorean theorem

$$W_1 = \sqrt{(a_1 M_{a1})^2 + (U_1 - C_{\theta_1})^2} \tag{92}$$

Inlet relative angle β_1 is

$$\beta_1 = \cos^{-1}[(a_1 M_{a1})/W_1] \tag{93}$$

Pressure ratio for the overall centrifugal compressor is known, but the impeller pressure ratio must be determined. Centrifugal compressors have the ability to produce high pressure ratio but, because of their high absolute rotor exit velocity, usually suffer a significant diffuser pressure loss. Not only does the flow exit at high speed, but it must be turned 90 deg to enter the combustor. Diffuser pressure loss is strongly driven by the rate of diffusion, size limitations, and impeller exit velocity. Generally a 10–12% pressure loss can be expected. Because of this pressure loss, the impeller must overpressurize the air to achieve an overall pressure ratio. Impeller pressure ratio can be expressed as

$$PR_{\text{imp}} = \frac{P_{T2}}{P_{T1}} = \frac{1}{[1 - (\Delta P/P)_{\text{diff}}]}(PR_{\text{ov}}) \tag{94}$$

Assuming an impeller exit tip speed, the tip radius is determined to be

$$r_2 = \frac{U_2(60)}{2\pi(\text{rpm})} \tag{95}$$

The Euler work equation is used to find the impeller exit swirl velocity $C_{\theta 2}$

$$\Delta H_{T(\text{comp})} = 1/gJ(U_2 C_{\theta_2} - U_1 C_{\theta_1}) \tag{96}$$

Substituting for $\Delta H_{T(\text{comp})} = Cp\Delta T_T$ and solving for C_{θ_2} gives

$$C_{\theta_2} = 1/U_2[Cp(T_{T2} - T_{T1})gJ + U_1 C_{\theta_1}] \tag{97}$$

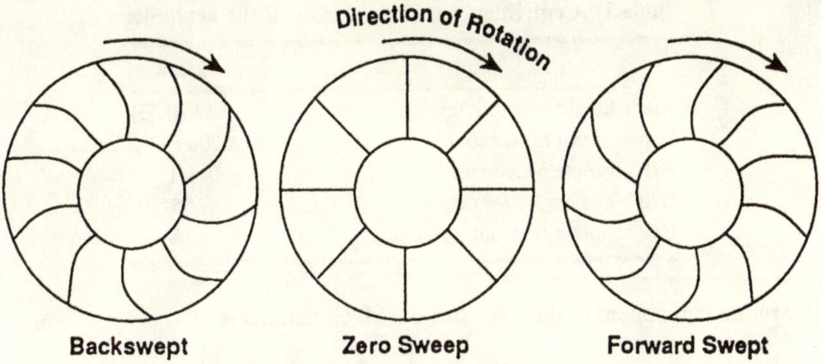

Fig. 20 Centrifugal exit blade shapes.

It is important that a sign convention be strictly followed. For the case where there is no preswirl ($C_{\theta 1} = 0$), this expression simplifies to

$$C_{\theta_2} = 1/U_2[Cp(T_{T2} - T_{T1})gJ] \tag{98}$$

With impeller exit tip speed and swirl known, the relative exit velocity can be found

$$W_2 = (U_2 - C_{\theta 2})/\sin(\beta_2) \tag{99}$$

where β_2 is the blade exit angle. Centrifugal compressors can have positive as well as negative backsweep angles (see Fig. 20). For the special case where $\beta_2 = 0$, U_2 is equal to $C_{\theta 2}$.

A term commonly used in centrifugal design is slip factor and is defined as

$$N_s = C_{\theta 2}/U_2 \tag{100}$$

By trigonometry,

$$\alpha_2 = \tan^{-1}(C_{\theta 2}/W_2 \cos\beta_2) \tag{101}$$

$$C_2 = C_{\theta 2}/\sin\alpha_2 \tag{102}$$

Expressed in terms of Mach number,

$$M_{c2} = C_2/\sqrt{\gamma g R T_2} \tag{103}$$

Since the static temperature is unknown, this expression in terms of total temperature becomes

$$M_{c2} = \frac{C_2\{1 + [(\gamma - 1)/2]M_{c2}^2\}^{\frac{1}{2}}}{\sqrt{\gamma g R T_{T2}}} \tag{104}$$

Solving for M_{c2}

$$M_{c2} = \sqrt{\frac{2C_2^2}{2\gamma g R T_{T2} - C_2^2(\gamma - 1)}} \tag{105}$$

Table 3 Centrifugal compressor state-of-the-art limits

	Limit
Blade height, *BHI*, in. (cm)	0.13 (0.33)
Impeller exit tip speed, U_2, ft/s (m/s)	2200 (670)
Exit absolute Mach number, M_{c2}	1.1
Exit blade angle, β_2, deg	55
Exit absolute flow angle, α_2, deg	75

The radial component of the exit absolute Mach number is

$$M_{c2_{\rm rad}} = M_{c2}(\cos\alpha_2) \tag{106}$$

Exit flow per annulus is determined using the exit conditions substituted into Eq. (86)

$$Wa_{\rm spec_2} = \sqrt{\frac{\gamma g}{RT_{\rm std}}} \frac{144(P_{\rm std})(M_{c2_{\rm rad}})}{\{1+[(\gamma-1)/2](M_{c2_{\rm rad}})^2\}^{\{(\gamma+1)/[2(\gamma-1)]\}}} \tag{107}$$

Annulus area is then

$$Aa_2 = Wa\frac{\sqrt{T_{T2}/T_{T\rm std}}}{Wa_{\rm spec_2}[P_{T2}/P_{\rm std}]} \tag{108}$$

Finally, the size of the blade at the impeller exit, or blade height (*BHI*) as it is commonly referred, can be determined by

$$BHI = Aa_2/2\pi r_2 \tag{109}$$

Although the term *state of the art* is subjective and changes over time, it is felt that some discussion of technology limiters would be helpful. Aerodynamic, structural, and manufacturing concerns all play a part in setting a centrifugal compressor design. Table 3 lists the constraints which have been determined based on existing design experience.

Blade height is limited by the need to control tip clearance losses and, therefore, manufacturing tolerances and aerodynamic passage size. The exit tip speed and backsweep angle are limited by the stress capability of titanium. If an aluminum rotor is to be used, the limits should be lowered to around 1600 ft/s U_2 and 10–20 deg backsweep. Exit absolute Mach number is not a problem for the impeller, per se, but can result in significant pressure losses in the diffuser. Entrance velocity into the vaned part of the diffuser should be subsonic. The small gap between the impeller and diffuser will provide enough diffusion to reduce the exit Mach from 1.1 to just below 1.0. The exit absolute flow angle should be limited to around 75 deg assuming that two sets of turning vanes would be available for flow straightening. These rules of thumb should be viewed as guidelines, not hard and fast limits. They are merely intended to help guide the reader in designing a centrifugal compressor.

B. Axial Compressor Design

Combined axial/centrifugal compressors are commonly used in expendable engines because front axial stages are easily linked to a centrifugal stage. Since increasing pressure ratio is key to improving performance, the axi/centrif is a good combination to enhance performance. All-axial compressors offer many advantages to axi/centrifs. Axial stages are easily ganged together without the need for a tortuous s-duct. They have better flow per frontal area since flow turning and diffusion occurs in the axial direction. Axial compressors do not typically have large diffusion losses because the pressure rise occurs in smaller increments. Historically, axial compressors have been viewed as more expensive because, for a given pressure ratio, the number of parts required to build up an all-axial multistage device is significantly larger than a single-stage centrifugal compressor. Recently, the manufacturing technology has matured to the point where multistage rotors can be cast in a single piece. This makes the all-axial compressor cost competitive with the axi/centrif.

The axial compressor can be examined as individual stages with the exit to the leading stage defining the inlet to the following stage. The exit vector diagram for one stage becomes the inlet triangle for the next. A similar nomenclature to the centrifugal compressor will be used for the axial (Fig. 21). The vector triangles should be examined at both the tip and hub. Preliminary design practice is to assume free-vortex flow, which means that the axial velocity at a given station is a constant from hub to tip. The tip is important because it represents the highest blade speed. The hub, on the other hand, will have the highest turning. Figure 22 illustrates the effect of moving from hub to tip on blade shape. Note that moving from hub to tip, the rotor speed increases, resulting in a change in blade turning ($\beta_1 - \beta_2$).

The methodology to generate the velocity triangles is essentially the same as that described in the centrifugal compressor design section and will not be repeated in the interest of brevity. However, the guidelines associated with axial compressor design are substantially different. Table 4 shows the suggested limits for axial compressors. As in the centrifugal compressor, the material selection plays a major role in setting the tip speed and temperature limit. This table relates

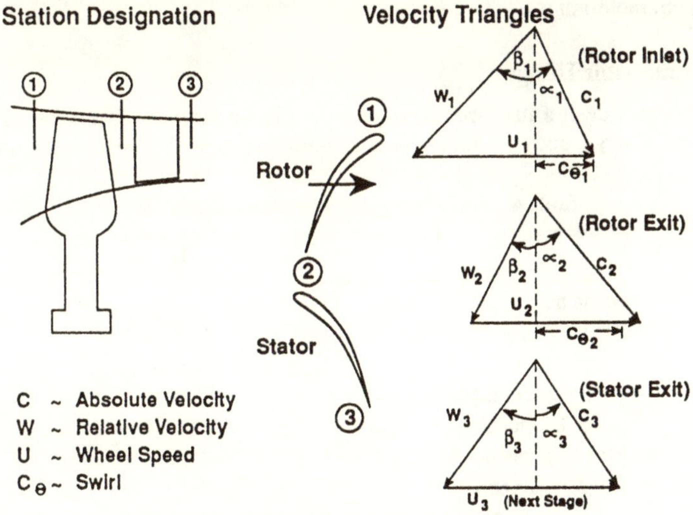

Fig. 21 Fan-axial compressor nomenclature.

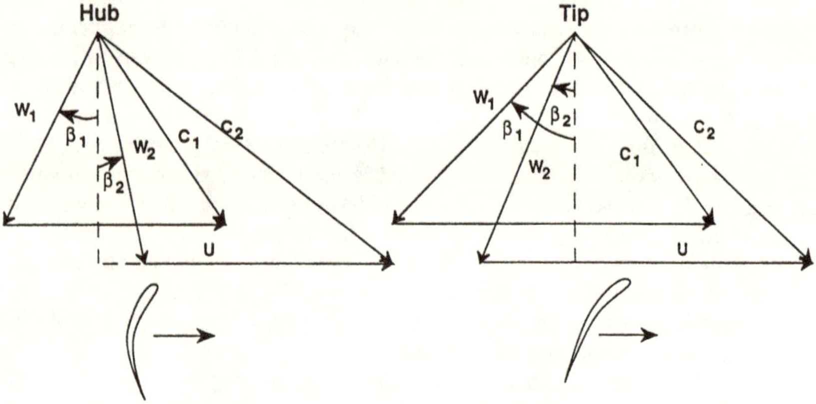

Fig. 22 Effect of span on blade shape.

to titanium-based materials. If aluminum was favored because of its manufacturing cost advantages, the tip speed would be limited to around 1000 ft/s and exit gas temperature to 300–400°F. Because of differences in the way an axial compressor is supported compared to a centrifugal, the tip-speed capability is significantly reduced. The fan tip speed and stator hub turning limit is higher since it operates in a lower temperature environment. The hub turning limit is based on allowable single stator turning without suffering large separation losses. The last stage exit Mach number limit is set to minimize combustor diffuser losses.

Combustor entrance Mach number should not exceed 0.2, so the Mach number entering the combustor prediffuser should be limited to 0.3. Many other factors are considered when designing rotating machinery including the choice of degree of reaction and diffusion factor. Information on how to calculate these parameters and how to interpret them can be found in Refs. 13 and 14. For preliminary analysis purposes, however, these guidelines provide a first-order estimate of compressor design characteristics.

C. Combustor Design

Combustor design has been and continues to be largely a black art. There is, however, a large body of experimental data which has established criteria for

Table 4 Axial compressor state-of-the-art limits

	Limit
Blade height, *BHI*, in. (cm)	0.25–0.4 (0.6–1.0)
Blade tip speed, U_{tip}, ft/s (m/s)	1500 (460)
	1600–1800 for a fan
Final stage exit Mach number, M_{a3}	0.30
Stator hub turning, α_3–α_2 deg	45–48, 50–55 for a fan
Max hub speed, U_{hub}, ft/s (m/s)	1200–1300 (366–396)
Compressor exit	
temperature, [T_{T3}, °F (°C)]	500 (260)

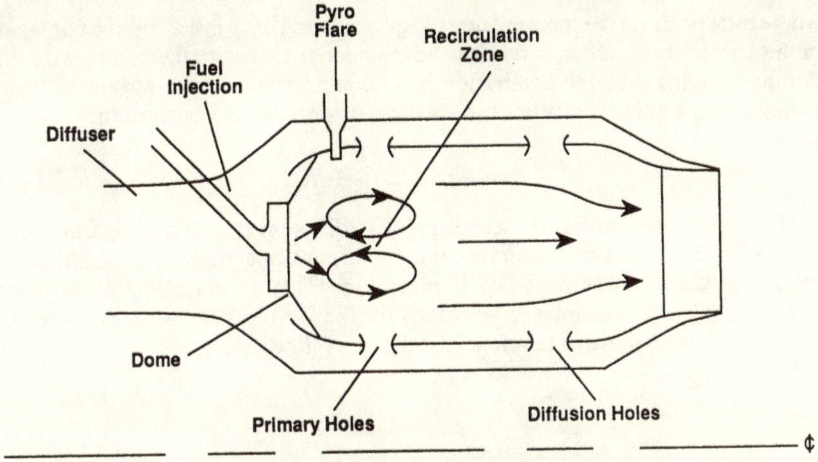

Fig. 23 Combustor key elements.

designing a combustor. The principle driver in combustor design is to minimize the size while maintaining acceptable performance.

Figure 23 shows the classic design features of a typical combustor. The gas turbine engine combustor is a subsonic burning device. In order to avoid blowout, the burning velocity is limited to around 15 ft/s. A recirculation zone is required to stabilize the flame and maintain low-burning velocity. Careful attention must be given to the air velocities around and through the dome. A portion of the incoming air is swirled and mixed with the fuel. High-shear stresses on the fuel atomize it to micron-size droplets. The mixture autoignites and is burned fuel rich in the dome. Fresh air enters through the primary holes, mixes with the remaining fuel, and is burned. This air helps maintain the recirculation zone as well. Downstream air entering the dilution holes reduces the temperature of the gases exiting the combustor to levels which the turbine can sustain.

Several combustor design parameters have been developed: 1) space heat release rate SR, 2) reference velocity V_{ref}, 3) dome velocity V_{dome}, 4) passage velocity V_{pass}, 5) residence time RT, and 6) length-to-height ratio L/H. Figure 24 shows pertinent dimensions for an annular combustor. Space heat rate is defined as follows.

$$SR = \frac{Wf(FHV)(\eta_{\text{burn}})}{P_{T3}(Vol_{\text{burn}})} \qquad [\text{BTU/atm-h-ft}^3] \tag{110}$$

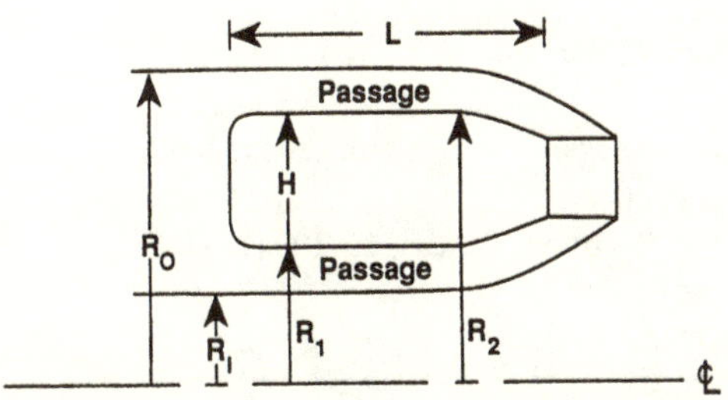

Fig. 24 Combustor dimensions.

where total pressure is expressed in atmospheres and Vol_{burn} is the volume contained in the burner itself. Often times it is advantageous to design the compressor and turbine first and then the combustor. With compressor exit and turbine entrance radii known, a guide is available to set combustor inner and outer radii.

$$Vol_{burn} = \pi L\left(R_2^2 - R_1^2\right) \tag{111}$$

Sometimes the combustor design point is different than the design point for other components. The corners of the flight envelope should be examined to determine where space heat rate is maximum. This occurs at the point where $Wf(\eta_{burn})/P_{T3}$ is maximized. Velocities for the various combustor passages are found using trigonometry relations and the mass flow equation.

Liner:

$$V_{pass} = Wa_{pass}/\rho_T A_{pass} \tag{112}$$

$$A_{pass} = \pi\left[\left(R_0^2 - R_2^2\right) + \left(R_1^2 - R_I^2\right)\right] \tag{113}$$

$$\rho_T = P_{T3}/T_{T3}R \tag{114}$$

Dome:

$$A_{dome} = \pi\left(R_2^2 - R_1^2\right) \tag{115}$$

$$Wg_{36} = Wa_3 + Wf - Wa_{pass} \tag{116}$$

$$V_{dome} = Wg_{36}/\rho_T A_{dome} \tag{117}$$

Reference:

$$A_{ref} = \pi\left[R_0^2 - R_I^2\right] \tag{118}$$

$$V_{ref} = (Wg_{36} + Wa_{pass})/\rho_T A_{ref} \tag{119}$$

Station 36 refers to the point in the burner where the fuel has been added, but burning has not yet occurred. For an annular burner, the length is found as follows.

$$L = Vol_{burn}/A_{burn} \tag{120}$$

Table 5 Combustor design constraints

	Limit
Space heat release rate, *SR*, Btu/atm-h-ft	10–12 × 10^6
(KJ/atm-h-m)	(3.5–4.2)
Passage velocity V_{pass}, ft/s (m/s)	100–200 (30–60)
Dome velocity, V_{dome}, ft/s (m/s)	15–30 (4–9)
Reference velocity, V_{ref}, ft/s (m/s)	60–100 (18–30)
Residence time, RT, ms	5–7
Burner length to height, L/H	1.8–2.5

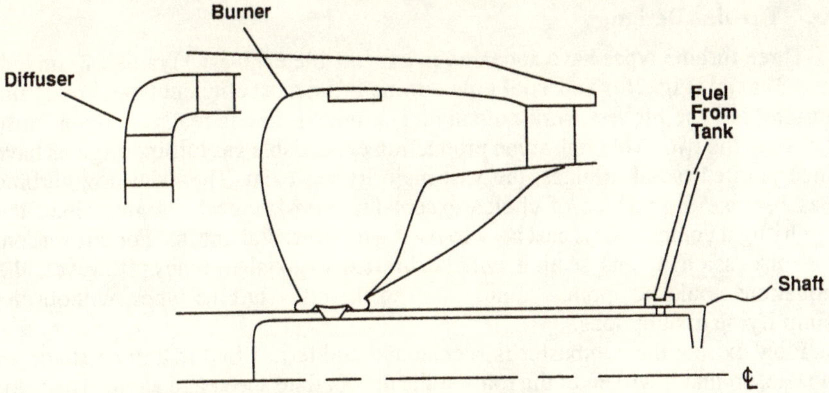

Fig. 25 Slinger combustor.

Residence time is

$$RT = L / V_{\text{dome}} \tag{121}$$

Design limits for the combustor are contained in Table 5. Depending on the cycle, different parameters determine the design. An iterative process is used to define the combustor dimensions. As in the compressor components, these are not hard and fast limits. Mission requirements can play a role in setting the combustor design. For example, if only a modest throttling range is required, the space heat rate limit could increase. Future technology developments such as advanced fuel/air swirlers could enhance burning and, therefore, reduce combustor size.

One other combustor type deserves mention because of its widespread usage in expendable gas turbine engines. The slinger combustor, illustrated in Fig. 25, has several advantages in that it is shorter and, because of its unique fuel delivery system, is less costly. Fuel enters into the center of the shaft and is pumped by the centrifugal force caused by the shaft's rotation. Centrifugally pressurized fuel is injected through holes in the shaft into the burner. Because the shaft is rotating, the fuel is readily atomized. The result is that a complex fuel injection system with air swirlers is not required.

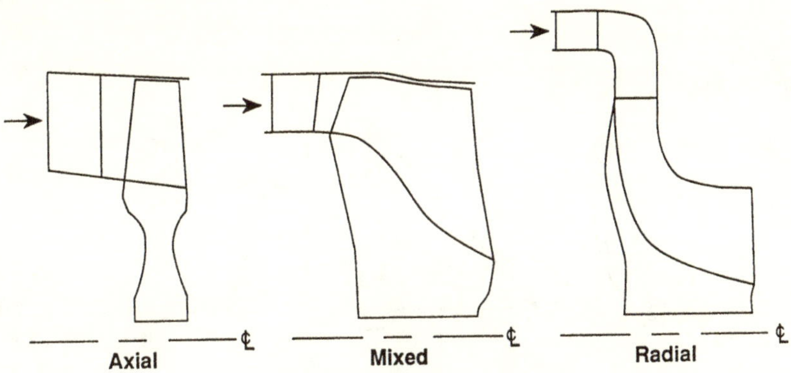

Fig. 26 Turbine types.

D. Turbine Design

Three turbine types have appeared in expendable engines: 1) radial, 2) mixed, and 3) axial (Fig. 26). The axial flow turbine is the most efficient type. The radial turbine has the highest work potential. The mixed flow is really a compromise between the two. Although some production expendable gas turbine engines have incorporated radial turbines, the vast majority are axial. The axial-flow turbine has become the turbine of choice except for ultra-low-cost designs where the centrifugal compressor is cast back to back with the radial turbine. For this reason, the discussion in this section will be limited to axial turbines. However, the equations could be applied equally well to the other turbine types, without the simplifying assumptions.

Flow exiting the combustor is accelerated and turned by the turbine stator, or nozzle, so that it will enter the rotor at the appropriate speed and angle. The rotor extracts the work necessary to drive the compressor. Vector triangles for the turbine are illustrated in Fig. 27. To minimize confusion, the nomenclature is similar to that which was used for the compressors.

Euler's equation can be applied equally well to the turbine.

$$\Delta H_{T\,(\mathrm{turb})} = 1/gJ(U_2C_{\theta 2} - U_3C_{\theta 3}) \tag{122}$$

Note that for the case illustrated in Fig. 27, the $C_{\theta 3}$ is in the negative direction.

Often times the turbine tip and hub radius are constant. With this assumption, the hub and tip speeds at the entrance and exit to the turbine are equal. This results in the following simplification:

$$\Delta H_{T\,(\mathrm{turb})} = U/gJ(C_{\theta 2} - C_{\theta 3}) \tag{123}$$

The turbine rpm is set equal to the compressor. Once again the tip streamline will be presented, but for an actual design, the mean (average) and hub should be examined. Free vortex flow can be assumed for the preliminary assessment. Examining the tip, mean, and hub streamlines will provide the designer with an

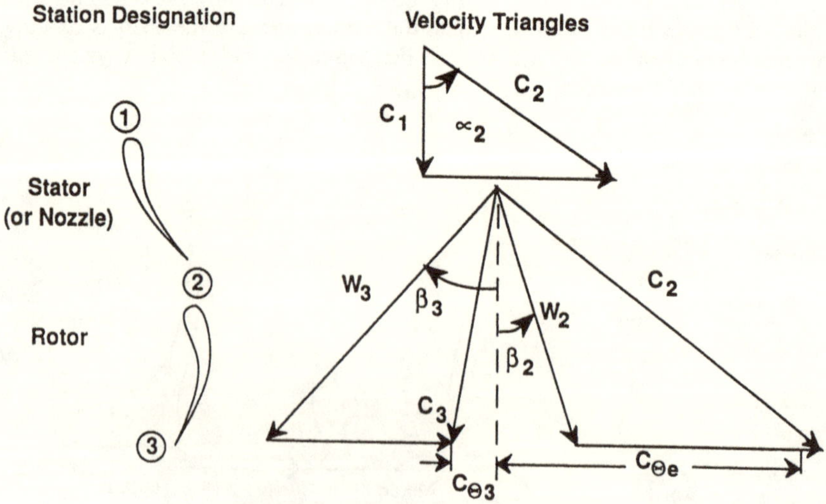

Fig. 27 Turbine design nomenclature.

understanding of the blade profiles as a function of span. With rpm and tip radius known, the tip speed is found using the following equation.

$$U_{\text{tip}} = 2\pi r_{\text{tip}}/60\ (\text{rpm}) \tag{124}$$

A parameter frequently used in turbine design is degree of reaction. It is defined as the static enthalpy drop across the rotor divided by the total enthalpy change across the stage.

$$DOR = (H_2 - H_3)/(H_{T1} - H_{T3}) \tag{125}$$

Using the Euler equation, degree of reaction for an axial machine can be expressed in velocity terms as

$$DOR = (W_3^2 - W_2^2)/[(C_2^2 - C_3^2) + (W_3^2 - W_2^2)] \tag{126}$$

Degree of reaction is useful in understanding the most common types of turbines. If the reaction is zero, the turbine is referred to as an "impulse" design. For reaction to be zero, the relative velocities entering and leaving the rotor are equal in magnitude and opposite in direction. The rotor blade shape is symmetric, i.e., $W_3 = W_2$ and $\beta_3 = -\beta_2$.

When the enthalpy drop across the stator and rotor are equal, the degree of reaction is 0.5. For this turbine type, the velocity triangles are symmetrical, i.e., $C_3 = W_2$, $C_2 = W_3$.

The final turbine type to be discussed is the zero exit swirl turbine. This type is recommended, if possible, because exit guide vanes are not required to straighten the exiting flow. Swirling flow entering the exhaust nozzle has a detrimental effect on the thrust performance. Swirl angles in excess of 15 deg have a significant thrust penalty. The zero exit swirl constraint produces less maximum work when compared to a turbine which allows for swirl. If possible, it is desired because it eliminates the need for exit guide vanes to deswirl the flow.

State-of-the-art limits for axial flow turbines are contained in Table 6. A good rule of thumb for determining required stage count is to examine the corrected work ($\Delta H_{T(\text{turb})}/\theta$). As already noted, if this value exceeds 32 Btu/lbm, more than one stage will likely be required. Angle and Mach number limits are set to minimize efficiency losses due to shocks and separation. Turbine materials are normally high-strength, nickel-based steel alloys because of the high-temperature

Table 6 Turbine state-of-the-art limits

	Limit
Stator exit, M	1.0–1.2
Stator exit angle, α_2, deg	75–80
Tip speed, U_{tip}, ft/s (m/s)	1600–1800 (488–549)
Hub rotor turning, β_3–β_2, deg	130
Rotor relative entrance, M	0.6
Extra swirl, deg	15
Turbine inlet temperature, °F (°C)	1900–1950 (1038–1066) (uncooled metallic)
AN^2, in.^{2}-rpm^2 (m^2-rpm^2)	$5\text{–}7\times10^9$ (1.3–1.8)

environment. These materials provide higher tip-speed capability compared to titanium. The temperature limit assumes the turbine nozzle is convectively cooled by compressor discharge air, a common practice across the industry. The rotor blades do not require cooling because the rotor is running away from the flow, resulting in a reduced relative speed. As a result, the relative stagnation temperature is lower than the absolute gas stagnation temperature. AN^2 (annulus area × rpm^2) is useful in defining the maximum structural limit for rim stress due to blade pull. The limit provided represents the state of the art for uncooled, expendable turbines.

E. Exhaust Nozzle Design

The principle driver in expendable engine nozzle design has historically been cost. Performance, although important, is not as critical and is often times sacrificed in favor of simplicity. Fixed nozzles have been used exclusively in all expendable engines to date. Therefore, this discussion will focus on fixed nozzle designs. The principle design parameters for the exhaust nozzle are velocity coefficient and flow coefficient. The velocity coefficient is defined as the actual jet velocity divided by the ideal isentropic, fully expanded jet velocity.

$$C_v = V_8/V_8' \tag{127}$$

Note that if the nozzle is fully expanded, $C_v = CFG$. The basic elements which contribute losses include exhaust flow angularity, friction, shocks, and separation. Flow coefficient C_f is defined as actual flow divided by theoretical flow.

$$C_f = Wa_8/Wa_8' \tag{128}$$

Flow coefficient is used to account for area constriction due to boundary-layer growth caused by friction and, in the worst case, separation. In terms of area, the flow coefficient can be expressed as

$$C_f = C_v(A_8'/A_8) \tag{129}$$

where A_8' is the effective area derived from the cycle calculation and A_8 is the actual area required. Hence, C_f is useful in defining the actual vs ideal predicted area required. Unfortunately, a simple method to analytically determine values for C_v and C_f does not exist. Experimental data, however, are available and are provided in the following sections.

Virtually all expendable gas turbine engines incorporate fixed converging nozzles. The fixed converging nozzle is the simplest, most cost effective nozzle. In addition, since all production expendable turbine engines to date have been used in subsonic applications, there has been no need for a converging-diverging nozzle. Figure 28 shows the effect of nozzle pressure ratio on C_v and C_f. The dotted line at 1.89 *NPR* corresponds to choked flow. The converging nozzle performance is excellent up to around 3.5 *NPR*. This is well within the *NPR* range for subsonic expendable turbines.

A key feature of a converging nozzle is its cone half-angle. Figure 29 (Ref. 11) shows the effect of cone half-angle on C_v and C_f. The absolute values are not important for this discussion. Test results illustrated in this figure indicate that the C_v line shown was within 2% for the various cone angles examined. Cone angle has little effect on C_v, but dramatically effects C_f. For large cone angles (greater than 30 deg), the C_f is very poor. Another major concern is that C_f, for the higher

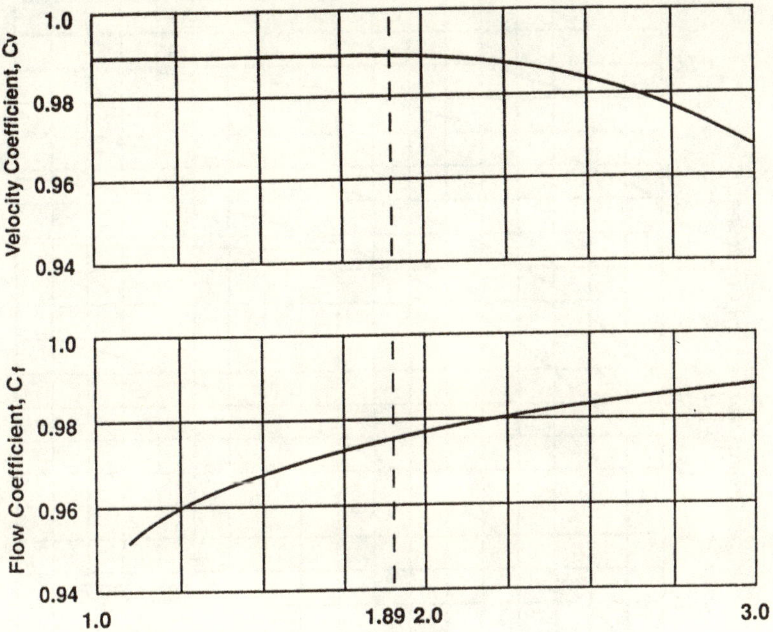

Fig. 28 Converging nozzle performance characteristics.

cone angles varies significantly with *NPR*. This has the net effect of causing the effective nozzle area to vary significantly with power setting. This will cause the engine to operate as if the nozzle area has changed, with the potential for loss in stall margin. For this reason, it is suggested that the nozzle half-cone angle not exceed 30 deg.

Future applications for gas turbine engines could potentially require *NPR* in excess of 3.5. For example, supersonic engines favor higher nozzle pressure ratios because of the need for greaterzz specific thrust. Consideration should be given to converging-diverging nozzles for high nozzle pressure ratios. Figure 30 shows the effect of nozzle area ratio (A_9/A_8) on C_v. Above 3.0 *NPR*, the converging nozzle ($A_9/A_8 = 1.0$) begins to roll off in terms of C_v. By 4.0 *NPR*, the 1.5 area ratio nozzle has a clear advantage. An interesting characteristic is that as nozzle pressure ratio goes below 2.5, the converging-diverging nozzle suffers a rather severe decrease in C_v, particularly for the 2.0 area ratio. This is primarily due to overexpansion. For man-rated engines, the need for very high nozzle pressure ratio for high specific thrust combined with a large throttling range drives the designer to a variable converging-diverging nozzle. To avoid the variability in an expendable engine nozzle, the performance penalty for a fixed nozzle needs to be considered. If a fixed converging-diverging nozzle is to be used, the area ratio is selected based on a compromise between maximum and cruise performance.

F. Controls/Accessories

No other component in an expendable gas turbine engine is driven as much by cost and reliability as the controls and accessories. Controls and accessories represent the most significant cost element of an expendable engine, in some cases in excess of 25%. Considerable effort is put into minimizing the complexity of this

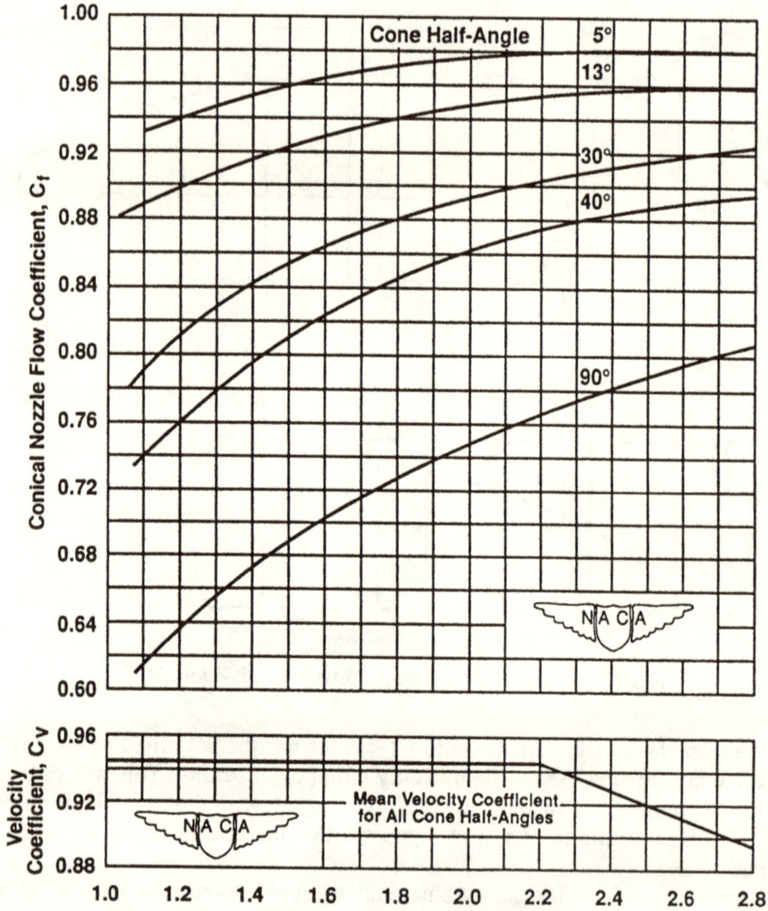

Fig. 29 Effect of cone angle on C_F and C_V.

part of the engine to keep costs down. Reliability is also required to be very high, usually greater than 0.995. The control system is the critical element which drives reliability as well. Cost and reliability tend to be at opposite ends of the spectrum, so these factors drive the design of expendable controls and accessories.

The vehicle requires power for guidance and control as well as warhead dispensing if required. If a missile/unmanned vehicle is to operate for a period any longer than a few minutes, a battery system is not adequate to provide systems power. The turbine engine is an excellent power source since power can be extracted directly from the shaft. Figure 31 shows a cut away of the J-402-CA-401 which is used to power the Harpoon. Built directly into the nose of the engine is a shaft mounted alternator and regulator which provides up to 2.66 kW of dc electric power. Other versions of the J-402 have been developed which produce up to 6 kW of power. This compact design has been used primarily for tactical applications. For longer range strategic vehicles such as Tomahawk and ALCM, power is extracted from a tower shaft. One horsepower is equal to 0.746 kW, a very small percentage of

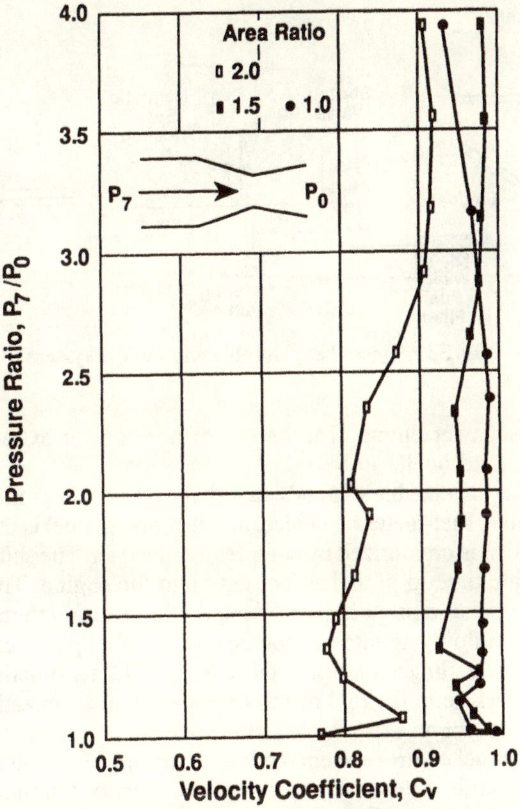

Fig. 30 ***C_v* comparison for various area ratio nozzles.**

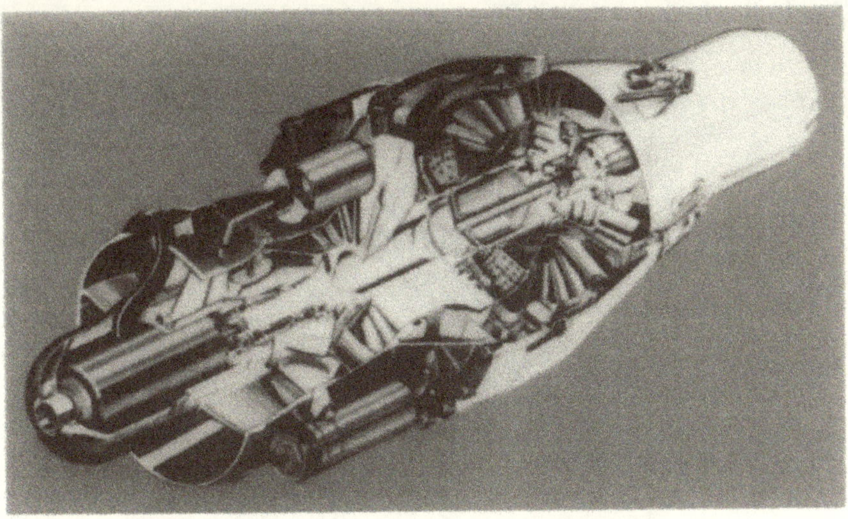

Fig. 31 **J402-CA-401 engine cutaway, courtesy of Teledyne.**

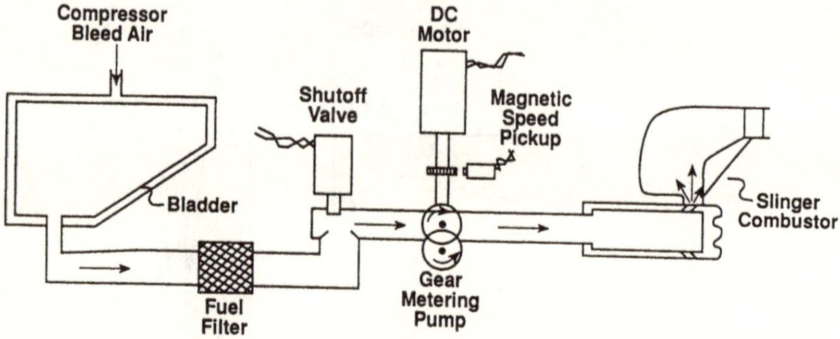

Fig. 32 Typical expendable engine fuel system.

a turbine's horsepower output. The impact of horsepower extraction on engine performance is quite small.

Figure 32 An expendable engine's fuel delivery system is made up of a fuel tank, shutoff valve, fuel-metering device, and fuel pump. Fuel is usually contained by a bladder which is pressurized by compressor bleed air. The shutoff valve opens at the appropriate time to initialize fuel flow into the engine. The fuel-metering device could be either a proportional solenoid or a metering pump. It is required to control the fuel flow rate into the combustor. The fuel pump can be a separate pump or an integral slinger as described in Sec. VI.C. Its function is to provide adequate fuel pressure to the fuel nozzles for good fuel atomization. illustrates a typical fuel system for an expendable turbine engine.

The electronic fuel control system regulates starting, acceleration, deceleration, and steady-state cruise. Because cost is critical, usually a minimum number of sensors are available. Typically, ram temperature T_{T2}, compressor discharge pressure P_{T3}, and rotor speed are input to the control unit. With this information the control uses a stored acceleration/deceleration schedule to regulate engine fuel flow to throttle the engine as required for a given mission (Fig. 33).

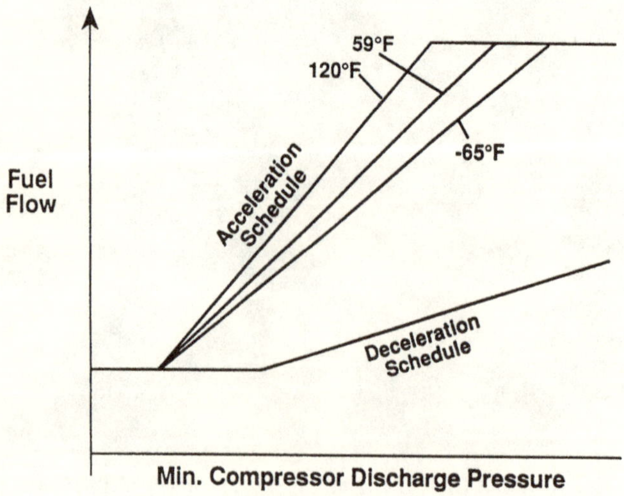

Fig. 33 Typical fuel control schedule.

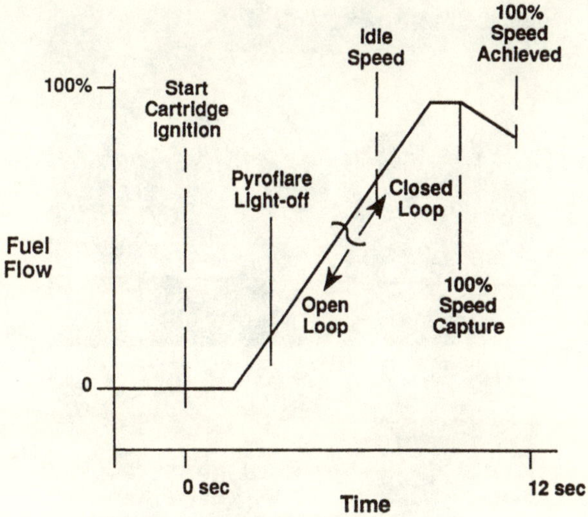

Fig. 34 Typical starting sequence.

The starting system is a key element of reliability for expendable turbine engines. Barring a major design/manufacturing flaw, if the engine starts it will more than likely complete the mission. Although some expendable engines are windmill started, most use a pyrotechnic device to accelerate the engine to lightoff speed. Windmill starting is not feasible for most applications because low-altitude launch requires starting and acceleration to maximum speed in a matter of seconds to avoid ground clobber. The problem of reliable starting is further aggravated by the need to provide rapid starting at a wide range of Mach numbers, altitudes, and ambient temperatures. The start cartridge impinges its exhaust gases on either the compressor or turbine, to rotate the engine to around 30% high-pressure-spool speed. Fuel flow is started and the burner is lit off by another pyroflare. Fuel flow is scheduled to provide stall-free acceleration to maximum speed. A great deal of development testing is dedicated to optimizing the appropriate pyroflare sizing, fuel initiation speed, lightoff speed, fuel scheduling, etc.

Start sequencing is controlled by the electronic control. From start of rotation to a predetermined sub-idle point, the fuel control operates in open loop with pyrotechnic sequencing. Fuel flow is determined by lapsed time from start initiation. When the engine reaches the specified sub-idle point, it switches to closed loop. Inputs from rotor speed, compressor discharge pressure, etc. are used to schedule the fuel flow up to maximum speed. Figure 34 shows the sequencing of the various elements as a function of time. Note that as the engine approaches 100% speed, the fuel flow is cut back. This is to minimize speed overshoot.

VII. Integration

It has been mentioned throughout this chapter how significantly the expendable turbine engine design is driven by missile size and weight constraints. The inlet and exhaust system are set by the overall system size constraints as well. The Harpoon, for example, is designed to fit in a round tube. As a result, a semiflush

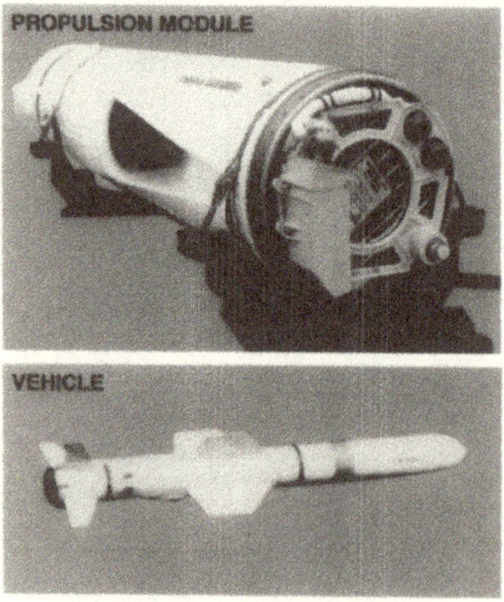

Fig. 35 Harpoon vehicle.

inlet is used (Fig. 35). The ALCM, on the other hand, must fit on a rotary rack, which has sufficient room for a top-mounted pitot inlet (Fig. 36). Low observable requirements can have a profound effect on inlet and exhaust designs also. This section will describe some of the tradeoffs associated with inlet and exhaust system selection.

A. Inlet System Integration

For purposes of discussion, the three inlet types shown in Fig. 37 will be examined. Each has its advantages and disadvantages. The pitot inlet has the best ram recovery but is most difficult to integrate; untreated, it has the worst radar cross section (rcs) characteristics. The distributed inlet has the worst performance

Fig. 36 ALCM vehicle courtesy of Williams International.

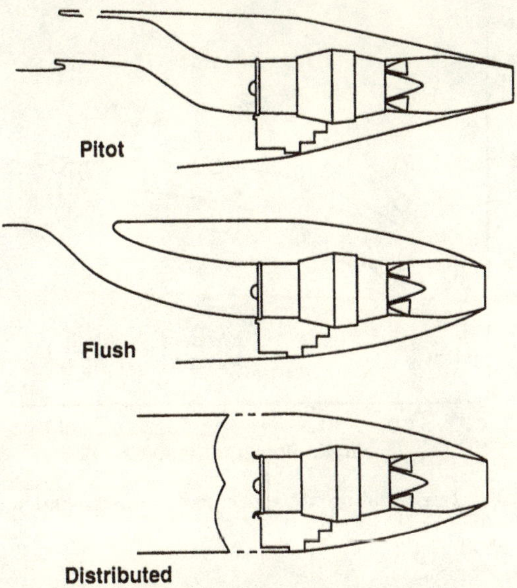

Fig. 37 Inlet types.[12]

but uses the minimal internal volume. In addition, the flush and distributed inlets have a more favorable rcs. Figure 38 shows the ram recovery for the three inlets as a function of percent airflow.[12]

Another interesting factor of inlet design is inlet distortion. Pressure distortion has a significant impact on the compressor stall margin. Stall margin for an engine can be seriously degraded by an inlet with a large amount of distortion. Different fans and compressors have varying degrees of sensitivity to distortion. They can also be sensitive to the location where the maximum distortion occurs. Some compressors stall as a result of high-hub distortion, whereas others are insensitive. Normally, the inlet for a given missile is wind-tunnel tested at a variety of critical mission conditions to determine anticipated distortion patterns. Launch, cruise,

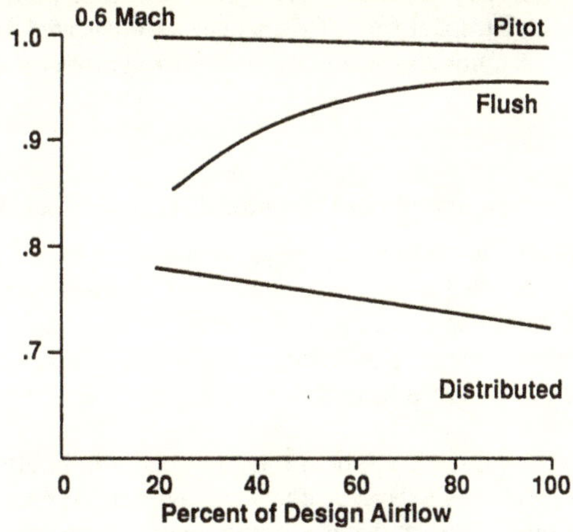

Fig. 38 Ram recovery for various inlet types.[12]

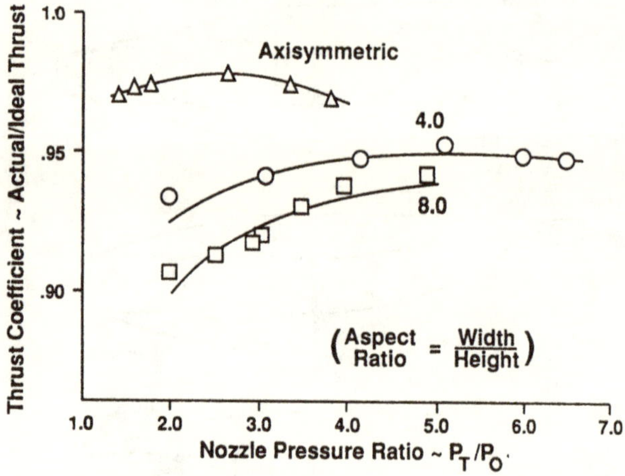

Fig. 39 Impact of nozzle aspect ratio on performance.[12]

and high-angle-of-attack operation are examined. These data are used to design distortion screens to simulate inlet characteristics during engine testing. As part of the flight qualification, the engine must demonstrate compressor stability at the critical mission conditions. A failure to perform as required can result in a redesign of the compression device.

B. Exhaust System Integration

The exhaust system performance characteristics have already been discussed at great length. Sometimes radar and infrared (IR) signature requirements place additional constraints on the designer. It is desirable to use a two-dimensional exhaust if reduced observables are required. The two-dimensional exhaust nozzle reduces the radar return to a narrow viewing aspect. In addition, a wide nozzle with a short height will have a tendency to more quickly dissipate the hot gasses with the freestream. The net result is a reduced IR signature compared to a round exhaust. There is a detrimental impact on performance, as illustrated in Fig. 39. Therefore, if observables are not a consideration, the axisymmetric nozzle is the best choice.

VIII. Future Trends in Expendable Gas Turbine Engines

Gas turbine engines have enjoyed a phenomenal growth in the expendable vehicle area. They have been used for a number of systems, some of which have been discussed in this chapter. Both subsonic tactical and subsonic strategic applications have been home to gas turbines. From straight turbojets to moderate bypass turbofans, gas turbines have gained acceptance in the unmanned vehicle community.

For subsonic tactical missile applications, low cost will likely continue to drive the engine design. Methods to reduce cost such as reduced parts count, advanced manufacturing technologies, and parts commonality with existing engines will strongly influence future subsonic tactical engines. Increased specific thrust capability is dependent on increased turbine rotor inlet temperature, and so advanced materials or cooling technologies which do not adversely affect cost are needed.

Subsonic strategic vehicles are not quite as driven by cost as are tactical vehicles. The engine designer for these applications can consider alternate cycles to enhance range capability. In addition, reconnaissance vehicles would benefit from improved fuel economy to allow for longer time on station. Significant improvements are possible by using innovative cycles such as high-bypass ratio turbofans ($BPR = 3–10$) or propfans.[10] Because of the importance of high component efficiencies for strategic engines, work is needed to develop the required aerodynamic technologies. Low observables is a major driver for long range or endurance, so close attention must be paid to integration issues.

Expendable gas turbine engines to date have been used exclusively for subsonic applications. Supersonic weapon systems have not used gas turbine engines primarily because extremely high turbine rotor inlet temperature is needed for adequate specific thrust. Recently, cooling technology has progressed to the point where the required turbine inlet temperature for supersonic applications is achieveable. Advanced materials such as ceramic composites and carbon-carbon appear ready for static parts application and may some day be able to be used on rotating parts as well. Supersonic weapon systems offer the potential to provide quick reaction capability required to attack mobile targets or for timely reconnaissance. Turbomachinery has the added benefit of longer range, compared to alternative propulsion systems, and greater mission flexibility. For example, a reconnaissance vehicle may desire the capability to reach its search area quickly, then slow down to provide adequate coverage.

Fig. 40 Allison J-102 supersonic expendable turbojet.

Much progress has been made towards the development of a supersonic expendable turbojet engine by the Allison company. The Allison J-102 (Fig. 40), developed and tested under contract to the U.S. Air Force's Aero Propulsion and Power Directorate of Wright Laboratory, has successfully operated at simulated 3.5 Mach number, 90,000 ft. This demonstration has greatly expanded the envelope of expendable gas turbine engines.

Appendix A: Gas Properties

Temp, °R Air, $\phi = 0$	Cp, Btu/lbm-°R	$\gamma(Cp/Cv)$
400	0.2393	1.402
600	0.2400	1.400
800	0.2434	1.392
1000	0.2486	1.381
1200	0.2547	1.368
1400	0.2610	1.356
1600	0.2671	1.345
1800	0.2725	1.336
2000	0.2773	1.329
Air plus fuel, $\phi = 0.25$		
1000	0.2543	1.370
1200	0.2609	1.357
1400	0.2678	1.344
1600	0.2743	1.333
1800	0.2803	1.324
2000	0.2855	1.316
2200	0.2900	1.310
2400	0.2938	1.305
2600	0.2972	1.300
2800	0.3001	1.296
3000	0.3028	1.293
Air plus fuel, $\phi = 0.50$		
1000	0.2597	1.359
1200	0.2670	1.346
1400	0.2743	1.334
1600	0.2813	1.323
1800	0.2878	1.313
2000	0.2934	1.305
2200	0.2983	1.299
2400	0.3026	1.293
2600	0.3063	1.289
2800	0.3096	1.285
3000	0.3125	1.281

Appendix B: Examples

Example 1

Consider a turbojet operating at $0.7Mn/200$ ft with the following cycle characteristics:

$$\eta_{ram} = 1.0 \qquad \eta_{polycomp} = 0.87 \qquad FHV = 18{,}100 \text{ Btu/lbm}$$
$$PR_{comp} = 6.0 \qquad \eta_{burn} = 0.98 \qquad P_{T4}/P_{T3} = 0.95$$
$$T_{41} = 1900°\text{F} \qquad \eta_{turb} = 0.85 \qquad P_{T7}/P_{T6} = 0.99$$
$$Wa = 15.0 \text{ lbs/s} \qquad CFG = 0.985$$

Calculate all component temperatures and pressures as well as overall performance.

Answer: The ambient temperature and pressure at 200 ft are 518.0°R and 14.59 psi, respectively. The γ are found in Appendix A. The inlet ram conditions are

$$T_{T2} = T_{amb}\left(1 + \frac{\gamma - 1}{2}M_0^2\right) = 518\left[1 + \frac{1.4 - 1}{2}(0.7)^2\right] = 568.8°\text{R}$$

$$P_{T2} = \eta_{ram}(P_{amb})\left(\frac{T_{T2}}{T_{amb}}\right)^{[\gamma/(\gamma-1)]} = 14.59\left(\frac{568.8}{518.0}\right)^{1.4/0.4} = 20.24 \text{ psi}$$

Compressor exit conditions are found using Eq. (33)

$$T_{T3} = T_{T2}(PR_{comp})^{[(\gamma-1)/\eta_{poly}(\gamma)]} = 568.8(6.0)^{[0.4/0.87(1.4)]} = 1024.5°\text{R}$$

$$P_{T3} = P_{T2}(PR_{comp}) = 20.24(6.0) = 121.4 \text{ psi}$$

The compressor adiabatic efficiency can be derived from the polytropic efficiency using Eq. (34):

$$\eta_{comp} = \frac{(6.0)^{(0.4/1.4)} - 1}{(6.0)^{[0.4/(0.87)(1.4)]} - 1} = 0.834$$

The compressor work, expressed in horsepower, is

$$\text{Work} = Wa_{comp}(Cp)(T_{T3} - T_{T2})(778 \text{ ft-lbs/Btu})\left(\frac{1 \text{ hp}}{550 \text{ ft-lbs/s}}\right)$$
$$= 15.0(0.24)(1024.5 - 568.8)778/550 = 2320.6 \text{ hp}$$

The fuel flow is found using Fig. 5 and Eq. (35) to correct for burner efficiency

$$FAR = \frac{(FAR')}{\eta_{burn}} = \frac{0.021}{0.98} = 0.0214$$

$$W_f = WA_{comp}(FAR)(3600 \text{ s/h}) = 1156 \text{ lbm/h}$$

The combustor exit pressure is

$$P_{T4} = 0.95(121.4) = 115.3 \text{ psi}$$

At the turbine inlet, the equivalence ratio is $\Phi = 0.0214/0.067 = 0.319$. Using Appendix A, $\Phi \approx 0.25$, the $\gamma = 1.30$, and $Cp = 0.293$ Btu/lbm-°F. The turbine expansion ratio is found with Eq. (38):

$$ER = \left[1 - \frac{\text{Work}_{\text{comp}}}{Wg_{\text{Turb}}(Cp)(T_{T4})(\eta_{\text{Turb}})}\right]^{[\gamma/(\gamma-1)]}$$

$$= \left[1 - \frac{2320.6(550/778)}{(15.0 + 0.3)(0.293)(2360)(0.85)}\right]^{(0.3/1.3)} = 0.411$$

Turbine exit temperature and pressure are

$$T_{T5} = T_{T4}[1 - \eta_{\text{Turb}}(1 - (ER)^{[(\gamma-1)/\gamma]})]$$

$$= 2360\{1 - 0.85[1 - (0.411)^{0.3/1.3}]\} = 1988°\text{R}$$

$$P_{T5} = 115.3(0.411) = 47.39 \text{ psi}$$

The nozzle is assumed to be converging. The total pressure entering the nozzle is $P_{T8} = 0.99(47.39) = 46.92$ psi. The nozzle pressure ratio is ($\gamma = 1.32$ and $C_p = 0.286$ Btu/lbm-°F)

$$NPR = \frac{P_{T5}}{P_{\text{amb}}} = \frac{47.39}{14.59} = 3.24$$

This is definitely choked, so from Eq. (17),

$$CNPR = \left[\frac{\gamma + 1}{2}\right]^{\gamma/(\gamma-1)} = \left[\frac{2.32}{2}\right]^{(1.32/0.32)} = 1.84$$

The nozzle exit static pressure and temperature are

$$T_8 = T_{T8}/[(\gamma + 1)/2] = 1988/(2.32/2) = 1713°\text{R}$$

$$P_8 = P_{T8}/[(\gamma + 1)/2]^{[\gamma/(\gamma-1)]} = 46.92/(2.32/2)^{(1.32/0.32)} = 25.44 \text{ psi}$$

The exit velocity is

$$V_0 = M_8\sqrt{\gamma g R T_8} = \sqrt{1.32(32.2)(53.35)(1713)} = 1971 \text{ ft/s}$$

The flight speed is

$$V_0 = M_0\sqrt{\gamma g R T_{\text{amb}}} = 0.7\sqrt{1.4(32.2)(53.35)(518.0)} = 781.3 \text{ ft/s}$$

The nozzle exit area is

$$A_8 = \frac{Wg_8}{P_8/RT_8(V_{\text{jet}})} = \frac{15.3}{25.44/[53.35(1713)](1971)} = 27.9 \text{ in.}^2$$

The gross thrust, from Eq. (47), is

$$FG = CFG\left[\frac{Wa(1 + FAR)}{9}V_{\text{jet}} + A_8(P_8 - P_{\text{amb}})\right]$$

$$= 0.985[15.3/32.2(1971) + 27.9(25.44 - 14.59)] = 1220.7 \text{ lbf}$$

The ram drag is

$$FR = \frac{WaV_0}{g} = \frac{15.0(781.3)}{32.2} = 364.0 \text{ lbf}$$

The net thrust is

$$FN = FG - FR = 856.7 \text{ lbf}$$

The specific thrust and thrust specific fuel consumption are

$$FN/WA = 856.7/15.0 = 57.1 \text{ lb}_f/\text{lb}_m/\text{s}$$

$$TSFC = 1157/856.7 = 1.35 \text{ lb}_m/\text{ lb}_f\text{-h}$$

Example 2

Calculate the exit conditions for a mixer with the following inlet conditions.
Bypass stream:

$$Wg_{16} = 10.0 \qquad T_{T16} = 698^\circ \text{ R}$$
$$P_{T16} = 37.2 \text{ psi} \qquad BPR = 1.0$$

Core stream:

$$Wg_5 = 10.2 \qquad Mn_5 = 0.175$$
$$P_{T5} = 36.4 \text{ psi} \qquad FAR_5 = 0.018$$
$$T_{T5} = 1650^\circ\text{R}$$

Answer: From Appendix A, the bypass $\gamma = 1.4$ and the core $\gamma = 1.3$. The entrance static pressure is

$$P_5 = P_{16} = \frac{P_{T5}}{\{1 + [(\gamma - 1)/2](M_5)^2\}^{[\gamma/(\gamma-1)]}}$$

$$= \frac{36.4}{[1 + (0.3/2)(0.175)^2]^{1.3/0.3}} = 35.7 \text{ psi}$$

The bypass stream Mach number is

$$M_{16} = \sqrt{\frac{2}{\gamma - 1}\left[\left(\frac{P_{T16}}{P_{16}}\right)^{[(\gamma-1)/\gamma]} - 1\right]}$$

$$= \sqrt{\frac{2}{0.4}\left[\left(\frac{37.2}{35.7}\right)^{(0.4/1.4)} - 1\right]} = 0.24$$

Using Eq. (67), the mixed flow temperature is

$$T_{T7} = \frac{T_{T16}(BPR) + T_{T5}(1 + FAR)}{BPR + 1 + FAR}$$

$$= \frac{698(1.0) + 1650(1.0 + 0.018)}{1.0 + 1 + 0.018} = 1178^\circ\text{R}$$

The mixed flow Mach number is found by using Eqs. (69–71)

$$f(M_{16}) = (0.24)^2[1 + (0.4/2)(0.24)^2][1 - 1.4(0.24)^2]^{-2} = 0.050$$

$$f(M_5) = (0.175)^2[1 + (0.3/2)(0.175)^2][1 + 1.3(0.175)^2]^{-2} = 0.028$$

$$F = (1 + BPR)\left[1 + BPR\left(\frac{T_{T16}}{T_{T5}}\right)\right]\left[\frac{1}{f\sqrt{(M_5)}} + BPR\sqrt{\frac{T_{T16}/T_{T5}}{f(M_{16})}}\right]^{-2}$$

$$= 2\left[1 + 1\left(\frac{698}{1650}\right)\right]\left[\frac{1}{\sqrt{0.028}} + 1\sqrt{\frac{698/1650}{0.050}}\right]^{-2} = 0.036$$

The FAR_7 is

$$FAR_7 = \frac{Wg_5}{Wa_{16}}(FAR_5) = \frac{10.2}{20.2}(0.018) = 0.009$$

From Appendix A, $\gamma_7 = 1.36$. The mixed flow Mach number is

$$M_7 = \left[\frac{2F}{1 - 2\gamma F + \sqrt{1 - 2(\gamma + 1)F}}\right]^{\frac{1}{2}}$$

$$= \left[\frac{2(0.036)}{1 - 2(1.36)(0.036) + \sqrt{1 - 2(2.36)(0.036)}}\right]^{\frac{1}{2}}$$

$$M_7 = 0.20$$

The core and bypass areas are

$$A_{16}(\text{bypass}) = \left(\frac{Wa_{16}}{P_{16}M_{16}}\right)\sqrt{RT_{T16}/(\gamma g\{1 + [(\gamma - 1)/2]M_{16}^2\}^{-(\gamma+1)/(\gamma-1)})}$$

$$= \frac{10}{35.7(0.24)}\sqrt{\frac{53.35(698)}{1.4(32.2)[1 + (0.4/2)(0.24)^2]^{-6}}}$$

$$= 34.7 \text{ in.}^2$$

$$A_5(\text{core}) = \frac{10.2}{35.7(0.175)}\sqrt{\frac{53.35(1650)}{1.3(32.2)[1 + (0.3/2)(0.175)^2]^{-7.67}}}$$

$$= 76.2 \text{ in.}^2$$

The exit area is

$$A_7 = A_{16} + A_5 = 110.9 \text{ in.}^2$$

Using Eq. (72)

$$P_7 = \frac{33.4(35.7)[1 + 1.4(0.24)^2] + 74.7(35.7)[1 + 1.3(0.175)^2]}{110.9(1 + 1.36(0.20))^2} = 34.7 \text{ psi}$$

$$P_{T7} = P_7\{1 + [(\gamma - 1)/2](M_7^2)\}^{[\gamma/(\gamma-1)]}$$

$$= 34.7[1 + (0.36/2)(0.2)^2]^{(1.36/0.36)} = 35.7 \text{ psi}$$

Example 3

Determine the rotor and stator velocity triangles at the tip for a single stage axial fan with the following characteristics.

Cycle: $(0.65Mn/SL)$

$$Wa = 16 \text{ lbs/s} \qquad \eta_{\text{comp}} = 0.87$$
$$PR = 1.64 \qquad Wa_{\text{spec}} = 40 \text{ lbs/s-ft}^2$$

Configurational:

$$r_H/r_T = .035 \qquad M_{a2}(\text{ rotor exit}) = 0.50$$
$$U_{\text{tip}} = 1600 \text{ ft/s} \qquad M_{a3}(\text{stator exit}) = 0.45$$

Assume a constant tip radius design, with a ram recovery of 1.0. There is no inlet guide vane, so the flow enters axially. The stator is to return the flow to axial.

Answer: The station numbers used in this example are 1 for the rotor inlet, 2 for the rotor exit, and 3 for the stator exit.

Rotor inlet: The inlet total temperature and pressure at $0.6Mn/SL$ are

$$T_{T1} = 519[1 + (1.4 - 1)/2(0.65)^2] = 562.9^\circ \text{ R]} \qquad (\theta = 1.085)$$

$$P_{T1} = 14.70(562.9/519)^{(1.4/0.4)} = 19.53 \text{ psi} \qquad (\delta = 1.329)$$

By iterating, Eq. (86) is used to determine that the inlet axial Mach number M_{a1} is equal to 0.565. The inlet speed of sound is

$$a_1 = 49.05\sqrt{\frac{T_{T1}}{\{1 + [(\gamma - 1)/2]M_{a1}^2\}}} = 49.05\sqrt{\frac{562.9}{[1 + 0.2(0.565)^2]}} = 1128 \text{ ft/s}$$

Since there are no inlet guide vanes, $\alpha_1 = 0$ deg. The absolute inlet velocity is

$$C_1 = a_1 M_{a1} = 1128(0.565) = 637.5 \text{ ft/s}$$

The annulus area and tip radius can be determined by rearranging Eq. (82):

$$Aa = \frac{Wa_{\text{corr}}}{Wa_{\text{spec}}} = (16)\frac{(\sqrt{1.085}/1.329)}{40} = 0.314 \text{ ft}^2$$

$$r_{\text{tip}} = \sqrt{\frac{Aa}{\pi[1 - (r_h/r_1)^2]}} = \sqrt{\frac{0.314}{\pi(1 - 0.35)^2}} = 0.337 \text{ ft} = 4.05 \text{ in.}$$

The RPM is found with Eq. (90) to be

$$\text{rpm} = \frac{U_{\text{tip}}(60)}{2\pi(r_{\text{tip}})} = \frac{1600(60)}{2\pi(0.337)} = 45{,}338 \text{ rpm}$$

The inlet relative velocity is

$$W_1 = \sqrt{(a_1 M_{a1})^2 + (U_{\text{tip}} - C_{\theta_1})^2}$$
$$= \sqrt{(637.5)^2 + (1600 - 0)^2} = 1722 \text{ ft/s}$$

The relative angle, or β_1, is

$$\beta_1 = \cos^{-1}\left(\frac{a_1 M_{a1}}{W_1}\right) = \cos^{-1}\left(\frac{637.5}{1722}\right) = 68.3 \text{ deg}$$

Rotor exit: Since this is a constant tip radius design, $U_{\text{tip1}} = U_{\text{tip2}}$ and $r_{\text{tip1}} = r_{\text{tip2}}$. The fan exit total temperature is

$$T_{T2} = T_{T1}\left[\frac{[(PR)(\gamma - 1)/\gamma - 1]}{\eta_{\text{comp}}} + 1\right]$$

$$= 562.9\left[\frac{(1.64)0.286 - 1}{0.87} + 1\right] = 661.2°\text{R}$$

The rotor exit swirl is found using Eq. (98)

$$C_{\theta_2} = 1/U_{\text{tip}}[Cp(T_{T2} - T_{T1})gJ]$$

$$= 1/1600[0.24(661.2 - 562.9)(32.2)(778)]$$

$$= 369.4 \text{ ft/s}$$

The rotor exit absolute velocity C_2 is found as follows:

$$a_2 = 49.05\sqrt{\frac{661.2}{1 + 0.2(0.5)^2}} - 1230.9 \text{ ft/s}$$

$$C_{a2} = a_2 M_{a2} = 1230.9(0.5) = 615.4 \text{ ft/s}$$

$$\alpha_2 = \tan^{-1}(369.4/615.4) = 31.0 \text{ deg}$$

$$C_2 = \sqrt{C_{\theta 2}^2 + C_{a2}^2} = \sqrt{369.4^2 + 615.4^2} = 717.8 \text{ ft/s}$$

The rotor exit relative angle β_2 and velocity W_2 is

$$\beta_2 = \tan^{-1}\left(\frac{U_{\text{tip}} - C_{\theta_2}}{C_{\theta_2}}\right) = \tan^{-1}\left(\frac{1600 - 369.4}{615.4}\right) = 63.4 \text{ deg}$$

$$W_2 = \frac{U_{\text{tip}} - C_{\theta_2}}{\sin\beta_2} = \frac{1600 - 369.4}{\sin(63.4)} = 1376 \text{ ft/s}$$

Stator exit: The stator for this design is to straighten the flow and diffuse it down to 0.4 Mn. Therefore, the exit velocity is found to be

$$a_3 = 49.05\sqrt{\frac{661.2}{1 + 0.2(0.45)^2}} = 1236.5 \text{ ft/s}$$

$$C_3 = 1236.5(0.45) = 556.4 \text{ ft/s}$$

Example 4

Determine the hub design characteristics for the turbine from example 1. Assume an axial, single-stage design with the following configurational constraints.

$$U_{tip} = 1600 \text{ ft/s} \qquad AN^2 = 35 \times 10^9$$
$$\text{rpm} = 32{,}000 \qquad C_{\theta 2} = 0 \text{ deg (zero exit swirl)}$$
$$C_{a3}/C_{a2} = 1.60$$

Answer: The tip radius is

$$r_{tip} = \frac{U_{tip}(60)}{2\pi(\text{rpm})} = \frac{1600(60)}{2\pi(\text{rpm})} = 0.477 \text{ ft} = 5.73 \text{ in.}$$

The annulus area is

$$Aa = \frac{AN^2}{N^2} = \frac{35 \times 10^9}{(32{,}000)^2} = 34.18 \text{ in.}^2$$

The hub radius and speed are

$$r_{hub} = \sqrt{r_+^2 - \frac{Aa}{\pi}} = 4.68 \text{ in.}$$

$$U_{hub} = U_{tip}(r_h/r_+) = 1600(4.68/5.73) = 1306.8 \text{ ft/s}$$

The exit velocity is found by first determining the exit specific flow.

$$Wa_{spec} = \frac{Wg\sqrt{\theta}}{\delta(Aa)} = \frac{15.3(1988/519)^{\frac{1}{2}}}{(47.39/14.7)(34.18/144)} = 39.132 \text{ lb}_m/\text{s} - \text{ft}^2$$

$$= 39.132 \text{ lb}_m/\text{s} - \text{ft}^2$$

Equation (86) can be used to iterate on a solution for the exit absolute Mach number, $M_{a3} = 0.571$. In this case, since $C_{\theta 3} = 0$ deg, $C_3 = C_{a3}$, exit velocity is

$$a_3 = \sqrt{\frac{\gamma g R T_{T3}}{\{1 + [(\gamma - 1)/2](M_{a3})^2\}}}$$

$$= \sqrt{\frac{1.3(32.2)(53.35)1988}{[1 + 0.15(0.571)^2]}}$$

$$a_3 = 2057.3 \text{ ft/s}$$

$$C_3 = C_{a3} = a_3(M_{a3}) = 0.571(2057.3) = 1174.7 \text{ ft/s}$$

The exit relative angle β_3 and velocity W_3 is

$$\beta_3 = \tan^{-1}(U_{hub}/C_{a3}) = \tan^{-1}(1306.8/1174.7) = 48.0 \text{ deg}$$

$$W_3 = \frac{U_{hub}}{\sin(\beta_3)} = \frac{1306.8}{\sin(48.0)} = 1758.5 \text{ ft/s}$$

To determine the rotor inlet conditions, annulus area is assumed to be constant. Equation (123) is used to determine the required inlet swirl $C_{\theta 2}$

$$C_{\theta 2} = \frac{\Delta H_{\text{turb}}(gJ)}{U_{\text{hub}}} + C_{\theta 3} = \frac{0.293(2360 - 1988)(32.2)(778)}{1306.8} + 0$$

$$C_{\theta 2} = 2089.5 \text{ ft/s}$$

The rest of the rotor inlet velocities and angles are found as follows.

$$C_{a2} = C_{a3} \Big/ \left(\frac{C_{a3}}{C_{a2}}\right) = \frac{1174.7}{1.60} = 734.2 \text{ ft/s}$$

$$C_2 = \sqrt{(C_{a2})^2 + (C_{\theta 2})^2} = \sqrt{(734.2)^2 + (2089.5)^2}$$

$$= 2214.7 \text{ ft/s}$$

$$\alpha_2(\text{stator exit angle}) = \cos^{-1}\left(\frac{C_{a2}}{C_2}\right) = \cos^{-1}\left(\frac{734.2}{2214.7}\right) = 70.6 \text{ deg}$$

$$W_{\theta 2} = C_{\theta 2} - U_{\text{hub2}} = 2089.5 - 1306.8 = 782.7 \text{ ft/s}$$

$$W_2 = \sqrt{(W_{\theta 2})^2 + (C_{a2})^2} = \sqrt{(782.7)^2 + (734.2)^2}$$

$$= 1073.2 \text{ ft/s}$$

$$\beta_2 = \cos^{-1}\left(\frac{C_{a2}}{W_2}\right) = \cos^{-1}\left(\frac{734.2}{1073.2}\right) = 46.8 \text{ deg}$$

The stator exit Mach number is found using Eq. (105)

$$M_{c2} = \sqrt{\frac{2C_2^2}{2\gamma g R T_{T2} - C_2^2(\gamma - 1)}}$$

$$= \sqrt{\frac{2(2214.7)^2}{2(1.3)(32.2)(53.35)(2360) - (2214.7)^2(0.3)}} = 1.04$$

The rotor relative entrance Mach number

$$M_{W2} = \sqrt{\frac{2W_2^2}{2\gamma g R T_{T2} - W_2^2(\gamma - 1)}}$$

$$= \sqrt{\frac{2(1073.2)^2}{2(1.3)(32.2)(53.35)(2360) - (1073.2)^2(0.3)}} = 0.475$$

Once again, the specific flow is used to determine the stator inlet axial velocity C_{a1}

$$Wg_{\text{spec}} = \frac{115.3(2360/519)^{\frac{1}{2}}}{(115.3/14.7)(34.18/144)} = 17.52 \text{ lb/s-ft}^2; \qquad Mn_{a1} = 0.219$$

$$a_1 = \sqrt{\frac{1.3(3.2)(53.35)(2360)}{[1 + 0.15(0.219)^2]}} = 1187.5 \text{ ft/s}$$

$$C_{a1} = 0.219(2287.5) = 501.0 \text{ ft/s}$$

Referring to Table 4, it can be seen that this turbine design is well within the design constraints listed.

References

[1]Benstein, E., "Small Flying Engines Are Different," AIAA/SAE 10th Propulsion Conf., AIAA Paper 74-1185, 1974.

[2]Boyce, M. P., *Gas Turbine Engineering Handbook,* Gulf, 1992.

[3]Cohen, H., Rogers, G. F. C., and Saravanamuttoo, H. I. H., *Gas Turbine Theory,* 3rd ed., Longman Scientific and Technical, 1987.

[4]Cropper, G. D., "Electronic Fuel Controls for Missile and RPV Gas Turbines," National Aerospace Engineering and Manufacturing Meeting, Society of Automotive Engineers, SAE Paper 751061, 1975.

[5]DeMeis, R., "Designing a Cruise Missile: General Dynamics BGM-109 Tomahawk," *Aerospace America,* Jan. 1985, pp. 110–114.

[6]Durham, F. P., *Aircraft Jet Power Plants,* Prentice-Hall, New York, 1961.

[7]Gunston, B., *Illustrated Guide to Modern Airborne Missiles,* Prentice-Hall, New York, 1986.

[8]Hesse, W. J., and Mumford, V. S., Jr., *Jet Propulsion for Aerospace Applications,* 2nd ed., Pitman, 1964.

[9]Hill, P. G., and Peterson, C. R., *Mechanics and Thermodynamics of Propulsion,* Addison-Wesley, Reading, MA, 1965.

[10]Hoy, W. D., "Long Range Subsonic Cruise Missile Propulsion Performance Comparison," AIAA/SAE/ASEE 25th Joint Propulsion Conference, AIAA Paper 89-2474, 1989.

[11]Anon., *Installation Handbook for Turbojet Engines,* General Electric Co., 1950.

[12]Kerkam, B. F., "Installation Effects on Cycle Selection for Small Turbofan Engines," AIAA 18th Aerospace Sciences Meeting, AIAA Paper 80-8106, 1980.

[13]Mattingly, J. D., Heiser, W. H., and Daley, D. H., *Aircraft Engine Design,* AIAA Education Series, AIAA, New York, 1987.

[14]Oates, G. C., *Aerothermodynamics of Gas Turbine and Rocket Propulsion,* AIAA Education Series, AIAA, New York, 1984.

[15]Schmidt, M. F., "Propeller Design Point Calculation Method for Comparing Turbofan/Propfan Engine Performance," Beijing International Gas Turbine Symposium and Exposition, ASME Paper 85-IGT-150, American Society of Mechanical Engeneers, 1985.

[16]Stricker, J. M., "Comparison Methodology For Long Range Subsonic Cruise Missile Propulsion Systems," AIAA/SAE/ASEE 25th Joint Propulsion Conf., AIAA Paper 89-2475, 1989.

[17]Wills, T. K., and Wise, E. P., "Development of a New Class of Engine: The Small Turbofan," AIAA/SAE 12th Propulsion Conf., AIAA Paper 70-618, 1976.

Chapter 11

Liquid-Fueled Ramjet Engine

Philippe Cazin*
ONERA, Chatillon sous Bagneux 92322, France
and
Jean-Marie Laurent†
Aérospatiale/CELERG, Clamart 92140, France

I. Introduction—Design of a Liquid-Fueled Ramjet Engine

COMPARISONS with other propulsion systems were made earlier in this volume, and this chapter deals more specifically with the choices and techniques applicable during the design of a liquid-fueled ramjet engine.

This type of engine can be broken down into three components that each have specific features that interact with each other: 1) the air supply system, 2) the fuel supply system, and 3) the combustion chamber and the ejection nozzle (see Fig. 1).

The air supply system has been covered in the previous chapter, and now this aspect will essentially be examined from the standpoint of calculating the ramjet engine's overall performance. However, the architecture of the chamber air intake and airflow distribution (pressure, speed, and turbulence) system is a decisive factor, along with the fuel injection, in obtaining correct operation of the combustion chamber and in determining whether or not instabilities appear.

The fuel supply system can be broken down into three main functions: 1) the storage of fuel in the missile and the systems used to transfer the fuel to the combustion chamber for use; 2) the regulation system that is used to vary the fuel flow according to preprogrammed values or, as is most often the case with recent missiles, according to values calculated aboard the missile as a function of the desired trajectories and flight conditions; and 3) the injection system that supplies the combustion chamber with fuel at the right place and at the correct rate for maximum efficiency.

The combustion chamber and ejection nozzle must take account of the specific requirements for use with tactical missiles; in particular, the compactness required for airborne applications calls for a booster to be built into the chamber. As a

*Director of Space Systems. Member AIAA.

†Director of Programs. Member AIAA.

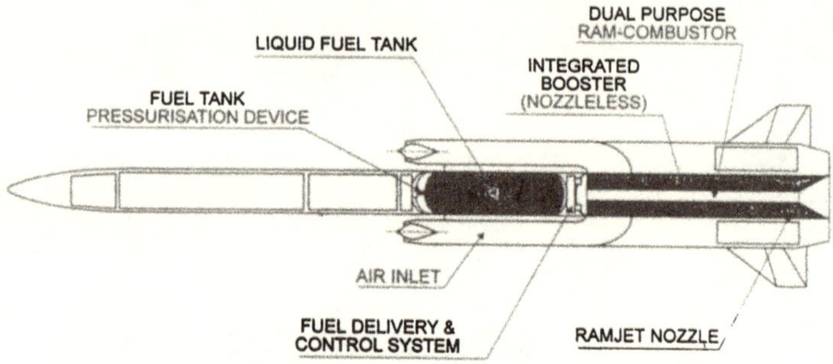

Fig. 1 Generic design for a liquid-fueled ramjet.

result, the design of combustion chambers must take into account this aspect of physical dimensions and also must be compatible with both the presence of air intake shutoff devices and an ejectable nozzle and the fact that there is no space inside for accommodating flame stabilization devices.

As the chamber itself is a complex acoustic resonator, control over combustion instabilities not only improves performance in certain cases but also is essential to avoid the destructive effects of the vibration generated, both for the engine and all of the missile's mechanical or electronic equipment.

Maximum performance throughout the operating range is obtained through a combination and overall optimization of these three subassemblies (air supply, fuel supply, and combustion chamber).

Consequently, this chapter will deal with the possible choices, from the technical and possibly technological aspects, according to the following plan: liquid fuels: available fuels, power performance, and selection criteria; fuel supply: storage, transfer, regulation, and injection; combustion chamber: design and size; combustion instabilities; and performance calculation.

Because all of those involved in the industrial and/or scientific aspects of this propulsion field do not adopt exactly the same approaches and methods, the following description does not pretend to cover every possible aspect; a consistent approach to the various subjects has, however, been adopted.

II. Liquid Fuels

For ramjet engine propulsion, liquid fuels offer a number of advantages over solid fuels, particularly in the following respects: 1) better specific thrust together with the fact that solid propellants must contain oxidizing products that gasify the product through combustion; 2) higher modulation ratios for liquids (ratio >20) because of their incompressibility, whereas a ratio of 7 would be a good result with solids, in addition to the requirement for gas generator operation to be stable between a few bar and a few hundred bar; and 3) more compactness and optimization of available space (possibility of tanks having any shape), whereas solid propellant gas generators also have to meet restrictive physical and ballistic dimension requirements in terms of the propellant block.

For all of these reasons, the use of a liquid fuel proves necessary when the specified missions involve long ranges or when the flight envelope entails considerable altitude changes and/or Mach number variations.

Although the best-known and most widely used fuel is kerosene, much research has been carried out and continues to be carried out to obtain industrial fuels providing an equivalent or greater calorific value per unit mass and, more important, a significant increase in the mass per unit volume, which would provide substantial gains in terms of range through improved compactness.

A. Main Liquid Fuels

The fuels currently developed can be classified into three main fuel categories: hydrocarbon-based fuels derived from oil or coal, synthetic fuels, and slurry fuels. The products described next are the most representative in these categories; Table 1 lists their main characteristics.

1. *Petroleum Distillate*

These are hydrocarbon-based fuels derived from oil or coal: TR0 kerosene, JP5, and so forth.

2. *Synthetic Fuels*

These fuels are essentially pure synthesized hydrocarbons or mixtures of pure and synthetic hydrocarbons. They are compounds of high fuel value per unit volume tailored to meet the operational requirements of their assigned system. These fuels are based on bridged ring hydrocarbon structures (which contain high carbon-to-hydrogen ratios) that increase both their density and volumetric heat of combustion. The simplest bridged structure that forms the building block for all of the hydrocarbon missile fuels is norbornane or bicyclo (2.2.1) heptane. Though never found as such in the final fuels, its structure serves as the building block for many of the fuels in use.

RJ-4. RJ-4 is a mixture of eight isometric forms produced by the hydrogenation of methylcyclopentadiene dimer, or tetrahydro (TH) dimer. RJ-4 is a

Table 1 Liquid fuels-main characteristics

	Petroleum distillate		Synthetic fuels					
	TR0	JP-5	RJ-4	JP-10	RJ-5	RJ-6	CSD	Slurry, BK
Chemical formula	$C_{10}H_{20}$	$C_{10}H_{19}$	$C_{12}H_{20}$	$C_{10}H_{16}$	$C_{10}H_{18.4}$	$JP_{10}RJ_5$	$C_{14}H_{18}$	TR0+B
Specific gravity	0.79	0.82	0.93	0.94	1.06	1.03	1.04	1.25
Flash point, °C	38	60	60	52	95	60	36	—
Freezing point, °C	−53	−46	−40	−79	−30	−54	−60	−40
Viscosity *CS* at −40°C	9	16	65	18	solid	130	105	—
Net heat of combustion								
Kcal/kg	10,350	10,150	9,990	10,050	9,860	9,910	10,000	9,600
Kcal/dm^3	8,180	8,234	9,290	9,450	10,450	10,200	10,400	12,000

high-density hydrocarbon having substantially more energy per unit volume than kerosene fuels. The mixture of isomers accounts for the variations in the boiling and freezing temperatures of the fuel.

JP-10. JP-10 has the single chemical structure exo-tetrahydro dicyclopentadiene. It is prepared by hydrogenating commercially available dicyclopendatiene, which yields the solid material endo-tetrahydro dicyclopentadiene. This intermediate endostructure is then isomerized in the presence of a catalyst to produce pure JP-10.

RJ-5. RJ-5 is a high-density mixture of hydrogenated dimers of norbornadiene, an unsaturated derivative of the bridged compound norbornane. It has the highest energy per unit volume of all of the missile fuels, but its high freezing point makes it unusable for many applications.

RJ-6. RJ-6 is made up of a mixture of RJ-5 and JP-10. Although RJ-6 has a lower volumetric heat of combustion, it has better low-temperature properties than RJ-5.

CSD. CSD is a mixture of hydrogenated dimers of norbornadiene and solvent (toluene) to decrease viscosity at low temperatures.

3. *Slurry Fuels*

Slurry or slush fuels are liquids similar to those in the previous family with the difference that very fine solid particles in suspension are introduced. The main additive that is used is boron, but others such as carbon, aluminum, or magnesium may be introduced as well. The greater the proportion of solid additive, the more energetic the mixture. It is, however, necessary to consider the implementation of operating parameters that limit their permissible loading levels.

1) The product obtained must always be capable of being injected within the specified environmental operating range.

2) The combustion efficiency in the chamber may significantly decrease because the solid particles require a minimum amount of energy to initiate their combustion and the residence time in the chamber may not permit complete combustion.

3) Because the mixture is more energetic, the final combustion temperatures are higher, and as a result, the thermal load on the chamber is increased.

B. Energy Performance During Combustion

To calculate the ramjet engine's overall performance, the intrinsic characteristics of fuels during their combustion in air are required. This methodology, which may be qualified as analytical, will gradually make way for numerical calculation methods (see Sec. IV.b.1) once some of the predictive calculation codes have reached a sufficiently reliable description of the aerochemical phenomena, on the one hand, and once the associated computer facilities allow their use on an industrial scale (calculation time and cost), on the other.

According to the temperature generating the flow of upstream air (Tt_0) and the richness of the fuel really burnt (Φ_b, the burnt equivalence ratio), the so-called Brinkley method (association of chemical equilibrium equations and mass conservation equations) allows the composition of gases after combustion and the total temperature at the end of combustion (Tt_4) to be determined.

The following physical parameters may be deduced from this: ratio of specific heats γ (Tt_4, Φ_b), molar mass of the mixture M_4 (Tt_4, Φ_b), and characteristic velocity C^* (Tt_4, Φ_b).

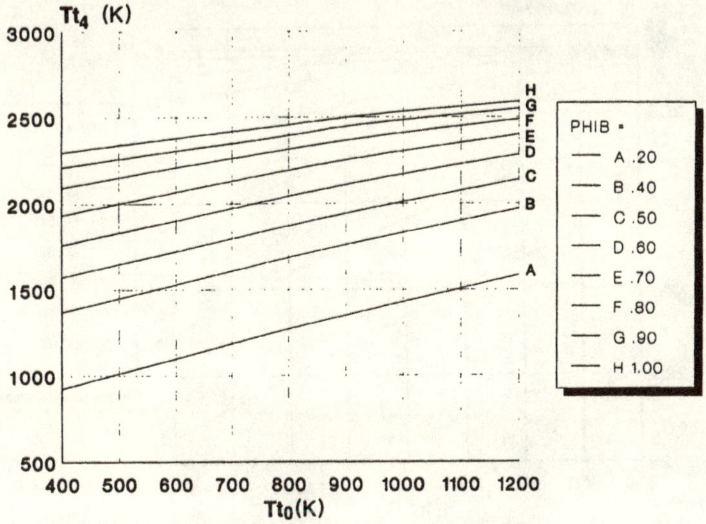

Fig. 2 Combustion temperature.

Note that the characteristic speed C^* taken here as the parameter representative of the air/fuel mixture is homogeneous with the method for calculating performances described in Sec. VI.

Other parameters may be taken, heating γ, etc., that are used to calculate the thermal load of the combustion chamber. For clarity, only the characteristic velocity is dealt with in this chapter:

$$C^* = \left(\frac{\gamma_4 + 1}{2}\right)^{\frac{\gamma_4+1}{2(\gamma_4-1)}} * \sqrt{\frac{R^* T t_4}{M_4^* \gamma_4}}$$

To establish orders of magnitude, the preceding parameters for air/kerosene combustion are given in Figs. 2–5.

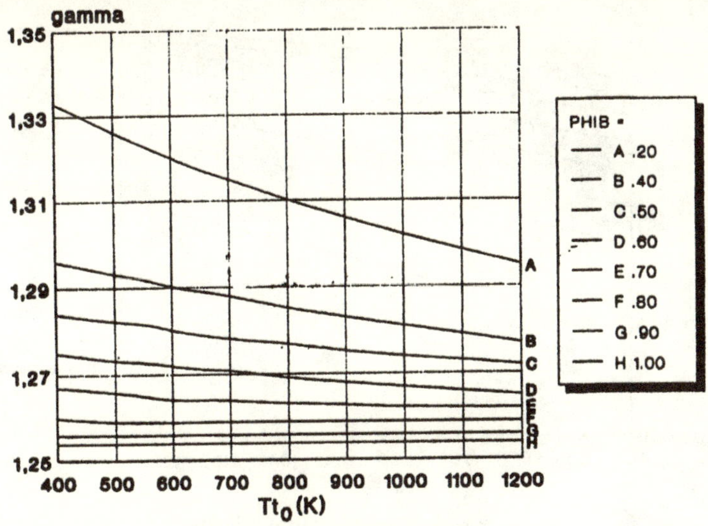

Fig. 3 Specific heat ratio.

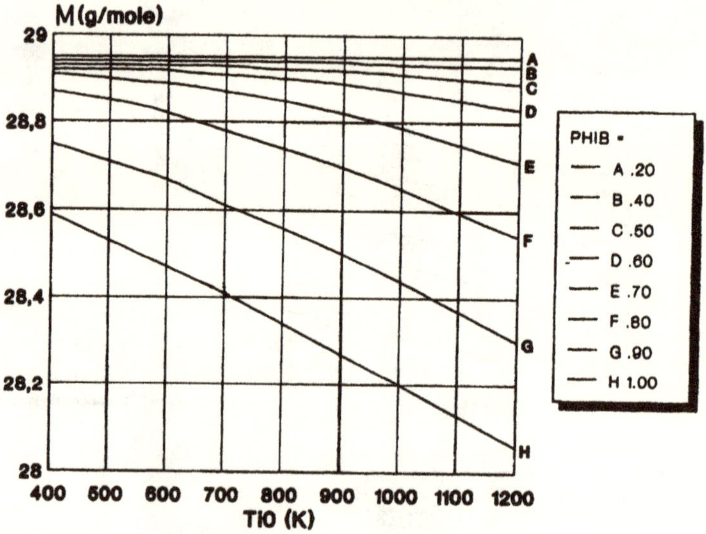

Fig. 4 Molar mass.

C. Choice Criteria

Depending on the mission, the choice of fuel will correspond to one or another priority criterion, such as the following.

1) Energy level: This requires a fuel with a high mass calorific value (or a high specific impulse), that is, having a lower calorific value of at least 10,000 kcal/kg.

2) Density and density-specific impulse: To enhance missile compactness (or for a given volume and improved range), the aim is to increase density-specific impulse. The most effective method is generally to increase the fuel density. Liquid

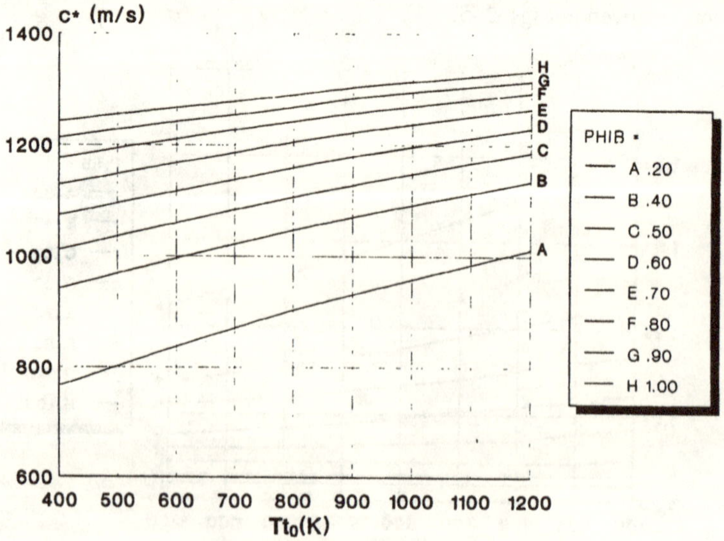

Fig. 5 Characteristic velocity.

fuels with densities of between 1000 and 1100 kg/m^3 have already been developed for tactical missile applications.

3) High-altitude operation: Certain slush fuels, such as those containing boron particles, are difficult to burn at low pressure and hence cannot be easily utilized for high-altitude missions.

4) Expansion: Density varies with temperature in a missile operating at extreme temperatures, and classic kerosene will expand by between 7 and 8% per 100°C. The fuel tank has to be oversized as a result.

5) Viscosity: Defining fuels with everincreasing densities is generally accompanied by increases in viscosity, in particular at low temperatures. This phenomenon is the limiting factor for very high density fuels on tactical missiles (in particular free-carried air-to-ground and air-to-surface variants) because of the practical consequences for fuel-forcing systems or pumps. For example, viscosities greater than 150 cs pose problems or otherwise necessitate special technology. Nevertheless, certain slush fuels can be used, despite their having even higher static viscosities, because of their highly thixotropic qualities; the high viscosity of these fuels when inert drops considerably once a shear force is applied (that is, under any form of movement).

6) Flash point: To facilitate ignition of the fuel, particularly when its temperature is very low (such as missiles externally carried on aircraft during prolonged high-altitude flight), the choice criterion dictates a low flash point. In some cases, this is not possible for safety reasons (such as limiting fire hazards aboard ships). Such a constraint may entail keeping the flash point above 60°C.

7) Freezing point: The freezing point must be lower than the lowest temperature likely to be encountered in storage or carriage.

8) Stealth: Certain operational missions require special fuels producing good exhaust discretion, from the electromagnetic and, above all, the optical viewpoints (visible and infrared spectrums). Such a constraint may necessitate hydrocarbon-based fuels only (that is, carbon and hydrogen alone), to the exclusion of metal or metalloid bases (aluminum and magnesium, and boron, respectively).

9) Storability: Fuels for tactical missiles must be suited to long-duration storage within the tank. However, the problem never concerns simple chemical products; kerosene, for example, despite being extremely common, is an oil derivative with a chemical composition that can vary widely depending on the refinery processes employed and the origin of the raw oil. For this reason, special standards have to be applied for long-duration fuels.

The main problems are the following.

1) Oxidization as a result of the presence of dissolved air: This can have a corrosive effect or result in deposition of gums. Hence, certain precautions are required when filling tanks, as well as the addition of antioxidizing agents to the fuel.

2) Corrosion of certain plastic or seal materials, which governs tank design: Additives are available to combat corrosion.

3) Presence of microorganisms (bacteria, fungi, mildew, mold, etc.) that can develop upon contact with the fuel. These organisms result in the formation of sludge and filamentary deposits liable to clog various parts and equipment. The presence of water in the fuel is an additional encouraging factor.

4) Settling: Certain high-density fuels are prepared from a suspension of particles of heavy solids ground into a liquid organic support. The stability of the suspension is assured by gelling the support, preventing any movement of the

particles. It is therefore necessary to ensure that a prolonged storage will not result in settling of the solid particles, with subsequent changes in the fuel characteristics (particularly at the injectors).

The preceding list is not exhaustive; depending on the application, other special properties may be required. These may, in particular, involve the presence of anti-icing additives to prevent formation of crystals likely to block filters or pipes; the addition of static dispersants to render the fuel more conductive; cost constraints for the fuel, which not only affect the cost of the serially produced missile but also the basic development costs as a result of the high number of tests performed; and the absence of toxicants, not only in the fuel itself, but also in the form of products of combustion discharged to the atmosphere. This constraint is associated more with missile development than with operational use.

III. The Supply, Regulation, and Injection System

Supplying fuel to the combustion chamber is, without doubt, the major function involved in ramjet engine operation. Its performance has a direct effect on combustion efficiency, on the appearance of instabilities, and, of course, on the engine's ability to operate within the specified flight envelope. As regards the latter aspect, the flow rate ratio to be provided, for fixed geometry air intakes, between an operating point at high altitude and low Mach number and an operating point at low altitude and high Mach number, may be greater than 20.

Its use in a tactical missile imposes significant, sometimes contradictory, constraints, which essentially involve the ability to provide a guaranteed service life of greater than 10 years (20 years desired) since the missile is considered as a munition, i.e., without specific maintenance (this aspect will be dealt with in a subsequent section), and satisfactory resistance to accidental attacks, principally fire and bullet impact. As far as fire is concerned, externally protecting the tank by means of a heat shield may be sufficient and delay the appearance of reactions sufficiently to meet the standards imposed. For bullet impact, vulnerability may be limited by segmenting the tank, which even increases the reliability of mission success when such an incident occurs in flight.

And finally, as for all moving vehicles containing a liquid, the sloshing around of such a liquid under the effect of varying accelerations (longitudinal or lateral) should be reduced as much as possible. In the case of tactical missiles, the lateral load factor may reach 20*g*, and under these conditions, the supply system must not generate significant variations either in relation to the position of the center of gravity or the pressure and flow rate of the fuel delivered.

To simplify its presentation, the fuel supply system may be divided into three subassemblies, the supply system, storage tank, and pressurization system; the regulation system, which sets the fuel flow rate to the desired value according to the flight conditions; and the injection system, which, as its name indicates, delivers the right quantity of fuel into the combustion chamber at the right place. All of these subassemblies will be described in detail, and the technical solutions considered will be mentioned.

A. Supply System

Many supply systems have been designed and developed as a result of research work in the aerospace field (see Fig. 6). The tank design depends directly on the choice adopted for the pressurization system since to provide a satisfactory

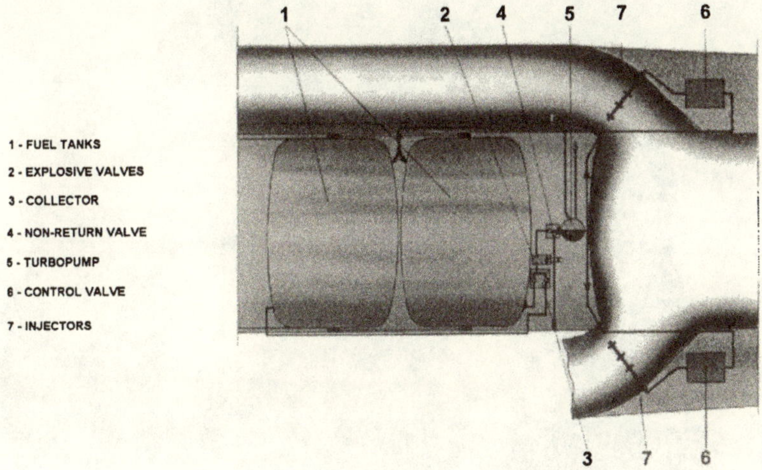

Fig. 6 General design for the supply system.

response to the previously mentioned constraints, the pressure at the output of this system must be greater than a value of about 10–20 bar. The pressure drop required to obtain the correct flow rate is produced in the regulation system.

The longer the flight times at high Mach numbers, the higher this minimum pressure must be to prevent kinetic heating resulting in the formation of vapor bubbles or nuclear boiling phenomena when the liquid is in contact with hot parts.

Of all of the technologies developed and tested, the choice comes down to two main technologies that are now operational: 1) a tank pressurized by a gas generator and 2) the use of a turbopump actuated by an air bleed in the air intake diffuser.

1. Supply by a Pressurized Tank

To pressurize a tank to high pressure requires the missile to contain either a gas tank at an even higher pressure or for gas to be generated by a pyrotechnic generator using liquid or solid ergols. The latter solution has currently been adopted because of its compactness and the fact that it is maintenance free (with solid ergols). In this case, the gases generated push a piston that slides directly in the tank's metal structure.

For this type of configuration (see Fig. 7), a considerable number of difficulties have to be overcome.

The generator must deliver gases that are sufficiently cold not to damage the structures. It must also provide a sufficient pressure (see previous paragraph) while preserving the tank's mechanical strength safety coefficient. For significant operating times, and to avoid having to fit a generator flow rate modulation system, such a generator may be divided into several small generators, the simultaneous or staggered ignition of which provides the best possible response to the required real instantaneous fuel flow rate.

The piston must be able to move with no significant frictional forces (as a result of the tank being distorted because of its internal pressure and flight loads) while providing for sealing between the fuel and the generator gases.

Fig. 7 Example of gas generator.

The tank must provide for long-term fuel storage. For this purpose, the best solution is to only allow the fuel to come into direct contact with metal parts. Under these conditions, accelerated aging tests have shown that if stainless steels are selected, a service life in excess of 100 years could be achieved (provided there is no decomposition of the fuel itself).

This architecture calls for the installation of a metal diaphragm between the piston and the fuel, which is sheared by the piston's stroke; the supply tube is closed off by a pyrotechnic valve that is opened when the missile is fired.

The fuel contained in the metal tank has an expansion coefficient per unit volume that is incompatible with the capabilities of the structure, and hence it is necessary to create a variable internal volume in the tank to allow the fuel to expand.

As with any system, the pressurized-tank supply system has advantages and drawbacks: the design is well suited to long-term storage, it eliminates all risk of fuel sloshing about during flight, it calls for a cylindrically shaped tank with a constant cross section produced with tight tolerances, but the solution is not very compatible with the use of multiple tanks.

2. *Supply by a Turbopump*

Since this technology has been widely developed for aviation needs (turbofan engine, liquid ergol rocket motors, etc.), its application to a ramjet engine for tactical missiles in principle poses no major problems. The basic restrictions lie in the little space available for its installation and in the need for maintenance-free operation.

The turbine driving the pump is supplied by an air bleed in the air duct, which is a solution offering better performance than the one involving the installation of a specific air bleed. The quantity of air required in relation to the total flow received is about 1%, and the turbine exhaust moreover can be used for the purpose of cooling the equipment housed in the same fairing.

The main technical difficulty lies in where to install the pump on the engine. Indeed, its position should be optimized to avoid pressure drops being created in the air supply and to minimize the fuel lines between the tank and the injectors.

If required, the turbopump turbine may be oversized to allow an ac generator to be connected to generate electrical power aboard the missile for long operating periods in particular. Figure 8 illustrates a possible layout.

As far as the tank is concerned, the adoption of a turbopump makes it possible to propose solutions that are better tailored to all types of architecture. The fuel is generally stored in a flexible bladder that collapses as the fuel is pumped out.

There are a number of problems to be solved with this type of tank.

1) For long-term storage, the bladder material must be chemically compatible with the fuel; metallizing the inside of the bladder may provide a satisfactory solution, especially as perfect sealing is required.

2) The sloshing around of the fuel must be limited to avoid sudden changes in the center-of-gravity position, on the one hand, and to provide a constant flow of fuel at sufficient pressure to prevent any likelihood of stalling, on the other. For this purpose, the gap between the tank structure and the bladder is pressurized by bleeding air at static pressure at the end of the air intake diffuser.

3) The architecture must make it possible to use all of the fuel stored; an expulsion efficiency objective of 97% is reasonable.

A few of the advantages of a turbopump supply are as follows: it is well adapted to long-duration missions, it can be used for tanks of any shape, and it is possible to divide up the storage.

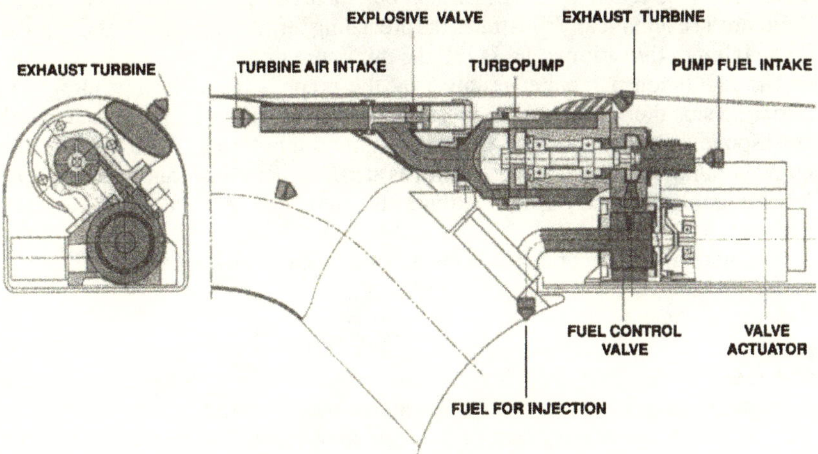

Fig. 8 Ramjet fuel conditioning with turbopump.

B. Regulation System

The regulation system has to provide the constant fuel flow required for the flight program, and hence throughout the flight envelope (Mach number, altitude, incidence) it must ensure correct engine operation; for this purpose, the flow rate must constantly lie between two limits: a maximum thrust limit compatible with the rich extinction limit and the stable operation of the air intakes and a minimum thrust limit compatible with the limit of lean extinction. The regulation system also must set the missile speed to the values requested by the flight program, as a function of time and altitude, using all of the values of thrust within the previous limits.

The problem of ramjet engine ignition and the initial few seconds of operation calls for special requirements: on the one hand, the whole fuel system must be rapidly and completely filled, and on the other, the availability of residual fuel increases the energy supplied in the combustion chamber. During this period, the fuel flow must provide the ramjet engine's maximum thrust while maintaining a sufficient margin with respect to air intake surges.

The regulation system includes a flow rate adjustment valve, a set of sensors, a computer, and a servosystem. The position of the regulation valve is servo-controlled in such a way that the value of a parameter characterizing the thrust (increasing function), calculated from the measurements made on the ramjet engine during operation, is equal to the control value *pco*. This same parameter, calculated as a function of the flight program conditions, should meet the requirements just defined, which can be summarized as follows:

For operating safety:

$$P_{\min} \leq P_{co} \leq P_{\max}$$

For missile speed regulation

$$P_{co} = P_c - k(M - Mc)$$

where K characterizes the servocontrol stiffness in terms of missile Mach number, Mc is the required Mach number, P_c is the value of the thrust parameter for the Mach number Mc, and $P_{\max}$ and $P_{\min}$ are functions of the missile flight parameters and ramjet engine operation (Mach number, incidence, reference pressure, etc.).

Figure 9 schematically illustrates the preceding formulas; points A B C represent the maximum thrust limit and D E F the minimum thrust limit.

The regulation parameter is given by the graph A B E F. Depending on its initial speed, the missile accelerates at maximum thrust to the Mach number corresponding to point B or decelerates at minimum thrust to the Mach number corresponding to point E; it then stabilizes at a point between B and E. The required Mach number may vary as a function of the time, altitude, and flight trajectories chosen.

Two parameters may be used. One is equal to the following ratio:

$$\frac{\text{fuel weight flow rate}}{\text{reference pressure}}$$

The fuel flow rate may be either measured directly by a flow meter or calculated from the measurements of one fuel temperature and two pressures, together with the position of the moving part of the regulation valve. The reference pressure is measured by one or more wall static pressure ports in the flow downstream of the air intakes, before the combustion chamber.

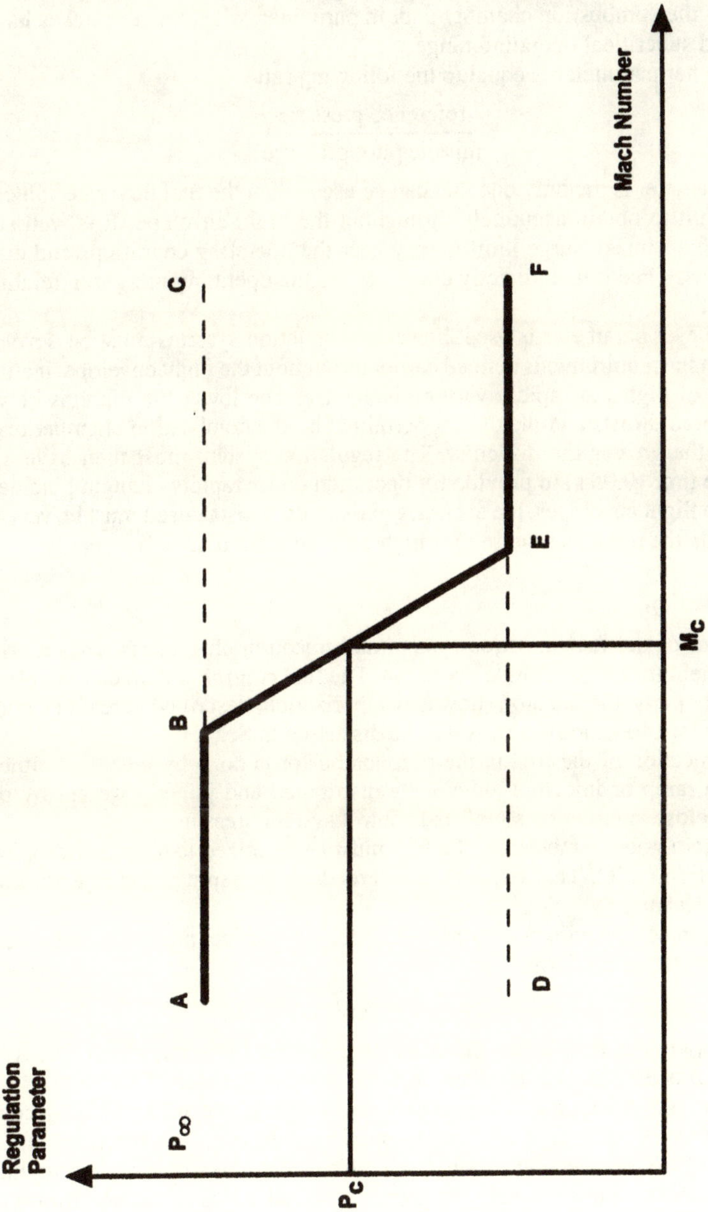

Fig. 9 Regulation parameter as a function of Mach number.

For a given ramjet engine, this parameter is then more especially a function of the combustion richness and depends little on Mach number and incidence. Its use is particularly well suited when the operating limits are determined by the stability range of the combustion chamber and, in particular, when the air intakes have an extended subcritical operating range.

The other parameter is equal to the following ratio:

$$\frac{\text{reference pressure}}{\text{missile pitot pressure}}$$

This parameter is the only one that can be used when the fuel flow rate value may be difficult to obtain accurately throughout the flight envelope. It is well suited when the air intake surge limit is very near the operating conditions and critical conditions, since it then directly characterizes the operating margin in relation to this limit.

Whichever parameter is used, however, regulation systems must be developed that meet the requirements defined earlier throughout the flight envelope, including the case of flights at rapidly varying incidence. The lower the margins between the required thrusts and the thrusts permitted by the combustion chamber and air intakes, the greater the difficulty. The regulation system must then have a low response time (0.05 s) to provide for operation under rapidly changing incidence. For wide flight envelopes, the accuracy class of the sensors used must be very high to provide the margins required for high-altitude operation.

C. Injection

Control of the fuel injection system and injection characteristics is a critical requirement for proper engine operation. The fuel control system can be split into two major parts: the injector, (flow and vaporization, described hereafter); and the injection location and type (this will be discussed in Sec. IV.)

The injection of the fuel in the ram combustor is done by means of either an injection ramp or injectors judiciously distributed and joined together by pipes that can close in case of a simplified control system pressure.

To obtain good combustion, the fuel must be atomized to very tiny droplets at the injection outlet. That requires a difference of pressure at the injector on the order of 10 bar.

For an injector having a constant area opening, the fuel mass flow per unit area can be written in the form

$$q = C_D S_c \sqrt{2\rho \Delta P}$$

where C_D is the flow ratio, Sc is the required fuel control which varies with altitude, ρ is the fuel density mass, and ΔP is the difference of pressure.

In this case, to obtain a fuel mass turn-down ratio between 15 and 20, necessary for trajectories with large changes in operating altitude, two solutions are conceivable: 1) to feed the injectors in a selective way, for example by interrupting or not the injector ramps this solution presents cost advantages, but not always at an optimum combustion efficiency; and 2) to have a feeding device able to produce pressures up to 200 bar. From the design characterized point of view, this solution is too restricting.

To improve this problem, a solution consists of implementing variable flow injectors by an injection area that can increase with pressure, providing a fuel mass flow history that is almost linear.

Fig. 10 Different kinds of injectors.

Different technical solutions (see Fig. 10) allow this result using the springiness deformation, diaphragm, or the injector itself. The localization of the injectors depend on the aerodynamic organization of the diffuser and on the combustion chamber. It is therefore logical to treat it in the chapter concerning that matter.

IV. Combustion Chamber

A. Generalities

The ramjet combustion chamber represents one of the longest development times of any part of the missile. It has to fulfill a large number of functions, particularly on modern systems employing integral boosters. Its functions are threefold.

1) It should ensure high-efficiency combustion, with minimum impact on the missile in terms of aerodynamic drag (entailing a reduced frontal section); hence the internal flux has a very high velocity (often up to 100 or 150 m/s in the intake duct and 300 m/s in the chamber itself). For this reason the flame must be stabilized by special methods, either mechanical or aerodynamic.

2) It should contain the propellant for any integral booster and withstand the resulting booster operating constraints [upstream port covers for boost phase, nozzle to maintain booster pressure (70–140 bar), then jettison itself after booster firing].

3) It should contribute to engine integrity, particularly during long-duration flights with very variable temperature and internal pressure conditions.

It should be noted that the preceding functions and the resulting constraints are practically independent of the physical nature of the fuel or its type.

The design of a combustion chamber depends on the aerodynamic configuration of the missile and the mission for which it is intended. In general terms, distinction is made between separate and integral booster approaches.

For the separate booster approach, the booster is mounted in tandem, laterally, or on a slide system within the combustion chamber. In this case, the chamber is not subjected to the constraints of an integral booster; that is, it can contain metal flame holder rings to facilitate combustion. It can operate at low pressure (less than 10 bar approximately) and has no requirement for jettisonable port covers. Its structure can be optimized for cruise flight conditions, allowing a much wider choice of technology; for example, liners may be film cooled by upstream bleed air (as employed for turbojet augmentors, etc.). The ramjet ignition sequence is not critical, and the engine can be lit even when the booster is firing. The chamber length is determined by the correct combustion efficiency; it can be shorter than an integral booster version.

The integral booster approach (where the booster is inside the combustion chamber) is currently the most favored for short- or medium-duration missions because of the excellent degree of compactness it affords. The constraints involved are as follows. The length is generally determined by the volume occupied by the booster propellant; this parameter is the determining factor, even if the missile is fired from a fast aircraft. The flame is stabilized essentially by purely aerodynamic means, since no flame-holder devices are allowed in the presence of the booster gases. Nevertheless, where the recirculation volumes are insufficient, combustion can be improved by adding small flame-holders upstream of the jettisonable part covers.

B. Development of a Combustion Chamber

As said previously, the development of a ram combustor is linked to the injection but also to the general architecture of the missile. Depending on the air inlet implantation, the aerodynamics of the combustor can be completely modified. Figure 11 shows some examples of general shapes to which can be added, in the cases of lateral or semilateral air inlet, the number of air inlets between 1 and 4.

1. *Modeling*

The working of a combustion chamber brings together many complex physical phenomena of which a modeling allows an optimization of the tuning tests and decreases the costs. Numerous numeric software are now available. They are in constant evolution to take into account the improvements, first, in the modeling of the physical and chemical phenomena and, second, in the scientific computers.

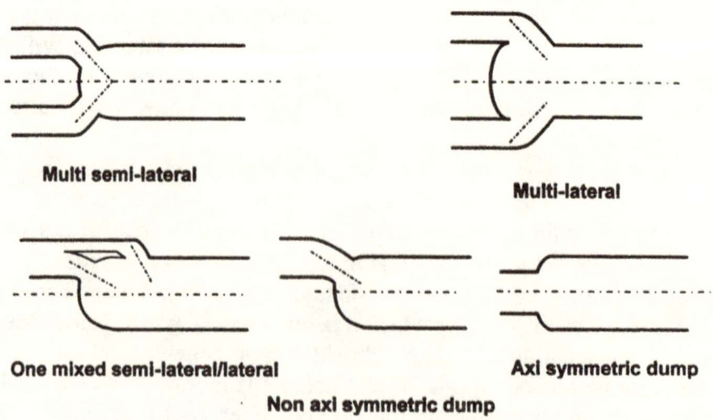

Fig. 11 Possibilities of combustor air entries.

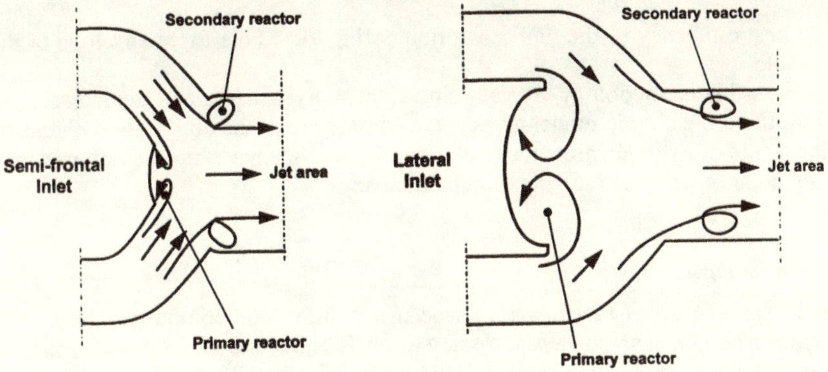

Fig. 12 Combustor modeling.

Nevertheless, semi-empirical methodology gives some results to understand how the combustor operates. It consists of doing mockup tests on a transparency in a hydraulic test bench for the aerodynamics of the combustor and of studying the fuel injection and distribution in an aerodynamic test bench.

The combustion chamber is then divided into different areas: jet stream and primary and secondary combustors, as indicated in Fig. 12.

The study in hydraulic simulation first gives the knowledge of the swirl areas (localization, volume, velocity, etc.), areas where most of the combustion is located, and second it provides help for determining the shape and location of aerodynamic obstacles so as to increase the swirls if necessary. (See Fig. 13.)

The study in aerodynamic simulation allows one to evaluate the fuel repartition in the different areas of the combustor, to measure the staying time of a fuel droplet in a reactor and its trajectory, and to optimize the location of the injectors to get

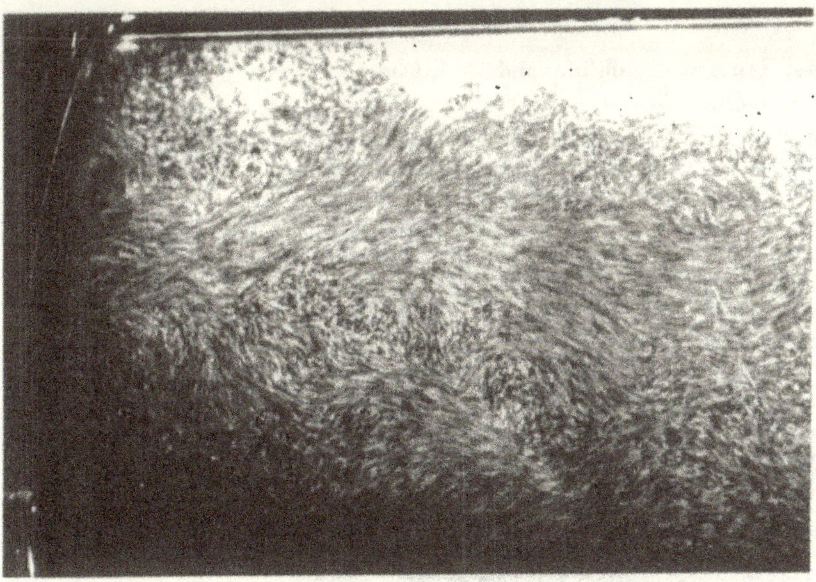

Fig. 13 Example of hydraulic simulation. (photo ONERA/CERT).

the best efficiency of the fuel repartition during the whole flight envelope of the missile.

With this methodology, by gathering elementary reactors and the interactions with the jet, a simple model gives some parameters of the combustion chamber: the limits of extinguishing, the evolution of the efficiency with the richness, and the range of richness for a maximum efficiency.

2. *Combustion Tests*

In spite of all of the progress made in modeling, the combustion tests on a scale one mockup are always necessary to verify and to measure the performances (pressure and combustion efficiency, limits of operating, thrust, etc.) in the whole domain of flight (Mach number, altitude, atmosphere, angle of attack, fuel temperature, etc.). In addition, these test are indispensable to bring to the fore the combustion instabilities and the cures.

V. Combustion Instabilities

Any heat engine that is expected to provide excellent performance throughout an extensive domain of operation runs the risk of encountering bothersome vibratory phenomena at a few points in this domain. This is not a special feature of the ramjet, and we know that serious disappointments happen in the form of combustion instabilities in solid as well as liquid-fueled rockets, when high performance is sought (e.g., instabilities in the Saturn V first stage, tangential instabilities in the Ariane engine, etc.).

Then again, the acuteness of the problems sometimes appears only during development, i.e., when the engine is being tested in all of its operational conditions, because prediction methods are still primitive.

A. Types of Instabilities and Their Consequences

Schematically, two types of combustion instabilities exist with clearly defined spectra. They produce either low-frequency or high-frequency vibrations.

Low-frequency vibrations (in the hundreds of Hertz) are caused by instabilities that appear at the thermodynamic limits of the combustion and that are sustained by resonance either of the jets entering the chamber or of the combustion chamber/air intake system. They are more sensitive to the low stagnation temperatures. If these vibrations are large, the low-frequency instabilities may fatigue and damage the missile structure and equipment and cause aerodynamic coupling phenomena with the air intake and thus limit the maximum thrust (see Fig. 14).

To know the influence of these low-frequency instabilities, one can carry out tests, for instance, on an axisymmetric inlet equipped with a mechanical device providing large pressure fluctuations between 0 and 200 Hz (see Fig. 15.) An important result of these experiments is that the internal boundary-layer bleed can reduce drastically the pressure fluctuations.[2]

High-frequency vibrations (screech, in the thousands of Hertz) are triggered at certain equivalence ratios. These vibrations may damage the heat shielding and apply high accelerations to the equipment on board.

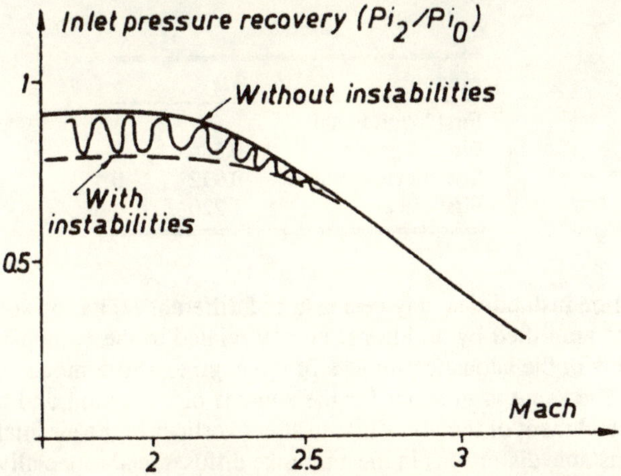

Fig. 14 Effect of pressure oscillations on inlet margin.

The frequency F of each acoustical mode can be calculated with following next relation, available for a cylindrical cavity:

$$F = a/2\sqrt{(2k/D)^2 + (n/L)^2}$$

where a is the acoustical velocity inside the combustor, D is the internal diameter, L is the length, and k and n are coefficients (see Table 2). For example, with numerical values of $a = 750\ \mathrm{mS^{-1}}$, $D = 0.40$, $L_1 = 1.20$, and $L_2 = 1.10$ m, the frequencies of the principal modes are as follows: for low frequencies, in the first longitudinal mode (combustor) $F = 340$ Hz and in the longitudinal mode (combustor + inlet) $F = 160$ Hz; for high frequencies, in the first tangential mode $F = 1100$ Hz, in the second tangential mode $F = 1820$ Hz, and in the first radial mode $F = 2290$ Hz.

Figure 16 shows a transverse acoustical mode.

B. Origin and Remedies

Like most unsteady phenomena, ramjet combustion instabilities are a result of complex physical phenomena. Although in-depth studies have been going on for many years, the phenomena are not yet fully understood.

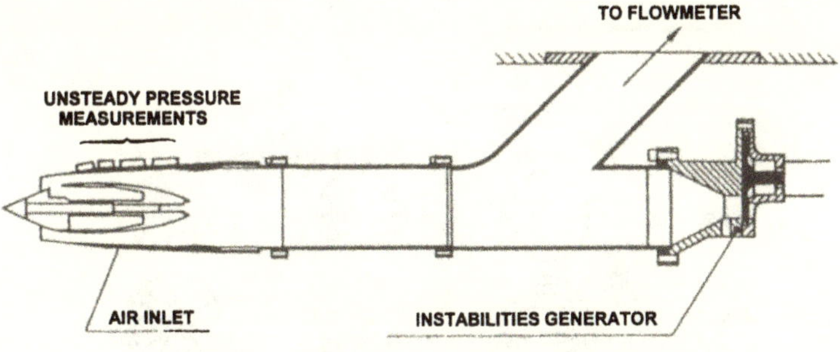

Fig. 15 Air inlet and instabilities generator.

Table 2

Mode	k	n
First longitudinal	0	1
First tangential	0.586	0
Second tangential	0.972	0
First radial	1.220	0

Combustion instabilities may be a result of different excitation source phenomena, may be amplified by additional energy related to the combustion, and may be coupled with the acoustical modes of the engine. Those modes may be easily identified. The same is not true for the sources of excitation and amplification: periodic detachment of the von Kármán street vortices in the air intake and chamber inlet, unstable distortions in the air intake diffuser, and especially the dynamic behavior of the reverse flow regions.

The complexity of these phenomena makes it difficult to understand and predict these instabilities scientifically, but it does allow for a great number of remedies that the engineer can provide at the experimental level.

A few concrete examples will illustrate the effectiveness of these remedies.

1. *First Example*

The air injection at the top of a flame holder makes it possible to raise the equivalence ratio limit at which the high-frequency screech instabilities appear (see Fig. 17).[3] In effect, air injection disturbs the shear layer where the vortices form that separate from the flame holder. This is the periodic aerodynamic phenomenon that, in this case, caused the instabilities.

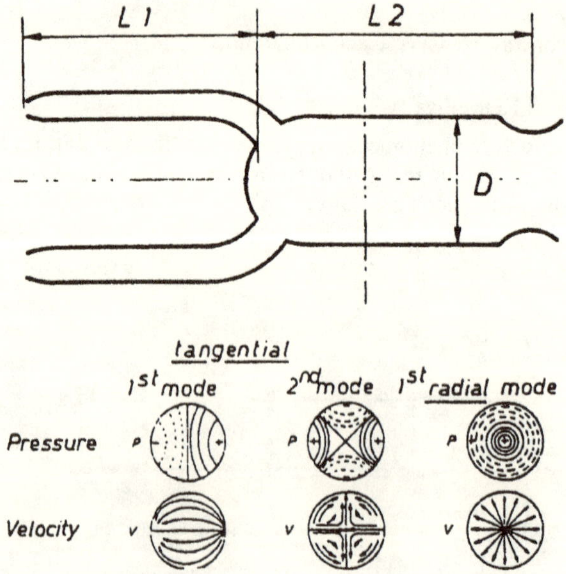

Fig. 16 Transverse acoustical mode.

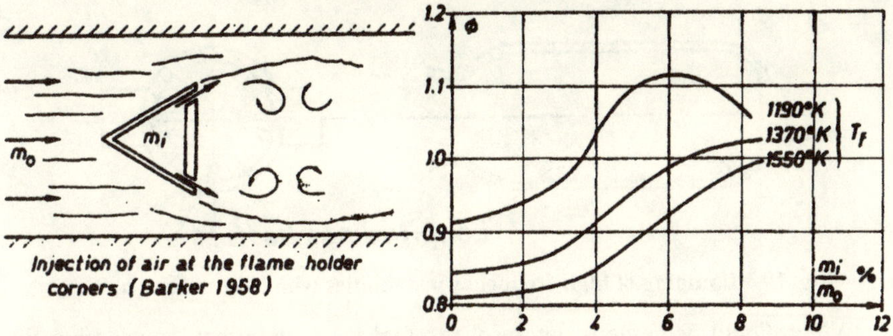

Fig. 17 Richness limit at which the instabilities appear vs injected air mass flow.

2. *Second Example*

The way the fuel is injected, vaporized, and carbureted in the different combustion chamber regions plays a very important role. Figure 19 shows that the flow rate of each injector can be varied to create or eliminate high-frequency instabilities.

The physical explanation of these phenomena is still difficult and controversial, although they are very much under study. It seems that, by modifying the injection, we modify the combustion in the reverse flow regions, which are the main sources of acoustical energy.[4,5]

3. *Third Example*

Other remedies may work, not by suppressing the cause of the instabilities, but by modifying the space structure of the pressure field and the frequencies. This is what the longitudinal partitions along the chamber wall do since they can damp certain (tangential) vibration modes.[4] This remedy is also well known by users of liquid and solid propellant rocket engines.

C. Some Reservations

It is undeniable that the risk of combustion instabilities appearing on long-duration missions must be considered seriously. This is an important stage in development. However, a few reservations should encourage prudence in a program manager.

1) There is no ideal "miracle" configuration that will get around these difficulties with certainty, as noted by Franco-American exchanges in 1982, on widely varying geometries (one, two, three, and four intakes).

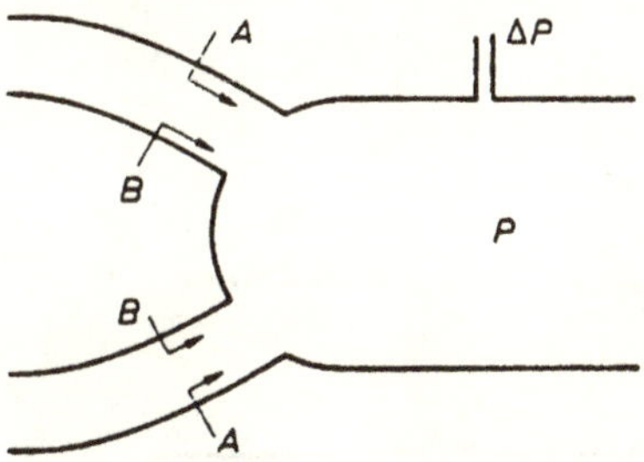

INJECTION REPARTITION %		HIGH FREQUENCY INSTABILITY LEVEL
A	B	Δp/p (%)
30	70	0
40	60	0 TO 2
50	50	8 TO 10
60	40	12 TO 30

Fig. 18 Level of instabilities vs the repartition of fuel flow rate.

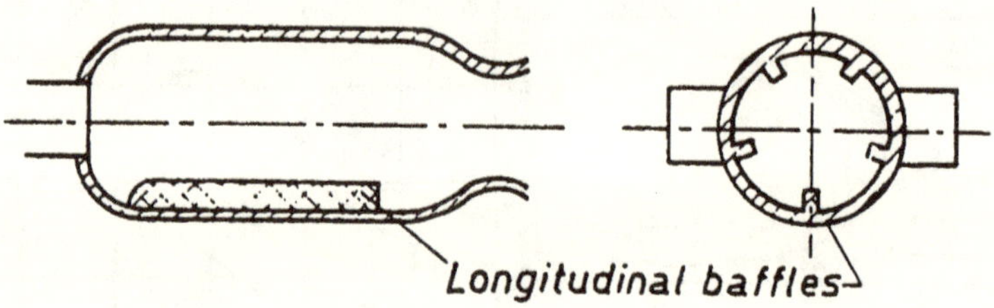

Fig. 19 Damping of high-frequency instabilities with longitudinal baffles.

2) The liquid or gaseous nature of the fuel has little effect, considering the importance of the injection mode, in ramjets. As an example, in 1982 ONERA bench tested an engine with four intakes supplied with kerosene, with a dense liquid fuel, with propane at 50°C and with gas from a gas generator (700°C). Roughly the same type of instability, and at equivalent levels, was observed in all four cases. As Fig. 21 shows, the instabilities can sometimes have catastrophic consequences.

VI. Performances

The performances of a ramjet depend on many parameters for which we cannot describe all of the influences in this chapter.

To calculate the thrust of the ramjet, the following parameters are necessary:

Geometry:

A_1/S_r : Captation (S_r : reference surface)
A_5/S_r : Nozzle throat
A_6/S_r : Nozzle exit

Air inlet:

$\eta_{02} = Pt_2/Pt_0$ pressure efficiency
$\varepsilon = A_1/A_0$ mass air flow coefficient

Fuel:

F_s : stoichiometric ratio
C_4^* : characteristic velocity
γ_4 : specific heat ratio

Combustion chamber:

$\eta c = \Phi_b/\Phi_i$ combustion efficiency
$\eta_{24} = Pt_4/Pt_2$ pressure efficiency
$\eta_j =$ nozzle efficiency

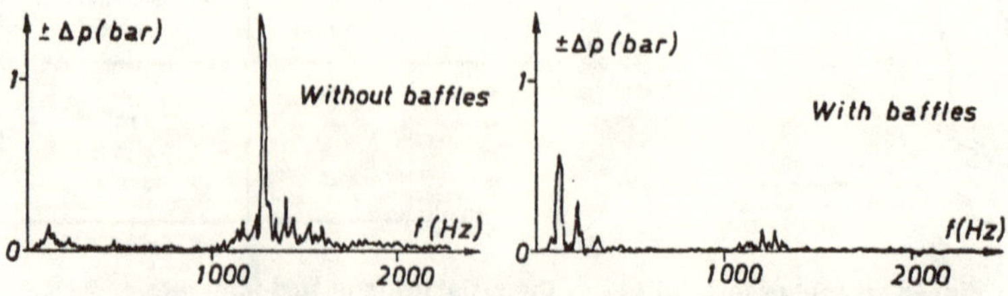

Fig. 20 Illustration of damping provided by longitudinal baffels.

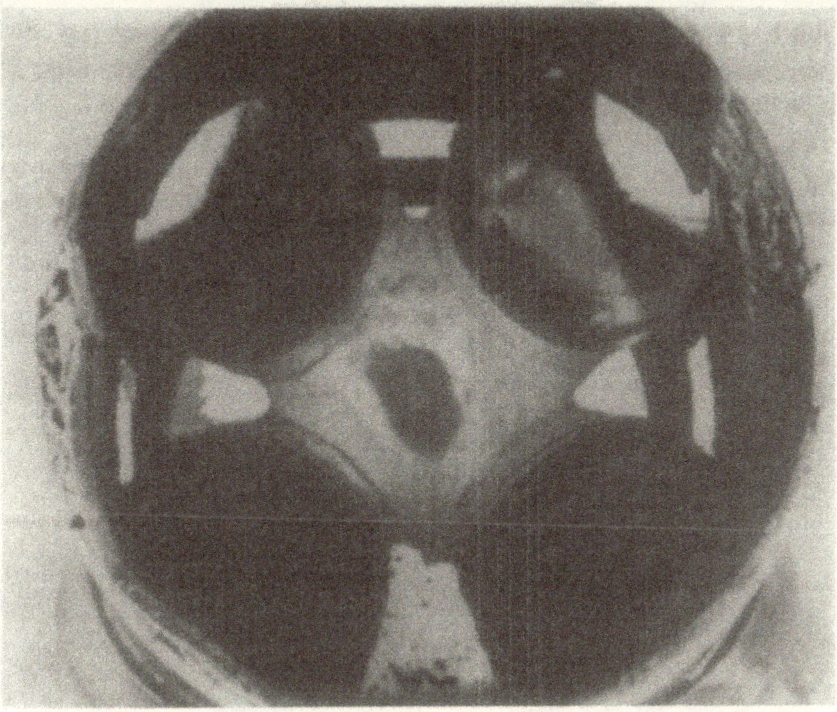

Fig. 21 Effect of high frequency instabilities on a ducted rocket combustor.

Flight characteristics:
Mach number, altitude, etc.
With all of these parameters, the thrust coefficient is given by the following:

$$C_F = \frac{2}{\gamma_0 M_0^2}\frac{A_6}{S_r}\left[\frac{\bar{\omega}(M_6)}{\bar{\omega}(M_0)}(1+\gamma_4 M_6^2)\eta_{02}\eta_{24}\eta_j - 1\right] - 2\frac{A_1}{S_r}\varepsilon$$

with

$$\frac{\eta_{02}}{\varepsilon} = \frac{1}{\sum(M_0)C_0^*}\frac{A_1}{A_5}\frac{C_4^*(1+f_s\Phi_i)}{\eta_{24}}$$

To adapt the solution to the requirements, some parameters have a large influence. Acceleration capacity will increase when the ratio A_5/A_1 is high. Fuel consumption will decrease (in cruise phase, for example) when the ratio A_5/A_1 is low. The range depends on the consumption but also on the choice of the fuel. If we take kerosene as a reference, a synthetic fuel like CSD will increase the range by about 13% and the use of a boron slush by about 50%. The performances of the air inlet are the most important factor for the ramjet.

VII. Conclusion

In conclusion, the use of a liquid-fueled ramjet gives a well-adapted response for supersonic missiles with a large diversity of atmospheric trajectories. This chapter treats only the technical and technological aspects of this propulsion type and the methodology for the design, which were the main topics of this book. Other publications on the liquid-fueled ramjet provide more explanations of performances.

References

[1]Cazin, P., "Liquid-Fueled Ramjets," AGARD/PEP Lecture Series 136, Sept. 1984.

[2]Sans, C., "Effets des instabilités de basse fréquence sur le fonctionnement des prises d'air de statoréacteur," AGARD/PEP Conference Proceedings 450, Oct. 1988.

[3]Baker, C. L., "Experiment Concerning the Occurring and Mechanism of High Frequency Combustion Instability," Ph.D. Thesis, California Inst. of Technology, Pasadena, CA, 1948.

[4]Laverdant, A., "Instabilités de combustion de haute fréquence dans les foyers de statoréacteurs. Etudes documentaire et expérimentale," ONERA Document, unpublished.

[5]Waugh, R. C., Brown, R. S., Hood, T. S., Flandro, G. A., Oates, G. C., and Reardon, F. H., "Ramjet Combustor Instability Investigation Literature Survey and Preliminary Design Study," United Technologies Corporation, Chemical Systems Division, Report UTC/CSD-2770-IT-1, San Jose, CA, 1982.

Chapter 12

Ducted Rockets

John L. Leingang* and Dean P. Petters†
Wright Laboratory, Wright–Patterson Air Force Base, Ohio 45433

Nomenclature

A	= cross-sectional area
b	= constant in burn rate equation
C_D	= discharge coefficient for fuel gas generator orifice, $C_D A_5 = A^*$
C_F	= nozzle gross thrust coefficient
$C_{F\text{vac}}$	= nozzle vacuum thrust coefficient
C_p	= specific heat at constant pressure
C_T	= ramjet thrust coefficient
C_v	= specific heat at constant volume
D	= drag force
D_{inlet}	= ram drag of inlet
E	= total energy stored in fuel
F	= force
$\bar{F}$	= stream thrust
f/a	= fuel–air ratio, equals fuel flow rate/air flow rate
f^*	= heating parameter
f_i^*	= ideal heating parameter
g_c	= 32.174, lbm ft/lbf s^2
I_{sp}	= fuel specific impulse, $\bar{T}/\dot{m}_f$, lbf/lbm/s
J	= mechanical equivalent of heat
M	= Mach number
$\boldsymbol{M}$	= molecular weight, lbm/lbm-mole
$\boldsymbol{M}_a$	= molecular weight of air
$\dot{m}$	= mass flow or weight flow, lbm/s
n	= gas generator burning rate exponent
P	= pressure
P_T^*/P_e	= nozzle pressure ratio
q	= dynamic pressure
R	= gas constant, $\boldsymbol{R}/\boldsymbol{M}$, ft lbf/lbm°R

*Supervisory Aerospace Engineer, Advanced Propulsion Division, Retired.
† Captain, U.S. Air Force, Aerospace Engineer, Aeropropulsion and Power Directorate.

R = universal gas constant
$\dot{r}$ = gas generator burning rate
S = specific stream thrust, $\bar{F}/\dot{m}$
$\boldsymbol{S}$ = gas generator burning surface area, in.^{2}
S_a = air specific stream thrust, $\bar{F}/\dot{m}_a$
S_a^* = critical/sonic air specific stream thrust, $\bar{F}^*/\dot{m}_a$
T = temperature
$\bar{T}$ = thrust force
$\bar{T}_{\text{net}}$ = net thrust force
$\bar{T}_{\text{net jet}}$ = net jet thrust force
$\bar{T}_{\text{gross}}$ = thrust of combustor nozzle
V = velocity
W = weight
X = mass flow function (see Appendix)
Y = constant area flow function (see Appendix)
Z = stream thrust function (see Appendix)
z = altitude
γ = specific heat ratio, C_p/C_v
ΔH_c = fuel heating value, Btu/lbm
ε = nozzle expansion ratio, A_e/A^*
$\eta_{c\Delta T}$ = temperature rise combustion efficiency
η_{cfa} = fuel–air ratio combustion efficiency
η_{cSa} = stream thrust combustion efficiency
η_{KE} = inlet kinetic energy efficiency
λ = flameholder/dump burner effective expansion ratio, A_4/A_{2a}
ρ = density

Subscripts

c = gas generator combustion chamber
e = nozzle exit
n = reference location
p = propellant
T = stagnation condition
0–6 = station numbers, Fig. 4
∞ = freestream conditions

Superscript

$*$ = nozzle throat

I. Introduction

THIS chapter describes the ducted rocket, its main advantages of simplicity and ruggedness for missile propulsion, types of fuels used, fuel flow throttling, and selection factors vis-a-vis a liquid ramjet. This chapter closes with a discussion of performance analysis and a sample calculation of performance.

II. Concept Description

A ducted rocket, Fig. 1, is defined herein as a ramjet in which the fuel is delivered to the ramjet combustor by preburning a solid fuel grain that contains a

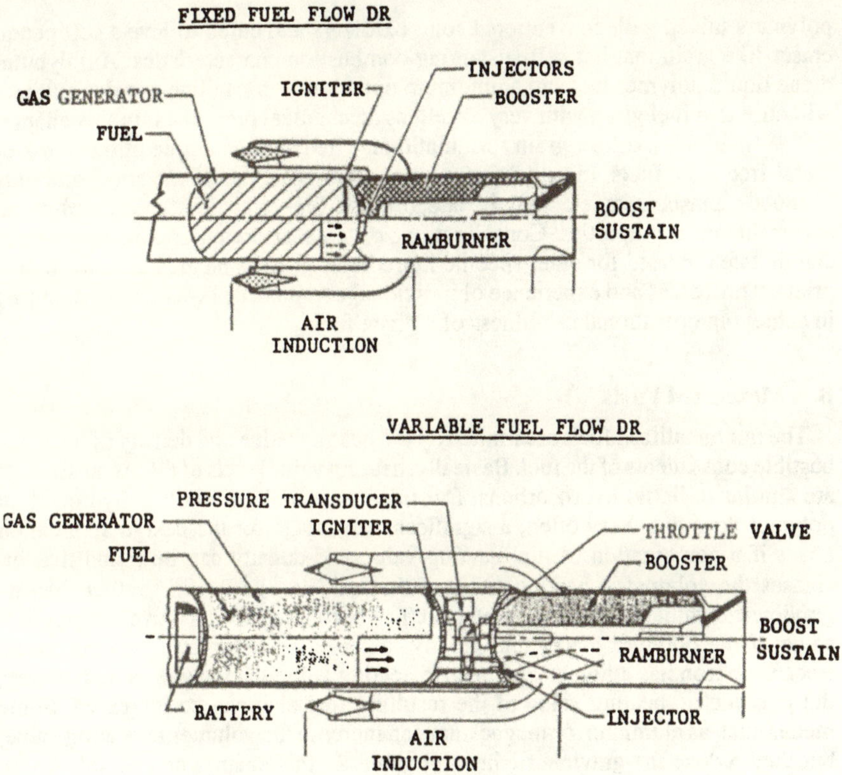

Fig. 1 Schematic of ducted rocket, from Ref. 1

small amount of oxidizer, thus gasifying it and expelling it from its containment chamber, termed a gas generator. Because the fuel is a solid, this engine is well suited to applications where ease of handling and ruggedness are highly important. The term ducted rocket has its origins in work sponsored by the Air Force Rocket Propulsion Laboratory (AFRPL), now part of the Air Force Phillips Laboratory, in the 1960s. A related term, air augmented rocket, has similar origins. Indeed, the ducted rocket grew out of work conducted by the AFRPL to increase the performance of solid rockets by adding external air to the rocket exhaust. Thus, a rocket may be considered to be ducted or to have its performance augmented. Another related propulsion cycle, the shrouded rocket, sought to increase the basic rocket performance by jet compression and combustion of air added to the supersonic fuel-rich rocket plume. The shrouded rocket seeks a modest increase of the basic rocket performance. Today, the ducted rocket derives very little rocket thrust from the fuel-rich gas generator. It could be thought of as a solid gas generator ramjet.

III. Fuel Characteristics

A. Nonmetallized Fuels

Ducted rocket fuels can consist of a wide variety of solid fuel ingredients mixed with the usual solid oxidizers. The most attractive fuels use liquid hydrocarbon polymers mixed with conventional solid oxidizers and cured to give a soft pencil eraser-like grain that has self-sustaining combustion characteristics. A polybutadiene liquid polymer fuel and ammonium perchlorate crystalline powder oxidizer will cure to a fuel grain with very excellent mechanical properties. In this chapter a 65% fuel/35% oxidizer grain formulation is cited as an example of this class of metal-free solid fuels. In actual formulations, burn rate modifiers, small amounts of metals, plasticizers, etc. may be added to the fuel to achieve specific physical and mechanical properties. Considerations of temperature environment extremes can dictate the need for other specific ingredients. This is an area where the proprietary processes and experience of individual companies plays an important part in achieving operational usefulness of a given fuel.

B. Metallized Fuels

The nonmetallized fuels are limited by the heating value and density of the combustible constituents of the fuel. Basically, heating value levels of these constituents are similar to liquid hydrocarbons. Densities are usually limited to hydrocarbon polymer densities. Very often, a significant advantage for the design application arises if a combination of fuel heating value and density can be found that increases the volumetric heating value of the fuel, i.e., its Btu/in.^{3} rather than its gravimetric value, Btu/lbm. A metal such as boron, in powder form, can replace some of the polymer binder and greatly improve the volumetric heating value. Because boron has a higher gravimetric heating value than any organic polymer, the gravimetric heating value of the resulting fuel also can be increased. Some metals such as aluminum or magnesium can increase the volumetric heating value, but they reduce the gravimetric heating value. In this chapter an example boron ducted rocket fuel is shown, having a composition of 50% boron, 10% binder, and 40% oxidizer. The boron content is not the maximum possible; rather, it is the maximum that permits the gas generator temperature to be greater than 2000 K (3600°R). Extensive combustion experience with boron dictates that the gas generator discharge temperature needs to exceed this value to guarantee ignition and consumption of the boron in the air stream.

C. Heating Values, Density, Ballistics

Table 1 shows the relevant characteristics of the two ducted rocket fuels used as examples in this chapter.

IV. Factors Affecting Selection of a Ducted Rocket

The principal attribute of a ducted rocket is the simplicity, ruggedness, and handling ease it enjoys by virtue of its solid fuel. In tactical missile applications this factor can weigh heavily and can favor selection of a ducted rocket even though its performance might be only one-half to two-thirds that of a liquid fuel ramjet. The ducted rocket also has reduced risk for flameout and combustion instability. The following discussion shows how one can quantify the selection factors between a solid ducted rocket and a liquid fuel ramjet to establish which candidate offers higher delivered performance. It is necessary to compare the total fuel feed systems of each in order to establish the complete weight penalty for delivering a given amount of fuel energy to the combustor. The components of the ducted rocket feed system include: the fuel, oxidizer, gas generator case and

Table 1 Ducted rocket fuels

	Hydrocarbon fuel	Boron fuel
Composition	65% HTBP 35% AP	50% boron 10% HTPB 40% AP
Density, lb/in.^{3}	0.0473	0.06818
Gravimetric heating value, Btu/lbm	12717	14944
Volumetric heating value, Btu/in.^{3}	601.5	1018.9
Gas generator combustion temp, °R	2276	4010
Gas generator discharge mol wt	29.068	32.514
Gas generator specific heat ratio	1.2469	1.1678
Stoichiometric fuel–air ratio	0.1182	0.1799

insulation system, and the throttle valve, injector, control, and any power source required for fuel delivery. Thus, the total weight of a ducted rocket fuel feed system can be expressed as

$$WDR_{\text{tot}} = W_F + W_{\text{ox}} + W_{\text{case}} + W_{\text{vcp}}$$

where the symbols are defined as follows:

W_F = fuel weight
W_{ox} = oxidizer weight
W_{case} = gas generator case and insulation system
W_{vcp} = valve, injector, control, power source

For a liquid fuel system the corresponding components are the fuel, fuel bladder or other internal fuel positioning devices, tank pressurant system, fuel pump, fuel control, injectors, and power source components specific to fuel delivery. Thus, its weight can be expressed by

$$WLRJ_{\text{tot}} = W_F + W_{\text{tank}} + W_{\text{pres}} + W_{\text{pump}} + W_{\text{contr}} + W_{\text{inj}}$$

where the symbols are defined as follows:

W_F = weight of fuel
W_{tank} = fuel tank weight
W_{pres} = tank pressurant equipment
W_{pump} = fuel pump and drive
W_{contr} = fuel flow controller
W_{inj} = fuel injectors and manifolds

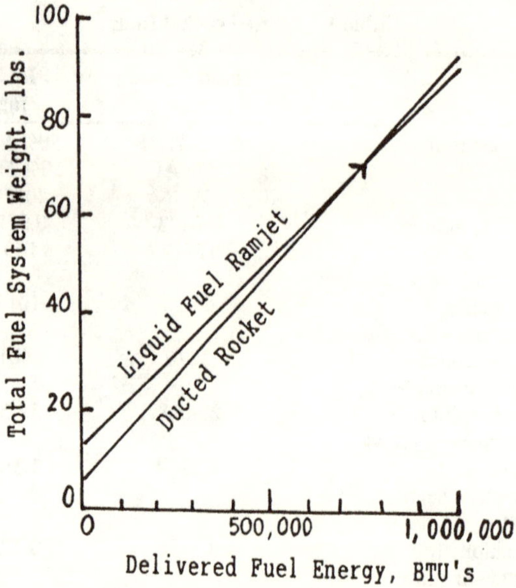

Fig. 2 Ducted rocket and liquid fuel ramjet delivered fuel energy comparison.

Figure 2 shows the result of an analysis conducted to compare the weight of a liquid fuel system with a ducted rocket. Figure 2 suggests that when the total energy to be delivered exceeds 765,000 Btu then the lightest weight approach is a liquid system. For all energy levels below this, the oxidizer in the propellant serves as the most efficient fuel "pressurant" and "delivery" approach. Even at energy delivery requirements of twice the crossover value shown in Fig. 2, the ducted rocket may still be selected because the benefits of its greater simplicity are more important than the performance loss.

Figure 2 was generated by applying the methods of Ref. 1 to define the weight of a ducted rocket fuel system and a liquid fuel system. The comparison was done for a 10-in.-diam missile using fuels with the following properties.

	Ducted rocket	Liquid
Fuel	Hydrocarbon	RJ5
Fuel density, lb/in.^{3}	0.0473	0.0386
Heating value, Btu/lbm	15,000	18,000

In addition to fuel density, heating value, and missile diameter, the method prescribes a slightly greater gas generator case weight for a boron ducted rocket than for a hydrocarbon ducted rocket.

For the example of Fig. 2, the Ref. 1 methods gave the following total fuel system plus fuel weights: hydrocarbon ducted rocket, $W_{\text{total}} = 1.312 \times W_F + 5.0$; and liquid fuel ramjet, $W_{\text{total}} = 13.726 + 1.355 \times W_F$. By noting that the total energy supplied by each is $E = \Delta H_c W_F$ we can express the fuel system weights as follows for $\Delta H_{c\text{DR}} = 15{,}000$ Btu/lbm and $\Delta H_{cLRJ} = 18{,}000$ Btu/lbm so that for the ducted rocket

$$W_{\text{total}} = 8.75 \times 10^{-5} E_{\text{DR}} + 5.0$$

and for the liquid fuel

$$W_{\text{total}} = 7.61 \times 10^{-5} E_L + 13.72$$

Solving these equations at the crossover point gives 765,000 Btu as the maximum delivered energy at which the ducted rocket will have the smallest fuel system weight. This corresponds to 51-lb ducted rocket fuel or 42.5-lb liquid fuel in a fuel system having a total weight of 71.9 lbs.

V. Performance of Fixed Flow Ducted Rocket

The simplest ducted rocket is one with a fixed flow of the fuel-rich gas generator. The fuel flow rate can be designed to be constant by having the grain maintain a constant burning surface area (usually an end-burning grain) and discharge through a port that maintains a constant flow area. The fuel flow rate is preset to a value expected to sustain or accelerate the missile after the booster has accelerated the missile to the ducted rocket takeover speed. This speed is usually in the range of Mach 2 to Mach 3. The fixed flow ducted rocket functions much as a boost-sustain rocket, i.e., a high-thrust boost phase is followed by a (approximately) constant-thrust sustain phase of flight. Of course, the sustain specific impulse will be several times that of a solid rocket sustainer.

Because the sustainer fuel flow is preset and invariant during flight, the missile trajectory must be constrained so that the sustainer fuel–air ratio does not fall below minimum limits or exceed maximum bounds. For this reason the launch envelope and the subsequent flight envelope may be more restrictive than an all-rocket missile. In addition, the fuel temperature at launch will affect its burn rate. Very cold temperatures will reduce the fuel flow rate and very high temperatures will cause an increase. If the missile maintains its end-of-boost speed and altitude, or climbs and accelerates along a constant air flow path, the sustainer fuel–air ratio will remain constant. A descending path will cause a reduction in fuel–air ratio, leading to lean blowout of the sustainer, in the extreme. A climbing path with or without acceleration that reduces air flow will produce higher fuel–air ratios. Carried to extreme, a rich limit fuel–air ratio will be exceeded causing rich blowout of the sustainer. Another consequence of increasing fuel–air ratio is the reduction of inlet operating margin. In the extreme, inlet unstart would occur and inlet "buzz" could be encountered. If the fuel gas generator discharge has significant axial momentum this will add to the inlet operating margin and tend to allow a higher fuel–air ratio to occur before the inlet operating margin is exceeded. The beneficial effect is that somewhat higher altitudes could be achieved with stable engine operation. For the fuels considered in this section, the momentum of the gas generator is small and so the additional altitude that can be achieved is not appreciable.

VI. Variable Flow Ducted Rocket

The variable flow ducted rocket uses a throttle valve to change the flow rate of the gas generator in accordance with commands from the missile flight control system. Basically, the throttle valve adjusts the exit area of a gas generator that is operating above the choking pressure ratio. For the simplest case we can say that the mass rate of flow from the gas generator is

$$\dot{m} = S\rho_p\dot{r}$$

and the burn rate is given by

$$\dot{r} = bP_c^n$$

Combining these expressions gives

$$\dot{m} = S\rho_p bP_c^n$$

We also know that the mass flow rate of the gas generator is given by

$$\dot{m} = C_D P_c A_c^* X^* \frac{\sqrt{g_c/R}}{\sqrt{T_{Tc}}}$$

where

$$X^* = \frac{\sqrt{\gamma}}{[(\gamma+1)/2)]^{[(\gamma+1)/2(\gamma-1)]}}$$

Combining the preceding equation by eliminating P_c leads to the following relation of mass flow and the flow area:

$$\dot{m} = [S\rho_P b]^{(n/(1-n))}, \left[\frac{\sqrt{T_{Tc}}}{X^*\sqrt{g_c/R}}\right]^{(n/(1-n))} [C_D A_c^*]^{-(n/(1-n))}$$

or

$$\dot{m} \propto A_c^{*(n/(1-n))}$$

for most fuels, n will be greater than 0.5 and less than 0.8. This means that the mass flow will vary as the inverse of flow area. Decreasing the flow area (closing the throttle valve further) increases the fuel flow, and increasing the flow area (opening the valve) reduces the fuel flow. Figure 3, from Ref. 1, shows the relation of fuel flow rate to chamber pressure and throttle valve flow area.

A variable flow ducted rocket retains the essential simplicity of an all-solid propulsion system but can achieve nearly the same flight envelope flexibility of operation of a liquid fuel ramjet if a throttling ratio of 5–10 is used. The maximum range of a variable flow ducted rocket can be several times that of a fixed flow ducted rocket because the fuel flow can be programmed to accelerate the missile to high altitude and then sustain efficient cruising flight at high speed. Typical maximum cruise conditions are Mach 3–3.5 at altitudes of 60,000–80,000 ft.

VII. Rustique Self-Throttling Ducted Rocket

Under development in France, the Rustique combines the mechanical simplicity of the fixed flow ducted rocket with most of the performance advantages of the variable flow ducted rocket. By having the fuel gas generator operate in an unchoked manner, pressure changes in the ramjet burner are communicated to the gas generator which then adjusts its pressure and fuel burn rate. Increases in ramjet burner pressure (increased speed, decreased altitude) will increase the gas generator pressure and provide the desired increase of fuel flow. Decreases in ramjet burner pressure (decreased speed, increasing altitude) will reduce the gas generator pressure and provide the desired reduction of fuel flow.

The Rustique can operate at a nearly constant fuel–air ratio over the range of ramjet burner pressure that keeps the gas generator flow unchoked. This self-regulation of fuel flow enables the concept to operate over a wider speed–altitude envelope than the fixed flow ducted rocket. Since gas generator operating pressures

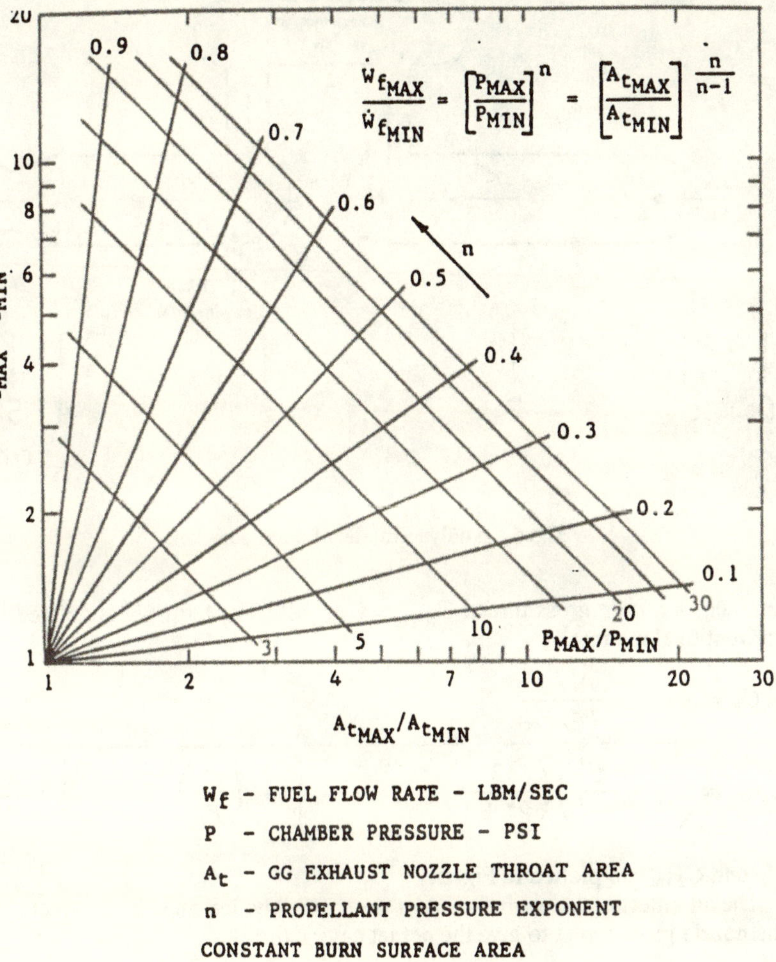

Fig. 3 Variable flow ducted rocket fuel flow modulation as function of gas generator throat area.

are low compared to a conventional gas generator, special attention must be paid to achieving desired fuel burn rates at these low pressures. Achieving a relatively high burn rate pressure exponent (up to 1.0) is important to achieving a nearly constant fuel–air ratio.

VIII. Performance Analysis

Figure 4 illustrates the analysis concept employed in this section for the ducted rocket. The fuel gas generator supplies combustible discharge to the ramjet combustor. Air to complete the fuel combustion enters the combustor from a "reservoir" where stagnation conditions exist. The air has been captured by the inlet and has been decelerated and fed to the reservoir. This performance analysis method is called the nozzle thrust coefficient method (C_F method) and has its roots in rocket motor performance analysis. The ducted rocket combustor-nozzle assembly

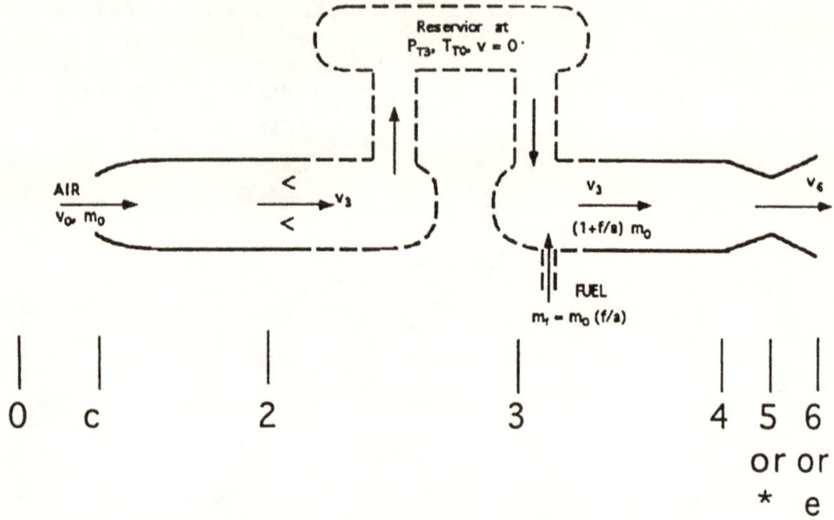

Fig. 4 Analysis model of ducted rocket.

produces a nozzle gross thrust: $\bar{T}_{\text{gross}} = C_F P_T^* A^*$. The term C_F is defined in the conventional manner as

$$C_F = C_{F\text{vac}} - \frac{A_e/A^*}{P_T^*/P_\infty}$$

$$C_{F\text{vac}} = \sqrt{\left(\frac{2\gamma^{*2}}{\gamma^* - 1}\right)\left(\frac{2}{\gamma^* + 1}\right)^{[(\gamma^*+1)/(\gamma^*-1)]}\left[1 - \left(\frac{P_e}{P_T^*}\right)^{[(\gamma^*-1)/\gamma^*]}\right]} + \frac{A_e/A^*}{P_T^*/P_e}$$

C_F and $C_{F\text{vac}}$ are plotted in Fig. 5.

The air entering the engine exerts a "ram drag" which must be subtracted from the nozzle gross thrust to give the net jet engine thrust

$$\bar{T}_{\text{net jet}} = \bar{T}_{\text{gross}} - D_{\text{inlet}}$$

The term net jet thrust refers to thrust based on freestream air inlet conditions, i.e., the engine inlet analytically extends forward to capture air at V_∞ at an inlet capture area of A_∞.

The expression for net jet thrust can be stated as

$$\bar{T}_{\text{net jet}} = \bar{F}_e - \bar{F}_\infty - P_\infty(A_e - A_\infty)$$

$$\bar{T}_{\text{net jet}} = \left[\frac{\dot{m}_e V_e}{g_c} + P_e A_e\right] - \left[\frac{\dot{m}_\infty V_\infty}{g_c} + P_\infty A_\infty\right] - P_\infty(A_e - A_\infty)$$

$$\bar{T}_{\text{net jet}} = \frac{\dot{m}_e V_e}{g_c} + (P_e - P_\infty)A_e - \frac{\dot{m}_\infty V_\infty}{g_c}$$

$$\bar{T}_{\text{net jet}} = P_T^* A^* C_F - (\dot{m}_\infty V_\infty / g_c)$$

A fuel net jet specific impulse is defined by

$$I_{\text{sp net jet}} = \bar{T}_{\text{net jet}}/\text{fuel flow rate}$$

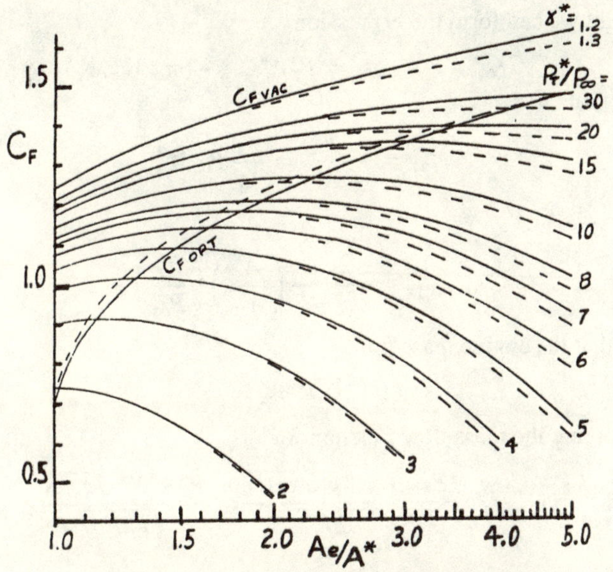

Fig. 5 Nozzle gross thrust coefficient.

The fuel flow rate is taken to be the total flow rate of the gas generator, i.e., its depletion rate. A thrust coefficient based on some reference area of the engine A_n can be defined from

$$C_{\text{Tn net jet}} = T_{\text{net jet}}/q_\infty A_n$$

Usually the combustor cross-sectional area (station 4) or the nozzle exit area (station 6 or e) is chosen as the reference area. In some cases, the freestream (station ∞) is chosen.

The specific impulse and the thrust coefficient can be related to each other by

$$I_{\text{sp net jet}} = \frac{C_{T\text{net jet}} A_n q_\infty}{(f/a)\dot{m}_\infty}$$

where the fuel flow is $(f/a)m_\infty$. Noting that

$$q_\infty = \rho_\infty V_\infty^2/2g_c \qquad \dot{m}_\infty = \rho_\infty A_\infty V_\infty$$

we obtain

$$I_{\text{sp net jet}} = \frac{V_\infty}{2g_c} \bullet \frac{C_{T\infty\,\text{net jet}}}{(f/a)(A_\infty/A_n)}$$

Note that the following are all equal:

$$C_{T\infty}A_\infty = C_{T4}A_4 = C_{Te}A_e = C_{Tn}A_n$$

With $A_n = A_\infty$, the I_{sp} can be expressed as

$$I_{\text{sp net jet}} = \frac{V_\infty}{2g_c} \bullet \frac{C_{T\infty\,\text{net jet}}}{(f/a)}$$

$$I_{\text{sp net jet}} = \frac{V_\infty}{2g_c} \bullet \frac{C_{T4\,\text{net jet}}}{(f/a)(A_\infty/A_4)}$$

$$I_{\text{sp net jet}} = \frac{V_\infty}{2g_c} \bullet \frac{C_{Te\,\text{net jet}}}{(f/a)(A_\infty/A_e)}$$

At this point we can form the expression

$$C_{T\infty\ \text{net jet}} A_\infty q_\infty = P_T^* A^* C_F - (\dot{m}_\infty V_\infty / g_c)$$

By noting that

$$q_\infty = \frac{\rho_\infty V_\infty^2}{2g_c} = \frac{\gamma_\infty}{2} P_\infty M_\infty^2$$

and $\dot{m}_\infty = \rho_\infty A_\infty V_\infty$, we arrive at

$$C_{T\infty\ \text{net jet}} = \frac{2C_F}{\gamma_\infty M_\infty^2}\left(\frac{A^*}{A_\infty}\right)\left(\frac{P_T^*}{P_\infty}\right) - 2$$

We know that the engine mass flow is

$$(1 + f/a)\dot{m}_\infty = \dot{m}^*$$

and by applying the mass flow function X

$$(1 + f/a)\frac{P_{T\infty} A_\infty X_\infty \sqrt{g_c/\mathbf{R}}}{\sqrt{T_{T\infty}/M_\infty}} = \frac{P_T^* A^* X^* \sqrt{g_c/\mathbf{R}}}{\sqrt{T_T^*/M^*}}$$

we get

$$\frac{A_\infty}{A^*} = \frac{(P_T^*/P_{T\infty})(X^*/X_\infty)}{(1 + f/a)\sqrt{(T_T^*/T_{T\infty})(M_\infty/M^*)}} = \frac{(P_T^*/P_{T\infty})(X^*/X_\infty)}{f^*}$$

where we have defined a "heating parameter" as

$$f^* = (1 + f/a)\sqrt{(T_T^*/T_{T\infty})(M_\infty/M^*)}$$

Substituting back into the thrust coefficient expression

$$C_{T\infty\ \text{net jet}} = \left[\frac{2(P_{T\infty}/P_\infty)(X_\infty/X^*)}{\gamma_\infty M_\infty^2}\right] f^* C_F - 2$$

The square bracketed term is plotted in Fig. 6. It is strongly a function of M_∞ and varies weakly with γ^* and T_∞. It is substantially independent of any engine parameters. The heating parameter f^* is a function of the fuel, the fuel–air ratio f/a, the flight condition $T_{T\infty}$, and the combustion efficiency η_c. The combustion efficiency affects the degree to which T_T^* can reach T_{Ti}^*, the ideal combustion temperature. A temperature rise combustion efficiency $\eta_{c\Delta T}$ is most often used to relate the achieved T_T^* to its ideal value

$$T_T^* = T_{T\infty} + \eta_{c\Delta T}(T_{Ti}^* - T_{T\infty})$$

An alternative definition of combustion efficiency, termed a heating parameter combustion efficiency, or a stream thrust efficiency is defined by

$$\eta_{cSa^*} = f_{\text{act}}^* / f_{\text{ideal}}^*$$

The term C_F is

$$C_F = \eta_{NC_F\text{vac}} C_{F\text{vac}} - \frac{\varepsilon}{pr}$$

Note that the nozzle pressure ratio is

$$pr = \frac{P_T^*}{P_\infty} = \frac{P_T^*}{P_{T2a}} \bullet \frac{P_{T2a}}{P_{T\infty}} \bullet \frac{P_{T\infty}}{P_\infty}$$

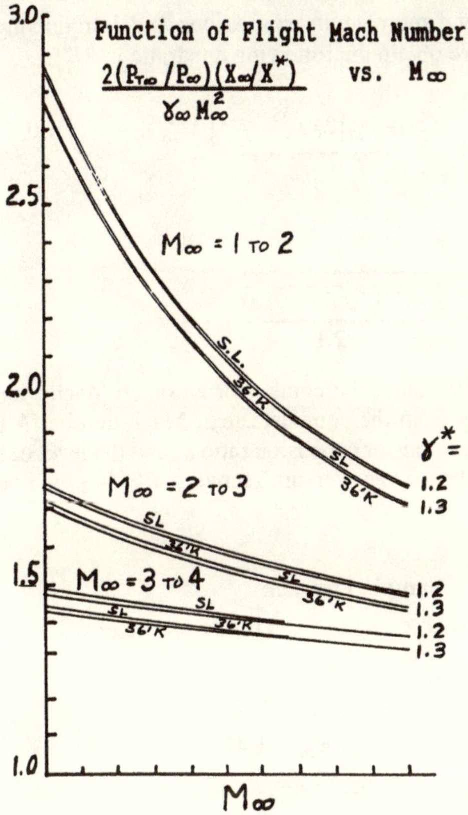

Fig. 6 Flight Mach number function.

Thus, the nozzle pressure ratio is the product of the combustor pressure recovery P_T^*/P_{T2a}, inlet pressure recovery $P_{T2a}/P_{T\infty}$, and the isentropic pressure ratio $P_{T\infty}/P_\infty$.

The combustor pressure recovery is established by considering continuity and momentum relations between the combustor entrance and exit

$$\dot{m}_4 = (1 + f/a)\dot{m}_{2a}$$

or

$$P_T^* A_4 X_4 \frac{\sqrt{g_c/R_4}}{\sqrt{T_T^*}} = (1 + f/a) P_{T2a} A_{2a} X_{2a} \frac{\sqrt{g_c/R_a}}{\sqrt{T_{T\infty}}}$$

then

$$\frac{P_T^*}{P_{T2a}} = \frac{A_{2a}}{A_4}\frac{X_{2a}}{X_4}(1 + f/a)\sqrt{\frac{T_T^*}{T_{T\infty}}\frac{M_\infty}{M^*}} = \frac{A_{2a}}{A_4}\frac{X_{2a}}{X_4} f^*$$

From momentum (stream thrust), for flow axially entering the dome of a constant area combuster and in the absence of friction, the net force acting on the dome area is equal to the change in stream thrust between stations 2a and 4. Thus,

$$F = P_{2a}(A_4 - A_{2a}) = P_T^* A_4 Z_4 - P_{T2a} A_{2a} Z_{2a}$$

By combining the continuity and momentum expressions and introducing the constant area flow function Y, we obtain the following quadratic in M_{2a}^4:

$$\underbrace{\gamma_{2a}\left[\gamma_{2a}Y_4^2 - \frac{\gamma_{2a}-1}{2}f^{*2}\right]}_{A} M_{2a}^4 + \underbrace{\gamma_{2a}\left[2\lambda Y_4^2 - f^{*2}\right]}_{B} M_{2a}^2 + \underbrace{\lambda^2 Y_4^2}_{C} = 0$$

which has the solution

$$M_{2a} = \sqrt{\frac{-B - \sqrt{B^2 - 4AC}}{2A}}$$

As the worked example will show later, the combustor entrance Mach number M_{2a} can be established uniquely from the combustor exit Mach number M_4, or its function Y_4, the flameholder/combustor expansion ratio λ, and the level of heat added f^*. Once M_{2a} is found, the burner pressure recovery P_T^*/P_{T2a} is readily calculated.

IX. Sample Problem

The objective is to calculate specific impulse and thrust coefficient of a ducted rocket defined as follows:

$M_\infty = 2.5$	$P_{T2a}/P_{T\infty} = 0.72$
$z_\infty = 2500$ ft	$A_c/A_{2a} = 1.477$
$f/a = 0.10$	$A_{2a}/A_4 = 0.303$
$\eta_{cSa^*} = 0.98$	$A^*/A_4 = 0.448$
$\eta_{nCf\,\text{vac}} = 0.975$	$A_e/A^* = 2.615$
hydrocarbon fuel	$A_\infty/A_c =$ to be calculated
65% HTPB	standard day atmosphere
35% AP	

Plots of f^*, γ^* and M^* for the hydrocarbon fuel are shown in Figs. 7–9. For the boron fuel (Table 1) they are shown in Figs. 10–12.

In this problem, operating conditions are prescribed, component efficiencies are prescribed, and engine cross section area ratios are prescribed. The inlet capture area is to be calculated. A standard day is assumed.

To start the calculation, an assumption of γ^* is required. At the end of the calculation, γ^* appropriate to the computed value of f^* can be compared to the initial assumption to see if the calculation should be repeated with a better assumption for γ^*.

From the stated conditions of $M_\infty = 2.5$, $z_\infty = 2500$ ft, and $f/a = 0.10$, it is established that $T_\infty = 510°\text{R}$, $X_\infty = 0.265$ (Fig. 14), and $\gamma^* = 1.17$ (Fig. 8). Using these values, calculations yield $T_{T\infty} = 1133.4°\text{R}$, $P_{T\infty}/P_\infty = 17.22$, $V_\infty =$ 2765 ft/s $q_\infty = 8479$ psf, and $f(M_\infty) = 1.5756$ (Fig. 6). Given the geometry data $A^*/A = 0.448$, $A_{2a}/A_4 = 0.303$, $\eta_{cSa^*} = 0.98$, and $f/a = 0.10$, the calculations give $Y_4 = 0.278$ (Fig. 13), $\lambda = A4/A2a = 3.3$, $fi^* = 2.18$ (Fig. 7), and $f^* = \eta_{cSa^*} fi^* = (0.98)(2.18) = 2.14$.

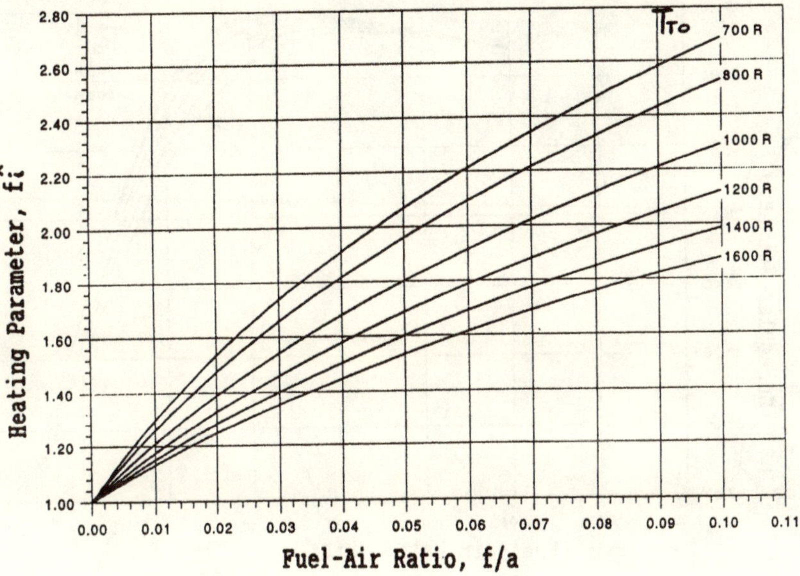

Fig. 7 Heating parameter for hydrocarbon ducted rocket fuel.

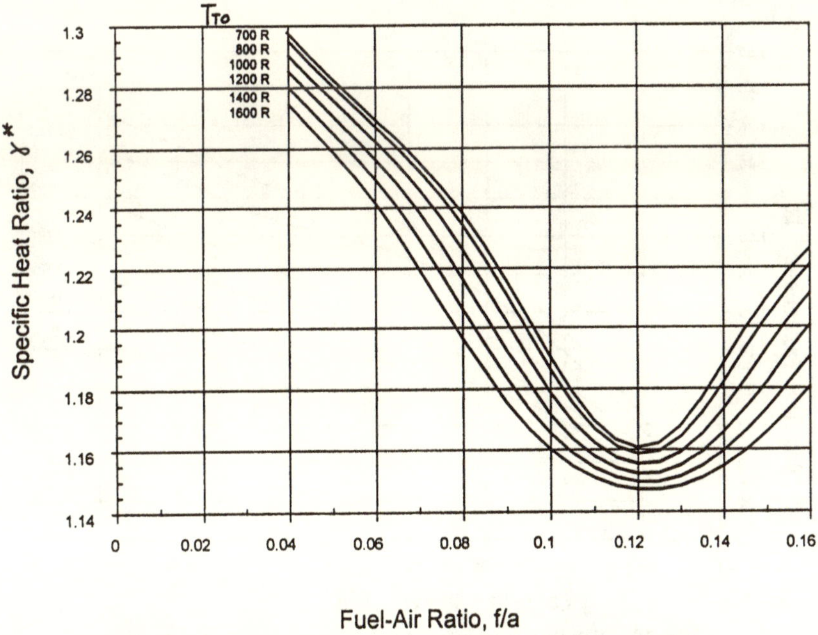

Fig. 8 Specific heat ratio for hydrocarbon ducted rocket fuel.

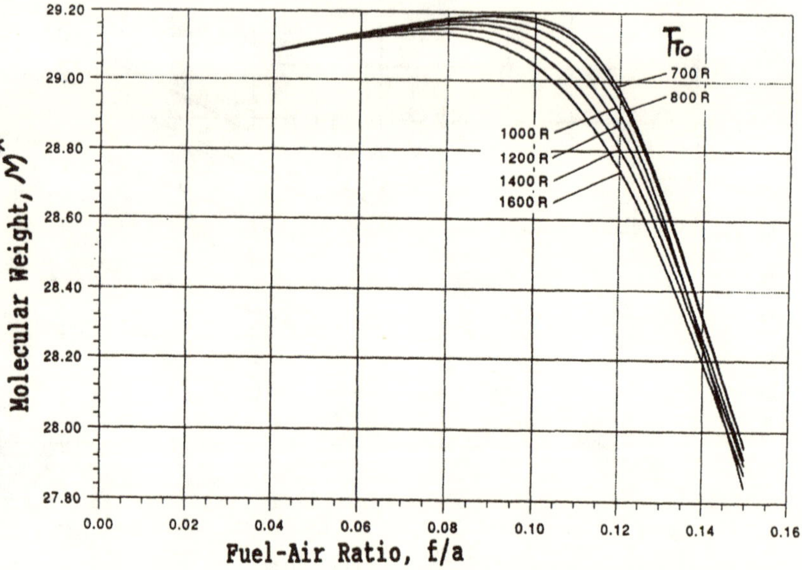

Fig. 9 Molecular weight of hydrocarbon ducted rocket fuel.

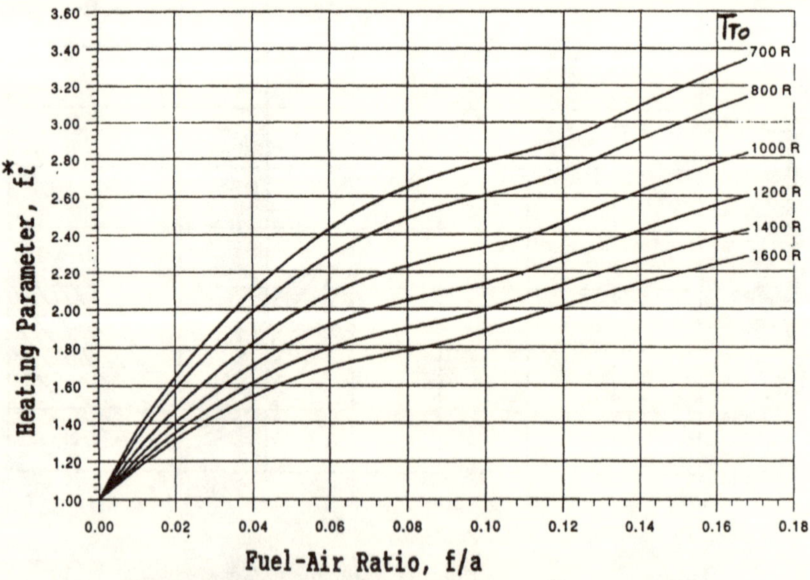

Fig. 10 Heating parameter for boron ducted rocket fuel.

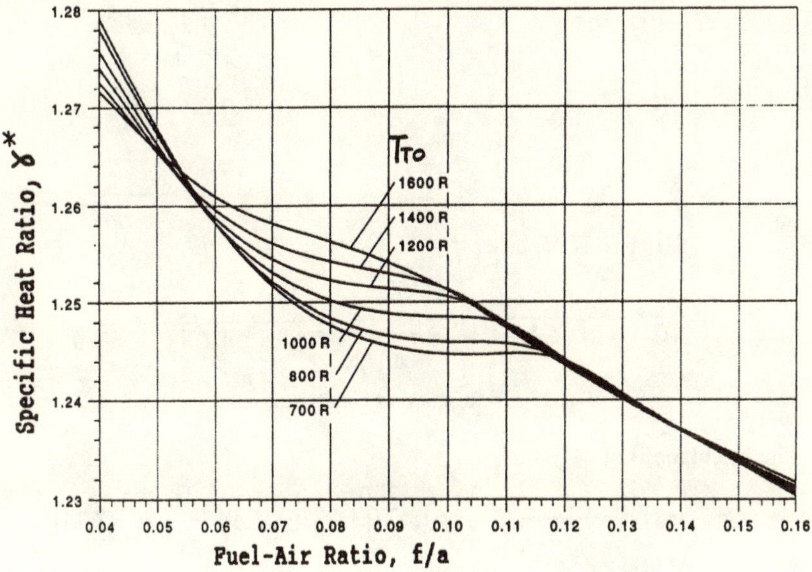

Fig. 11 Specific heat ratio for boron ducted rocket fuel.

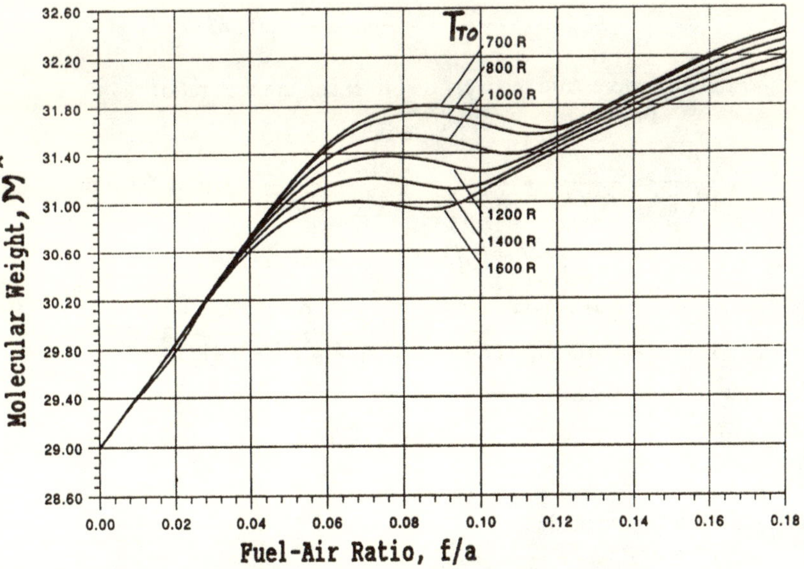

Fig. 12 Molecular weight for boron ducted rocket fuel.

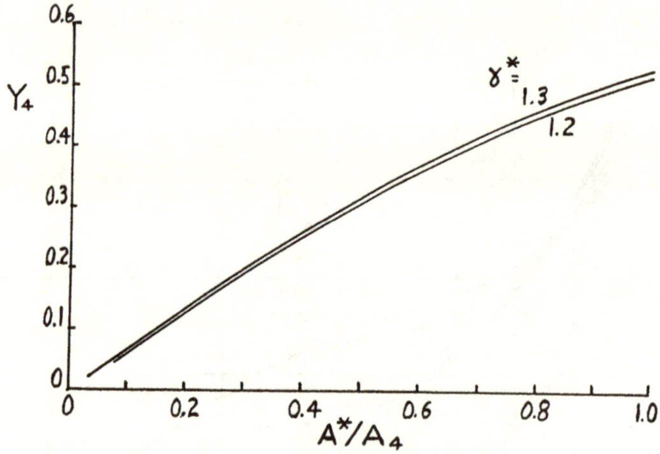

Fig. 13 *Y* function at combustor exit.

The calculations of M_{2a} are

$$A = \gamma_{2a}\left[\gamma_{2a}Y_4 - \frac{\gamma_{2a}-1}{2}f^{*2}\right] = 1.372\left[1.372(0.278)^2 - \frac{0.372}{2}(2.14)^2\right]$$
$$= -1.0232$$

$$B = \gamma_{2a}\left[2\lambda Y_4^2 - f^{*2}\right] = 1.372[2(3.3)(0.278^2) - (2.14)^2] = -5.5834$$

$$C = \lambda^2 Y_4^2 = (3.3)^2(0.278)^2 = 0.8416$$

$$M_{2a} = \sqrt{\frac{-B - \sqrt{B^2 - 4AC}}{2A}} = 0.3831$$

For the combustor pressure recovery and inlet capture area ratio:

$$X_{2a} = 0.413$$

$$\frac{P_T^*}{P_{T2a}} = \frac{A_{2a}/A_4}{A^*/A_4} \bullet \frac{X_{2a}}{X^*}f^* = \frac{0.303}{0.448} \bullet \frac{0.413}{0.65} \bullet 2.14 = 0.9196$$

$$\frac{P_{T2a}}{P_{T\infty}} = 0.72 \qquad \frac{A_c}{A_{2a}} = 1.477$$

$$\frac{A_\infty}{A_c} = \frac{P_{T2a}/P_{T\infty}}{(A_c/A_{2a})(X_\infty/X_{2a})} = \frac{0.72}{(1.477)(0.255/0.413)} = 0.7895$$

Using the following data

$$\varepsilon = A_e/A^* = 2.615$$
$$\eta_{NCf\,\mathrm{vac}} = 0.975$$

these calculations are made

$$\frac{P_T^*}{P_\infty} = \frac{P_T^*}{P_{T2a}} \bullet \frac{P_{T2a}}{P_{T\infty}} \bullet \frac{P_{T\infty}}{P_\infty} = (0.9196)(0.72)(17.22) = 11.40$$

$$\frac{A_\infty}{A_4} = \frac{A_\infty}{A_c} \bullet \frac{A_c}{A_{2a}} \bullet \frac{A_{2a}}{A_4} = (0.7895)(1.477)(0.303) = 0.3533$$

$$\frac{A_\infty}{A_4} = \frac{A_\infty}{A_c} \bullet \frac{A_c}{A_{2a}} \bullet \frac{A_{2a}}{A_4} \bullet \frac{1}{A^*/A_4} \bullet \frac{1}{A_e/A^*}$$
$$= (0.7895)(1.477)(0.303)(1/0.448)(1/2.615) = 0.3016$$

$$C_{F\text{vac}} = 1.53 \text{ (Fig. 5)}$$

$$C_F = \eta_{NCf\text{vac}} C_{F\text{vac}} - \frac{\varepsilon}{pr} = (0.975)(1.53) - \frac{2.615}{11.40} = 1.2624$$

$$C_{T\infty} = f(M_\infty) f^* C_f - 2 = (1.5756)(2.14)(1.2624) - 2 = 2.2566$$

$$I_{SP} = \frac{V_\infty}{2g_c} \bullet \frac{C_{T\infty}}{f/a} = \frac{2766}{2(32.174)} \bullet \frac{2.2566}{0.10} = 970$$

$$C_{T4} = C_{T\infty} \frac{A_\infty}{A_4} = 2.2566(0.3533) = 0.797$$

Appendix: Compressible Flow Functions

Mass flow continuity (conservation) and momentum (stream thrust) relations used in compressible flow replace the primitive flow variables, P, ρ, V, and T with Mach number functions X, Y, Z, and total (stagnation) conditions for total temperature T_T and total pressure P_T. Mass flow is expressed as

$$\dot{m} = \rho A V = P_T A X \frac{\sqrt{g_c/R}}{\sqrt{T_T}}$$

where

$$X = \frac{\sqrt{\gamma} M}{\{1 + [(\gamma - 1)/2] M^2\}[(\gamma + 1)/(2(\gamma - 1))]}$$

Stream thrust is expressed as

$$\bar{F} = \dot{m} V / g_c + PA = P_T A Z$$

where

$$Z = \frac{1 + \gamma M^2}{\{1 + [(\gamma - 1)/2] M^2\}^{(\gamma/(\gamma - 1))}}$$

In situations where continuity and momentum relations must be solved simultaneously, such as a combustor, the continuity equation is divided by the stream thrust equation leading to the factor X/Z. Another compressible flow function, termed the constant area flow function is defined as

$$Y = \frac{X}{Z} = \frac{M\sqrt{\gamma\{1 + [(\gamma - 1)/2] M^2\}}}{1 + \gamma M^2}$$

The functions X, Y, and Z are plotted in Fig. 14. Figure 15 shows how the gas law, the definition of Mach number, and the definitions of total pressure and total temperature are used to express continuity and stream thrust in terms of Mach number functions and total conditions.

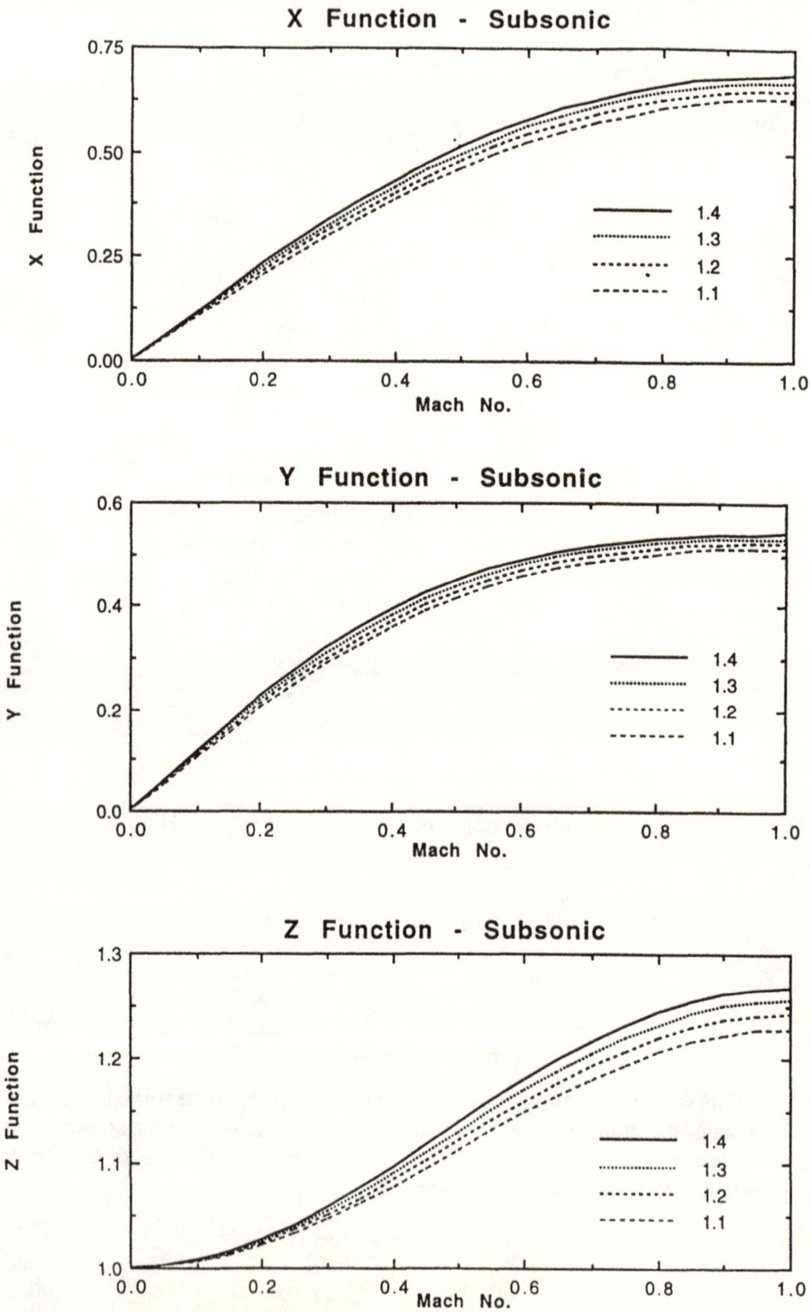

Fig. 14a Subsonic compressible flow functions.

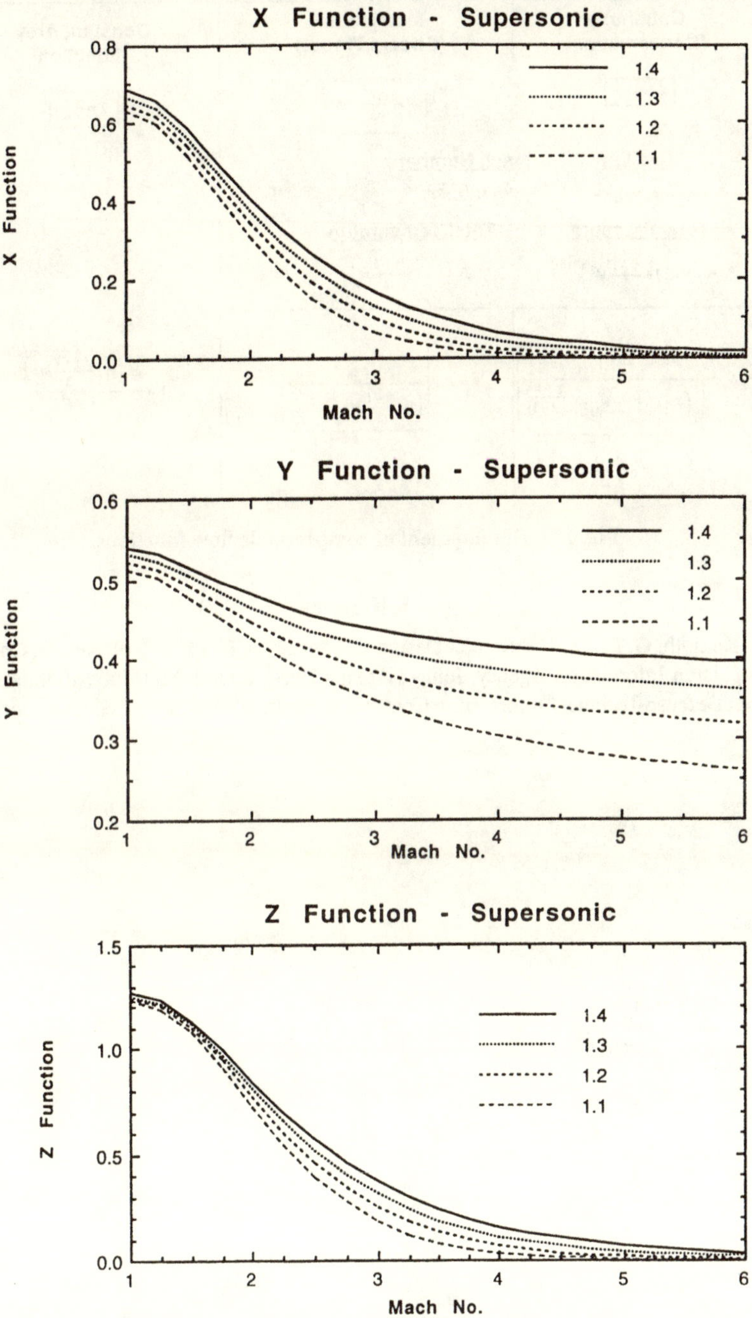

Fig. 14b Supersonic compressible flow functions.

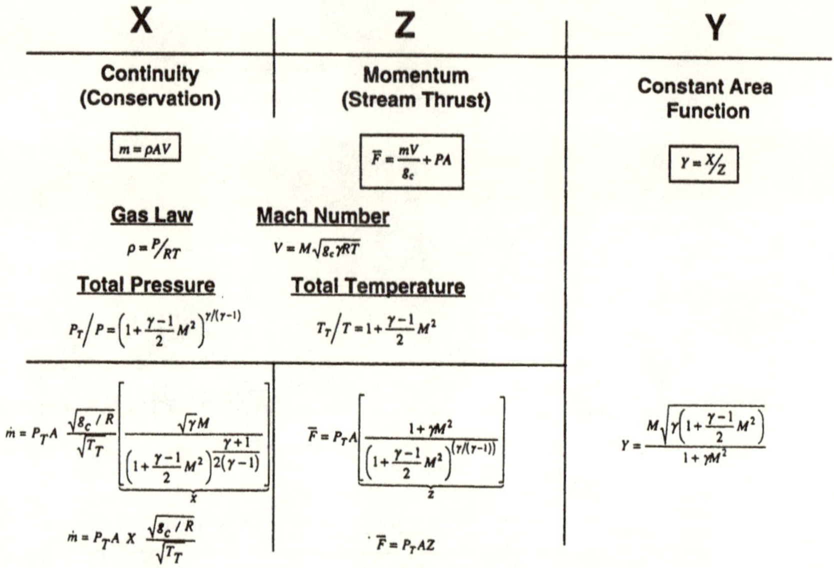

Fig. 15 Development of compressible flow functions.

Reference

[1]Kinhoth, G. D., et al., "Ramjet Design Handbook," CPIA Pub. 319, Rev. B, Chemical Propulsion Information Agency, Johns Hopkins Univ., Laurel, MD, July, 1988; available from Defense Technical Information Center, Alexandria, VA, to qualified requestors.

Chapter 13

Solid Fuel Ramjets

Steven Leisch*
United Technologies Corporation, San Jose, California 95151
and
David W. Netzer†
Naval Postgraduate School, Monterey, California 93943

Nomenclature

A	= area
A_b	= fuel surface area
A_c	= inlet capture area
A_{fp}	= fuel port area
A_p	= propellant port area
A_{ref}	= reference area for vehicle drag
a	= acoustic speed
C_D	= vehicle drag coefficient
C_d	= nozzle discharge coefficient
C_F	= nozzle thrust coefficient
C^*	= characteristic exhaust velocity
F	= thrust
f	= fuel–air ratio, m_f/m_{air}
G	= air mass flux, m_{air}/A_{fp}
g_c	= units proportionality factor, 32.174 lbm ft/lbf s^2, 1 kg m/N s^2
g_0	= acceleration of gravity, 32.174 ft/s^2, 9.8067 m/s^2
I_{sp}	= specific impulse
L	= length
M	= Mach number
$\dot{m}_{\text{air}}$	= mass flow rate of air
$\dot{m}_f$	= mass flow rate of fuel
P	= pressure
PF	= perimeter factor
PLC	= empirical loss factor
$PLCDM$	= empirical loss factor

*Project Engineer, Chemical Systems Division.

†Professor, Department of Aeronautics and Astronautics.

pr	= perimeter of fuel grain
q	= dynamic pressure, $p_\infty + \frac{1}{2}\rho_\infty v_\infty^2$
R	= gas constant
r	= fuel regression rate
t_{res}	= residence time
T	= temperature
TM	= thrust margin
v	= velocity
v_∞	= vehicle flight velocity
W_l	= vehicle launch weight
Wp	= weight of propellant
W_{pm}	= weight of propellant in mixing section
W_{poc}	= weight of propellant overcast on fuel grain
γ	= process from 4 to 5, C_p/C_v
Δv	= ideal vehicle velocity change
$\eta_{\Delta T}$	= temperature-rise combustion efficiency
ρ	= density
ϕ	= equivalence ratio, $f/f_{\text{stoichiometric}}$

Subscripts

c	= chamber
ca	= capture
f	= fuel
g	= fuel grain
m	= mixer
t	= stagnation property
1, . . . ,	= vehicle station numbers, Fig. 5

I. Introduction

A. Background

THE solid fuel ramjet (SFRJ) has been considered as a propulsion device for missiles since the 1930s. Much of the early work in the United States is contained in limited distribution corporate and JANNAF reports. Early designs[1,2] used metallized fuels with high-bypass ratios to obtain high-combustion efficiencies and various amounts of oxidizer in the fuel grain in order to improve flammability and ignition characteristics. It is a very simple device, storable, and has the potential for high reliability at a cost close to that of an equivalent solid rocket motor. The SFRJ can provide specific impulse in the 900–1400 s range and density impulse in the 1150–2300 gm-s/cm^3 range. This results in a 200–400% increase in range over that of a solid rocket motor with comparable size and weight.[3]

A typical configuration for the SFRJ is shown in Fig. 1. Also shown for comparison are configurations for a liquid fuel ramjet and a ducted rocket. Figure 2 illustrates the major combustion regions.[3–5] The configurations include an air inlet diffuser which slows the supersonic air to low-subsonic speeds. This, in turn, can raise the air to high-combustor inlet static temperatures and pressures. Some means for flame stabilization is also required under most flight conditions. The heat release rates of hydrocarbon fuels with air are considerably less than for a similar hybrid rocket in which oxygen or some other liquid oxidizer would be

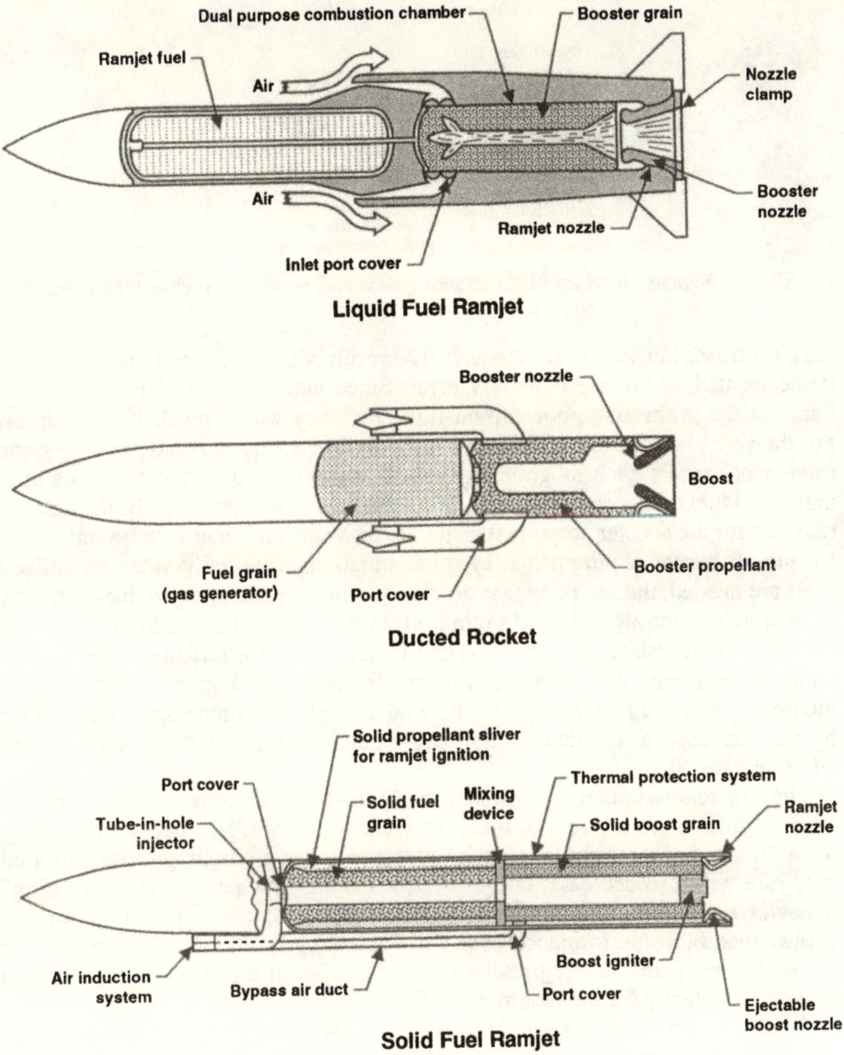

Fig. 1 Ramjet engine configurations.

used. Thus, boundary-layer flame stabilization, which is satisfactory for the hybrid motor design, is not applicable to the SFRJ except for very high inlet air temperatures. The SFRJ is normally considered to be part of an integral rocket ramjet (IRR) with a solid propellant booster cast inside the SFRJ motor cavity. Thus, gas turbine type bluff-body or can-type flame stabilizers cannot be used. This requires the use of an aerodynamic flame stabilization device, such as the rearward facing step or side-dump geometry, in which fuel rich recirculation zones are generated.[5,6] The energy released within or along the edge of the recirculation zone has to be sufficient to sustain combustion in the rest of the combustor.

Downstream of flow reattachment a turbulent boundary layer develops (see Fig. 2). Fuel pyrolysis products diffuse outward from the wall, and oxygen from

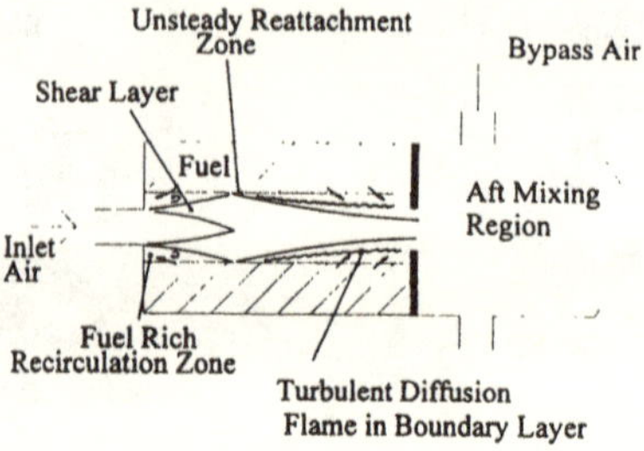

Fig. 2 Schematic of an SFRJ showing flow and combustion characteristics.

the air diffuses inward toward the wall. The result is an annular turbulent diffusion flame located within the boundary layer. Since unburned fuel exists below the flame at the grain exit, poor combustion efficiency would result if the exhaust nozzle were placed at that location. An aft chamber is required (and possibly some mixing enhancers such as grain exit swirl generators) to permit the remaining unreacted fuel to mix and burn with the available oxygen. Fortunately, the volume required for the booster propellant generally provides sufficient volume and length for the aft burner requirements. In some situations, especially when metallized fuels are needed, the use of bypass air can be quite beneficial. By reducing the air mass flow rate through the fuel grain port, more fuel can be placed in the ramjet for the same length restriction. This can increase the cruise range of the missile while at the same time providing a means for better mixing and combustion of the products leaving the fuel port. These advantages are somewhat compromised by the increased weight and complexity required for the multiple inlets and/or additional ducting.

The fuel regression rate depends primarily on the inlet air temperature and the air mass flux (G_{air}) through the fuel port. Generally $r \sim G_{\text{air}}^x$, where x is typically 0.4–0.7. Thus, if the air mass flux is decreased (as in a climbing mode) the fuel flow rate will also decrease, but to a lesser extent. This partial "self-throttling" behavior is advantageous, but it generally is not adequate to provide optimum equivalence ratios/performance over wide operating limits. Variable bypass can be used to provide some regression rate control but at the expense of increased system complexity for the required valving.

Since the energy release is distributed along the entire length of the fuel grain, there is no high-energy release zone which can readily couple with the acoustic or vortex shedding frequencies present in the motor. Thus, the SFRJ provides a combustion system which does not experience the combustion instability problems so often observed in liquid fueled ramjets.

Because of the interaction of the fuel regression rate and the combustion efficiency with the air mass flux and inlet air temperature, the design of an SFRJ for a particular tactical missile application requires considerable cycle analysis to insure that the required thrust can be provided over the entire flight envelope. The SFRJ is well suited for air-to-ground applications but can also be a competitive propulsion system for air-to-air missions. Typical ranges are 35–500 km with Mach numbers from 2 to 5. It is also particularly well suited for tubular projectiles since it can withstand the high-acceleration environments.

The primary data needed by the SFRJ designer are expressions for combustion efficiency and fuel regression rate as a function of operating conditions and knowledge of the flammability limits. Most often these data are obtained experimentally using subscale hardware. Scaling laws[7–10] are then needed to guide the development of the full-scale hardware. In addition, cold-flow-visualization studies and/or measurements are often used to better understand the mixing processes, especially for complex geometries such as side-dump inlets.[11]

B. Fuels

After the initial utilization of metallized fuels[1,2] most efforts turned to hydrocarbons. Plexiglas® has also been used in many research investigations due to its availability and machinability. However, it is not a particularly good fuel due to the large percentage (32%) of oxidizer contained in the solid form. Most of the hydrocarbon studies have utilized variants of HTPB. This fuel has very good physical properties,[12] which are needed to provide grain integrity and performance when subjected to wide ranges of environmental conditions.

The desire to increase volumetric performance has in recent years led to the development of highly metallized fuels that can deliver high-combustion efficiencies.[13] Typically, 50% of a metal such as boron or boron carbide is mixed with HTPB. Small amounts of catalysts or oxidizers[1,14–16] can be used to enhance surface heating of the metals, resulting in improved combustion efficiency, although the specific impulse decreases due to the use of an oxidizer. However, the surface behavior of the fuel grain can be significantly altered by the presence of metals.[17,18] The use of metals in the fuel may also effect the ignition characteristics.[19] In addition, their use can increase the plume signature, which may restrict their selection for specific missions.

C. Combustion

Figure 2 showed the major combustion regions and flow phenomena in the SFRJ. The flame located either along the shear layer of the inlet flow or within the developing boundary layer downstream of flow reattachment conducts heat to the fuel surface by convection and radiation.[3,4,20] Convection often dominates the heat transfer. However, for fuels such as HTPB utilized with equivalence ratios near unity, considerable soot is produced. As the concentration of soot increases toward the aft end of the grain, the contribution of radiation to the total heat flux increases. Without significant radiation and/or flow acceleration in the port, the fuel regression will most often be a maximum at the flow reattachment zone, where stagnation heat transfer conditions are approached. This typically occurs between 6 and 9 inlet step heights downstream of the combustor inlet. With significant radiation present (soot or metals) or when significant flow acceleration occurs in long grains, the fuel regression may be highest at the aft end of the fuel grain.[21]

When metals are included, the combustion process can be significantly altered[22] as shown in Fig. 3.[23,24] Metal particles emerging on the fuel surface must be heated to their ignition temperature before they can ignite and burn when they reach the oxygen rich region above the diffusion flame. The latter is much weaker when metals are present, since fewer hydrocarbon vapors are produced by pyrolysis of the fuel surface. Metallized fuels can produce significantly increased radiative heat transfer to the fuel surface from the burning metal particles. However, a high-number density of particles below the flame can also block radiation to the surface.[19] The ignition and combustion requirements may not be mutually

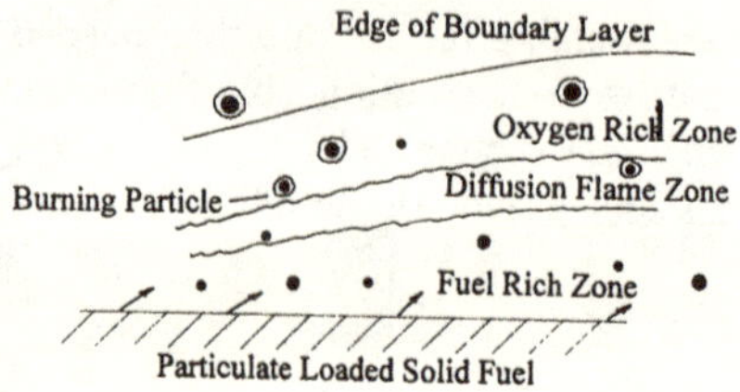

Fig. 3 SFRJ combustion regime for metallized fuels, adapted from Ref. 23.

compatible due to the effects of particle size on the speed and trajectory of the particles as they leave the fuel surface.[23]

The benefit of controlled mixing of the fuel pyrolysis products[25] and the air, either within the fuel port or the aft mixing region, has been demonstrated. Noncircular inlets can increase mixing,[26] but may affect the uniformity of the fuel regression rate. Inlet air swirl may also provide increased fuel regression rates but may affect the flame stabilization and the uniformity of the fuel regression rate.[27,28] The use of bypass air is almost always beneficial and appears to be necessary for obtaining high-combustion efficiency with metallized fuels.[2,29–31]

D. Combustion Efficiency

Without bypass air the combustion efficiency of HTPB-based fuels generally decreases as an equivalence ratio of one is approached. This results from the lack of mixing of the available fuel. For example, the combustion efficiency of HTPB can be given approximately as follows:

$$\eta_{\Delta T} = \frac{T_{t4,\mathrm{exp}} - T_{t2}}{T_{t4,\mathrm{th}} - T_{t2}} = 1 - 0.35\phi^{0.92} \qquad \text{for} \quad \phi \leq 1$$
$$= 0.3 + 0.35\phi^{0.92} \qquad \text{for} \quad \phi > 1$$

where $T_{t4,\mathrm{exp}}$ is the actual nozzle inlet stagnation temperature, $T_{t4,\mathrm{th}}$ the theoretical inlet stagnation temperature, and T_{t2} the combustor inlet stagnation temperature. $T_{t4,\mathrm{th}}$ is found using an equilibrium combustion code such as the NASA SP273 or the NAWC PEPCODE.

In contrast, the combustion efficiency of metallized fuels often increases with equivalence ratio.[22] With bypass the sensitivity of the combustion efficiency to equivalence ratio can be significantly reduced.

E. Flammability Limits

Flammability limits are most often expressed for a specific fuel composition as shown in Fig. 4.[3,4,32]

Large values of A_3/A_5 result in low-fuel port Mach numbers and improve flammability limits. Large values of A_3/A_{2a} increase the size and energy of the recirculation zone. This improves flammability but reduces the allowable fuel loading for a specified outside diameter. Increased T_{t2} results in higher fuel regression rates and higher heat release, which are conducive to improved flammability. It is generally only the initial geometry which is critical for flammability since A_3 increases as the fuel regresses. Different fuels can have significantly different

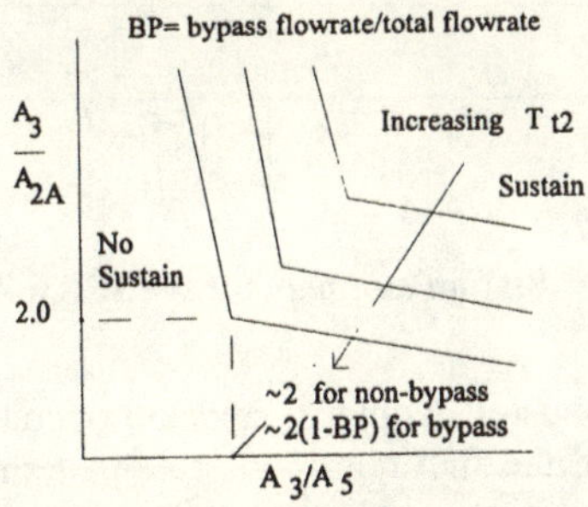

Fig. 4 Typical flammability limits for the SFRJ.

flammability limits, and this may be due to the fuel regression rate (surface blowing rate), the fuel pyrolysis product composition and the amount of subsurface heating that is permitted.[33–35]

At higher pressures most investigations have shown that the SFRJ combustion is diffusion limited. However, regions of the recirculation zone and the entire combustor at very low pressures may be dominated by kinetic rates.

The requirements for A_3/A_{2a} can be improved by using a tube-in-hole injector[4] in which much of the high-velocity inlet air is injected into the port downstream of the inlet. This isolates the recirculation region from much of the inlet air flow.

F. Fuel Regression Rate

The original efforts to develop regression rate expressions[3,4] were based on the earlier hybrid rocket motor modeling work. These early models considered coupled convective and radiative heat transfer to the fuel surface from a diffusion flame located with a turbulent boundary layer.[36] Corrections then had to be made to account for the effects of the recirculation zone and the flow reattachment region. Many experimental and analytical research efforts[15,32,37–42] have been directed at developing fuel regression rate expressions or at determining the variables which most directly affect the regression rate. Typically,

$$\dot{r} = fn[G_{\text{air}}, T_{t2}, P_c, L_g\,(\text{or }\phi),\ \text{other geometric effects}]$$

For example, an approximate expression for HTPB is

$$\dot{r}\,(\text{cm/s}) = 4.44 \times 10^{-5} G_{\text{air}}^{0.53} P_c^{0.33} T_{t2}^{0.71}$$

where $G_{\text{air}} = \text{kg/m}^2\text{s}$, $P_c = \text{MPa}$ and T_{t2} in degrees Kelvin.

The expressions are generally obtained for higher combustor pressures (0.13–1.03 MPa) and may not be applicable at the very low pressures (0.03–0.07 MPa) experienced at high altitudes.

G. Numerical Modeling

A numerical model for the turbulent, recirculating combustion process with wall mass addition that is applicable to the SFRJ was first published in 1977.[5] Subsequently, improved models have been developed which include kinetic effects and three-dimensional geometries.[21,42–54] A review of these models has been published.[55] The numerical models can provide valuable insight into the effects

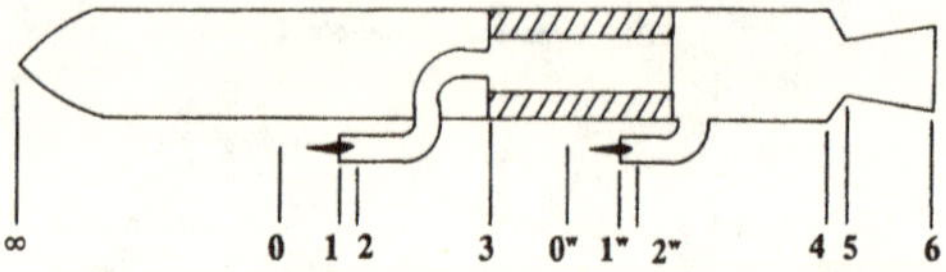

Fig. 5 Station numbers for the SFRJ, Ref. 56.

of fuel properties, combustor geometry, and the operating environment on the combustion process and the fuel regression rate pattern. However, they are not currently adequate for accurate predictions of combustion efficiency, flammability limits, or fuel regression rate, and the required computational times remain excessive.

H. Design Equations

Figure 5 shows the station numbers for the SFRJ which have been agreed upon by the NATO community.[56] The equations in Table 1 are a summary of the major ones used in the SFRJ design process. They will be discussed subsequently.

II. Combustor Design

A. Overview

Design of the ramjet combustor begins with a consideration of the flight and physical requirements imposed by the intended application. Flight requirements may include altitude(s) of operation, vehicle range and velocity, operational temperature limits, and flight trajectories. Physical requirements may include vehicle length, diameter, and weight restrictions; payload weight and volume; and other

Table 1 Equation summary

$P_{t6}/P_{t\infty}$	$= (P_{t0}/P_{t\infty})(P_{t2}/P_{t0})(P_{t2a}/P_{t2})(P_{t3}/P_{t2a})(P_{t4}/P_{t3})(P_{t6}/P_{t4})$
$0/\infty$	= missile forebody
$2/0$	= inlet shocks and flow separation
$2a/2$	= inlet turn and dump
$3/2a$	= combustor inlet
$4/3$	= combustor
$6/4$	= exhaust nozzle
P_{t3}/P_{t2a}	$= \exp\left(-PLCM_{2a}^2\,\gamma_{2a}/2\right)$
P_{t2a}/P_{t2}	$= 1 - PLCDM\, M_2^2\,(P_2/P_{t2})\gamma_2/2$
$\dot{m}_f$	= fuel flow rate $= \rho_f A_b r$
$\dot{m}_5^a$	$= \dot{m}_{\mathrm{air}} + \dot{m}_f = C_d P_{t4} A_5 \left[(\gamma/RT_{t4})\{2/(\gamma+1)\}^{(\gamma+1)/(\gamma-1)}\right]^{\frac{1}{2}}$
$\eta_{\Delta T}$	$= (T_{t4} - T_{t2})/(T_{t4,\mathrm{th}} - T_{t2})$
$F_{\mathrm{req'd}}$	= required thrust $= TMC_D q A_{\mathrm{ref}}$
F_{int}	= internal thrust $= \dot{m}_6 V_6 - \dot{m}_{\mathrm{air}} V_\infty + (p_6 - p_\infty) A_6 = C_F P_{t4} A_5 - \dot{m}_{\mathrm{air}} V_\infty$
I_{sp}	$= F/\dot{m}_f g_0 = [C_F C^*(1+f) - M_\infty a_\infty]/f$
C_F^a	$= \left\{[2\gamma/(\gamma-1)][2/(\gamma+1)]^{(\gamma+1)/(\gamma-1)}\left[1 - (p_6/P_{t4})^{(\gamma-1)/\gamma}\right]\right\}^{\frac{1}{2}}$
	$= \quad + [(p_6 - p_\infty)/P_{t4}]\,(A_6/A_5)$
C^*	$= P_{t4} A_5 C_d / \dot{m}_5$
$C_{\mathrm{th}}^{*\,a}$	$= \left\{R_4 T_{t4,\mathrm{th}} (1/\gamma)[(\gamma+1)/2]^{(\gamma+1)/(\gamma-1)}\right\}^{\frac{1}{2}}$

[a]Value of γ in these expressions is the process value (Ref. 56).

Table 2 Typical engine parameters determined during mission study

Ratio of grain port to nozzle throat area, A_3/A_5
Air bypass ratio
Inlet "shock-on-lip" Mach number
Mach number at ramjet ignition
Equivalence ratio at ramjet ignition
Thrust margin (TM) at ramjet ignition
Missile body diameter
Fuel weight
Propellant weight
Inlet capture area

possible restrictions related to the airframe. These requirements are generally imposed based on launcher size and limitations.

The length and diameter of the combustor are typically established at the beginning of the design phase based on vehicle restrictions. An engine that is to propel an air-launched missile, for example, will be restricted in length and diameter by the capability and dimensions of the launch aircraft. Because the mission goals are generally to maximize range and/or velocity, the largest engine that will meet the vehicle size restrictions is usually selected, subject to any weight limitations.

The ramjet combustor typically includes all components of the engine aft of the inlet diffuser. However, achieving an optimum combustor design for the intended application requires consideration of all major engine components, as well as the airframe design, during the preliminary design stage. The problem is to perform a balanced integration of the air induction system, combustor, fuel, propellant, airframe, and mission characteristics to define the best overall vehicle design. Such an integration, sometimes called a mission study or mission analysis, is best performed by allowing certain key design variables to adjust simultaneously during the optimization to the demands of each other as well as to the performance requirements of the mission. To be performed efficiently, this requires the use of a sophisticated computer model. Table 2 lists some of the key engine parameters that could be determined through this study. The results from this process are specific design values which will provide optimum engine performance for the intended mission. What remains of the design process then is to define each engine component in detail based on the mission study results as well as thermal and structural analyses.

In spite of the assertion that the best way to design the combustor is to adjust key design variables simultaneously, the design process can be looked at in a linear manner (i.e., *a* sets *b* which determines *c*, etc.) as long as one remembers that many iterations may be required for optimization. Each loop in the design process proceeds as given in the following steps.

B. Design Procedure

1. Thrust

The altitude, velocity, and vehicle size requirements will dictate a certain thrust level. The required thrust can be expressed as

$$F = (TM) \times C_D \times q \times A_{\text{ref}}$$

For the solid fuel ramjet, (TM) is generally taken to be 1.2 at ramjet takeover. At other points in the trajectory it may be more or less, but the takeover value is usually the most critical. The drag coefficient C_D and reference area A_{ref} may not necessarily be defined at this point which is why an iterative procedure is required. Nevertheless, the physical requirements will allow the drag and reference area to be bracketed so that a reasonable number may be selected to start off the design procedure.

2. *Specific Impulse and Pressure Losses*

The specific impulse at the design conditions(s) may be calculated next. The equation is given by

$$I_{\text{sp}} = [(C_F \times C^*)(1+f) - M_\infty a_\infty]/fg_0$$

Selection of the fuel–air ratio f will depend on the fuel used, the range of altitudes to be flown, and the design point. In the absence of air flow controls, the SFRJ will move toward more fuel rich conditions as the vehicle climbs. Because operation at values above stoichiometric are undesirable from the point of view of maximum range, a vehicle with a climbing trajectory will generally be sized with a less than stoichiometric fuel–air ratio at ramjet takeover. A descending or constant altitude trajectory would result in an operating equivalence ratio near one or less, depending on the mission requirements.

The thrust coefficient C_F is a function of the engine pressure ratio, specific heat ratio for the nozzle expansion process, and the nozzle area ratio. The engine pressure ratio can be written as the product of individual pressure ratios within the engine as follows:

$$P_{t6}/P_{t\infty} = (P_{to}/P_{t\infty})(P_{t2}/P_{to})(P_{t2a}/P_{t2})(P_{t3}/P_{t2a})(P_{t4}/P_{t3})(P_{t6}/P_{t4})$$

The internal pressure ratios are largely empirical values obtained from existing databases. The freestream ratio is a function of the vehicle Mach number and freestream specific heat ratio. The combustor pressure ratio is strongly dependant on the amount of heat addition within the combustor and the Mach number of the flow. Figure 6 shows the relationship between pressure ratio and total temperature ratio across a constant area combustor for different internal Mach numbers. It is usually desirable to keep the forward end Mach number M_3 less than 0.3.

The injector pressure ratio represents the flow as it passes from the injector face and dumps into the forward end of the fuel grain. The ratio depends on the flow Mach number at the injector, the specific heat ratio, and the geometry of the dump. An empirical expression that has been used by designers is

$$P_{t3}/P_{t2a} = \exp\left(-PLC\, M_{2a}^2 \gamma_{2a}/2\right)$$

where PLC is an empirical factor describing the loss effect associated with the particular dump geometry used. Injector Mach numbers should be kept below 0.3 to minimize losses. A plot of PLC as a function of dump area ratio is shown in Fig. 7. Flameholding considerations will usually limit the area ratio to values less than 0.7 whereas fuel volume loading and loss considerations will dictate it be greater than 0.5. A typical value for PLC would thus be approximately 0.2.

The inlet turn-and-dump ratio represents those losses associated with turning the flow from the inlet duct into the plenum and then on to the injector. It also includes

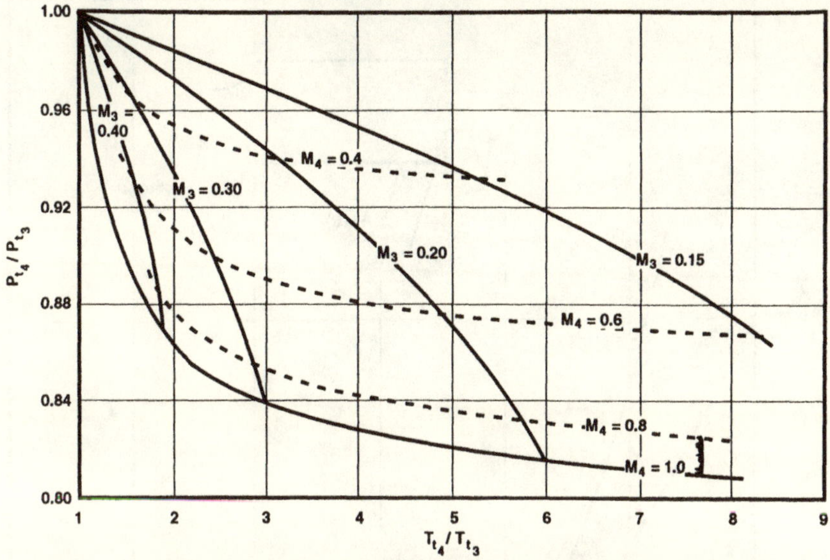

Fig. 6 Stagnation pressure loss due to heat addition.

losses associated with friction, expansion, and contraction within that same region. Another empirical equation that has been used by designers is given by

$$P_{t,2a}/P_{t2} = 1 - PLCDM\, M_2^2 (P_2/P_{t2})\gamma_2/2$$

where M_2 is the diffuser Mach number, P_2 is the pressure in the diffuser, and *PLCDM* is a loss factor similar to *PLC* just discussed. The numerical value of *PLCDM* depends on the turn section and injector used. A plot of $P_{t,2a}/P_{t2}$ as a function of diffuser Mach number and *PLCDM* is shown in Fig. 8. As earlier, Mach numbers in this section are typically kept at levels below 0.3. In volume limited designs a sharp turn is generally required, resulting in *PLCDM* values in the 1.2–1.5 range. If an S-turn can be used, lower values of *PLCDM* can be combustor.

Pressure recovery of the inlet is highly dependant on the flight Mach number and the inlet design. For example, a normal shock inlet will have higher losses than an internal contraction design, and the recovery will decrease significantly with the strength of the shock. At typical takeover Mach numbers a normal shock inlet may have a value of 0.4, whereas an internal contraction type inlet would have better recovery.

Calculation of the characteristic velocity C^* in the combustor is made based on the selected fuel–air ratio and the expected combustion efficiency. The following equation describes the theoretical characteristic velocity:

$$C^*_{\text{th}} = \left\{R_4 T_{t4,\text{th}}\,(1/\gamma)[(\gamma+1)/2]^{(\gamma+1)/(\gamma-1)}\right\}^{\frac{1}{2}}$$

The theoretical temperature ($T_{t4,\text{th}}$), γ, and R_4 are obtained from fuel tables or a calculation assuming equilibrium conditions. The values are primarily a function of inlet air temperature and fuel-air ratio. There is a minor dependance on combustor pressure, but this can usually be ignored during the first few iterations. To obtain the actual C^*, the theoretical combustor temperature should be reduced to account for

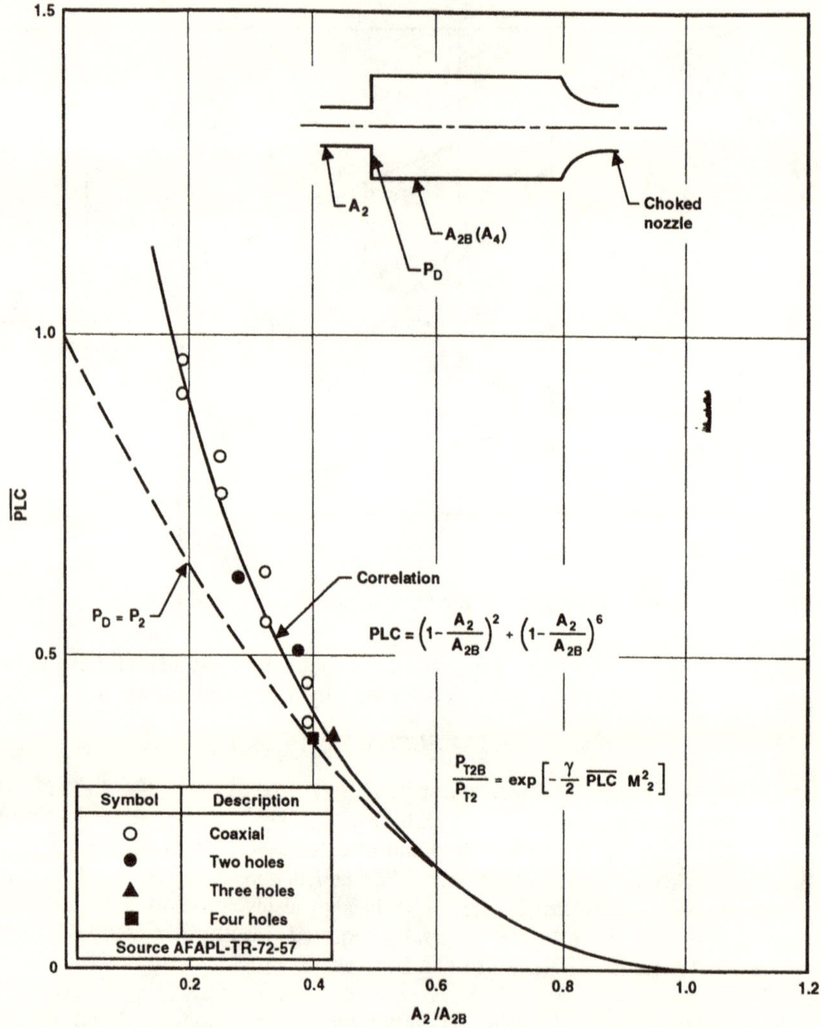

Fig. 7 Coaxial dump pressure loss coefficient as a function of area ratio.

combustion inefficiencies ($\eta_{\Delta T}$). The amount of reduction is based on a database established for the desired fuel.

3. *Flow Rates*

The fuel flow rate can be written in the following form:

$$\dot{m}_f = F/I_{sp}g_0$$

where F is the thrust from step 1 and I_{sp} is the specific impulse from step 2. Having selected a fuel–air ratio in step 2, the air flow rate can now simply be calculated as

$$\dot{m}_{air} = \dot{m}_f/f$$

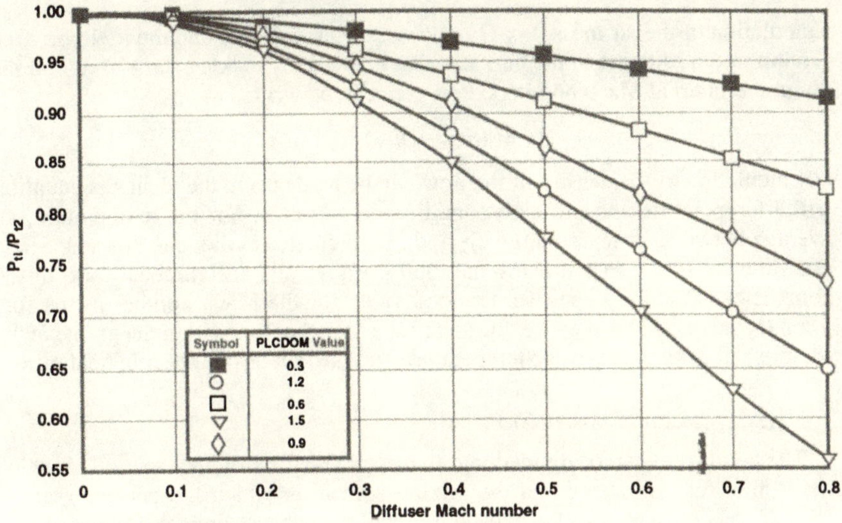

Fig. 8 Injector pressure ratio as a function of inlet Mach number.

4. *Throat Area*

The required ramjet throat area A_5 may be calculated from the continuity equation as follows:

$$A_5 = (\dot{m}_{\text{air}} + \dot{m}_f)C^*/(P_{t4}C_d)$$

Typical values for C_d are 0.97–1.0. It has been assumed in this equation that $P_{t5} = P_{t4}$.

5. *Capture Area*

The required inlet capture area A_{ca} is calculated from the mass continuity equation:

$$A_{\text{ca}} = \dot{m}_{\text{air}}/\rho_o v_o$$

The subscript o refers to the location outside the inlet entrance but within the vehicle bow shock.

6. *Fuel Surface Area*

The required burning surface area of the fuel is given as

$$A_b = \dot{m}_f/\rho_f r$$

However, before this calculation can be performed, an estimate of the regression rate at the design point must be made. Regression rate, besides being fuel dependant, is strongly a function of air mass flux and air temperature. It is to a lesser extent dependent on pressure and certain other engine parameters. For the first iteration, only air mass flux and temperature need be considered. The form of the equation generally takes the following form (see the expression for HTPB):

$$\dot{r} = aG^n T_a^m$$

Calculation of the air mass flux G requires the air flow rate and the fuel port area A_3. An initial estimate of the port area can be made by making some assumptions about the internal Mach numbers. Port area can be written as

$$A_3 = (\dot{m}_{\text{air}} + \dot{m}_f)/\rho_3 M_3 a_3$$

Calculation of the density of the flow can be made using the ideal gas equation if the temperature and pressure conditions are known. For the temperature, an average between air temperature T_{t2} and T_{t4} calculated earlier can be used. Static pressure can be estimated using the total pressure and specific heat ratio in the combustor (P_{t4} and γ) and an assumed value for the Mach number in the fuel grain (typically 0.3). Likewise, the sound speed in the flow within the grain can be established using the average temperature and the combustor specific heat ratio.

7. *Mixing Section/Booster Grain*

The combustor can be divided into two sections: the sustainer section in which the ramjet fuel is located and the mixing section in which the booster grain is located. There may also be booster propellant overcast onto the fuel grain in the sustainer section. In setting the length of the mixing section several factors must be considered: residence time available for adequate combustion, amount of booster propellant required to achieve takeover velocity, and amount of fuel required to meet range requirements. An optimum balance must be met and it is not guaranteed at the outset that the range, velocity, and physical requirements can all be met simultaneously.

The mixing section length needed to provide the minimum residence time (determined through fuel studies, but typically 2–4 ms) can be calculated by making an estimate of the Mach number in the mixing section. Mach number is a function of the mixing section-to-nozzle throat area ratio and the specific heat ratio. Nozzle throat area was calculated in step 4 and the mixing section area A_4 is set by the diameter of the engine, one of the physical restrictions. The mixing section length is given by

$$L_m = M_4 a_4 t_{\text{res}}$$

A check must be made next to determine if this length is adequate for the amount of booster propellant needed. The required booster propellant weight W_p is determined from the required Δv for the vehicle and the maximum permissible initial vehicle weight, both of the latter being system restraints. An initial estimate can be made neglecting drag and gravity effects. Thus,

$$\Delta v = g_0 I_{\text{sp}}\, \ell n\, [W_l/(W_l - W_p)]$$

The amount of propellant that can be overcast onto the fuel grain is calculated first. The total volumetric loading of fuel and propellant within the fuel grain portion of the motor is limited to a maximum of about 88% $[(A_4 - A_p)/A_4 \leq 0.88]$. At loading levels beyond this, bore strain due to temperature changes becomes unacceptably high, and cracking of the grain is likely to occur. Since port area of the fuel grain has been established, the minimum port area of the overcast propellant can be calculated. The length of the sustainer section has in effect also been established (total combustor length minus mixing section length), and so the maximum amount of propellant that can be overcast is given by

$$W_{\text{poc}} = (A_3 - A_p) L_g \rho_p$$

The amount of propellant that can be cast into the mixing section is given by

$$W_{\text{pm}} = (A_4 - A_p)L_m\rho_p$$

The total of W_{pm} and W_{poc} must be greater than or equal to W_p calculated from the Δv equation. If not, the fuel grain must be shortened enough to accommodate the required propellant. This, of course, affects the amount of fuel in the sustainer section and will reduce the range capability. A check must be made to see if there will be sufficient fuel to meet the range requirement after adjusting the length. If not, it may mean that the requirements are not entirely compatible and that some compromise will be required. It may be necessary to reduce the port area of the fuel grain, thus adding fuel, and bypass some of the air around it to keep the fuel port Mach number within the desired range. More discussion of bypass vs nonbypass configurations will be presented in the next section.

Selection of specific propellant properties and booster grain design are generally reserved until after the preliminary ramjet design is completed. It is generally sufficient at this stage to specify propellant I_{sp} and calculate how much propellant is required to size the ramjet. When the time comes to prepare the specific booster grain design, general rules used in solid rocket motors apply. Some of the other considerations associated with integration of the booster are discussed in a later section.

8. *Bypass*

Up to now, it has been assumed that all of the air captured by the inlet was passed through the fuel grain before reaching the mixing section. Depending on the physical restrictions placed on the vehicle, it may be advantageous to bypass a portion of the incoming air around the fuel grain and dump it into the forward end of the mixing section. Doing so allows a smaller port to be used in the fuel grain nonetheless maintaining acceptable Mach numbers within that section. This generally results in a shorter, more compact fuel grain, leaving more length in the mixing section for booster loading and residence time during ramjet operation. It also can provide additional mixing of the fuel and air streams, thus promoting higher combustion efficiency. The disadvantages are that the bypass ducting adds weight and complexity to the vehicle.

9. *Fuel Grain Design*

Having established port area, burn surface area, and length of the fuel grain, the remaining variable to be determined is grain perimeter,

$$pr = A_b/L_g$$

Another way to represent the perimeter is through the perimeter factor PF. This is simply the ratio of the perimeter of the grain port to that of a circular port with equal area,

$$PF = pr/(4\Pi A_3)^{0.5}$$

The shape of the grain port may now be established based on the required perimeter factor. This is largely a geometric problem, however, stress and fuel utilization concerns must also be considered. Some examples of port shapes with differing perimeter factors are shown in Fig. 9. After selecting a pre-

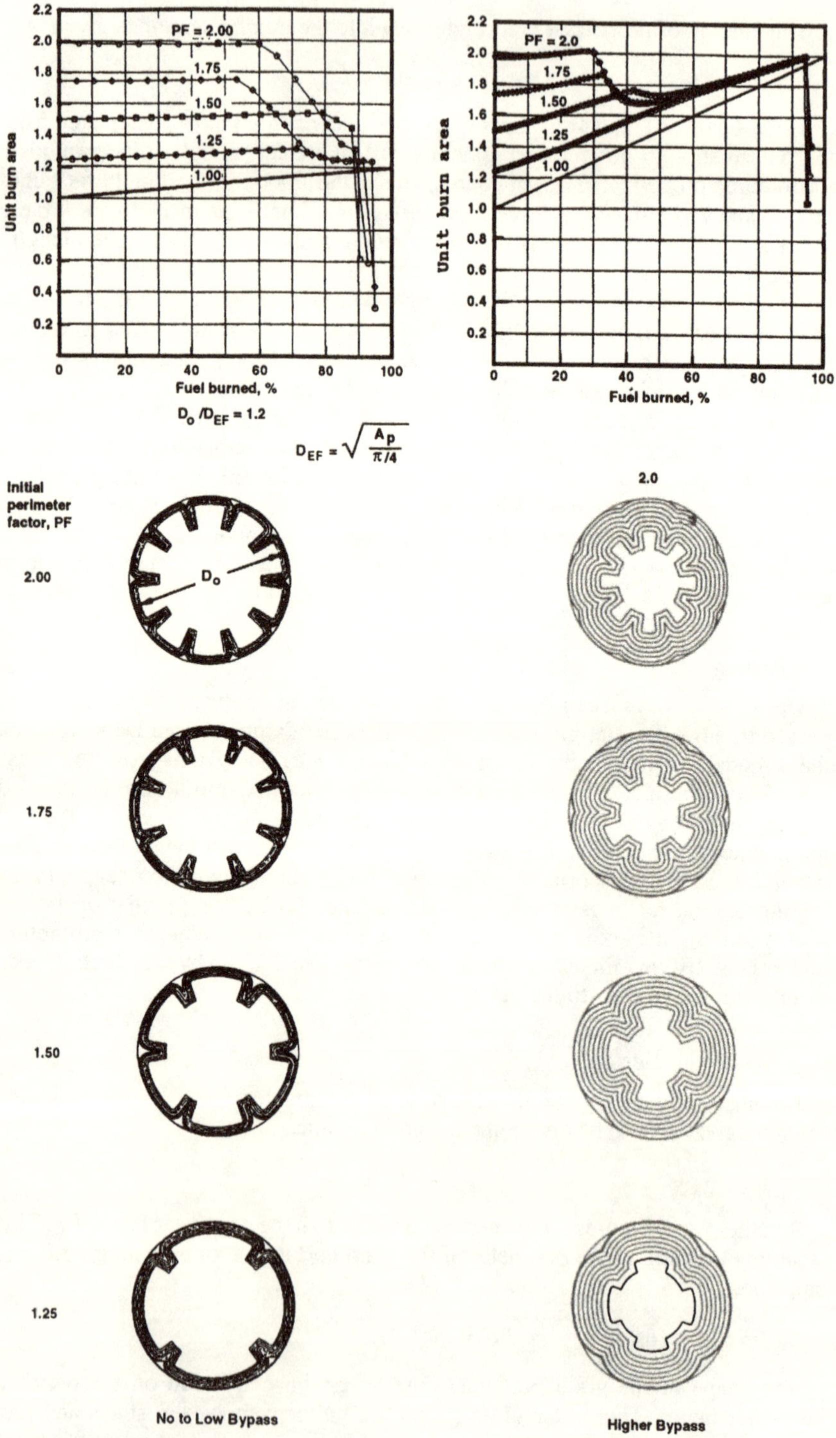

Fig. 9 Typical Boeing SFRJ-AMPP real fuel grain design matrix.

liminary candidate, the burn surface area as a function of port area or web thickness should be calculated and used in a trajectory study to see how the grain design matches the thrust demands associated with the required trajectory. As mentioned previously, some iteration will normally be required for optimization.

10. Air Injector

The area available for the air injector is limited on the lower end by the maximum allowable pressure loss and on the high end by flame blowoff limits. As the engine operates and the port opens up, the port-to-injector area ratio (A_3/A_{2a}) gets larger, and flameholding improves (i.e., moves away from the blowoff limit). It is, therefore, generally desirable to make the injector area as large as possible but still within the blowoff limits at takeover. Figure 10 shows a typical relationship between air temperature and port-to-injector area ratio, showing blowoff limits. Figure 4 also shows the effects of port-to-nozzle throat area ratio (A_3/A_5) on blowoff limits.

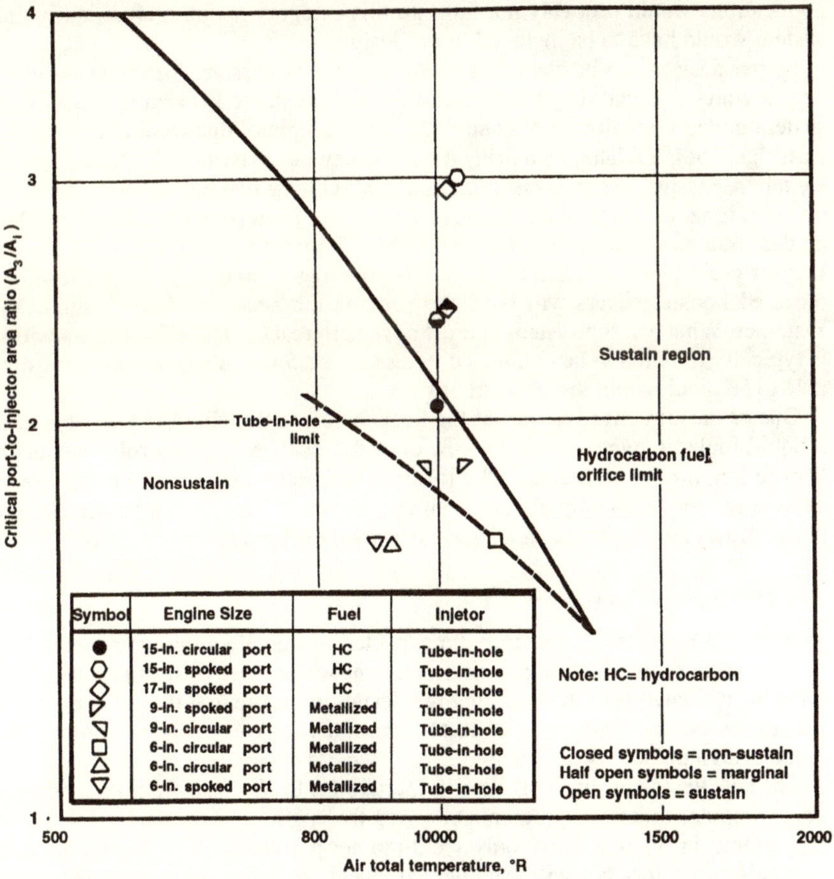

Fig. 10 Effect of temperature on critical port-to-injector area ratio.

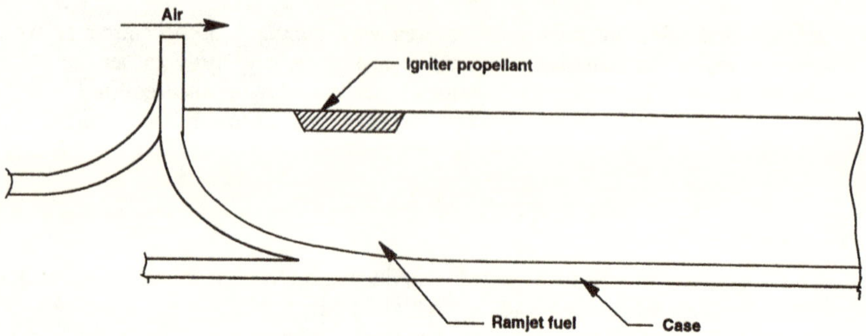

Fig. 11 Passive ignition concept for the SFRJ.

11. Ramjet Igniter

Several options exist for ignition of the ramjet combustor, the selection of which will depend in part on engine operating conditions. If the engine were to be boosted to a high enough Mach number such that the inlet air temperature was in excess of the autoignition temperature (fuel dependent), no igniter would be required. Most applications would probably not fall into this category and, therefore, an igniter system would have to be included in the design.

Igniter designs can be classified as either active or passive. An active system is one in which a signal (electrical, optical, mechanical, etc.) is sent to the ignition system under a definite set of conditions. An example of this would be an igniter cartridge that is initiated by a firing device which has been signaled by a pressure switch. A passive system is one that is integral with the combustor and requires no outside signal to initiate the ignition (Fig. 11). An example of this would be slots or depressions placed in the fuel into which booster propellant is cast. As the booster grain burns out, and the engine begins transition to ramjet operation, the recessed booster slivers will burn and provide the necessary heat to ignite the sustainer. Whatever the system, it must be able to react quickly. Ramjet transition is typically required to be completed in less than 0.5 s, and so the igniter must be able to respond within this amount of time.

One of the attractive features of the SFRJ is its relative simplicity compared to a liquid fueled ramjet or variable flow ducted rocket. Active control systems are kept to a minimum to maintain the inherent simplicity and, therefore, the use of passive igniters is usually a more desirable choice. A sketch of a passive igniter design using booster propellant slivers is shown in Fig. 11.

C. Off-Design Operation

An SFRJ engine will generally be expected to operate at altitudes and Mach numbers other than the design point. Performance at these off-design conditions must be evaluated prior to finalizing the design to ensure that the vehicle will be able to operate under all expected flight conditions. If it will not, some compromise at the design point may be required.

The best way to evaluate off-design operation is to "fly" the preliminary design on a computer trajectory program. Allowing the engine components to interact as they would in flight will not only yield the net performance of the engine on a particular trajectory but will usually also reveal what changes might be made if the performance is unsatisfactory.

Some general trends may be noted here as to the effects of off-design operation. A passively controlled SFRJ (i.e., no airflow control) will tend to higher equivalence ratios as the vehicle climbs to higher altitudes. Operation at an equivalence ratio greater than one is always an inefficient use of the fuel and should be avoided. For this reason, a vehicle which is to be launched at a low altitude and then climb to a higher cruise altitude generally has the engine sized to operate at fuel lean conditions at takeover so that it does not exceed an equivalence ratio of one at the cruise condition.

The effect of increasing equivalence ratio with Mach number may lead to an overspeed condition. Overspeed refers to a condition in which the operating temperature of the combustor or the airframe skin exceeds acceptable levels due to excessive speed. This can happen, for example, if a vehicle is designed to climb to a high-altitude cruise, but instead is operated on a level flight at a lower altitude. If such deviations in trajectory are possible, the design must be such that overspeed cannot occur.

One way to alleviate the problems associated with varying equivalence ratio is to include an active airflow control system. As mentioned previously, one of the attractive features of the SFRJ is its relative simplicity and, therefore, the addition of active control diminishes this feature. However, if it is essential that the vehicle operate over a very wide range of conditions and trajectories, air flow control can provide this capability. There are various schemes to provide airflow control, but the most effective ones are those that vary the amount of air that is bypassed around the sustainer section. This allows direct control of the engine equivalence ratio, within limits, regardless of altitude and can also be used to prevent overspeeding.

A nonbypass engine configuration may also include an air flow control device if necessary. The nature of such a device is different than in the previous case since now there is no airbypass. In one concept, rather than controlling the amount of air that flows over the grain, the level of swirl imparted to the incoming flow can be varied. Studies show that increasing the swirl number of the flow increases the heat transfer coefficient from the air to the fuel, and correspondingly increases regression rate and equivalence ratio. The range over which the equivalence ratio can be altered is more limited with this type of device compared to a variable bypass ratio system.

D. Insulation

With the preliminary design established, thermal considerations must now be addressed. The first concern is protecting the case internally from the heat associated with both booster and sustainer combustion. Any internal metal surface that could be exposed to combustion gases must be protected. The largest area to be protected is the inside wall of the mixing section. To protect this surface during combustion a layer of insulating material is usually applied directly to the case wall. The most prevalent material used to date is a silicone rubber developed by Dow-Corning designated 93–104. Among its advantages are that it is castable, can be cured at ambient or elevated temperatures, transitions to a tough, low conductivity char, and can be used in high shear, high heat-flux environments. Other materials such as ceramics or metal matrices are being studied, but more work is currently needed.

Another alternative is to incorporate an air cooled liner into the mixing section. This could be a perforated metal cylinder with an annular region between it and the case into which some inlet air is directed as a cooling gas. This is generally a more complex design compared to the rubber insulation method.

Another area of concern is the bond interface between the fuel grain and the case. Depending on the flight duration and speed, this interface could reach unacceptably high temperatures, thereby degrading the bond. At those conditions insulation must be provided either between the fuel and the case or on the external surface of the case.

E. Inlet Considerations

Many factors must be considered in the design of the inlet, not the least of which is performance. A design which meets all of the restrictions imposed on it with the highest possible performance, not just of the inlet, but of the entire propulsion unit, is of course the final design sought. It is beyond the scope of this chapter to explore detailed inlet design procedures, and many other texts exist on the subject. A few of the design considerations unique to the SFRJ will be mentioned here.

The overall size of the inlet(s) will be strongly influenced by the capture area required to provide the necessary air flow at the design point. Whether a single inlet or multiple inlets are used to produce this capture area will depend on vehicle type and size restrictions, degree of warhead shielding, use of the bypass configuration, use of active airflow control device, type of turn guidance, and sensitivity of combustion to flow distortions.

The inlet of an SFRJ (or forward inlet if multiple inlets are used) is typically located farther forward on the airframe as compared to liquid fuel ramjets or ducted rockets (see Fig. 1 for comparison). This is because the inlet provides air to the fuel section and, therefore, must be located forward of the sustainer. In a liquid fuel ramjet (LFRJ) or ducted rocket, air is introduced into the combustor downstream of the fuel tank or gas generator.

The forward position of the inlet on an SFRJ is more likely to pose problems of warhead shielding in the case of a missile, and must, therefore, be considered in the inlet design. The use of multiple inlets, with the bypass inlets farther aft, may be considered if this problem exists. Vehicle envelope restrictions, as may occur for a missile to be launched from a tube, must always be kept in mind when locating the inlets. The use of rectangular inlets as opposed to concentric may also provide some relief to the warhead shielding problem.

If warhead shielding is not a problem, placing both primary and bypass inlets forward on the airframe has the advantage that there is less aerodynamic shielding of the inlet by the body of the vehicle and, generally, a more uniform flow profile entering the inlet. Taken to the limit, a concentric normal shock or centerbody inlet at the most forward position of the vehicle may be the simplest of all designs. However, performance of this type of inlet is limited, and the design is only applicable to vehicles with limited forward payloads and no use of bypass ducting. Tubular projectiles are the most common form of vehicle employing this type of inlet.

The type of turn guidance to be used on the vehicle (if any) must be considered, especially if a skid-to-turn method is used. In this case, side mounted inlets may be shielded excessively on the side opposite the turn, causing very poor performance or flameout of the engine. A bottom mounted inlet would be a better choice in this case.

F. Booster Integration

The amount of booster propellant required and the relative split between sustainer and mixing sections was calculated during the preliminary engine design, as discussed previously. Other considerations include thrust, burn time of the propellant, maximum allowable acceleration, allowable ejecta, storage and operational temperature requirements, and propellant composition and mechanical properties.

Factors influencing the thrust output as a function of time include maximum allowable acceleration of the vehicle, desired boost trajectory, weight (mass) and drag of the vehicle, and practical considerations of propellant burn times and grain design. The design process is essentially the same as used for a solid propellant motor, with the exception that trades between booster propellant weight and ramjet fuel weight must be made, as discussed earlier. The design process for solid propellant motors can be found in other texts and will not be duplicated here.

One aspect of ramjet engine design related to the booster is the subject of ejectable vs nonejectable components. The nozzle throat sizes for the ramjet and the booster portions of the flight are normally different, with the booster throat size being smaller. One method of accomplishing this is to position an ejectable booster nozzle within the ramjet nozzle and expel the booster nozzle by some means during the transition from rocket to ramjet flight. However, in some applications, particularly air launched missiles, it is not desirable to have large components being fired out the back of the vehicle. In such cases there will be a requirement that no ejecta larger than some selected size be released from the engine. One solution that has been used in these circumstances is a "nozzleless" booster. In this design, the aft section of the propellant grain port is shaped to act as a nozzle, accelerating the exiting reaction products. A sketch of each type of booster is shown in Fig. 12.

Another component related to the booster and associated with the ejectable vs nonejectable design considerations is the port cover. During boost, the inlet duct(s) must be sealed against the booster gases, and this is done by means of a cover over any airflow ports into the combustor. Like the nozzle, they may be solid components that are simply ejected during transition if there is not a requirement against it. Otherwise, a means of consuming or fragmenting the covers must be incorporated into the design. One method is to fabricate the covers from a reactive

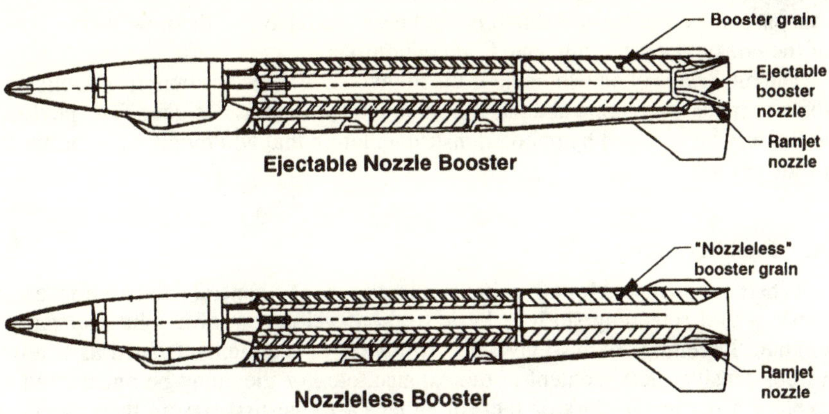

Fig. 12 Ejectable nozzle vs nozzleless booster.

material, such as a strong propellant or intermetallic that will be consumed during transition. Another method is to make the cover from a strong but brittle material such as ceramic or glass that can be fractured into many small pieces of acceptable size.

Bonding the propellant to the ramjet fuel is a consideration unique to the solid fuel ramjet. Migration of liquid species, such as plasticizers, from either the propellant or the fuel to the other must also be considered and prevented. Special coatings are generally used to prevent migration and promote bonding at the propellant/fuel interface.

III. Typical Development Process

The development of an SFRJ engine proceeds through a series of studies and tests. Work typically begins with the investigation of fuels identified for specific or general applications. A preliminary flight design is then established based on certain requirements, and a test engine is designed to simulate the key components and dimensions of the flight version. Often, this first test engine is a subscale version of the flight engine, but it need not be. Inlet and combustor tests are normally conducted separately during the early stages of development to minimize cost and allow separate study of each system. For this reason, combustor tests are generally conducted in "direct-connect," also called "connected-pipe," configurations. As the names imply, air is ducted to the combustor directly via a pipe, rather than being captured by an inlet immersed in a supersonic flowfield. In later stages of direct-connect testing, an inlet simulator may be used that produces a flowfield entering the combustor closely matching that in the real inlet.

Results of the direct-connect testing are used to modify the engine design as necessary to meet requirements. Air induction system (inlet) testing is performed concurrent with the combustor tests in a similar manner, usually starting with subscale and then proceeding to full scale. On successful completion of both inlet and combustor developments, the two systems are combined in a series of semifreejet and/or freejet tests. A freejet test is one in which the entire airframe/engine is situated in a supersonic wind tunnel that is flowing air at the desired velocity and temperature. In the semifreejet arrangement, the test engine consists only of the engine and inlet systems, and air is provided through a nozzle positioned at the entrance to the inlet. Duplication of external airframe contours is not necessary in this case since the entire vehicle is not being modeled and the flow downstream of the nozzle does not duplicate flight conditions.

The final stage of development is the flight test. A vehicle is built that duplicates the air frame and engine and has the same total weight as designed. The payload will likely be replaced by onboard instrumentation that will monitor performance of the engine.

A. Fuel Screening

Whether conducted for a specific application or as a general database for future work, a fuel screening study is basic to the development of a solid fuel ramjet engine. Screening studies may start with a set of requirements such as heating value, density, metal content in the exhaust, etc., or they may be undertaken to explore a particular class of fuel. In either case, the first step in the process is to identify a matrix of candidate formulations that are expected to fall into the range of interest. Calculations are likely to accompany this process to establish theoretical properties of the candidates.

Hand mixes are initially made of all or some of the candidates to evaluate compatibility of the ingredients and ease of processing. Viscosity measurements are taken as a measure of processability. Successful hand mixes are generally followed by pint mixes in a motorized mixer. Again, viscosity of the mix is measured, and samples are cast that can be used for preliminary structural testing.

Selected formulations are then taken into a 1-gal mixer, and laboratory-scale (2.5–3.0 in. diam.) fuel grains are cast. These are simple grains usually cast into a steel tube having a circular port. Combustion tests are conducted with these grains to evaluate regression rate and combustion efficiency at conditions of interest. Samples are also cast from the 1-gal mixes for additional mechanical property testing. Modulus and stress and strain capabilities are some of the mechanical properties of interest.

B. Subscale Testing

Depending on the size of the full-scale engine, a series of subscale tests may be performed. The larger the size of the engine, the more often will subscale testing be performed. The reason for this is almost entirely lower cost. Initial testing will reveal the largest scope of changes required, and it is less costly to make these changes in subscale hardware. The subscale design should duplicate the key internal flow contours, area ratios, and fuel geometry of the full-scale engine.

C. Scaling

Design of subscale hardware requires similarity scaling of three major parameters that characterize the solid fuel ramjet: regression/fuel flow rate, flameholding parameters, and combustion efficiency. The fuel regression rate is primarily dependent on air mass flux, air temperature, and pressure; and, consequently, these variables must be equal to the design values to achieve results representative of the full-scale engine. In addition to regression rate, the fuel-to-air ratios must equal the flight values. Because the air flow rate will be dictated by the engine scale and the desired air mass flux, the proper fuel-to-air ratio must be set by selecting the correct burn surface area. It is important that the port cross section be geometrically similar to the full-scale design and, therefore, adjustment of the burn surface area is controlled through the grain length. Ideally, the grain length scaling should be equal to the overall engine scaling, however, there is generally a diameter effect on regression rate that must be accounted for when selecting the correct grain length.

The second area of concern is the flameholding of the engine. The key parameters that dictate flameholding limits and characteristics are the port-to-injector and the port-to-nozzle area ratios (see Fig. 4). These values should be initially set to the design values with the capability of adjustment if necessary.

The third item of interest is the combustion efficiency. The two variables that have the greatest effect on combustion efficiency are equivalence ratio and combustor residence time. Equivalence ratio has already been addressed and, therefore, efficiency can be modified by changing the mixing section length. Booster testing is not a concern at this stage of the development, and so maintaining similarity here is not of importance. Typically, mixing sections of different length will be tested to evaluate the correlation between residence time and combustion efficiency. These data can be used to modify the flight design if necessary.

D. Full-Scale Testing

There are four general types of full-scale testing; direct connect, semifreejet, freejet, and flight. These four configurations are depicted in Fig. 13. Depending on resources available to the development effort and the size and type of engine to be tested, all or only some of these test types may be used. Each one is discussed in the following text.

1. Direct Connect

If a test series with subscale hardware was performed, the results will be used to make any design modifications to the flight vehicle before proceeding to full-scale testing. The first series of full-scale tests will usually be performed in the direct-connect configuration, again due to lower cost. Any unexpected changes in

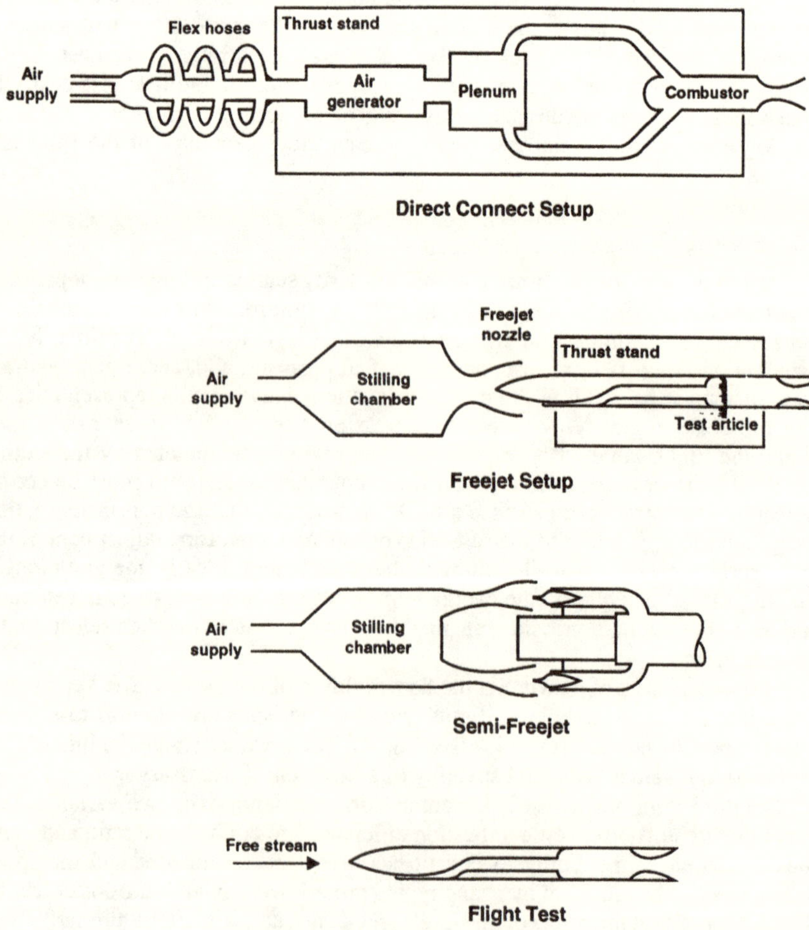

Fig. 13 Test configurations.

performance after scaling up can usually be modified for less cost with this type of hardware. As with subscale hardware, all internal contours, area ratios, and key components are duplicated in the test engine. Testing is conducted at conditions spanning the range of operation.

2. *Semifreejet*

During the time that direct-connect testing is proceeding, inlet testing is usually also being performed. In the semifreejet configuration the two systems are integrated without having to duplicate the entire external airframe. This allows interactions between the inlet and combustor to be defined, along with overall engine performance, at a cost lower than that required for freejet or flight testing. It also allows testing in a facility with smaller airflow capacity than required for freejet testing. In this configuration, an air nozzle (or nozzles) is located directly upstream of the inlet(s) and there is no attempt to duplicate flow conditions over the external portion of the test engine (airframe).

3. *Freejet*

This configuration of ground testing comes closest to flight testing. All internal and external contours, area ratios, materials, and engine components are duplicated and evaluated in the freejet test. The test takes place in a wind tunnel large enough to accommodate the entire test vehicle and can simulate flight conditions (temperature and pressure) over the entire range of interest. The test vehicle is usually heavily instrumented with pressure and temperature sensors, as well as other measurement devices, to document flow profiles and performance.

4. *Flight*

The ultimate test and demonstration of the design is made by flight testing. This is normally done only after extensive testing and modification in one or more of the other configurations to minimize the risk of failure. As in the freejet test, the engine contains numerous sensors to document the performance and flight characteristics.

References

[1]Bulban, E. J., "Light Rocket Features Solid-Fuel Ramjet," *Aviation Week,* July 31, 1961, pp. 87–88, 909.

[2]Jenkins, H. P., Jr., "The Solid-Fuel Ramjet," *Astronautics,* Dec. 1959, pp. 44–45, 92, 94.

[3]Meyers, T. D., "Special Problems of Ramjet with Solid Fuel," Ramjet and Ramrocket Propulsion Systems for Missiles, AGARD PEP Lecture Series 136, Sept. 1984.

[4]Anon., "Solid Fuel Ramjet," United Technologies Chemical Systems Div., United Technologies Corp., San Jose, CA.

[5]Netzer, D. W., "Modeling Solid-Fuel Ramjet Combustion," *Journal of Spacecraft and Rockets,* Vol. 14, No. 12, 1977, pp. 762–766.

[6]Schulte, G., Pein, R., and Hogl, A., "Temperature and Concentrations Measurements in a Solid Fuel Ramjet Combustion Chamber," *Journal of Propulsion and Power,* Vol. 3, No. 2, 1987, pp. 114–120.

[7]Netzer, D. W., and Gany, A., "Burning and Flameholding Characteristics of a Miniature Solid Fuel Ramjet Combustor," AIAA Paper 88-3044, July 1988.

[8]Zvuloni, R., Gany, A., and Levy, Y., "Geometric Effects on the Combustion in Solid Fuel Ramjets," *Journal of Propulsion and Power,* Vol. 5, No. 1, 1989, pp. 32–37.

[9]Ben-Arosh, R., and Gany, A., "Similarity and Scale Effects in Solid-Fuel Ramjet Combustors," *Journal of Propulsion and Power,* Vol. 8, No. 3, 1992, pp. 615–623.

[10]Zvuloni, R., Levy, Y., and Gany, A., "Investigation of a Small Solid Fuel Ramjet Combustor," *Journal of Propulsion and Power,* Vol. 5, No. 3, 1989, pp. 269–275.

[11]Binn, B. A., Scott, W. E., and Netzer, D. W., "Combustion Behavior of Solid Fuel Ramjets, Vol. I, Correlation of Reacting and Non-Reacting Flow Characteristics," Naval Postgraduate School Rept., NPS-67-81-010, Monterey, CA, July 1981.

[12]Lee, T.-H., and Netzer, D. W., "Temperature Effects on Solid-Fuel Ramjet Fuel Properties and Combustion," *Journal of Propulsion and Power,* Vol. 8, No. 3, 1992, pp. 721–723.

[13]Gany, A., and Netzer, D. W., "Fuel Performance Evaluation for the Solid-Fuel Ramjet," *International Journal of Turbo and Jet-Engines,* Vol. 2, 1985, pp. 157–168.

[14]Chen, D. M., Hsieh, W. H., Snyder, V. Y., and Kuo, K. K., "Study of the Thermophysical Properties and Combustion Behavior of Metal-Based Solid Fuels," 24th Joint Propulsion Conf., July 11–13, 1988.

[15]Mady, C. J., Hickey, P. J., and Netzer, D. W., "Combustion Behavior of Solid-Fuel Ramjets," *Journal of Spacecraft and Rockets,* Vol. 15, No. 3, 1978, pp. 131–132.

[16]Raghunandan, B. N., Ravichandran, E. R., and Marathe, A. G., "Combustion Related to Solid-Fuel Ramjets," *Journal of Propulsion and Power,* Vol. 1, No. 6, 1985, pp. 502–504.

[17]Gany, A., and Netzer, D. W., "Combustion Studies of Metallized Fuels for Solid-Fuel Ramjets," *Journal of Propulsion and Power,* Vol. 2, No. 5, 1986, pp. 423–427.

[18]Karadimitris, A., Scott, C., III, Gany, A., and Netzer, D. W., "Regression and Combustion Characteristics of Boron Containing Fuels in Solid Fuel Ramjets," *Journal of Propulsion and Power,* Vol. 7, No. 3, 1991, pp. 341–347.

[19]Snyder, T. S., Jarymowycz, T., Pace, K. K., and Kuo, K. K., "Solid Fuel Ignition and Combustion Characteristics Under High-Speed Crossflows," AIAA Paper 90-2075, July 1990.

[20]Vinnemeier, F. M., and de Wilde, J. P., "Heat Transfer in a Solid Fuel Ramjet Combustor," AIAA Paper 90-1783, 1990.

[21]Metochianakis, M. E., and Netzer, D. W., "Modeling Solid Fuel Ramjet Combustion Including Radiation to the Fuel Surface," *Journal of Spacecraft and Rockets,* Vol. 2, No. 4, 1983, pp. 405–406.

[22]Nabity, J., Lee, T.-H., Natan, B., and Netzer, D., "Combustion Behavior of Boron Carbide Fuel in Solid Fuel Ramjets," *Combustion of Boron-Based Solid Propellants and Solid Fuels,* edited by K. K. Kuo and R. Pein, CRC Press/Begell House, Boca Raton, FL, 1992, pp. 287–302.

[23]Natan, B., and Gany, A., "Ignition and Combustion Characteristics of Individual Boron Particles in the Flowfield of a Solid Fuel Ramjet," 23rd Joint Propulsion Conf., AIAA Paper 87-2034, June/July 1987.

[24]Gany, A., "Combustion of Boron-Containing Fuels in Solid Fuel Ramjets," *Combustion of Boron-Based Solid Propellants and Solid Fuels,* edited by K. K. Kuo and R. Pein, CRC Press/Begell House, Boca Raton, FL, 1992.

[25]Schadow, K. C., Cordes, H. F., and Chieze, D. J., "Experimental Studies of Combustion Process in a Tubular Combustor with Fuel Addition Along the Wall," *Combustion Science and Technology,* Vol. 19, 1978, pp. 51–57.

[26]Wilson, K. J., Schadow, K. C., Gutmark, E., Korting, P. A. O. G., and Dijkstra, F., "Effect of Inlet Geometry on the Flow and Combustion in a Solid Fuel Ramjet," Proceedings of the 31st Annual Israel Conf. on Aviation and Astronautics, Tel Aviv, Israel, Feb. 1990.

[27]Campbell, W. H., Jr., Ko, B. N., Lowe, S. R., and Netzer, D. W., "Solid-Fuel Ramjet Fuel Regression Rate/Thrust Modulation," *Journal of Propulsion and Power,* Vol. 8, No. 3, 1992, pp. 624–629.

[28]Duesterhaus, D. A., and Hogl, A., "Measurements in a Solid Fuel Ramjet with Swirl," Joint Propulsion Conf., AIAA Paper 88-3045, July 1988.

[29]Natan, B., and Gany A., "Combustion Characteristics of a Boron-Fueled SFRJ with Aft-Burner," Ninth International Symposium on Air Breathing Engines, ISABE 89-7013, Athens, Greece, Sept. 3–8, 1989.

[30]Natan, B., and Gany, A., "Effects of Bypass Air on the Combustion of Boron Particles in a Solid Fuel Ramjet," 25th Joint Propulsion Conf., AIAA Paper 89-2886, July 1989.

[31]Natan, B., and Netzer, D. W., "Experimental Investigation of the Effects of Bypass Air on Boron Combustion in a Solid Fuel Ramjet," *Combustion of Boron-Based Solid Propellants and Solid Fuels,* edited by K. K. Kuo and R. Pein, CRC Press/Begell House, Boca Raton, FL, 1992, pp. 427–437.

[32]Schulte, G., "Fuel Regression and Flame Stabilization Studies of Solid Fuel Ramjets," *Journal of Propulsion and Power,* Vol. 2, No. 4, 1986, pp. 301–304.

[33]Hewett, M. E., and Netzer, D. W., "Light Transmission Measurements in Solid Fuel Ramjet Combustors," *Journal of Spacecraft and Rockets,* Vol. 18, No. 2, 1981, pp. 127–132.

[34]Wooldridge, R. C., and Netzer, D. W., "Ignition and Flammability Characteristics of Solid Fuel Ramjets," *Journal of Propulsion and Power,* Vol. 7, No. 5, 1991, pp. 846–848.

[35]Metochianakis, M. E., Goodwin, W. V., Katz, U., and Netzer, D. W., "Combustion Behavior of Solid Fuel Ramjets, Vol. II, Effects of Fuel Properties and Fuel Air Mixing on combustion Efficiency," Naval Postgraduate School Rept., NPS-67-81-011, Monterey, CA, Aug. 1981.

[36]Hadar, I., and Gany, A., "Fuel Regression Mechanism in a Solid Fuel Ramjet," *Prop., Expl., Pyrotech.,* Vol. 17, 1992, pp. 70–76.

[37]Korting, P. A. O. G., van der Geld, C. W. M., Wijchers, T., and Schoyer, H. F. R., "Combustion of Polymethylmethacrylate in a Solid Fuel Ramjet," *Journal of Propulsion and Power,* Vol. 6, No. 3, 1991, pp. 263–270.

[38]de Wilde, J. P., "The Pyrolysis of Polymethylmethacrylate During Combustion of A Solid Fuel Combustion Chamber," Delft Univ. of Technology, Rept. LR-531, Delft, The Netherlands, Dec. 1987.

[39]van der Geld, C. W. M., Korting, P. A. O. G., and Wijchers, T., "Combustion of PMMA, PE and PS in a Ramjet," *Combustion and Flame,* Vol. 79, No. 4, 1990, pp. 299–306.

[40]Boaz, L. D., and Netzer, D. W., "An Investigation of the Internal Ballistics of Solid Fuel Ramjets," Naval Postgraduate School Rept., NPS-57Nt73031A, Monterey, CA, March 1, 1973.

[41]Dijkstra, F., Korting, P. A. O. G., and van den Berg, R. P., "Ultrasonic Regression Rate Measurement in Solid Fuel Ramjets," AIAA Paper 90-1963, 1990.

[42]Elands, P. J. M., Dijkstra, F., and Zandbergen, B. T. C., "Validation of the Flow and Combustion Processes in a Solid Fuel Ramjet," AIAA Paper 91-1869, June 1991.

[43]Netzer, D. W., "Model Applications to Solid-Fuel Ramjet Combustion," *Journal of Spacecraft and Rockets,* Vol. 15, No. 5, 1978, pp. 263–264.

[44]Stevenson, C. A., and Netzer, D. W., "Primitive Variable Model Applications to Solid Fuel Ramjet Combustion," *Journal of Spacecraft and Rockets,* Vol. 18, No. 1, 1981, pp. 89–94.

[45]Milshtein, T., and Netzer, D. W., "Three-Dimensional, Primitive-Variable Model for Solid Fuel Ramjet Combustion," *Journal of Spacecraft and Rockets,* Vol. 23, No. 1, 1986, pp. 113–117.

[46]Elands, P. J. M., Korting, P. A. O. G., Dijkstra, F., and Wijchers, T., "Combustion of Polyethylene in a Solid Fuel Ramjet—A Comparison of Computational and Experimental Results," *Journal of Propulsion and Power,* Vol. 6, No. 6, 1990, pp. 732–739 (AIAA Paper 88-3043, July 1988).

[47]Richardson, J., deGroot, W. A., Jagoda, J. J., Walterick, R. E., Hubbartt, J. E., and Strahle, W. C., "Solid Fuel Ramjet Simulator Results: Experimental and Analysis in Cold Flow," *Journal of Propulsion and Power,* Vol. 1, No. 6, 1985, pp. 488–493.

[48]Vos, J. B., "The Calculation of Turbulent Reacting Flows with a Combustor Model Based on Finite Chemical Kinetics," Ph.D. Thesis, Delft Univ. of Technology, Delft, The Netherlands, April 1987.

[49]De Wilde, J. P., "Fuel Pyrolysis Effects on Hybrid Rocket and Solid Fuel Ramjet Combustor Performance," Ph.D. Thesis, Delft Univ. of Technology, Delft, The Netherlands, Feb. 1991.

[50]Jarymowycz, T., Yang, V., and Kuo, K., "A Numerical Study of Solid Fuel Combustion Under Supersonic Crossflows," AIAA Paper 90-2076, July 1990.

[51]Elands, P. J. M., Dijkstra, F., and Zandbergen, B. T. C., "Experimental and Computational Flammability Limits in a Solid Fuel Ramjet," AIAA Paper 90-1964, July 1990.

[52]Elands, P. J. M., "The Prediction of the Flow and Combustion in a Solid Fuel Combustion Chamber by Means of Two Combustion Models Based on the Diffusion Flame Concept," AIAA Paper 87-1702, July 1987.

[53]De Wilde, J. P., "A Fuel Pyrolysis Model for Combustion Calculations," AIAA Paper 91-1868, June 1991.

[54]Elands, P. J. M., "The Modelling of Heat and Mass Transfer Near Solid Boundaries and Comparison with Experimental Results," Delft Univ. of Technology Rept. LR-567, Delft, The Netherlands, Nov. 1988.

[55]Currie, T. C., and Atkinson, J. C., "Computational Fluid Dynamics for Solid-Fuel Ramjet Flowfield Analysis—A State of the Art Review," ECS-Power Systems, Inc., Canada, Feb. 1991.

[56]Anon., "Experimental and Analytical Methods for the Determination of Connected-Pipe Ramjet and Ducted Rocket Internal Performance," NATO/AGARD Rept. AGARD-AR-323, July 1994.

Chapter 14

High Mach Number Applications: Combined Cycles

Lawrence E. Fink*
The Boeing Company, Seattle, Washington 98124

Nomenclature

A = area
m = mass
M = Mach number
P = pressure

Subscript and Superscript

t = stagnation condition
$\cdot$ = rate

I. Introduction

A. Definition

COMBINED-CYCLE propulsion refers to the combination of different propulsion elements in a single engine with cycle-process interactions between elements. This is in contrast to missiles that employ different types of engines in combination (combination engines), such as BOMARC (Boeing Michigan Air Research Center) which used a rocket booster in the fuselage and pod mounted ramjet engines for sustained propulsion. Another example that contrasts with combined-cycle propulsion is the integral rocket ramjet engine that combines a solid rocket motor with a ramjet engine, sharing the same combustion chamber in sequence. In other words, in combined-cycle engines, engine cycles interact with one another. Sometimes the distinction between combined cycles and combination engines is obscure, such as when two physically separate and different engines share the same inlet. However, under the broad sense of the term, this is treated as a combined-cycle engine in this discussion because the airflow to one engine is affected by the operation of the other engine.

*Principal Engineer, Boeing Defense & Space Group.

The term composite propulsion has caught on to refer to combined-cycle engines. To add to the confusion of terms, composite engine is used more and more to describe engines that use the new metal-matrix and ceramic reinforced composite materials to lower cost or decrease weight of advanced technology engines. The term multicycle or combined cycle is better than composite to avoid confusion. Consequently, the term combined cycle will be used in this discussion.

B. Combined-Cycle Engine Features

The motivation to combine engine cycles is to take advantage of the superior engine performance (high thrust or specific impulse) offered in specialized flight regimes, or to broaden the regime over which acceptable performance is delivered by a propulsion system. For example, as shown in Fig. 1, different types of air-breathing engines offer superior specific impulse in different Mach number regimes. One could design a missile with separate engines to operate in each Mach number range, but a weight and volume savings is realized when common elements between the engines are used to as great an extent as possible. For example, a common inlet, combustor, or nozzle might be used. Another benefit is sometimes realized when two cycles are operated simultaneously. Better performance can be obtained from a synergistic effect than if each cycle were operated alone. In tactical missile applications, elimination of a separate engine, such as a booster, can be an overwhelming engine design driver to lower cost, weight, and volume, even at the expense of sustain propulsion specific impulse.

There are some relative liabilities associated with combined-cycle engines compared to combination engines which relate mainly to technical and programmatic risk. The engine and inlet technology making up combination propulsion systems exist, and the technical risk of using them in combination on a missile is relatively low. On the other hand, combined-cycle propulsion systems are basically new, and present the missile designer with a set of challenging technical uncertainties

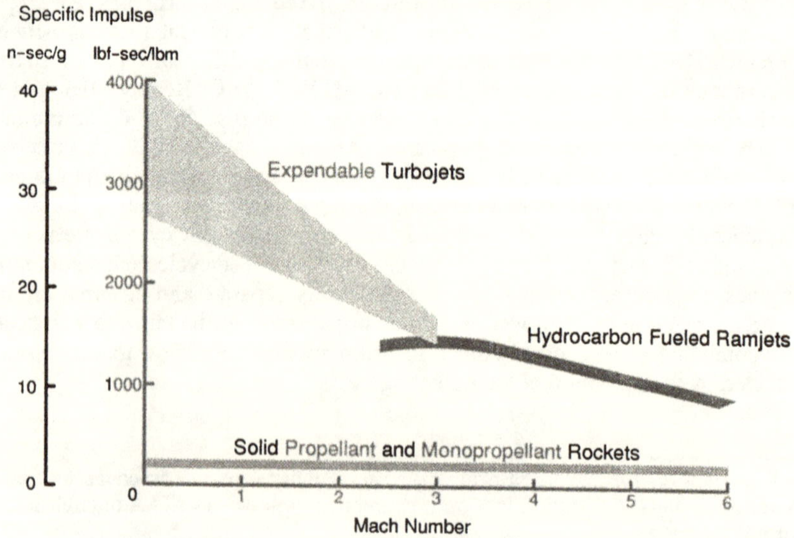

Fig. 1 Specific impulse trends over the tactical missile speed regime.

and associated developmental program risks. For example, can the different engine cycles be made to smoothly and reliably interact as flight conditions change and thrust is modulated during flight? Engineering development programs have demonstrated the operation and control of various types of combined-cycle engines. It is likely that these challenges will be met by propulsion companies and government agencies, and combined-cycle engines will be used more in future missiles or aircraft.

The weight savings offered by combined-cycle engines are especially important in hypersonic missiles (Mach 5+). An incremental increase in propulsion system weight has an amplified effect on takeoff gross weight which can be as high a factor as six for a Mach 5 vehicle which cruises at approximately 27 km (90,000 ft).

As implied by Fig. 1, the greater the spread of flight speed, the more important it is to use combined-cycle engines. Combined-cycle engines have received the greatest attention as propulsion for space launch vehicles and aircraft. To date, the expendable nature of tactical missiles and their shorter range have not required combined-cycle engines, and missile programs are reluctant to accept the higher risk of fielding new engine concepts. However, as the need for tactical missiles to counter faster, technologically advanced threats requires greater speed and range, combined-cycle engines offer a unique capability. Future mission studies may show that only combined-cycle engines can meet weapon system requirements.

The combination of basic engine types (turbine/compressor, ramjet, and rocket engines) in various ways, and in some cases adding other components such as heat exchangers, results in a number of possible combined-cycle engines. These range from obvious combinations, such as the turbojet engine and the ramjet being combined to form the turboramjet, to a unique class of engines cycles that employ heat exchangers using liquid hydrogen to liquify air and burn the two in a rocket engine.[1] However, combined-cycle engines using liquid hydrogen will not be described as they require special handling and are better suited outside the realm of tactical missiles for longer range and higher speeds.

The most-studied combined-cycle engine for tactical missiles has been the air turborocket (ATR). Other combined cycles have received more attention for space launch vehicles, usually using liquid hydrogen fuel for at least one of the cycles.[2] However, there is historical evidence for application of all the following engine cycles to tactical missiles using storable fuel.

II. Engine Cycles

State-of-the-art combined-cycle engines are composed of two or more of the elements listed in Fig. 2 in addition to the air inlet and nozzle. The challenge to the propulsion engineer is to combine these elements in a way that meets the application requirements while trading factors of cost, weight, thrust, size, and specific impulse. The more popular combinations of elements have acquired their own names. Some of these systems will be discussed and compared.

A. Turboramjet

One popular combined-cycle engine cycle is the turboramjet. It takes advantage of the high-specific impulse of the turbojet at the lower speeds, and operates as a ramjet at the higher speeds. It has been studied in various forms for over 40 years. The French flew a turboramjet powered aircraft (Le Griffon) in the mid-1950s which briefly held the world 100-km closed-course speed record and was

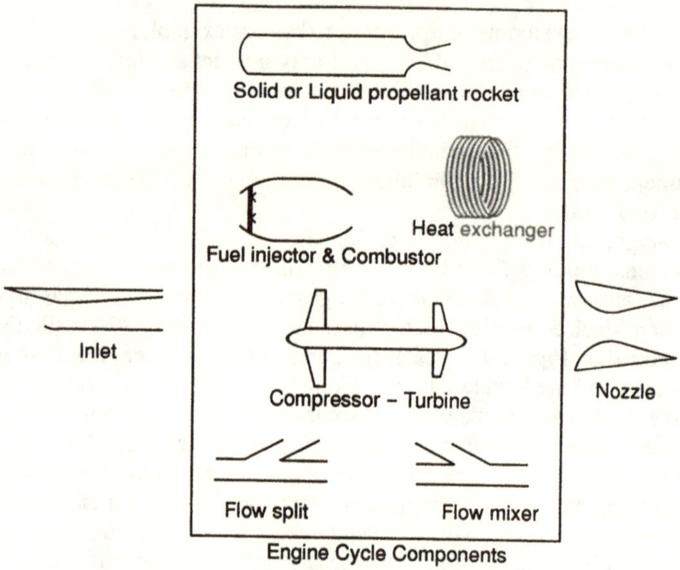

Fig. 2 Candidate composite propulsion elements.

able to accelerate in a climb at Mach 2.1 (Ref. 3). Improvements in materials and both turbojet engine and ramjet engine technology allow a new generation of turboramjet to be considered which has significantly greater potential for operating at higher speeds and with good fuel economy.

Usually the change from turbojet, or simultaneous turbojet/ramjet operation, to all-ramjet operation, occurs at about Mach 3 due to thrust requirements and temperature constraints. Sufficient nozzle pressure ratio is provided by inlet compression alone at these speeds, and further air compression only aggravates an already acute compressor discharge temperature or turbine inlet temperature problem.[4] In the first case, the engine speed (rotation rate) must be constrained to limit compressor pressure ratio (hence, the temperature due to thermodynamic relations). In the second case, fuel/air ratio must be constrained to limit the gas temperature entering the turbine stage. In either case, thrust is limited by these temperature constraints. As small engine technology progresses to allow higher turbine temperatures, the fuel/air ratio will approach stoichiometric proportions. An essential feature of turboramjet engines is a door or shut-off vanes to completely block air to the turbojet engine at the higher Mach numbers (greater than Mach 3) to protect the turbojet and to use all available air to produce thrust from the ramjet.

The liquid fuel ramjet can operate up to Mach 5, or to Mach 6 with more advanced technology. This flight velocity range can be achieved with liquid hydrocarbon fuels, thus providing safe handling, transportation, and storage features conducive to logistically supportable operational systems.

Generally, there are two types of turboramjets, parallel and tandem. In the parallel type, a flow splitter directs air from the inlet to either a turbojet or ramjet engine, or both, as depicted in Fig. 3. The turbojet may or may not have an afterburner and may or may not use a variable angle compressor or variable area turbine. However, the weight and complexity associated with these variable cycle components generally make them unsuitable for tactical missiles. The parallel

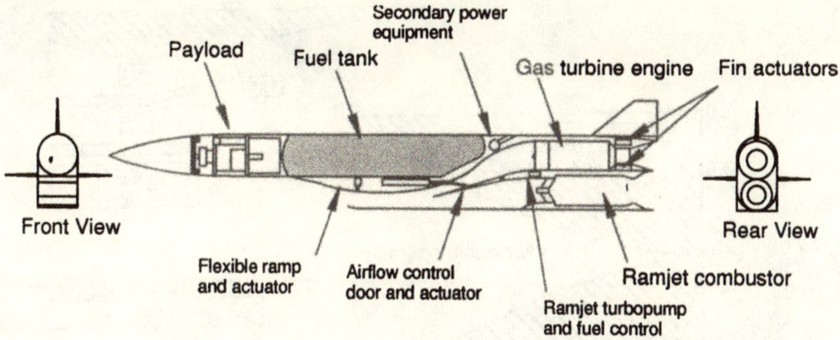

Fig. 3 Over/under turboramjet missile configuration.

type of turboramjet is also known as the over/under or wrap-around turboramjet depending on the configuration of the ramjet combustor. The over/under turboramjet, Fig. 3, has an advantage over the wrap-around in that it is easier to employ a flow splitter to divert flow to the ramjet and completely close off flow to the turbojet at high speed to protect the engine from high air temperatures. However, the wrap around, Fig. 4, is more compact and lighter weight.

If the flow split in an over/under turboramjet is made in supersonic flow, a key operating parameter is the inlet internal contraction ratio. On one hand, it is a liability by limiting how much flow can be diverted, as shown in Fig. 5. On the other hand, it can be an asset by increasing inlet total pressure recovery to the ramjet as speed increases, shown in Fig. 6. These limits and benefits of flow split are not applicable if the split occurs in subsonic flow.

Flow through the turbojet and ramjet engines in the over/under arrangement exhaust through separate nozzles, but the wrap-around engine can use either combined or separate annular nozzles. The combined nozzle is desirable for its potentially lower weight; however, it significantly restricts ramjet operating flexibility in simultaneous turbojet and ramjet operation which may be desirable for inlet/engine airflow matching.

The other type of turboramjet is the tandem turboramjet, shown in Fig. 7. It operates as an afterburning turbojet at low speed. At high speed (greater than Mach 3), flow to the turbojet core is shut off with doors, and the air flow is ducted into the afterburner section. It can be designed to incorporate an exhaust ejector so that it can operate in both modes at low- and medium-flight speeds. If so equipped,

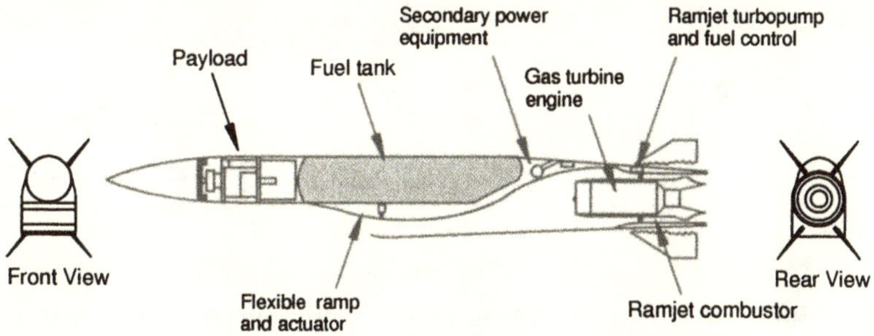

Fig. 4 Wrap-around turboramjet missile configuration.

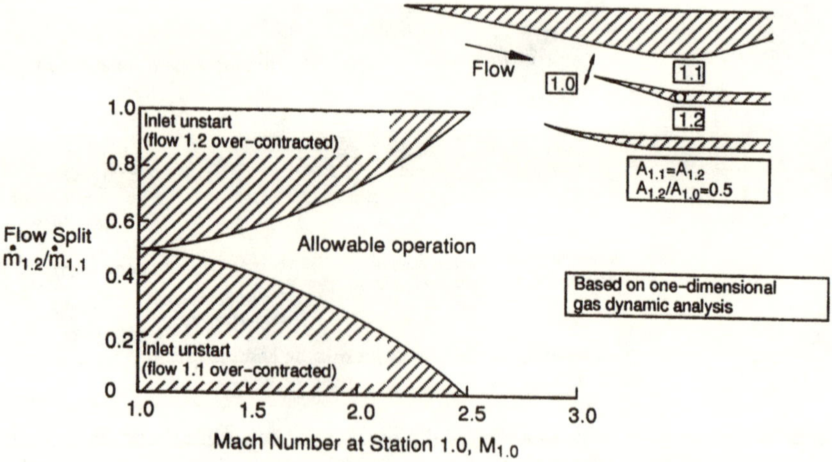

Fig. 5 Supersonic flow split limits; station number 1.0 is the cowl lip; 1.1 and 1.2 are the inlet throats for the respective flows.

most of the excess air supplied by the inlet (over what is demanded by the turbojet) can be ducted to the ejector/afterburner to avoid the drag penalty associated with dumping the bypass air overboard. Unlike the parallel type, which has independent ramburners, the tandem type has less combustor wall surface that must be cooled. The tandem type has the smallest cross-sectional area which implies a smaller fuselage cross-section requirement and lower drag. However, the length required for the afterburning section results in substantially greater engine length than in the wrap-around design.

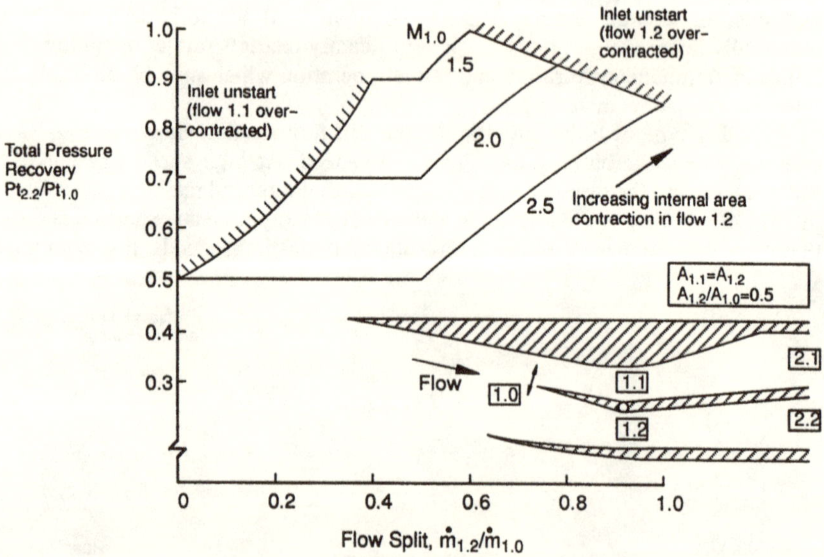

Fig. 6 Increased critical total pressure recovery with supersonic flow split; station number 1.0 is the cowl lip; 1.1 and 1.2 are the inlet throats for the respective flow; 2.1 and 2.2 are the diffuser exits for the respective flows.

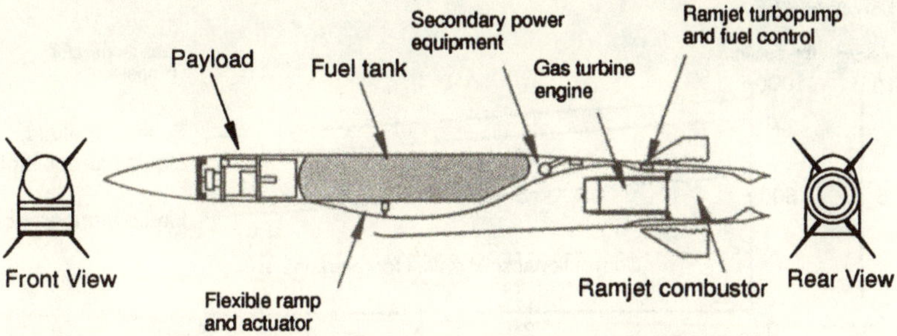

Fig. 7 Tandem turboramjet missile configuration.

A design variation that can be applied to either parallel or tandem turboramjet designs is a through-flow concept, where the air to the ramjet passes through part of the gas turbine compressor. The advantage of the concept is lower engine weight and slightly lower (better) specific fuel consumption at transonic Mach numbers but at the expense of higher (worse) specific fuel consumption at cruise Mach numbers and a larger inlet than the split-flow concepts shown in Figs. 3, 4, and 7.

B. Air Turborocket

The air turborocket is an air-breathing engine cycle similar to a turbojet. However, the compressor is driven by a gas generator exhausting through a turbine, Fig. 8. In a sense, it is actually a combination of turbojet, ramjet, and rocket. The ATR has also been known as the air turboramjet and supercharged ramjet. However, to avoid confusion with the turboramjet, the air-turborocket name is usually preferred.

After mixing with the air discharge from the compressor, the fuel-rich gas generator exhaust burns in a combustor downstream of the turbine. The gas generator, a fuel-rich rocket engine, can use either liquid monopropellants pumped through a catalyst bed or bipropellants. Alternatively, a fuel-rich solid propellant gas generator grain is preferred for some tactical missiles because of its logistic advantages

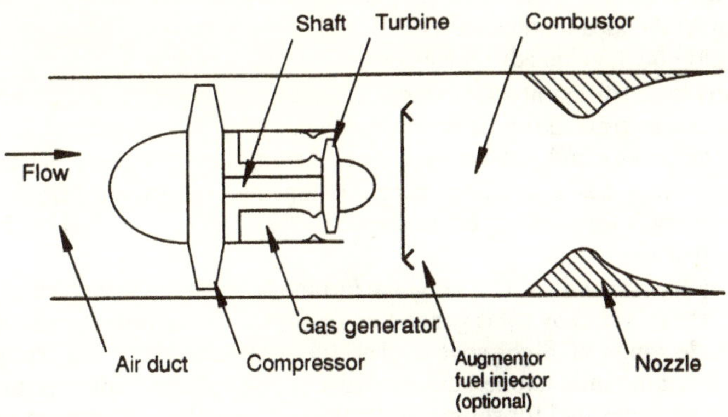

Fig. 8 ATR engine concept.

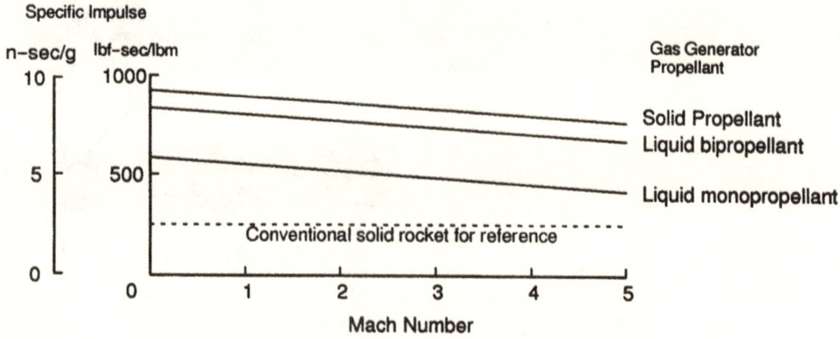

Fig. 9 Representative ATR specific impulse.

over liquid propellants but with the attendant problems of either erosion or deposition on the turbine vanes. The solid propellant gas generator flow can be throttled by a hot-gas valve.[5]

Figure 9 shows some typical performance values for ATRs with different fuels. OTTO II (propylene glycol dinitrate) torpedo liquid monopropellant, although mildly toxic, is relatively safe and easy to handle and has been an often-used propellant for the ATR, but its specific impulse is relatively low. Up to 20% of a kerosene type fuel can be mixed with the OTTO II fuel to increase the density, to increase the gravemetric heating value, and to lower the temperature of the gas generator exhaust to increase turbine life. Alternatively, a hydrocarbon fuel such as kerosene, with a higher heating value than a monopropellant, can be discharged into the combustor (mixing with the gas generator exhaust) to operate as an afterburner.

The history of the ATR can be traced back to the Aerojet patent by House filed in 1950 and granted in 1963.[6] In a related concept, Goddard was reported as working on a "turbine rocket plane" as early as 1932, in which a rocket-driven turbine was used to drive two propellers.[2] In the early 1950s, a 16.700-N (3750-lb-f) thrust, 66-cm- (26-in.) diam demonstration ATR using ethylene oxide fuel was demonstrated. Experiments were performed with secondary injection of JP-4. In 1956, the U.S. Navy (with Texaco Experiment, Inc.) demonstrated a 9900-N (2000-lb-f) thrust propyne fueled, monopropellant engine. In the early 1980s an ATR, using Otto II and methanol propellants with 5300-N (1200-lb-f) thrust, was studied with the idea it could be used in a U.S. Navy drone.[7]

To exploit the logistic advantages of solid propellant over liquid propellant, the U.S. Army tested different ATR designs using a solid propellant gas generator.[5,8] A U.S. patent was granted for the concept shown in Fig. 10. An unusual feature was that the compressor and turbine were ejected at high speed to make a transition to ramjet operation. Later studies predicted that a solid propellant ATR would have approximately 3 times the specific impulse of a comparable solid rocket motor and 2.5 times the range.

The key feature of the ATR is that the turbine is not in the main airflow path but is separately powered by a gas generator. This gives the ATR the ability to operate over a wide range of flight speeds. The fuel-rich exhaust of the gas generator keeps the turbine inlet temperatures relatively low and constant over the throttle range and provides fuel for subsequent combustion in the downstream ramburner section. A gas temperature of 816°C (1500°F) for hydrazine monopropellant is

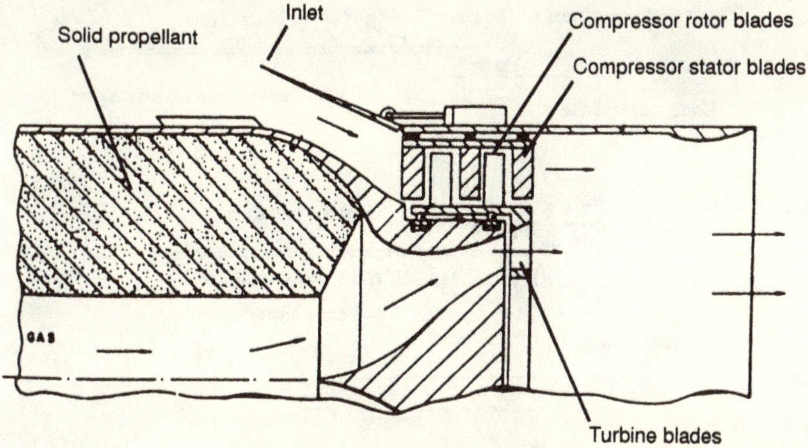

Fig. 10 Solid propellant ATR.

typical, 1100°C (2000°F) for a solid propellant gas generator. Unlike the turbojet, the turbine inlet temperature is not a function of flight speed. Consequently, for comparable technology and materials, it can operate at high Mach numbers where a turbojet could not operate due to high air stagnation temperature. Unlike the ramjet, the ATR is capable of producing sufficient thrust for subsonic (as low as Mach 0.3) through supersonic (Mach 3+) flight speeds. A rocket booster would still be an option for a ground-launched missile to eliminate the need for a ground-roll or launching rail takeoff but would be unnecessary for an air launch.

Although Fig. 8 depicts an ATR with axial compressor and turbine, centrifugal compressor and radial inflow turbine designs are applicable.[5] A compact design that leads to low cost is combination of the compressor and turbine into one part. Figure 11 illustrates two monorotor concepts that incorporate this feature. An engine integration option especially suitable for a podded engine, and implied by the top diagram, Fig. 11a, is the location of the gas generator external to the engine with a hot-gas duct to the turbine to shorten the engine. In the tip-driven compressor concept, the turbine is part of the compressor wheel, consisting of a ring shrouding the compressor blade tips onto which are machined short-span turbine blades. One advantage of the tip-turbine concept is that the turbine blade speed is high by virtue of its large radius from the axis of rotation which eliminates the need of a turbine-to-compressor speed-reducing gearbox, but stresses are high.

Another advantage of the ATR is high thrust-to-weight ratio combined with moderately high-specific impulse and a simpler engine than a turbojet. Particularly on accelerating missiles, high thrust-to-weight ratio is of paramount importance, with specific impulse of secondary importance. The cost and specific impulse of the ATR is between simple turbojets and ramjets (not considering the cost of a rocket booster however, which is required for a ramjet). The isolation of the gas generator from the airflow path allows primary combustion to occur at a pressure independent of altitude. The compression provided by the fan and hot-fuel gases that mix with the air allow the ramburner to sustain combustion at higher altitudes than the ramjet.

Like turbojets, the compressor determines the amount of airflow through the engine rather than the inlet as with a ramjet. Consequently one must be concerned

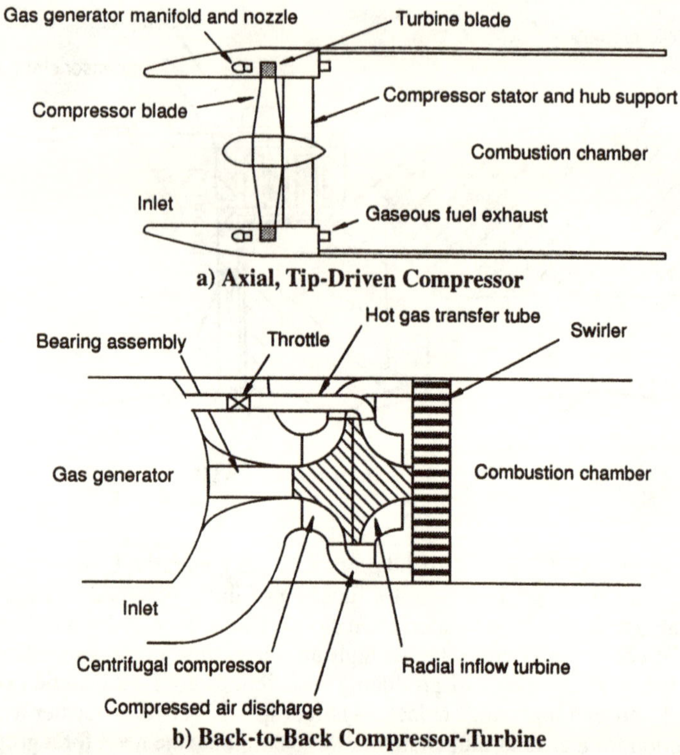

Fig. 11 ATR monorotor concepts.

with inlet and engine airflow matching throughout the flight regime and throttle range.

At high speeds (Mach 3+), the need to limit compressor rotation speed to prevent structural failure at high temperatures and to prevent combustor fuel/air ratio from exceeding stoichiometric due to the high compressor power requirement causes the compressor pressure ratio to approach unity. When this happens, the mechanical compressor is ineffective and the engine operates essentially as a ramjet. The specific impulse of the ramjet-mode ATR is less than that of a ramjet because the heating value of the gas generator propellant (monopropellant, bipropellant, or solid) is at best approximately one-half that of a hydrocarbon (kerosene) ramjet fuel with air. ATR-ramjet combinations similar to the arrangement of turbojet and ramjet to form different turboramjet systems could be used.

A helpful concept in analyzing the capability of an ATR design to accelerate a missile through the transonic regime is air specific impulse. Typically, a value of air specific impulse of 1.5 N-s/g (150 lb-f-s/lbm) is required. Depending on the propellant, this implies a compressor pressure ratio of at least 3:1 is required. In missile applications, this also requires that the thrust at Mach 1.5 be approximately 2.5 times the thrust at Mach 0.5.

If the inlet size is too small, thrust will be limited by the air delivered to the engine. If too large, the excess air bypassed overboard will cause a drag penalty. The additive and boundary-layer bleed drag will also be higher.

A variable geometry inlet and nozzle significantly improve the performance of a propulsion system using an ATR that is required to operate over a wide range of flight Mach numbers. However, the complexity and cost of variable geometry components for tactical missiles is not attractive. Consequently, it is not surprising that other propulsion systems using a combination of engines, each designed for a particular flight Mach number range, can sometimes out-perform an ATR on the basis of lower fuel consumption. The choice is a tradeoff of weight, size, fuel consumption, and costs associated with development and production of several engines rather than a single engine.

In application studies with launch at low altitude followed by climb, acceleration, and cruise at high altitude, the ATR seems to perform well while climbing at constant dynamic pressure until the cruise Mach number is reached. The Mach number is maintained to the cruise altitude. This has the additional advantage of providing a design dynamic pressure for the inlet and structure.

Maximum thrust is limited by an rpm structural limit and by lack of air if the inlet does not supply a sufficient quantity. Minimum thrust is limited by a maximum allowable bypass flow rate determined by bypass duct or nozzle size (usually up to around 20% of captured airflow).

C. Ejector Ramjet

The ejector ramjet (also called shrouded rocket) usually refers to an air-breathing engine with a rocket situated at the head end of the combustion chamber. Its supersonic exhaust has a momentum high enough that it increases the cycle pressure ratio and overall engine performance over that obtainable in the conventional ramjet. In contrast to the ducted rocket, the rocket in the ejector ramjet has a near stoichiometric exhaust rather than a fuel-rich exhaust as in a ducted rocket. Its stoichiometric operation precludes excess fuel combustion with the air during mixing and diffusion. Fuel energy is later added to the mixed flow following full diffusion by means of an afterburner type combustor. This maximizes cycle performance, since heat is added at the highest static pressure and lowest subsonic Mach number, minimizing Fanno flow heat-addition total pressure loss.[1] To minimize mixing length, multiple rocket nozzles can be used, as illustrated by Fig. 12.

The primary benefit of the ejector ramjet over the ducted rocket is its ability to produce significant thrust at low speed. In fact, sufficient thrust can be produced

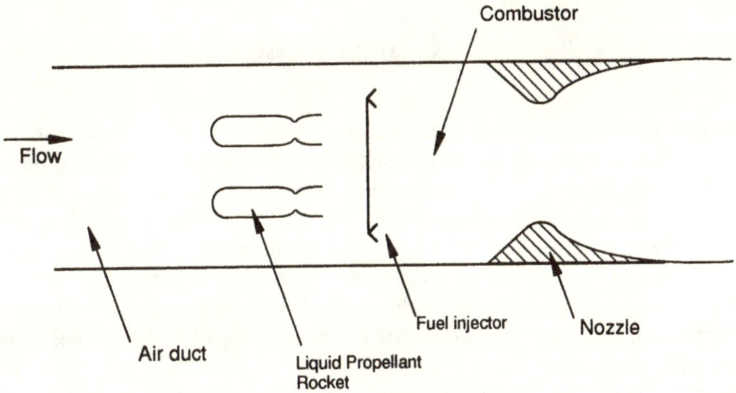

Fig. 12 Ejector ramjet concept.

to allow operation at subsonic flight velocity but usually not enough thrust to eliminate the need for a separate rocket booster for ground-launched applications.

At subsonic flight speed, the rocket primary flow acts as an ejector and draws additional air into the engine. It is the momentum of this primary jet that increases the cycle total pressure ratio over that of a conventional ramjet. At higher speeds, the rocket can be turned off and the engine can function as a ramjet. If the rocket is restartable, the rocket can provide additional thrust at supersonic speeds by increasing total pressure in the flow stream through jet compression. This enables higher ramjet fuel–air ratios and avoids the need for a variable nozzle throat area to maintain a specified inlet supercritical margin.

This is a classic example of synergism offered by two different propulsion cycles operating simultaneously. The ejector acts as a jet compressor. It resembles conventional ejectors as used in industrial applications, such as in steam locomotives where steam exhaust nozzles are precisely located in the smokebox to interact with the smokestack to draw air through the firebox. Similarly, steam ejector systems are used to reduce the nozzle exit back pressure to simulate high-altitude operation in rocket and ramjet engine test facilities.

Compared to a mechanical compressor turbine, ejectors require longer length and are not as efficient. On the other hand, they are lightweight and tolerant of flow distortion and ingested foreign objects.

The ejector ramjet term is also sometimes used to describe engines using vaporized fuel in the ejector with the design objective of providing an ejector effect.[9] This is different from a ducted rocket in that the ejector primary flow is 100% fuel vapor and no rocket oxidizer is involved. This discussion will reserve the term vapor-ejector ramjet for this engine, and rocket-ejector ramjet for engine using a near-stoichiometric rocket.

In the vapor-ejector ramjet, the fuel is vaporized by some of the heat of combustion and expanded through the primary nozzle.[8] It is important to use a fuel that does not coke during vaporization. Generally, a pure hydrocarbon fuel like propane, rather than a cracked hydrocarbon fuel like JP-4 or gasoline, can be used. A synthetic hydrocarbon fuel might be used and offer the benefit of higher fuel density compared to propane. Gluharref[10] patented an engine using this principle (and acoustic waves) with propane fuel and has offered plans for a home-built jet engine he calls a pressure jet. Its combustor and long tailpipe resembles a pulsejet but no valves are used.[10–12]

III. Component Design

A. Inlets

Efficient propulsion system performance requires that the inlet and engine airflow be closely matched. However, on typical climbing trajectory profiles using turbojet or ATR engines, the change in inlet airflow usually decreases more rapidly than the airflow demanded by the engine. This results in excessive airflow at lower Mach numbers, or part throttle at higher Mach numbers if the inlet is sized for the higher Mach number cruise condition. The excess flow must be "bypassed" around the engine. The effects of angle of attack may also need to be considered as two-dimensional inlets, or axisymmetric inlets in a wing or body flowfield increase airflow with angle of attack.

A reasonable approach to sizing the inlet and engine is to size the inlet to provide enough airflow for the engine at the cruise condition. In a turboramjet, the ramjet

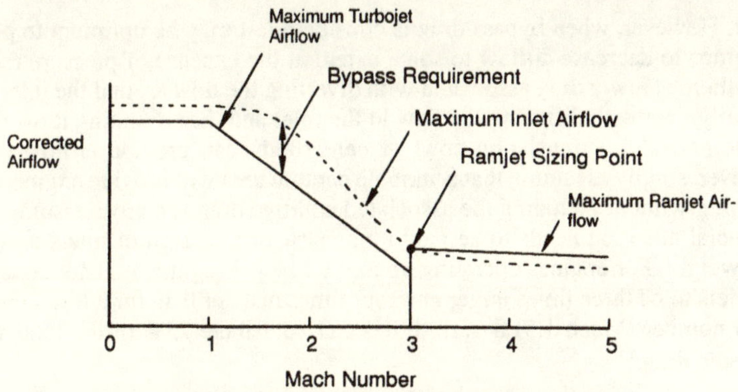

Fig. 13 Engine and inlet airflow matching.

is then sized to accommodate the captured flow at the turbojet/ramjet transition point, Fig. 13, and the turbojet is sized for thrust at the transonic pinch-point, Fig. 14.

The excess air flow can be handled by external inlet spillage, diverted overboard, ducted to an ejector nozzle, or diverted to a ramjet engine. Diverting flow to a ramjet engine or to an ejector nozzle results in the lowest drag, particularly if it is positioned to fill in separated flow regions in the nozzle and, thus, decrease nozzle drag. Bypassing flow to the ramjet lowers the base drag associated with the otherwise unused ramjet nozzle and can provide increased thrust when fuel is added and burned (above approximately Mach 1.1). Bypass to an ejector nozzle can be impractical if the mass flow is large because of the space required for the ducts. For missiles operating between Mach 1 and Mach 2 a reasonable operating scheme is to design the system to operate with a subcritical inlet (external spillage). Since the engine then takes as much airflow as needed, the excess is spilled around the inlet cowl, eliminating the need for bypass doors in the inlet diffuser. Of course, the disadvantage is the drag penalty and lower pressure recovery, but the penalty is not too great below Mach 2.

The performance of variable geometry inlets (movable ramps) are usually based on a ramp angle schedule that maximizes inlet total pressure recovery, hence,

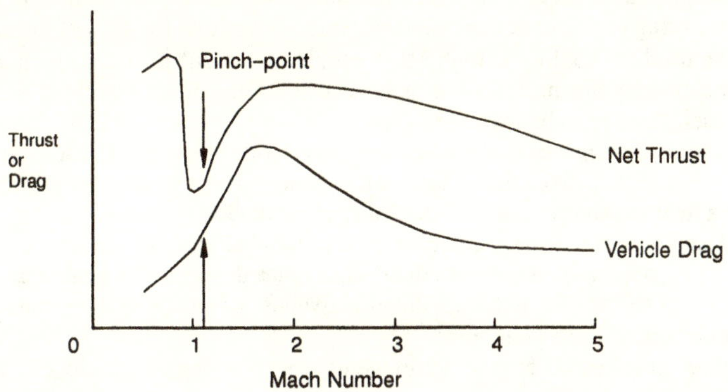

Fig. 14 Turbojet sized for thrust margin over drag at the pinch point.

thrust. However, when bypass drag is considered, it may be optimum to position the ramps to decrease airflow to some extent at the expense of pressure recovery since there is lower drag associated with diverting the flow around the inlet supersonically than with diffusing the flow in the inlet and then dumping it overboard.

Alternatively, a translating cowl or centerbody can provide airflow control. However, simply assuming that a variable capture area will provide engine/airflow matching without including the associated additive drag is a gross assumption.

Special attention needs to be paid to the structural design of inlets to manage their weight. For engines operating up to the Mach 4 regime, it is not unusual for the inlets to be three times larger and four times heavier than for a low supersonic Mach number (Mach 1.5) design. Inlets can contribute 25% of the total vehicle transonic drag.[13]

B. Engine and Nozzle Cooling

Maintenance of structural integrity for high-speed engines (Mach 4+) is a substantial challenge. Air cooling of the engine and airframe can be used below moderate supersonic Mach numbers (Mach 2). However, as flight speed increases, fundamental air thermodynamics causes the air to be at such a high temperature that it is unsuitable for cooling. At high speed, fuel is the logical choice to cool leading edges, engine, and/or nozzle. The cooling challenge is especially difficult using conventional liquid hydrocarbon fuels because of their inherently limited heat sink capability. One approach to the cooling problem with hydrocarbon fuels is to use endothermic fuels [such as methylcyclohexane (MCH)] which decompose during the process of heat addition, thereby taking on substantially more heat energy than conventional hydrocarbon fuels heated to the same temperature. This decomposition occurs at approximately 370°C (700°F) for MCH in the presence of a catalyst. However, at too high surface temperatures, undesirable chemical reactions may occur. These reactions occur at temperatures greater than approximately 480°C (900°F) for MCH, and yield insoluble products, coke, which build up on the cooling passage walls and reduce cooling effectiveness. Figure 15 illustrates the properties of some candidate endothermic fuels.

Most of the heat load is from the engine, but other components require cooling as shown in Fig. 16. The catalytic reaction of an endothermic fuel can absorb over one-half the total heat load in this example.

When air is used for cooling, it is important to account for the momentum loss in the thrust/drag performance calculations. It is not unusual for 15% of the captured air to be used for cooling at high Mach numbers.[13] A substantial reduction in the required cooling air mass flow is realized when it is cooled by a heat exchanger using fuel, thereby reducing drag.

One study determined that the best way to provide cooling is to use a secondary fluid to transfer heat from the fuel to cooling panels. The argument for this approach is that a heat exchanger is a very design specific device. The changes due to flight profile and fuel flow rate cannot be accommodated by a single heat exchanger design. Consequently, an intermediate fluid is used to transfer heat from the hot components to the fuel, although the effectiveness of such a system must include its weight and volume penalties.[14]

Higher temperature engine components improve engine efficiency, decrease size or increase thrust for a given size, allow operation at greater speeds, and simplify propulsion system design when cooling features are reduced or eliminated.

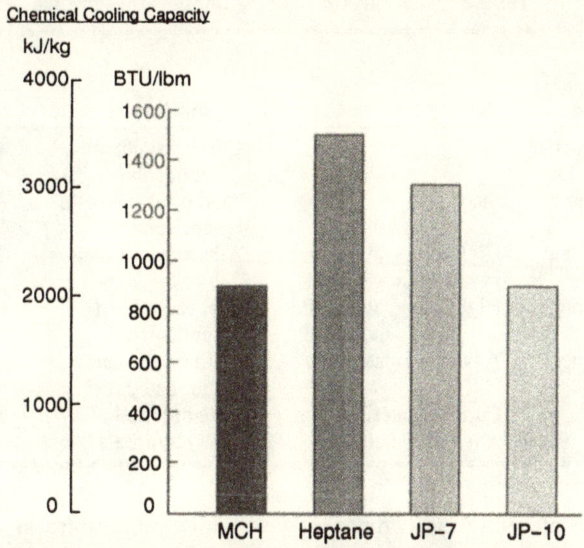

Fig. 15 Chemical heat sink capacity of some endothermic fuels (from Ref. 7).

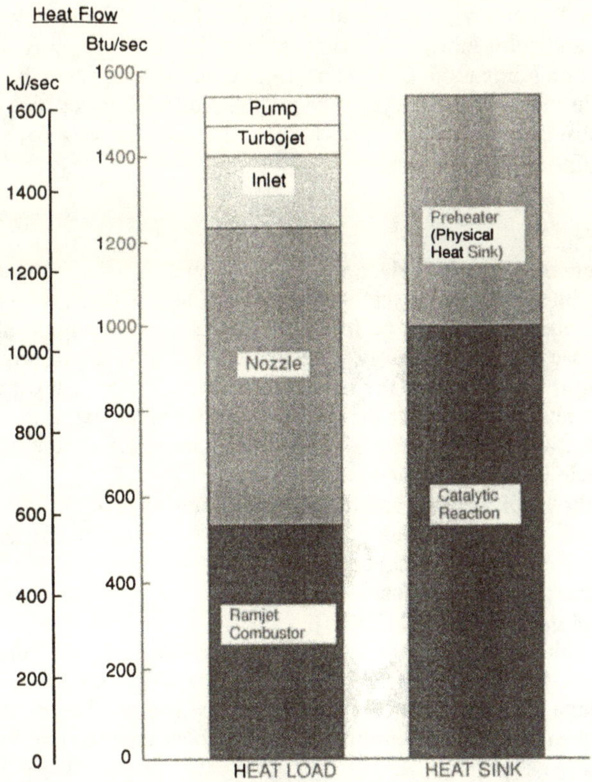

Fig. 16 Heat load and available heat sink for turboramjet at Mach 5 (from Ref. 8).

Table 1 Gas turbine bearing temperature limits

Bearing type	Bearing material	Lubrication	Maximum temperature °F	Maximum temperature °C
Rolling element	Steel	Current synthetic lubricants	450	230
Rolling element	Steel	Advanced synthetic lubricants	600	320
Rolling element	High-temperature alloy	Advanced synthetic lubricants	750	400
Rolling element	High-temperature alloy	Fuel cooled and lubricated	1000	540
Rolling element	Advanced materials	Solid or powder lubricated	1500	820
Gas foil	Current materials	none required	1300	700
Gas foil	Advanced materials	none required	2000	1100

Bearings are presently the component that limits temperature in gas turbines. Table 1 lists maximum temperature limits for different bearing types, materials, and lubricants. As shown, there exists potential for significant temperature increase over the more common steel/current synthetic lubricant rolling element bearing combination. Relatively short-life missile engines are particularly well suited for fuel-cooled and solid-lubricated bearings, in which the bearing retainer is worn away to provide lubrication. Gas foil bearings have demonstrated the potential for use in missile engines but have yet to be fully exploited. Besides high-temperature capability, foil bearings offer an advantage in higher operating speed and longer life than rolling element bearings.[15]

C. Nozzles

A key element in nozzle selection for high-speed missile applications is nozzle/airframe integration. An excellent performing nozzle (high-thrust coefficient) may not provide good system performance if integrated poorly, resulting in high boattail or base drags. Conversely, a well-integrated nozzle will provide the best overall design, usually when the nozzle is designed to be overexpanded at the lower Mach numbers. Finding the best overall nozzle/airframe integration concept requires trades between nozzle types, dimensions, shape, cost, and impact on overall vehicle weight, drag, and performance.

The net thrust of an air-breathing engine operating at high speed is the difference between two quantities of large magnitude, gross thrust and ram drag. Consequently, engine performance is more sensitive to nozzle efficiency as flight velocity increases. It is not unusual for just 1% change in nozzle thrust coefficient to change net thrust by 8% (Ref. 13).

Like the inlet, design of the nozzle must be carefully considered in high-speed engines. Higher speeds result in greater nozzle pressure ratios, and higher expansion area ratios are required for thrust to be greater than drag. The nozzle for a Mach 4 engine can be four times as large and as heavy as one for a Mach 1.5 engine.[13]

Nozzles for tactical missiles must be low cost, implying simplicity, and must operate at high temperatures. Figure 17 compares the characteristics of five nozzle

Shape	Advantages	Disadvantages
#1 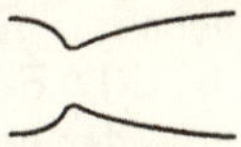 Axisymmetric Convergent-Divergent	• Low risk • Low weight	• Limited capability for are variation
#2 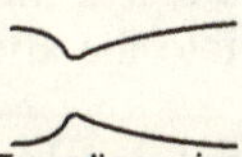 Two-dimensional Convergent-Divergent	• Easier to incorporate variable area capability than #1 • Hot parts shielding	• Highest weight
#3  Plug (axisymmetric or 2-D)	• Shortest length • Good operation at off-design pressure ratio • Variable area capability • Hot parts shielding	• Plug and plug support cooling
#4 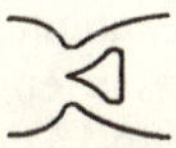 Expansion-Deflection	• Short length • Good operation at off-design pressure ratio • Hot parts shielding	• Centerbody cooling
#5 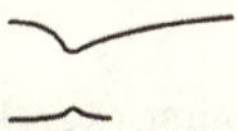 Two-dimensional Single Expansion Ramp (SERN)	• Lower weight and smaller size than #2 • Large expansion ratio by using fuselage as nozzle surface	• Slightly lower performance than #1 or #2 • Thrust vector changes with pressure ratio
#6 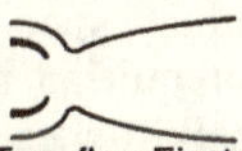 Two-flow Ejector (axisymmetric or 2-D)	• Well-suited to engine cycles with two flowpaths	• May require weight and expense of variable geometry if operated over wide range of Mach numbers and altitudes

Fig. 17 Nozzle concept comparison.

types. The simplest designs are convergent-divergent (C-D) types, but they are also the longest. The expansion-deflection (E-D) and plug designs may require cooling, however, they have the shortest length. These two designs, as well as the design flexibility of the two-dimensional C-D design, can be used to shield hot engine parts from infrared seekers. The single expansion ramp (SERN) has good integration characteristics when the propulsion flow path is offset from the vehicle centerline and a large exit area is needed, but it is longer and has a thrust vector bias unless compensated with variable geometry surfaces. Length is often a limiting design constraint for missiles, requiring a compromise in nozzle design for optimum system performance.

Some nozzles are better suited to some engines than others. The three parallel flow engines require different nozzle installation concepts. Simultaneous operation of the turbojet and ramjet engines may require separate, dual nozzle systems. Dual nozzles maintain good flow control, maximum performance, and short length. Dual nozzles allow separate nozzle control over both turbojet and ramjet propulsion systems, which enables better optimization of the nozzle performance. The ejector nozzle, shown at the bottom of Fig. 17, combines the flow from parallel flow paths through a single nozzle. A variable throat area, however, may be necessary to accommodate a wide range of flight conditions and throttle settings.

The wrap-around turboramjet could use the combination of concentric, dual nozzles or a single ejector nozzle. Dual nozzle combinations would be C-D, E-D, or plug for the turbojet and C-D nozzles for the ramjet. The over/under concept could use of different types of combination nozzle concepts, particularly the two-dimensional types. The tandem turboramjet engine is best suited to more conventional single nozzle designs since all of the flow is ducted into the afterburner/ramjet.

A shorter, potentially lower weight alternative to two-dimensional nozzles is the use of multiple, side-by-side axisymmetric nozzles. However, the nozzle integration with the vehicle must be carefully considered to avoid a drag penalty associated with the area in the cusps between nozzles.

Nozzle performance parameters such as length, divergence angle, and weight affect overall performance. It is usually advantageous to design the divergence angle a little larger than optimum since little penalty in thrust coefficient occurs compared to the benefits of decreased nozzle weight and length. For example, if peak nozzle performance occurs at a divergence angle of 10 deg, a savings in length and weight of approximately 33% can be made by increasing the angle to 15 deg.

IV. Applications

The objective of combined-cycle engines is to broaden the flight regime over that obtained by a single engine cycle alone. A requirement to operate at both subsonic and supersonic speeds is enough of a requirement to at least require consideration of a combined-cycle engine. Low weight, small size, high thrust, and low specific fuel consumption are the mutually exclusive desired parameters, requiring the missile designer to tradeoff one characteristic for another. The smallest size, or highest thrust for a given size, is offered by the ejector rocket, with the penalty of the greatest specific fuel consumption. The turboramjet has the largest size and lowest specific fuel consumption, with the ATR between the two. The vapor-ejector ramjet has been studied for subsonic applications, having the lowest thrust and theoretically, a moderate specific fuel consumption. However, some experimental results have not demonstrated performance close to its theoretical potential.[12] At high speed (Mach 2+), however, it could perform like a conventional liquid fuel ramjet.

If the mission requirement is only acceleration from subsonic to a supersonic cruise, the conventional integral rocket/ramjet (liquid fuel, solid fuel, or ducted rocket) may be the best choice for tactical missiles. However, one of the combined-cycle engines would be attractive for missions that require a subsonic end game after supersonic cruise because of munitions constraints (dispensing submunitions), guidance requirements (acquire and track), or a requirement to loiter. An ejector rocket or ATR could be used to advantage in an accelerating mission with limited or no cruise, and it could be better in some applications than an integral rocket/ramjet because no rocket booster may be needed (or at least only a small one for ground launch). ATRs also have the greatest altitude capability of these combined-cycle engines because the gas generator-driven compressor enables higher combustor pressure than a ramjet or turbojet.

Turboramjets would be the best for long-range missions with speeds from subsonic to Mach 4–6 in cruise. ATRs would be best for range shorter than suitable for turboramjets but longer than that might be provided by a conventional solid rocket motor. One study concluded that, in general, a solid propellant ATR will provide significant mission or vehicle benefits over conventional solid propellant rocket motors when the mission range is at least 16 km[6] (10 miles). The large Mach number operating range, and high-thrust capability of an ATR would be useful with a standoff range requirement where a subsonic cruise gives good fuel economy for most of the mission, and a fast accelerating, supersonic dive or dash through defenses at the target helps survivability.

References

[1]Escher, W. J. D., Teeter, R. R., and Rice, E. E., "Airbreathing and Rocket Propulsion Synergism: Enabling Measures for Tomorrow's Orbital Transports," AIAA Paper 86-1680, June 1986.

[2]Curran, E. T., "Introduction," *High-Speed Flight Propulsion Systems,* edited by S. N. B. Murthy and E. T. Curran, Vol. 137, Progress in Astronautics and Aeronautics, American Institute of Aeronautics and Astronautics, Washington, DC, 1991, pp. 1–20.

[3]Dugger, G. L., *Ramjets,* Selected Reprints, Vol. 6, American Institute of Aeronautics and Astronautics, New York, June 1969, pp. 89–101.

[4]Blevins, G., Hartsel, J., and Powel, T., "Variable Cycle Concepts for High Mach Applications," AIAA Paper 87-2103, June 1987.

[5]Thomas, M. E., and Christensen, K., "Air-Turbo-Ramjet Propulsion for Tactical Missiles," AIAA Paper 94-2719, June 1994.

[6]House, W. C., "Gas Generator Turbojet Motor," U.S. Patent 3,110,153, filed Sept. 5, 1950, issued Nov. 12, 1963.

[7]Briggs, M. M., Campbell, J. V., Andrus, S. R., and Burgner, G. R., "Synthesis and Performance of an Air-Turboramjet-Propelled Supersonic Target Vehicle," AIAA Paper 84-0075, Jan. 1984.

[8]Kesting, L., "Solid Propellant Air Turbo Rocket," U.S. Patent 4,096,803, filed Dec. 27, 1976, issued June 27, 1978.

[9]Supp, W. E., Watson, K. A., and Miller, J. H., "Subsonic Performance Potential of Ramjets and Ejector Ramjets," AFAPL-TR-72-7, Air Force Aero Propulsion Lab., May 1972.

[10]Gluhareff, E. M., "Valveless Jet Engine with Inertia Tube," U.S. Patent 3,093,962, June 18, 1963.

[11]Thoms, W., "A Jet You can Build in Your Own Shop," *Mechanix Illustrated,* Jan. 1975.

[12]Oldson, J., "The Gluhareff Pressure Jet," AIAA Paper 94-2825, June 1994.

[13]Rhode, J., "Airbreathing Combined Cycle Engine Systems," *Rocket-Based Combined-Cycle (RBCC) Propulsion Technology Workshop,* NASA Conf. Pub. 10090, March 1992.

[14]Petley, D., and Jones, S., "Thermal Management for a Mach 5 Cruise Aircraft Using Endothermic Fuel," AIAA Paper 87-2103, Sept. 1990.

[15]O'Connor, L., "Fluid-Film Foil Bearings Control Engine Heat," *Mechanical Engineering,* May 1993, pp. 72–75.

[16]Mechanical Technology Inc., Lathan, New York.

Author Index

PROGRESS IN ASTRONAUTICS AND AERONAUTICS SERIES VOLUMES

***1. Solid Propellant Rocket Research (1960)**
Martin Summerfield
Princeton University

***2. Liquid Rockets and Propellants (1960)**
Loren E. Bollinger
Ohio State University
Martin Goldsmith
The Rand Corp.
Alexis W. Lemmon Jr.
Battelle Memorial Institute

***3. Energy Conversion for Space Power (1961)**
Nathan W. Snyder
Institute for Defense Analyses

***4. Space Power Systems (1961)**
Nathan W. Snyder
Institute for Defense Analyses

***5. Electrostatic Propulsion (1961)**
David B. Langmuir
Space Technology Laboratories, Inc.
Ernst Stuhlinger
NASA George C. Marshall Space Flight Center
J. M. Sellen Jr.
Space Technology Laboratories, Inc.

***6. Detonation and Two-Phase Flow (1962)**
S. S. Penner
California Institute of Technology
F. A. Williams
Harvard University

***7. Hypersonic Flow Research (1962)**
Frederick R. Riddell
AVCO Corp.

***8. Guidance and Control (1962)**
Robert E. Roberson
Consultant
James S. Farrior
Lockheed Missiles and Space Co.

***9. Electric Propulsion Development (1963)**
Ernst Stuhlinger
NASA George C. Marshall Space Flight Center

***10. Technology of Lunar Exploration (1963)**
Clifford I. Cumming
Harold R. Lawrence
Jet Propulsion Laboratory

***11. Power Systems for Space Flight (1963)**
Morris A. Zipkin
Russell N. Edwards
General Electric Co.

***12. Ionization in High-Temperature Gases (1963)**
Kurt E. Shuler, Editor
National Bureau of Standards
John B. Fenn
Associate Editor
Princeton University

***13. Guidance and Control–II (1964)**
Robert C. Langford
General Precision Inc.
Charles J. Mundo
Institute of Naval Studies

***14. Celestial Mechanics and Astrodynamics (1964)**
Victor G. Szebehely
Yale University Observatory

***15. Heterogeneous Combustion (1964)**
Hans G. Wolfhard
Institute for Defense Analyses
Irvin Glassman
Princeton University
Leon Green Jr.
Air Force Systems Command

***16. Space Power Systems Engineering (1966)**
George C. Szego
Institute for Defense Analyses
J. Edward Taylor
TRW Inc.

***17. Methods in Astrodynamics and Celestial Mechanics (1966)**
Raynor L. Duncombe
U.S. Naval Observatory
Victor G. Szebehely
Yale University Observatory

***18. Thermophysics and Temperature Control of Spacecraft and Entry Vehicles (1966)**
Gerhard B. Heller
NASA George C. Marshall Space Flight Center

***19. Communication Satellite Systems Technology (1966)**
Richard B. Marsten
Radio Corporation of America

***20. Thermophysics of Spacecraft and Planetary Bodies: Radiation Properties of Solids and the Electromagnetic Radiation Environment in Space (1967)**
Gerhard B. Heller
NASA George C. Marshall Space Flight Center

*Out of print.

***21. Thermal Design Principles of Spacecraft and Entry Bodies (1969)**
Jerry T. Bevans
TRW Systems

***22. Stratospheric Circulation (1969)**
Willis L. Webb
Atmospheric Sciences Laboratory, White Sands, and University of Texas at El Paso

***23. Thermophysics: Applications to Thermal Design of Spacecraft (1970)**
Jerry T. Bevans
TRW Systems

24. Heat Transfer and Spacecraft Thermal Control (1971)
John W. Lucas
Jet Propulsion Laboratory

25. Communication Satellites for the 70's: Technology (1971)
Nathaniel E. Feldman
The Rand Corp.
Charles M. Kelly
The Aerospace Corp.

26. Communication Satellites for the 70's: Systems (1971)
Nathaniel E. Feldman
The Rand Corp.
Charles M. Kelly
The Aerospace Corp.

27. Thermospheric Circulation (1972)
Willis L. Webb
Atmospheric Sciences Laboratory, White Sands, and University of Texas at El Paso

28. Thermal Characteristics of the Moon (1972)
John W. Lucas
Jet Propulsion Laboratory

***29. Fundamentals of Spacecraft Thermal Design (1972)**
John W. Lucas
Jet Propulsion Laboratory

***30. Solar Activity Observations and Predictions (1972)**
Patrick S. McIntosh
Murray Dryer
Environmental Research Laboratories, National Oceanic and Atmospheric Administration

31. Thermal Control and Radiation (1973)
Chang-Lin Tien
University of California at Berkeley

32. Communications Satellite Systems (1974)
P.L. Bargellini
COMSAT Laboratories

33. Communications Satellite Technology (1974)
P.L. Bargellini
COMSAT Laboratories

***34. Instrumentation for Airbreathing Propulsion (1974)**
Allen E. Fuhs
Naval Postgraduate School
Marshall Kingery
Arnold Engineering Development Center

35. Thermophysics and Spacecraft Thermal Control (1974)
Robert G. Hering
University of Iowa

36. Thermal Pollution Analysis (1975)
Joseph A. Schetz
Virginia Polytechnic Institute
ISBN 0-915928-00-0

***37. Aeroacoustics: Jet and Combustion Noise; Duct Acoustics (1975)**
Henry T. Nagamatsu, Editor
General Electric Research and Development Center
Jack V. O'Keefe, Associate Editor
The Boeing Co.
Ira R. Schwartz, Associate Editor
NASA Ames Research Center
ISBN 0-915928-01-9

***38. Aeroacoustics: Fan, STOL, and Boundary Layer Noise; Sonic Boom; Aeroacoustics Instrumentation (1975)**
Henry T. Nagamatsu, Editor
General Electric Research and Development Center
Jack V. O'Keefe, Associate Editor
The Boeing Co.
Ira R. Schwartz, Associate Editor
NASA Ames Research Center
ISBN 0-915928-02-7

39. Heat Transfer with Thermal Control Applications (1975)
M. Michael Yovanovich
University of Waterloo
ISBN 0-915928-03-5

***40. Aerodynamics of Base Combustion (1976)**
S.N.B. Murthy, Editor
J. R. Osborn, Associate Editor
Purdue University
A.W. Barrows
J. R. Ward,
Associate Editors
Ballistics Research Laboratories
ISBN 0-915928-04-3

---
*Out of print.

41. Communications Satellite Developments: Systems (1976)
Gilbert E. LaVean
Defense Communications Agency
William G. Schmidt
CML Satellite Corp.
ISBN 0-915928-05-1

***42. Communications Satellite Developments: Technology (1976)**
William G. Schmidt
CML Satellite Corp.
Gilbert E. LaVean
Defense Communications Agency
ISBN 0-915928-06-X

***43. Aeroacoustics: Jet Noise, Combustion and Core Engine Noise (1976)**
Ira R. Schwartz, Editor
NASA Ames Research Center
Henry T. Nagamatsu, Associate Editor
General Electric Research and Development Center
Warren C. Strahle, Associate Editor
Georgia Institute of Technology
ISBN 0-915928-07-8

***44. Aeroacoustics: Fan Noise and Control; Duct Acoustics; Rotor Noise (1976)**
Ira R. Schwartz, Editor
NASA Ames Research Center
Henry T. Nagamatsu, Associate Editor
General Electric Research and Development Center
Warren C. Strahle, Associate Editor
Georgia Institute of Technology
ISBN 0-915928-08-6

***45. Aeroacoustics: STOL Noise; Airframe and Airfoil Noise (1976)**
Ira R. Schwartz, Editor
NASA Ames Research Center
Henry T. Nagamatsu, Associate Editor
General Electric Research and Development Center
Warren C. Strahle, Associate Editor
Georgia Institute of Technology
ISBN 0-915928-09-4

***46. Aeroacoustics: Acoustic Wave Propagation; Aircraft Noise Prediction; Aeroacoustic Instrumentation (1976)**
Ira R. Schwartz, Editor
NASA Ames Research Center
Henry T. Nagamatsu, Associate Editor
General Electric Research and Development Center
Warren C. Strahle, Associate Editor
Georgia Institute of Technology
ISBN 0-915928-10-8

***47. Spacecraft Charging by Magnetospheric Plasmas (1976)**
Alan Rosen
TRW Inc.
ISBN 0-915928-11-6

48. Scientific Investigations on the Skylab Satellite (1976)
Marion I. Kent
Ernst Stuhlinger
NASA George C. Marshall Space Flight Center
Shi-Tsan Wu
University of Alabama
ISBN 0-915928-12-4

49. Radiative Transfer and Thermal Control (1976)
Allie M. Smith
ARO Inc.
ISBN 0-915928-13-2

***50. Exploration of the Outer Solar System (1976)**
Eugene W. Greenstadt
TRW Inc.
Murray Dryer
National Oceanic and Atmospheric Administration
Devrie S. Intriligator
University of Southern California
ISBN 0-915928-14-0

***51. Rarefied Gas Dynamics, Parts I and II(two volumes) (1977)**
J. Leith Potter
ARO Inc.
ISBN 0-915928-15-9

***52. Materials Sciences in Space with Application to Space Processing (1977)**
Leo Steg
General Electric Co.
ISBN 0-915928-16-7

***53. Experimental Diagnostics in Gas Phase Combustion Systems (1977)**
Ben T. Zinn, Editor
Georgia Institute of Technology
Craig T. Bowman, Associate Editor
Stanford University
Daniel L. Hartley, Associate Editor
Sandia Laboratories
Edward W. Price, Associate Editor
Georgia Institute of Technology
James G. Skifstad, Associate Editor
Purdue University
ISBN 0-915928-18-3

*Out of print.

54. Satellite Communication: Future Systems (1977)
David Jarett
TRW Inc.
ISBN 0-915928-18-3

***55. Satellite Communications: Advanced Technologies (1977)**
David Jarett
TRW Inc.
ISBN 0-915928-19-1

***56. Thermophysics of Spacecraft and Outer Planer Entry Probes (1977)**
Allied M. Smith
ARO Inc.
ISBN 0-915928-20-5

***57. Space-Based Manufacturing from Nonterrestrial Materials (1977)**
Gerald K. O'Neill, Editor
Brian O'Leary, Assistant Editor
Princeton University
ISBN 0-915928-21-3

***58. Turbulent Combustion (1978)**
Lawrence A. Kennedy
State University of New York at Buffalo
ISBN 0-915928-22-1

***59. Aerodynamic Heating and Thermal Protection Systems (1978)**
Leroy S. Fletcher
University of Virginia
ISBN 0-915928-23-X

60. Heat Transfer and Thermal Control Systems (1978)
Leroy S. Fletcher
University of Virginia
ISBN 0-915928-24-8

***61. Radiation Energy Conversion in Space (1978)**
Kenneth W. Billman
NASA Ames Research Center
ISBN 0-915928-26-4

***62. Alternative Hydrocarbon Fuels: Combustion and Chemical Kinetics (1978)**
Craig T. Bowman
Stanford University
Jorgen Birkeland
Department of Energy
ISBN 0-915928-25-6

***63. Experimental Diagnostics in Combustion of Solids (1978)**
Thomas L. Boggs
Naval Weapons Center
Ben T. Zinn
Georgia Institute of Technology
ISBN 0-915928-28-0

***64. Outer Planet Entry Heating and Thermal Protection (1979)**
Raymond Viskanta
Purdue University
ISBN 0-915928-29-9

***65. Thermophysics and Thermal Control (1979)**
Raymond Viskanta
Purdue University
ISBN 0-915928-30-2

***66. Interior Ballistics of Guns (1979)**
Herman Krier
University of Illinois at Urbana-Champaign
Martin Summerfield
New York University
ISBN 0-915928-32-9

***67. Remote Sensing of Earth from Space: Role of "Smart Sensors" (1979)**
Roger A. Breckenridge
NASA Langley Research Center
ISBN 0-915928-33-7

68. Injection and Mixing in Turbulent Flow (1980)
Joseph A. Schetz
Virginia Polytechnic Institute and State University
ISBN 0-915928-35-3

***69. Entry Heating and Thermal Protection (1980)**
Walter B. Olstad
NASA Headquarters
ISBN 0-915928-38-8

***70. Heat Transfer, Thermal Control, and Heat Pipes (1980)**
Walter B. Olstad
NASA Headquarters
ISBN 0-915928-39-6

***71. Space Systems and Their Interactions with Earth's Space Environment (1980)**
Henry B. Garrett
Charles P. Pike
Hanscom Air Force Base
ISBN 0-915928-41-8

***72. Viscous Flow Drag Reduction (1980)**
Gary R. Hough
Vought Advanced Technology Center
ISBN 0-915928-44-2

***73. Combustion Experiments in a Zero-Gravity Laboratory (1981)**
Thomas H. Cochran
NASA Lewis Research Center
ISBN 0-915928-48-5

***74. Rarefied Gas Dynamics, Parts I and II (two volumes) (1981)**
Sam S. Fisher
University of Virginia
ISBN 0-915928-51-5

*Out of print.

***75. Gasdynamics of Detonations and Explosions (1981)**
J. R. Bowen
University of Wisconsin at Madison
N. Manson
Universite de Poitiers
A. K. Oppenheim
University of California at Berkeley
R. I. Soloukhin
Institute of Heat and Mass Transfer, BSSR Academy of Sciences
ISBN 0-915928-46-9

76. Combustion in Reactive Systems (1981)
J. R. Bowen
University of Wisconsin at Madison
N. Manson
Universite de Poitiers
A. K. Oppenheim
University of California at Berkeley
R. I. Soloukhin
Institute of Heat and Mass Transfer, BSSR Academy of Sciences
ISBN 0-915928-47-7

***77. Aerothermodynamics and Planetary Entry (1981)**
A. L. Crosbie
University of Missouri-Rolla
ISBN 0-915928-52-3

***78. Heat Transfer and Thermal Control (1981)**
A. L. Crosbie
University of Missouri-Rolla
ISBN 0-915928-53-1

***79. Electric Propulsion and Its Applications to Space Missions** (1981)
Robert C. Finke
NASA Lewis Research Center
ISBN 0-915928-55-8

***80. Aero-Optical Phenomena (1982)**
Keith G. Gilbert
Leonard J. Otten
Air Force Weapons Laboratory
ISBN 0-915928-60-4

***81. Transonic Aerodynamics (1982)**
David Nixon
Nielsen Engineering & Research, Inc.
ISBN 0-915928-65-5

***82. Thermophysics of Atmospheric Entry (1982)**
T. E. Horton
University of Mississippi
ISBN 0-915928-66-3

***83. Spacecraft Radiative Transfer and Temperature Control (1982)**
T. E. Horton
University of Mississippi
ISBN 0-915928-67-1

84. Liquid-Metal Flows and Magneto-hydrodynamics (1983)
H. Branover
Ben-Gurion University of the Negev
P. S. Lykoudis
Purdue University
A. Yakhot
Ben-Gurion University of the Negev
ISBN 0-915928-70-1

***85. Entry Vehicle Heating and Thermal Protection Systems: Space Shuttle, Solar Starprobe, Jupiter Galileo Probe (1983)**
Paul E. Bauer
McDonnell Douglas Astronautics Co.
Howard E. Collicott
The Boeing Co.
ISBN 0-915928-74-4

***86. Spacecraft Thermal Control, Design, and Operation (1983)**
Howard E. Collicott
The Boeing Co.
Paul E. Bauer
McDonnell Douglas Astronautics Co.
ISBN 0-915928-75-2

***87. Shock Waves, Explosions, and Detonations (1983)**
J. R. Bowen
University of Washington
N. Manson
Universite de Poitiers
A. K. Oppenheim
University of California at Berkeley
R. I. Soloukhin
Institute of Heat and Mass Transfer, BSSR Academy of Sciences
ISBN 0-915928-76-0

88. Flames, Lasers, and Reactive Systems (1983)
J. R. Bowen
University of Washington
N. Manson
Universite de Poitiers
A. K. Oppenheim
University of California at Berkeley
R. I. Soloukhin
Institute of Heat and Mass Transfer, BSSR Academy of Sciences
ISBN 0-915928-77-9

***89. Orbit-Raising and Maneuvering Propulsion: Research Status and Needs (1984)**
Leonard H. Caveny
Air Force Office of Scientific Research
ISBN 0-915928-82-5

---
*Out of print.

90. Fundamental of Solid-Propellant Combustion (1984)
Kenneth K. Kuo
Pennsylvania State University
Martin Summerfield
Princeton Combustion Research Laboratories, Inc.
ISBN 0-915928-84-1

91. Spacecraft Contamination: Sources and Prevention (1984)
J. A. Roux
University of Mississippi
T. D. McCay
NASA Marshall Space Flight Center
ISBN 0-915928-85-X

92. Combustion Diagnostics by Nonintrusive Methods (1984)
T. D. McCay
NASA Marshall Space Flight Center
J. A. Roux
University of Mississippi
ISBN 0-915928-86-8

93. The INTELSAT Global Satellite System (1984)
Joel Alper
COMSAT Corp.
Joseph Pelton
INTELSAT
ISBN 0-915928-90-6

94. Dynamics of Shock Waves, Explosions, and Detonations (1984)
J. R. Bowen
University of Washington
N. Manson
Universite de Poitiers
A. K. Oppenheim
University of California at Berkeley
R. I. Soloukhin
Institute of Heat and Mass Transfer, BSSR Academy of Sciences
ISBN 0-915928-91-4

95. Dynamics of Flames and Reactive Systems (1984)
J. R. Bowen
University of Washington
N. Manson
Universite de Poitiers
A. K. Oppenheim
University of California at Berkeley
R. I. Soloukhin
Institute of Heat and Mass Transfer, BSSR Academy of Sciences
ISBN 0-915928-92-2

96. Thermal Design of Aeroassisted Orbital Transfer Vehicles (1985)
H. F. Nelson
University of Missouri-Rolla
ISBN 0-915928-94-9

97. Monitoring Earth's Ocean, Land, and Atmosphere from Space–Sensors, Systems, and Applications (1985)
Abraham Schnapf
Aerospace Systems Engineering
ISBN 0-915928-98-1

98. Thrust and Drag: Its Prediction and Verification (1985)
Eugene E. Covert
Massachusetts Institute of Technology
C. R. James
Vought Corp.
William F. Kimzey
Sverdrup Technology AEDC Group
George K. Richey
U.S. Air Force
Eugene C. Rooney
U.S. Navy Department of Defense
ISBN 0-930403-00-2

99. Space Stations and Space Platforms–Concepts, Design, Infrastructure, and Uses (1985)
Ivan Bekey
Daniel Herman
NASA Headquarters
ISBN 0-930403-01-0

100. Single- and Multi-Phase Flows in an Electromagnetic Field: Energy, Metallurgical, and Solar Applications (1985)
Herman Branover
Ben-Gurion University of the Negev
Paul S. Lykoudis
Purdue University
Michael Mond
Ben-Gurion University of the Negev
ISBN 0-930403-04-5

101. MHD Energy Conversion: Physiotechnical Problems (1986)
V. A. Kirillin
A. E. Sheyndlin
Soviet Academy of Sciences
ISBN 0-930403-05-3

102. Numerical Methods for Engine-Airframe Integration (1986)
S.N.B. Murthy
Purdue University
Gerald C. Paynter
Boeing Airplane Co.
ISBN 0-930403-09-6

103. Thermophysical Aspects of Re-Entry Flows (1986)
James N. Moss
NASA Langley Research Center
Carl D. Scott
NASA Johnson Space Center
ISBN 0-930430-10-X

---

*Out of print.

***104. Tactical Missile Aerodynamics (1986)**
M. J. Hemsch
PRC Kentron, Inc.
J. N. Nielson
NASA Ames Research Center
ISBN 0-930403-13-4

105. Dynamics of Reactive Systems Part I: Flames and Configurations; Part II: Modeling and Heterogeneous Combustion (1986)
J. R. Bowen
University of Washington
J.-C. Leyer
Universite de Poitiers
R. I. Soloukhin
Institute of Heat and Mass Transfer, BSSR Academy of Sciences
ISBN 0-930403-14-2

106. Dynamics of Explosions (1986)
J. R. Bowen
University of Washington
J.-C. Leyer
Universite de Poitiers
R. I. Soloukhin
Institute of Heat and Mass Transfer, BSSR Academy of Sciences
ISBN 0-930403-15-0

***107. Spacecraft Dielectric Material Properties and Spacecraft Charging (1986)**
A.R. Frederickson
U.S. Air Force Rome Air Development Center
D. B. Cotts
SRI International
J. A. Wall
U.S. Air Force Rome Air Development Center
F. L. Bouquet
Jet Propulsion Laboratory, California Institute of Technology
ISBN 0-930403-17-7

***108. Opportunities for Academic Research in a Low-Gravity Environment (1986)**
George A. Hazelrigg
National Science Foundation
Joseph M. Reynolds
Louisiana State University
ISBN 0-930403-18-5

109. Gun Propulsion Technology (1988)
Ludwig Stiefel
U.S. Army Armament Research, Development and Engineering Center
ISBN 0-930403-20-7

110. Commercial Opportunities in Space (1988)
F. Shahrokhi
K. E. Harwell
University of Tennessee Space Institute
C. C. Chao
National Cheng Kung University
ISBN 0-930403-39-8

111. Liquid-Metal Flows: Magnetohydrodynamics and Application (1988)
Herman Branover, Michael Mond, and Yeshajahu Unger
Ben-Gurion University of the Negev
ISBN 0-930403-43-6

112. Current Trends in Turbulence Research (1988)
Herman Branover, Micheal Mond, and Yeshajahu Unger
Ben-Gurion University of the Negev
ISBN 0-930403-44-4

113. Dynamics of Reactive Systems Part I: Flames; Part II: Heterogeneous Combustion and Applications (1988)
A. L. Kuhl
R & D Associates
J. R. Bowen
University of Washington
J.-C. Leyer
Universite de Poitiers
A. Borisov
USSR Academy of Sciences
ISBN 0-930403-46-0

114. Dynamics of Explosions (1988)
A. L. Kuhl
R & D Associates
J. R. Bowen
University of Washington
J.-C. Leyer
Universite de Poitiers
A. Borisov
USSR Academy of Sciences
ISBN 0-930403-47-9

115. Machine Intelligence and Autonomy for Aerospace (1988)
E. Heer
Heer Associates, Inc.
H. Lum
NASA Ames Research Center
ISBN 0-930403-48-7

116. Rarefied Gas Dynamics: Space Related Studies (1989)
E. P. Muntz
University of Southern California
D. P. Weaver
U.S. Air Force Astronautics Laboratory (AFSC)
D. H. Campbell
University of Dayton Research Institute
ISBN 0-930403-53-3

---
*Out of print.

117. Rarefied Gas Dynamics: Physical Phenomena (1989)
E. P. Muntz
University of Southernn California
D. P. Weaver
U.S. Air Force Astronautics Laboratory (AFSC)
D. H. Campbell
University of Dayton Research Institute
ISBN 0-930403-54-1

118. Rarefied Gas Dynamics: Theoretical and Computational Techniques (1989)
E.P. Muntz
University of Southernn California
D. P. Weaver
U.S. Air Force Astronautics Laboratory (AFSC)
D. H. Campbell
University of Dayton Research Institute
ISBN 0-930403-55-X

119. Test and Evaluation of the Tactical Missile (1989)
Emil J. Eichblatt Jr.
Pacific Missile Test Center
ISBN 0-930403-56-8

120. Unsteady transonic Aerodynamics (1989)
David Nixon
Nielsen Engineering & Research, Inc.
ISBN 0-930403-52-5

121. Orbital Debris from Upper-Stage Breakup (1989)
Joseph P. Loftus Jr.
NASA Johnson Space Center
ISBN 0-930403-58-4

122. Thermal-Hydraulics for Space Power, Propulsion and Thermal Management System Design (1989)
William J. Krotiuk
General Electric Co.
ISBN 0-930403-64-9

123. Viscous Drag Reduction in Boundary Layers (1990)
Dennis M. Bushnell
Jerry N. Hefner
NASA Langley Research Center
ISBN 0-930403-66-5

***124. Tactical and Strategic Missile Guidance (1990)**
Paul Zarchan
Charles Stark Draper Laboratory, Inc.
ISBN 0-930403-68-1

125. Applied Computational Aerodynamics (1990)
P. A. Henne
Douglas Aircraft Company
ISBN 0-930403-69-X

126. Space Commercialization: Launch Vehicles and Programs (1990)
F. Shahrokhi
University of Tennessee Space Institute
J. S. Greenberg
Princeton Synergetics Inc.
T. Al-Saud
Ministry of Defense and Aviation Kingdom of Saudi Arabia
ISBN 0-930403-75-4

127. Space Commercialization: Platforms and Processing (1990)
F. Shahrokhi
University of Tennessee Space Institute
G. Hazelrigg
National Science Foundation
R. Bayuzick
Vanderbilt University
ISBN 0-930403-76-2

128. Space Commercialization: Satellite Technology (1990)
F. Shahrokhi
University of Tennessee Space Institute
N. Jasentuliyana
United Nations
N. Tarabzouni
King Abulaziz City for Science and Technology
ISBN 0-930403-77-0

129. Mechanics and Control of Large Flexible Structures (1990)
John L. Junkins
Texas A&M University
ISBN 0-930403-73-8

130. Low-Gravity Fluid Dynamics and Transport Phenomena (1990)
Jean N. Koster
Robert L. Sani
University of Colorado at Boulder
ISBN 0-930403-74-6

131. Dynamics of Deflagrations and Reactive Systems: Flames (1991)
A. L. Kuhl
Lawrence Livermore National Laboratory
J.-C. Leyer
Universite de Poitiers
A. A. Borisov
USSR Academy of Sciences
W. A. Sirignano
University of California
ISBN 0-930403-95-9

*Out of print.

132. Dynamics of Deflagrations and Reactive Systems: Heterogeneous Combustion (1991)
A. L. Kuhl
Lawrence Livermore National Laboratory
J.-C. Leyer
Universite de Poitiers
A. A. Borisov
USSR Academy of Sciences
W. A. Sirignano
University of California
ISBN 0-930403-96-7

133. Dynamics of Detonations and Explosions: Detonations (1991)
A. L. Kuhl
Lawrence Livermore National Laboratory
J.-C. Leyer
Universite de Poitiers
A. A. Borisov
USSR Academy of Sciences
W. A. Sirignano
University of California
ISBN 0-930403-97-5

134. Dynamics of Detonations and Explosions: Explosion Phenomena (1991)
A. L. Kuhl
Lawrence Livermore National Laboratory
J.-C. Leyer
Universite de Poitiers
A. A. Borisov
USSR Academy of Sciences
W. A. Sirignano
University of California
ISBN 0-930403-98-3

135. Numerical Approaches to Combustion Modeling (1991)
Elaine S. Oran
Jay P. Boris
Naval Research Laboratory
ISBN 1-56347-004-7

136. Aerospace Software Engineering (1991)
Christine Anderson
U.S. Air Force Wright Laboratory
Merlin Dorfman
Lockheed Missiles & Space Company, Inc.
ISBN 1-56347-005-0

137. High-Speed Flight Propulsion Systems (1991)
S.N.B. Murthy
Purdue University
E.T. Curran
Wright Laboratory
ISBN 1-56347-011-X

138. Propagation of Intensive Laser Radiation in Clouds (1992)
O. A. Volkovitsky
Yu. S. Sedenov
L. P. Semenov
Institute of Experimental Meteorology
ISBN 1-56347-020-9

139. Gun Muzzle Blast and Flash (1992)
Gunter Klingenburg
Fraunhofer-Institut fur Kurzzeitdynamik, Ernst-Mach-Institut (EMI)
Joseph M. Heimerl
U.S. Army Ballistic Research Laboratory (BRL)
ISBN 1-56347-012-8

140. Thermal Structures and Materials for High-Speed Flight (1992)
Earl. A. Thornton
University of Virginia
ISBN 1-56347-017-9

141. Tactical Missile Aerodynamics: General Topics (1992)
Michael J. Hemsch
Lockheed Engineering & Sciences Company
ISBN 1-56347-015-2

142. Tactical Missile Aerodynamics: Prediction Methodology (1992)
Michael R. Mendenhall
Nielsen Engineering & Research, Inc.
ISBN 1-56347-016-0

143. Nonsteady Burning and Combustion Stability of Solid Propellants (1992)
Luigi De Luca
Politecnico di Milano
Edward W. Price
Georgia Institute of Technology
Martin Summerfield
Princeton Combustion Research Laboratories, Inc.
ISBN 1-56347-014-4

144. Space Economics (1992)
Joel S. Greenberg
Princeton Synergetics, Inc.
Henry R. Hertzfeld
HRH Associates
ISBN 1-56347-042-X

145. Mars: Past, Present, and Future (1992)
E. Brian Pritchard
NASA Langley Research Center
ISBN 1-56347-043-8

146. Computational Nonlinear Mechanics in Aerospace Engineering (1992)
Satya N. Atluri
Georgia Institute of Technology
ISBN 1-56347-044-6

147. Modern Engineering for Design of Liquid-Propellant Rocket Engines (1992)
Dieter K. Huzel
David H. Huang
ISBN 1-56347-013-6

---
*Out of print.

148. Metallurgical Technologies, Energy Conversion, and Magnetohydrodynamic Flows (1993)
Herman Branover
Yeshajahu Unger
Ben-Gurion University of the Negev
ISBN 1-56347-019-5

149. Advances in Turbulence Studies (1993)
Herman Branover
Yeshajahu Unger
Ben-Gurion University of the Negev
ISBN 1-56347-018-7

150. Structural Optimization: Status and Promise (1993)
Manohar P. Kamat
Georgia Institute of Technology
ISBN 1-56347-56-X

151. Dynamics of Gaseous Combustion (1993)
A. L. Kuhl
Lawrence Livermore National Laboratory
J.-C. Leyer
Universite de Poitiers
A. A. Borisov
USSR Academy of Sciences
W. A. Sirignano

152. Dynamics of Heterogeneous Gaseous Combustion and Reacting Systems (1993)
A. L. Kuhl
Lawrence Livermore National Laboratory
J.-C. Leyer
Universite de Poitiers
A. A. Borisov
USSR Academy of Sciences
W. A. Sirignano
University of California
ISBN 1-56347-058-6

153. Dynamic Aspects of Detonations (1993)
A. L. Kuhl
Lawrence Livermore National Laboratory
J.-C. Leyer
Universite de Poitiers
A. A. Borisov
USSR Academy of Sciences
W. A. Sirignano
University of California
ISBN 1-56347-057-8

154. Dynamic Aspects of Explosion Phenomena (1993)
A. L. Kuhl
Lawrence Livermore National Laboratory
J.-C. Leyer
Universite de Poitiers
A. A. Borisov
USSR Academy of Sciences
W. A. Sirignano
University of California
ISBN 1-56347-059-4

155. Tactical Missile Warheads (1993)
Joseph Carleone
Aerojet General Corporation
ISBN 1-56347-067-5

156. Toward a Science of Command, Control, and Communications (1993)
Carl R. Jones
Naval Postgraduate School
ISBN 1-56347-068-3

157. Tactical and Strategic Missile Guidance Second Edition (1994)
Paul Zarchan
Charles Stark Draper Laboratory, Inc.
ISBN 1-56347-077-2

158. Rarefied Gas Dynamics: Experimental Techniques and Physical Systems (1994)
Bernie D. Shizgal
University of British Columbia
David P. Weaver
Phillips Laboratory
ISBN 1-56347-079-9

159. Rarefied Gas Dynamics: Theory and Simulations (1994)
Bernie D. Shizgal
University of British Columbia
David P. Weaver
Phillips Laboratory
ISBN 1-56347-080-2

160. Rarefied Gas Dynamics: Space Sciences and Engineering (1994)
Bernie D. Shizgal
University of British Columbia
David P. Weaver
Phillips Laboratory
ISBN 1-56347-081-0

161. Teleoperation and Robotics in Space (1994)
Steven B. Skaar
University of Notre Dame
Carl F. Ruoff
Jet Propulsion Laboratory, California Institute of Technology
ISBN 1-56347-095-0

162. Progress in Turbulence Research (1994)
Herman Branover
Yeshajahu Unger
Ben-Gurion University of the Negev
ISBN 1-56347-099-3

---
*Out of print.

163. Global Positioning System: Theory and Applications Volume I (1996)
Bradford W. Parkinson
Stanford University
James J. Spilker Jr.
Stanford Telecom
Penina Axelrad, Associate Editor
University of Colorado
Per Enge, Associate Editor
Stanford University
ISBN 1-56347-107-8

164. Global Positioning System: Theory and Applications Volume II (1996)
Bradford W. Parkinson
Stanford University
James J. Spilker Jr.
Stanford Telecom
Penina Axelrad, Associate Editor
University of Colorado
Per Enge, Associate Editor
Stanford University
ISBN 1-56347-106-X

165. Developments In High-Speed Vehicle Propulsion Systems (1996)
S.N.B. Murthy
Purdue University
E. T. Curran
Wright Laboratory
ISBN 1-56347-176-0

166. Recent Advances in Spray Combustion I (1996)
Kenneth K. Kuo
Pennsylvania State University
ISBN 1-56347-175-2

167. Fusion Energy in Space Propulsion (1996)
Terry Kammash
University of Michigan
ISBN 1-56347-184-1

168. Aerospace Thermal Structures and Materials for a New Era (1995)
Earl A. Thornton
University of Virginia
ISBN 1-56347-182-5

169. Liquid Rocket Engine Combustion Instability (1995)
Vigor Yang
William E. Anderson
Pennsylvania State University
ISBN 1-56347-183-3

170. Tactical Missile Propulsion (1996)
Gordon E. Jensen
United Technologies
David W. Netzer
Naval Postgraduate School
ISBN 1-56347-118-3

*Out of print.

www.ingramcontent.com/pod-product-compliance
Lightning Source LLC
Chambersburg PA
CBHW020809140125
20212CB00001B/1